Protein Reviews

Series Editor:
M. Zouhair Atassi

For further volumes:
http://www.springer.com/series/6876

Mark O.J. Olson
Editor

The Nucleolus

 Springer

Editor
Mark O.J. Olson
University of Mississippi Medical Center
Department of Biochemistry
2500 North State Street
Jackson, Mississippi 39216
USA
molson@umc.edu

ISBN 978-1-4614-0513-9   e-ISBN 978-1-4614-0514-6
DOI 10.1007/978-1-4614-0514-6
Springer New York Dordrecht Heidelberg London

Library of Congress Control Number: 2011935031

© Springer Science+Business Media, LLC 2011
All rights reserved. This work may not be translated or copied in whole or in part without the written permission of the publisher (Springer Science+Business Media, LLC, 233 Spring Street, New York, NY 10013, USA), except for brief excerpts in connection with reviews or scholarly analysis. Use in connection with any form of information storage and retrieval, electronic adaptation, computer software, or by similar or dissimilar methodology now known or hereafter developed is forbidden.
The use in this publication of trade names, trademarks, service marks, and similar terms, even if they are not identified as such, is not to be taken as an expression of opinion as to whether or not they are subject to proprietary rights.

Printed on acid-free paper

Springer is part of Springer Science+Business Media (www.springer.com)

Preface

The Nucleolus: A Nuclear Body Full of Surprises

The deeper we delve into nature, the more surprises we find. The nucleolus is no exception; as we learn more about the structure and functions of the nucleolus, the more surprising it becomes. It has taken almost two centuries to reach this point. In fact, well over a century passed between the first description of the nucleolus (Wagner 1835) and the publication of definitive experiments that established its primary function as a factory for ribosome biogenesis during the 1960s (summarized by Hadjiolov 1985). In the past four to five decades, research has been largely focused on investigating its structure and ribosome assembly process, defining its component parts and determining how it does and what it does. Still ongoing, these efforts are now at a relatively mature level, taking us out of the "black box" era. The picture that has emerged is a highly complex, multistep vectorial process that utilizes a large number of components. Although there is still much to be learned about the mechanisms of ribosome biogenesis, the field has moved into structural and functional analyses of individual components and larger sub-complexes as well as studies on integration and regulation within the system and by the cell.

With the primary focus of research during the second half of the twentieth century on the elucidation of the role of the nucleolus in ribosome assembly, most researchers did not expect that it could do much else. Consequently, the nucleolus managed to keep its other functions hidden. However, within the past two decades something extraordinary happened; new functions for the nucleolus began to appear. In many cases, some of these were met with skepticism, but several of the new roles have now become established and even found in textbooks. Others are under active investigation. These novel tasks for the old factory have given the field a new vitality, generating renewed excitement and interest. Moreover, the findings have attracted researchers who had little or no previous interest in the nucleolus.

The surprising features of the nucleolus are not limited to its newly discovered functions; they also include aspects of its conventional role. Consequently, almost two-thirds of this volume is devoted to traditional functions of the nucleolus. There has been a near-explosion of progress in elucidating nucleolar structures, functions, and mechanisms during the past decade. How do we account for these developments? It is best explained by a synergistic effect between the renewed interest in the subject and the continuous development and improvement of technology. As an example, advances in mass spectrometry allowed researchers to identify virtually every protein molecule in the nucleolus. Even with highly sensitive instrumentation, this was only possible because of the availability of genome sequences from several species. To the surprise of most researchers, several thousand polypeptides were found in the nucleolus (see Chap. 2), many of which have no apparent function in ribosome biogenesis. With this finding, questions about the dynamics of these polypeptides arose. Mass spectrometry coupled with isotopic methods has allowed researchers to analyze the dynamics of multiple molecules moving in and out of the nucleolus under various physiological conditions. Complementing this is the availability of laser scanning confocal microscopy coupled with photobleaching techniques to measure the dynamics of individual molecules in living cells. Not only has recent research provided us with new information, but it has changed our perception of the nucleolus; we are now forced to change our mental image of the nucleolus as the static structure shown in textbooks to one in which the components are constantly in motion. Although the details are important, the changing big picture may be more significant. To quote Sir William Bragg, "The important thing in science is not so much to obtain new facts as to discover new ways of thinking about them."

This subject has matured to the point where every subtopic cannot be covered in one volume. Therefore, we have focused on recent progress in specialized topics within the general subject. We apologize to those researchers whose work is not covered.

The Complex Nucleolus

The acquisition of greater knowledge about the nucleolus has also brought more complexity. Is the complexity surprising? Probably not, if we consider the complex products it assembles, the things it does and how it does them. Prokaryotes get along quite well with a relatively simple system of ribosome assembly. A superficial examination of the general features eukaryotic ribosome biogenesis (Fig. 0.1) suggests that the process is relatively simple. However, when one delves into the details described in the chapters of this volume, the eukaryotic ribosome production system turns out to be exceedingly complex. As eukaryotes evolved the complexity increased and so arose the need for a nucleolus. This came about for a number of reasons. Important insights into this issue occur when the compositions and structures prokaryotic and eukaryotic ribosomes are compared. Eukaryotic ribosomes are about 40% larger than their bacterial counterparts; their RNAs are longer and they have about 25 more proteins. Recent progress in X-ray crystallography also helps us make the comparison. The crystal structure of the prokaryotic ribosome became available about a decade ago (Ramakrishnan and Moore 2001), but recently the structures of the yeast 80S ribosome (Ben Shem et al. 2010) and the Tetrahymena

Preface

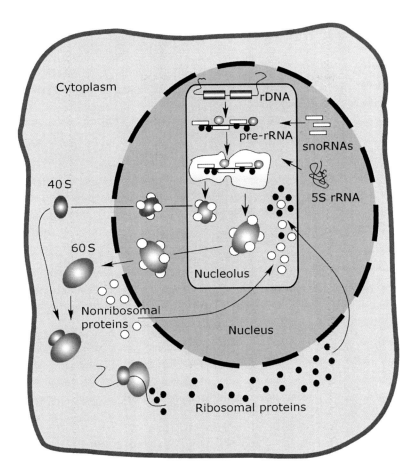

Fig. 0.1 Major steps in eukaryotic ribosome biogenesis. The process starts with transcription of preribosomal RNA (pre-rRNA) from multiple copies of the genes for pre-rRNA (rDNA). Nonribosomal proteins (*open circles*) and small nucleolar RNAs (snoRNAs; *open rectangles*) associate with the nascent transcript. The pre-rRNA is methylated and pseudouridylated under the guidance of the snoRNAs. 5S rRNA, a component of the 60S subunit, is added to the maturing complex. The pre-rRNA undergoes a series of cleavages ultimately resulting in 18S, 5.8S and 28S (25S in yeast) rRNAs. The complex is split into the two precursor particles for the small (40S) and large (60S) ribosomal subunits. Ribosomal proteins (*black circles*) are added to the precursor complexes at various stages of assembly. The nearly mature subunits are exported to the cytoplasm through the nuclear pore complexes with the aid of adaptor proteins. The small and large subunits are eventually incorporated into ribosomes in the cytoplasm. Figure modified from Olson (2004)

40S subunit in complex with initiation factor 1 (Rabl et al. 2011) were published. Although the core structure of the ribosome is conserved across all organisms, the additional components lie at the periphery. The added segments of rRNA and extra proteins appear to play a role in the regulation of translation. Hence, the assembly system had to evolve and become more complex to accommodate these regulatory components. Moreover, longer RNAs offer more opportunities for misfolding and a precise order of assembly is required to prevent this from happening during ribosome assembly. Of particular importance is the pseudoknot in the 18S rRNA that is required for ribosome function. The formation of this structure is delayed until later in the assembly process by utilization of base pairing with small nucleolar RNAs (snoRNAs) (Hughes 1996). Thus, timing of events and precision in the assembly process adds more complexity.

A second factor is related to the hundreds of RNA modifications found in eukaryotic rRNA, which are largely absent in prokaryotes. The relatively few modifications in the latter are performed by freestanding enzymes. For performing these operations in eukaryotes, the nucleolus contains a multitude of small (snoRNAs), which serve as guides, along with their modifying enzymes and other associated proteins. This system contributes hundreds of components to the nucleolar machinery that are not seen in prokaryotes.

Although ribosome production is regulated in prokaryotes, it is more tightly controlled in eukaryotes to meet the needs of the cell. The various levels of regulation are described in Chaps. 4, 6, 8, 12, and 13. The number of regulatory factors is growing; to a large extent these interact with the transcriptional machinery. This introduces additional components into the nucleolus, many of them only transiently. In addition, there is control of virtually every step of ribosome biogenesis, thereby adding more proteins and RNAs to the mixture.

A major difference between prokaryotes and eukaryotes is that the latter contain multiple copies of the genes for rRNA (rDNA), numbering in the hundreds. This brings us to the fourth reason for the complexity. The genes are tandem repeats, which in themselves increase the complexity. In addition, for efficient utilization of the transcription, processing and assembly machinery the rDNA repeats are carefully packaged within the compact structure of the nucleolus. This is likely to be the primary factor in the development of the nucleolus.

Finally, the nucleolus has several other functions in addition to ribosome production (covered in Part 3 of this volume). These include routine housekeeping tasks e.g., signal recognition particle (SRP) assembly and nucleolar participation in regulation of cell growth and the cell cycle e.g., nucleostemin. These novel functions add another layer of complexity to an already complex nuclear body.

Nucleolar Structure and Organization

Within the cell nucleus, individual chromosomes tend to occupy preferred territories, which form clusters of genes for efficient use of the transcription machinery (Misteli 2011). One of these territories is the nucleolus, which evolved to be an organized structure for efficient production of ribosomal RNA and ultimately, ribosomes. Were it not for the multiple copies of the genes for ribosomal RNA (rDNA) and their clustering at the nucleolar organizer regions (NORs) on chromosomes, the nucleolus would not exist. Without the gene clustering, eukaryotic cells might go about making ribosomes the way that prokaryotes do, in a less organized

Preface ix

manner. However, as described in Chap. 1, nucleolar structure is not just due to gene organization, but is closely related to the process of assembly of pre-ribosomal particles. This is similar to the structural role of RNA in the biogenesis of other nuclear bodies (Shevtsov and Dundr 2011). It remains essentially correct that the nucleolus is "an organelle formed by the act of building a ribosome" (Mélèse and Xue 1995). This phenomenon accounts for at least two of the major components of nucleoli of higher eukaryotes: the dense fibrillar components (DFCs) and the granular components (GCs), which contain pre-ribosomal RNP particles at various stages of assembly. Ribosome assembly flows from transcription at the border between the fibrillar centers (FCs) and the DFCs, continues in the DFCs, and nears completion in the GCs. Curiously, lower eukaryotes and anamniote higher eukaryotes; e.g., turtles, do not have FCs (Thiry and Lafontaine 2005). The FCs are the interphase equivalent to the NORs, which contain the rDNA. The difference appears to be due to the fact that amniotes have much longer spacer regions in the rDNA than anamniotes. How ribosome biogenesis differs with or without FCs is not clearly understood. As also discussed in Chap. 1, the size of the nucleolus depends on the activity of the cell, with rapidly growing cells having larger nucleoli than cells that are less active.

One of the most unexpected findings has been the identification of more than 6,000 polypeptides in nucleoli (see Chap. 2). Only about 30% of these are related to the process of ribosome biogenesis, including ribosomal proteins and the machinery for producing ribosomes. The diverse identities and functions of the remaining 70%, supports the idea that nucleolus engages in many functions other than ribosome assembly. However, many of these polypeptides have no known functions, leaving the field open for further study. Nucleolar proteomics has moved a step further in being able to quantitatively analyze alterations in protein content under changing physiological conditions. For example, it is possible to monitor changes in the nucleolar protein content following inhibition of transcription or DNA damage. This will further our understanding of changes in nucleolar function in response to chemotherapy or the stress response.

To meet the enormous demand for proteins, growing cells have as many as ten million ribosomes (Alberts et al. 2007). Consequently, the nucleolus must have sufficient capacity to produce large numbers of ribosomal subunits at a rapid pace. The process of evolution has scaled up the first source of raw materials by providing multiple copies of the genes for rRNA. The numbers vary from a few hundred in birds and mammals to several thousand in amphibians. These are contained in tandem repeats connected by spacer regions, whose lengths vary according to the species from which they are derived. Chapter 3 provides a detailed description of how these genes are organized at the DNA level and in chromatin. Complexed with histones and other proteins, the rDNA chromatin can adopt at least three different functional states. The genes that are generally permanently inactive are in the form of condensed, heterochromatic chromatin. Of the two other forms, one is less condensed, but inactive and the other is completely active and fully decondensed. With the aid of labeling techniques, these forms can be identified microscopically. As might be expected, the most active forms are found in the DFC and completely inactive rDNA appears as buds on the nucleolar periphery. McKeown and Shaw also describe in Chap. 3 the kinds of proteins associated with the various forms of rDNA.

The different forms of rDNA chromatin have evolved to be responsive to the needs of the cell and at the same time to facilitate conservation of cellular resources. RNA levels can be modulated either by controlling the rate of transcription or by regulating the number of genes available for transcription; cells obviously use both mechanisms. Although regulation of the transcription machinery has been extensively studied over the past three decades, what accounts for switching on and off of individual genes has become an active area of study. This introduces us to a relatively new area of molecular biology, epigenetics, which is the study of heritable changes in gene expression caused by mechanisms other than changes in the underlying DNA sequence. These chromatin alterations may be carried through multiple cell divisions or they may be perpetuated through numerous generations. The epigenetic process involves the placement of "marks" on histones (acetylation and methylation) and DNA (methlyation). In addition, in vertebrates, the positioning of the nucleosome seems to determine whether an rRNA gene is active or silent. Chapter 4 focuses on the three forms of rDNA chromatin: active, reversible silent, and stable silent rRNA genes and nicely complements and extends the information in Chap. 3. Central to the silencing process is the nucleolar remodeling complex (NoRC), which associates with newly replicated silent rRNA genes. This complex attracts an assortment of enzymes, which modify the histone and DNA components of chromatin. A surprising aspect of the silencing process is that it requires a noncoding RNA that originates in the intergenic spacer region (IGS). Santoro also discusses the intriguing idea that the silencing of large blocks of rDNA results in their heterochromatinization, which not only contributes to the architecture of the nucleolus, but it is also important in maintaining genomic stability.

It has long been known that the rRNA genes are organized in NORs and that these regions of chromosomes can be identified by silver staining (Goodpasture and Bloom 1975). But what facilitates this organization and other than active epigenetic marks, what signals transcriptional competence? Using Xenopus IGSs, McStay and colleagues (Chap. 5) were able construct what are called pseudo-NORs. These have essentially the characteristics of true NORs including silver staining and recruitment of the transcriptional apparatus. However, the pseudo-NORs are not transcriptionally active, because they lack promoter sequences. Thus, it follows that the IGS region and not the transcribed sequences of the rDNA are responsible for NOR formation. One important protein that is involved in NOR formation and rDNA organization is the upstream binding factor (UBF), which is an abundant transcription factor for RNA polymerase I (Pol I). UBF should be considered to be a multifunctional protein in that it not only plays a major role in enhancing transcription, but it also is an architectural factor that participates in the decondensation of active rDNA chromatin. Because of the manner in which UBF acts, it seems likely that these two roles are not separable.

In summary, a variety of factors contribute to the structure and organization of the nucleolus. Although multiple genes for rRNA may exist in a given cell type, chromatin programming at the DNA and protein levels determines whether they are active in nucleoli. Once that commitment is made, the final structure of the nucleolus depends on the cell type in which it is located and the rate of ribosome production that is required by that cell.

Preface   xi

The Role of the Nucleolus in Ribosome Biogenesis

The complex journey of ribosomal RNA on its way to becoming an essential component of a new ribosome begins with transcription by RNA Pol I. Although the transcription of 5S rRNA by RNA Pol III is of equal importance to the cell, it occurs in the nucleoplasm of higher eukaryotes and it is not covered in the volume. At the foundation of Pol I transcription is an elaborate apparatus containing ten catalytic core and four associated subunits in the mammalian enzyme (Chap. 7). By itself, the enzyme is not really functional; it needs nearly a dozen additional factors for initiation, elongation, and termination to operate at optimal efficiency. A surprising feature of the initiation process is that it is highly dynamic; i.e., the individual components move in and out of the nucleolus very rapidly until they become stabilized in the initiation complex (Dundr et al. 2002). Once the polymerase machinery has been assembled it must rapidly move along the rDNA. Although this process is poorly understood, there are several candidate factors, including chromatin remodeling proteins that clear the path for the polymerase to progress down the template. More intriguing is the finding that the apparent driving force for the movement is the combination of nuclear actin and myosin, which function together as a molecular motor. Because ribosome biogenesis is an energy-intensive process, nature has devised multiple mechanisms to conserve energy, but still meet the needs of the cell. Consequently, the activities of nearly all Pol I transcription factors are altered by posttranslational modifications, which in turn, are regulated by numerous signaling pathways. These are triggered in response to metabolic stress, growth factors, nutrient availability, oncogenesis, and phases of the cell cycle. It is now abundantly clear that the level of ribosome biogenesis does not simply depend on the number of rRNA genes available, but that the rate of transcription is fine-tuned to meet the changing conditions in which the cell finds itself.

The steps taken by pre-rRNA during and after transcription are numerous and complex. They have been reviewed in detail recently by Henras et al. (2008); therefore, they are not covered in depth in this volume. However, it is important to highlight a few salient features of the process. How does pre-rRNA make its way from a very long precursor to the 18S, 5.8S, and 28S rRNAs found in ribosomes? Obviously, nucleases are required to do the job, but what determines their ability to precisely generate the ends of the three ribosomal RNAs? It turns out that a subset of the numerous snoRNAs are essential for cleavage. These are not nucleases themselves, but they seem to serve as chaperones or anchors to recruit processing factors and their associated nucleases to the sites to be cleaved. The best known of these is U3 snoRNA as part of a snoRNP complex, which associates with the nascent transcript during transcription. In addition to being an essential factor for pre-rRNA cleavage, U3 also participates in base pairing that facilitates the accurate formation of a pseudoknot in the 18S rRNA.

For ribosomes to function optimally, ribosomal RNA needs to be posttranscriptionally modified. Approximately, 200 sites are modified in vertebrate rRNA with a combination of base methylation, 2′-*O*-methylation, and pseudouridylation. These modifications are believed to stabilize secondary and tertiary structures of the RNA; cell growth and viability are optimal when most or all sites are modified. In Chap. 7, Bleichert and Baserga describe the modification process and the machinery that performs this task. Again, the $2'-O$-methylation and pseudouridylation, but not the base methylation modifications, are precisely directed by snoRNPs. As the multitude of snoRNAs began to be discovered, researchers were surprised to find so many of them, numbering into the hundreds of unique snoRNAs in some species. Now that we know the number of modifications, their locations and the mechanism by which they take place it is clear why the number of snoRNAs is large. Most of these are well characterized and there are crystallographic structures available for a few of the snoRNP complexes (Reichow et al. 2007). In addition, we are beginning to understand how the proteins of these complexes affect the RNA components, facilitating the positioning of the RNA substrates into the active site of the modifying enzymes (Hamma and Ferré-D'Amaré 2010).

As indicated in Chap. 6, the level of transcription by RNA Pol I is adjusted to the cellular growth rate and is also dependent on the phase of the cell cycle of a given cell. But do alterations in ribosome production also affect the cell cycle? There is now evidence for communication between the ribosome biogenesis apparatus and the cell cycle. Chapter 8 provides us with insights into how ribosome biogenesis is monitored during G1 phase and how this influences the G1/S transition. Several studies show that when ribosome biogenesis components are depleted in yeast, the cells accumulate in the G1 phase, although the molecular mechanisms for this have not been determined. In multicellular organisms, deficiencies in certain ribosomal proteins or in factors required for ribosome assembly cause G1 arrest. For these organisms, the G1 arrest is largely mediated by the p53 response (see Chap. 12 for more details on this topic). Depletion of other factors; e.g., nucleophosmin/NPM or nucleolin, causes defects in progression through mitosis. More importantly, several defects in ribosome biogenesis result in diseases, including those labeled as "ribosomeopathies." These are now beginning to be understood, but much work is needed before treatment strategies can be developed.

The numerous steps in ribosome biogenesis require a multitude of different proteins. Some of the best characterized of these are the abundant proteins nucleolin, nucleophosmin/NPM/B23, and NOPP140, which are covered in Chaps. 9–11, respectively. A surprising common feature of these proteins is that they contain what might be considered extremes in the distribution of positively and negatively charged regions. The already highly acidic segments are also phosphorylated by kinase CK2, which contributes to their characteristically low isoelectric points (pIs around 5). Another unusual feature is that the positively charged segments are interspersed with basic segments. These proteins are also heavily modified by additional posttranslational modifications too numerous to mention. So, if these polypeptides have structural features in common, are their functions also similar? The answer to this is mixed. Although the sequences of these proteins became available several decades ago, the functions have been difficult to elucidate. The one apparently universal function of these three proteins is that they all have chaperone activities of one form or another. Nucleolin is able to assist in nucleosome assembly and

Preface xiii

chromatin remodeling through a kind of chaperone activity. NPM also is capable of aiding in nucleosome assembly and it has characteristics very similar to traditional molecular chaperones. NOPP140 acts as a different kind of chaperone by delivering snoRNPs to the nucleolus. However, chaperoning seems to be only part of what these proteins do. For example, nucleolin is essential for Pol I transcription and its RNA binding activity is needed for ribosome assembly. NPM is a ribonuclease that is essential for cleavage of pre-rRNA and it is also involved with centrosome duplication. NOPP140 is a component of Cajal bodies and is also a transcription factor for RNA Pol II. Thus, these are multifunctional proteins that are utilized for many cellular activities.

Novel Functions of the Nucleolus

In the early 1990s clues began to appear that suggested that the nucleolus did other things besides assemble ribosomes. Researchers were surprised to find proteins and RNAs in the nucleolus that had no apparent function in ribosome biogenesis. This idea has been especially reinforced by proteomic studies, which have revealed that a minority of proteins in the nucleolus are involved with its traditional role (see Chap. 2). The list of new functions for the nucleolus is growing and the nucleolus is now established as "plurifunctional" as proposed by Pederson (1998).

Why do multiple functions not related to ribosome biogenesis cluster in the nucleolus? We have a poor understanding of this but there are a few clues that might point us in the right direction. Organisms have evolved to utilize what is available to them and the nucleolus provides an abundance of molecular machinery of which to take advantage. The most obvious example is one involving spliceosomal RNAs, which traffic through the nucleolus to be modified by 2'-O-methylation and pseudouridylation (Lange 2004). The nucleolus contains the enzymes and guide snoRNAs to accomplish that task. In the case of another RNP, the SRP, it is less obvious why assembly is partially performed in the nucleolus (see Chap. 15). Although the SRP is a RNP, there is no evidence that it utilizes ribosome biogenesis components for the assembly process and the SRP RNA is not modified in the way that spliceosomal RNAs are. Furthermore, the SRP components are not found in the same locations as are pre-ribosomal particles. We are left with the presumption that in assembling the SRP, the nucleolus provides a platform that is separate from the rest of the cell. What anchors the SRP components in the nucleolus has not been determined.

The primary mediator of the response to cell stress is the tumor suppressor protein p53, which can trigger either apoptosis or inhibition of cell growth cell cycle arrest when its cellular levels are increased (Ryan et al. 2001). p53 is normally kept at low levels by a continuous cycle of syntheses and degradation. About a decade ago, it was discovered that the nucleolus is the location of a few proteins linked to p53 regulation, operating by some poorly understood mechanism (Zhang and Xiong 2001). An important advance in our understanding of the role of the nucleolus in this process came through the work of Rubbi and Milner (2003) who showed that a number of agents that cause p53 stabilization also disrupt the nucleolus. This suggested that the nucleolus acts as a general stress sensor for the cell. How it performs this task is not entirely clear, but Chap. 12 provides us with an overview of the machinery involved. A key player is the tumor suppressor ARF, which is primarily a nucleolar protein. ARF is an inhibitor of the ubiquitin ligase, MDM2, which marks p53 for degradation by the proteasomal system. This inhibition of MDM2 appears to take place in the nucleoplasm, so it seems possible that ARF is released from the nucleolus by its disruption, although this remains a debatable issue. There is also an intriguing relationship between ARF and ribosome biogenesis; ARF interacts with NPM/B23, which serves as a ribonuclease for the cleavage of at least one site in pre-rRNA. It is interesting that overexpression of ARF stimulates the degradation of NPM, which would obviously cause a defect in the processing of pre-rRNA. As indicated in Chap. 8, unproductive ribosome synthesis can lead to cell cycle arrest. This illustrates the intricate relationships among the cell stress response, ribosome biogenesis, and the cell cycle.

Other nucleolar proteins aid in controlling cell cycle progression. The most well characterized of these are the proteins belonging to the nucleostemin family (see Chap. 13). The protein was named such because of its enrichment in embryonic stem cells (Tsai and McKay 2005). Nucleostemin (NS) is a major factor in controlling cell cycle progression; low levels of it inhibit, intermediate levels promote, and overexpression inhibits progression. These effects are channeled through the p53 system. NS is a GTP-binding protein, with the GTP-bound form preferring the nucleolar location. Conversely, the GTP-unbound form of NS has a nucleoplasmic location, where it interacts with MDM2. This has a stabilizing effect on MDM2 by preventing its ubiquitylation, which ultimately results in a lower transcriptional activity of p53. The current knowledge of NS reinforces the idea that the nucleolus is not only itself regulated by cell growth and division, but that it actively participates in their control.

The nucleolus seems to need additional nucleoplasmic actors to play supporting roles. One of these is the Cajal body (CB). Because of its proximity to and occasional physical association with the nucleolus, it was originally called the nucleolar accessory body by its discoverer, Santiago Ramón y Cajal in 1903. In the 1960s, when electron microscopists examined the CBs, they found that they were composed of aggregates of tangled threads and named the structure the coiled body. Consequently, the major protein component of CBs was given the name, coilin. However, about 10 years ago the name of the CB was changed to Cajal body to honor its discoverer. Although it has been over a hundred years since this nuclear body was first observed, its functions were poorly understood until recently. As described in Chap. 15 maturation of snoRNPs occurs in the CBs; this is part of the supply chain for providing tools to build ribosomes in the nucleolus. Additionally, there is exchange of some nonribosomal proteins between the CBs and the nucleolus. Finally, the nucleolus and CBs share a similar response to stress, probably as a means of coordinating the levels of ribosome production with the availability of snoRNPs.

Preface

We have seen that cells take advantage of the nucleolus for performing a variety of functions not related to ribosome biogenesis. In the same vein, invading organisms utilize the nucleolus for crucial parts of their life cycles. This is especially the case with viruses, many of which have components that locate in the nucleolus (see Chap. 14). Viral proteins of many different types, including those from RNA and DNA viruses can be found in the nucleolus. Because viruses carry limited amounts of genetic information, one can understand why they need to hijack cellular structures and components for replication. However, in many cases, it has yet to be determined what the nucleolar locations of these components do for the virus. Of special importance is HIV-1, which has two proteins that are found in nucleoli of infected cells. One of these is the Rev protein whose function is to facilitate the transport of unspliced or partially spliced HIV-1 mRNA to the cytoplasm. The nucleolar location is essential for that function. The second HIV protein that locates partially in the nucleolus is the Tat protein, which binds the HIV-1 mRNA TAR element. As with Rev, the nucleolar trafficking of Tat is essential for HIV-1 replication. The nucleolar location of these viral components is interesting in itself, but even more appealing is the possibility that the nucleolar machinery can be utilized for treatment of HIV-1 infections (Chap. 17). Rossi and his colleagues have developed ribozymes based on snoRNAs that cleave HIV RNA, which results in inhibition of replication (Unwalla et al. 2008). Taking this approach one step further, Rossi and colleagues used siRNA in a TAR decoy to inhibit viral replication (Unwalla and Rossi 2010). What is more important about these pioneering studies is that they are now being translated into clinical trials (DiGiusto et al. 2010) and they offer hope for the development of new therapeutic modalities.

The Future of the Nucleolus

Does the nucleolus hold more surprises or is the field at a level at which major discoveries will be few and far between? As Niels Bohr once said, "Prediction is very difficult, especially about the future." Thus, we can only speculate about the outlook for new discoveries in this subject. Future directions are discussed in most chapters of this volume, but a few issues should be highlighted and expanded. The first of these deals with mechanism at several levels. Although, the component parts of the ribosome biogenesis process have been defined, we are only beginning to understand how they do what they do. For example, we do not really understand how the transcription machinery is propelled along the template and how this is coordinated with the vectorial process of ribosome assembly. In another example, we have a general idea of how the snoRNPs operate to modify rRNA and to aid in the cleavage of pre-rRNA, but our understanding of the mechanism by which it takes place is limited. Expanding this knowledge will require difficult and painstaking work utilizing genetic engineering, enzymology, more X-ray crystallography of complexes, and possibly technologies that have not yet been invented.

A second issue is related to regulation. Much regulation is at the level of transcription and involves communication with the rest of the cell. This is beginning to be understood, but much more detail is needed. But what about regulation of the ribosome biogenesis process itself? Ribosome assembly requires a precise order and timing of events. How are these controlled? Another issue concerns the feedback of ribosome biogenesis with the cell cycle.

For decades, researchers have been attempting to correlate ultrastructure with function in the nucleolus. One of the puzzling features of nucleolar ultrastructure is the fibrillar center. It appeared later in evolution and is not present in some lower eukaryotes. It contains rDNA and RNA Pol I, but transcription occurs only at its periphery. So what is it and what does it do for the cell? This and other poorly understood ultrastructural questions should be answered in the future. Relating ultrastructure to function can be taken a step further by doing it in three dimensions. Ongoing studies using electron tomography are aimed at understanding the three-dimensional organization of nucleolar components (Tchelidze et al. 2008). We look forward to advances in this area.

Although the potential for surprises in the area of ribosome biogenesis may have reached its apex, the chances for finding more novel functions in the nucleolus remain high. Proteomics has shown us that there are hundreds of proteins of unknown function in the nucleolus; these are likely to keep researchers busy for many years. In addition, the roles of many viral components in the nucleolus will continue to intrigue us and hopefully, move beyond the phenomenology that is now the case with many viral components in the nucleolus. More importantly, there is already evidence that we can take advantage of our knowledge of the nucleolus to develop therapeutic strategies. Hopefully, this approach will be extended to viruses in addition to HIV and to other diseases. We may even see a new era of nucleolar translational medical research.

References

Alberts B, Johnson A, Lewis J, Raff M, Roberts K, Walter P (2007) Molecular biology of the cell, 5th edn. Garland Publishing, New York, p 360

Ben Shem A, Jenner L, Yusupova G, Yusupov M (2010) Crystal structure of the eukaryotic ribosome. Science 330:1203–1209

DiGiusto DL, Krishnan A, Li L, Li H, Li S, Rao A, Mi S, Yam P, Stinson S, Kalos M, Alvarnas J, Lacey SF, Yee JK, Li M, Couture L, Hsu D, Forman SJ, Rossi JJ, Zaia JA (2010) RNA-based gene therapy for HIV with lentiviral vector-modified CD34(+) cells in patients undergoing transplantation for AIDS-related lymphoma. Sci Transl Med 2:36ra43

Dundr M, Hoffmann-Rohrer U, Hu QY, Grummt I, Rothblum LI, Phair RD, Misteli T (2002) A kinetic framework for a mammalian RNA polymerase in vivo. Science 298:1623–1626

Goodpasture C, Bloom SE (1975) Visualization of nucleolar organizer regions in mammalian chromosomes using silver staining. Chromosoma 53:37–50

Hadjiolov AA (1985) The nucleolus and ribosome biogenesis, vol 12. Cell biology monographs. Springer, Wien

Hamma T, Ferré-D'Amaré AR (2010) The box H/ACA ribonucleoprotein complex: interplay of RNA and protein structures in post-transcriptional RNA modification. J Biol Chem 285:805–809

Henras AK, Soudet J, Gerus M, Lebaron S, Caizergues-Ferrer M, Mougin A, Henry Y (2008) The post-transcriptional steps of eukaryotic ribosome biogenesis. Cell Mol Life Sci 65:2334–2359

Hughes JM (1996) Functional base-pairing interaction between highly conserved elements of U3 small nucleolar RNA and the small ribosomal subunit RNA. J Mol Biol 259:645–654

Lange TS (2004) Trafficking of spliceosomal small nuclear RNAs through the nucleolus. In: Olson MOJ (ed) The nucleolus. Landes Bioscience, Austin, pp 329–342

Mélèse T, Xue Z (1995) The nucleolus: an organelle formed by the act of building a ribosome. Curr Opin Cell Biol 7:319–324

Misteli T (2011) The inner life of the genome. Sci Am 304:66–73

Olson MOJ (2004) Introduction. In: Olson MOJ (ed) The nucleolus. Landes Bioscience, Austin, pp 1–9

Rabl J, Leibundgut M, Ataide SF, Haag A, Ban N (2011) Crystal structure of the eukaryotic 40S ribosomal subunit in complex with initiation factor 1. Science 331:730–736

Ramakrishnan V, Moore PB (2001) Atomic structures at last: the ribosome in 2000. Curr Opin Struct Biol 11:144–154

Reichow SL, Hamma T, Ferre-D'Amare AR, Varani G (2007) The structure and function of small nucleolar ribonucleoproteins. Nucleic Acids Res 35:1452–1464

Rubbi CP, Milner J (2003) Disruption of the nucleolus mediates stabilization of p53 in response to DNA damage and other stresses. EMBO J 22:6068–6077

Ryan KM, Phillips AC, Vousden KH (2001) Regulation and function of the p53 tumor suppressor protein. Curr Opin Cell Biol 13:332–337

Shevtsov SP, Dundr M (2011) Nucleation of nuclear bodies by RNA. Nat Cell Biol 13:167–173

Tchelidze P, Kaplan H, Beorchia A, O'Donohue MF, Bobichon H, Lalun N, Wortham L, Ploton D (2008) Three-dimensional reconstruction of nucleolar components by electron microscope tomography. Methods Mol Biol 463:137–158

Thiry M, Lafontaine DL (2005) Birth of a nucleolus: the evolution of nucleolar compartments. Trends Cell Biol 15:194–199

Tsai RY, McKay RD (2005) A multistep, GTP-driven mechanism controlling the dynamic cycling of nucleostemin. J Cell Biol 168:179–184

Unwalla HJ, Li H, Li SY, Abad D, Rossi JJ (2008) Use of a U16 snoRNA-containing ribozyme library to identify ribozyme targets in HIV-1. Mol Ther 16:1113–1119

Unwalla HJ, Rossi JJ (2010) A dual function TAR Decoy serves as an anti-HIV siRNA delivery vehicle. Virol J 7:33

Wagner R (1835) Einige bemerkungen un fragen über das keimbläschen (vesicular germinativa). Müller's Arch Anat Physiol U Wiss Med 268

Zhang YP, Xiong Y (2001) Control of p53 ubiquitination and nuclear export by MDM2 and ARF. Cell Growth Differ 12:175–186

Acknowledgments

When I was first introduced to the nucleolus in the late 1960s, the state of knowledge on the subject was such that it could be covered adequately in one book. At that time, Harris Busch and Karel Smetana had just completed such a volume (Busch H, Smetana K (1970) The nucleolus. Academic Press, New York). The timing of the publication of the latter volume was important in that it followed the decade in which researchers determined that the nucleolus does, in fact, make pre-ribosomal RNA and that the genes for such are located in the nucleolus organizer regions on chromosomes. This book was also the first comprehensive overview of the subject. The next compilation was published in 1982, when Jordan and Cullis gathered material from a symposium at the 200th meeting of the Society for Experimental Biology in 1980 in Oxford, UK, and assembled it into a volume also entitled *The Nucleolus* (Jordan EG, Cullis CA (1982) The nucleolus. Cambridge University Press, New York). The latter book covered research from the previous decade concerning the locations and multiplicity of ribosomal RNA and protein genes, transcription and maturation of pre-rRNA, assembly of pre-ribosomal particles, and regulation of ribosome biogenesis. Another comprehensive volume on the nucleolus (Hadjiolov AA (1985) The nucleolus and ribosome biogenesis. Springer, New York) is still a useful reference for many aspects of the nucleolus. A book by Thiry and Goessens focuses on nucleolar ultrastructure and also includes a very useful historical overview, presented in outline form (Thiry M, Goessens G (1996) The nucleolus during the cell cycle. R. G. Landes Company, Austin). A more specialized volume on rDNA transcription was compiled in 1998 by Marvin Paule (Paule MR (ed) (1998) Transcription of ribosomal RNA genes by eukaryotic RNA polymerase I. R.G. Landes Company, Georgetown). Each year, several reviews provide updates on general aspects of the nucleolus as well as one with specialized topics within the subject area. Many of these are cited in the chapters of this volume.

The latest attempt at comprehensive coverage of the field was a book edited by the editor of the current volume (Olson MOJ (ed) (2004) The nucleolus. R.G. Landes Company, Georgetown). However, the body of knowledge has exploded to the point where several volumes would be needed to cover all topics of this field. Therefore, the current collection of chapters focuses on well-established components, systems, and mechanisms operating in the nucleolus. These are covered in greater depth than in previous compilations. This volume is published under the umbrella of *Protein Reviews*; thus, it is reasonable to expect that the emphasis would be on nucleolar proteins, especially those that have been studied extensively. This is not to diminish the crucial importance of nucleic acids, which are covered indirectly in various chapters. A guiding theme of this volume is that the structure and function of the nucleolus ultimately is the result of individual macromolecules interacting with other macromolecules. Thus, we have included chapters on nucleolar ultrastructure, the locations and dynamics of nucleolar components, and on regulatory systems. Because of recent discoveries on novel functions of the nucleolus, about a third of the book is devoted to that topic.

Having worked in this field for about four decades, it has been a pleasure to see it mature to where it stands today. The subject has always been intriguing, but it is even more so now and full of surprises. Therefore, I thank Professor M. Z. Atassi, the Editor of this series, for giving me the opportunity to compile and edit this volume. I also thank all of the authors for their enthusiasm and cooperation in submitting their chapters. I have made numerous friends in this field over the years, too many to name individually; I dedicate this book to all of them.

Contents

Part I Nucleolar Structure and Organization

1 Structural Organization of the Nucleolus as a Consequence of the Dynamics of Ribosome Biogenesis 3
Danièle Hernandez-Verdun

2 The Dynamic Proteome of the Nucleolus 29
François-Michel Boisvert, Yasmeen Ahmad, and Angus I. Lamond

3 The Structure of rDNA Chromatin 43
Peter J. Shaw and Peter C. McKeown

4 The Epigenetics of the Nucleolus: Structure and Function of Active and Silent Ribosomal RNA Genes 57
Raffaella Santoro

5 UBF an Essential Player in Maintenance of Active NORs and Nucleolar Formation 83
Alice Grob, Christine Colleran, and Brian McStay

Part II Role of the Nucleolus in Ribosome Biogenesis

6 The RNA Polymerase I Transcription Machinery 107
Renate Voit and Ingrid Grummt

7 Small Ribonucleoproteins in Ribosome Biogenesis 135
Franziska Bleichert and Susan Baserga

8 Crosstalk Between Ribosome Synthesis and Cell Cycle Progression and Its Potential Implications in Human Diseases 157
Marie Gérus, Michèle Caizergues-Ferrer, Yves Henry, and Anthony Henras

xxi

Contents

9 The Multiple Properties and Functions of Nucleolin 185
Rong Cong, Sadhan Das, and Philippe Bouvet

10 The Multifunctional Nucleolar Protein Nucleophosmin/NPM/B23 and the Nucleoplasmin Family of Proteins 213
Shea Ping Yip, Parco M. Siu, Polly H.M. Leung, Yanxiang Zhao, and Benjamin Y.M. Yung

11 Structure and Function of Nopp140 and Treacle 253
Fang He and Patrick DiMario

Part III Novel Functions of the Nucleolus

12 The Role of the Nucleolus in the Stress Response 281
Laura A. Tollini, Rebecca A. Frum, and Yanping Zhang

13 New Frontiers in Nucleolar Research: Nucleostemin and Related Proteins 301
Robert Y.L. Tsai

14 Viruses and the Nucleolus 321
David Matthews, Edward Emmott, and Julian Hiscox

15 Assembly of Signal Recognition Particles in the Nucleolus 347
Marty R. Jacobson

16 Relationship of the Cajal Body to the Nucleolus 361
Andrew Gilder and Michael Hebert

17 Role of the Nucleolus in HIV Infection and Therapy 381
Jerlisa Arizala and John J. Rossi

Index 403

Contributors

Yasmeen Ahmad Wellcome Trust Centre for Gene Regulation and Expression, College of Life Sciences, University of Dundee, Dundee, UK

Jerlisa Arizala Division of Molecular Biology, Beckman Research Institute of The City of Hope, Duarte, CA, USA

Susan Baserga Departments of Molecular Biophysics and Biochemistry, Genetics, and Therapeutic Radiology, Yale University, New Haven, CT, USA

Franziska Bleichert Department of Molecular and Cell Biology, University of California, Berkeley, CA, USA

François-Michel Boisvert Wellcome Trust Centre for Gene Regulation and Expression, College of Life Sciences, University of Dundee, Dundee, UK

Philippe Bouvet Université de Lyon, Ecole Normale Supérieure de Lyon, CNRS USR 3010, Laboratoire Joliot-Curie, Lyon, France

Michèle Caizergues-Ferrer Centre National de la Recherche Scientifique, Laboratoire de Biologie Moléculaire Eucaryote, Toulouse, France

Université de Toulouse, UPS, Toulouse, France

Christine Colleran Centre for Chromosome Biology, School of Natural Sciences, National University of Ireland Galway, Galway, Ireland

Rong Cong Université de Lyon, Ecole Normale Supérieure de Lyon, CNRS USR 3010, Laboratoire Joliot-Curie, Lyon, France

The Institute of Biomedical Sciences and School of Life Sciences, East China Normal University, Shanghai, China

Sadhan Das Université de Lyon, Ecole Normale Supérieure de Lyon, CNRS USR 3010, Laboratoire Joliot-Curie, Lyon, France

Patrick DiMario Department of Biological Sciences, Louisiana State University, Baton Rouge, LA, USA

Edward Emmott Institute of Molecular and Cellular Biology, Biological Sciences, and Astbury Centre for Structural Molecular Biology, University of Leeds, Leeds, UK

Rebecca A. Frum Department of Radiation Oncology, University of North Carolina at Chapel Hill, Chapel Hill, NC, USA

Marie Gérus Centre National de la Recherche Scientifique, Laboratoire de Biologie Moléculaire Eucaryote, Toulouse, France,

Université de Toulouse, UPS, Toulouse, France

Andrew Gilder Department of Biochemistry, The University of Mississippi Medical Center, Jackson, MS, USA

Alice Grob Centre for Chromosome Biology, School of Natural Sciences, National University of Ireland Galway, Galway, Ireland

Ingrid Grummt Division of Molecular Biology of the Cell II, German Cancer Research Center, Heidelberg, Germany

Fang He Department of Biological Sciences, Louisiana State University, Baton Rouge, LA, USA

Michael Hebert Department of Biochemistry, The University of Mississippi Medical Center, Jackson, MS, USA

Anthony Henras Centre National de la Recherche Scientifique, Laboratoire de Biologie Moléculaire Eucaryote, CNRS, Toulouse Cedex 9, France

Université de Toulouse, UPS, Toulouse, France

Laboratoire de Biologie Moléculaire Eucaryote, CNRS, Toulouse Cedex, France

Yves Henry Centre National de la Recherche Scientifique, Laboratoire de Biologie Moléculaire Eucaryote, CNRS, Toulouse Cedex 9, France

Université de Toulouse, UPS, Toulouse, France

Laboratoire de Biologie Moléculaire Eucaryote, CNRS, Toulouse Cedex, France

Danièle Hernandez-Verdun Institut Jacques Monod-UMR 7592, CNRS, Université Paris Diderot, Paris Cedex 13, France

Julian Hiscox Institute of Molecular and Cellular Biology, Faculty of Biological Sciences, and Astbury Centre for Structural Molecular Biology, University of Leeds, Leeds, UK

Marty R. Jacobson Saccomanno Research Institute, St. Mary's Hospital & Medical Center, Grand Junction, CO, USA

Angus I. Lamond Wellcome Trust Centre for Gene Regulation and Expression, College of Life Sciences, University of Dundee, Dundee, UK

Contributors

Polly H.M. Leung Department of Health Technology and Informatics, The Hong Kong Polytechnic University, Hung Hom, Kowloon, Hong Kong SAR, China

David Matthews School of Cellular and Molecular Medicine, University of Bristol, Bristol, UK

Peter C. McKeown Department of Botany and Plant Science, Aras de Brun, NUI Galway, Galway, Ireland

Brian McStay Centre for Chromosome Biology, School of Natural Sciences, National University of Ireland Galway, Galway, Ireland

John J. Rossi Division of Molecular Biology, Beckman Research Institute of The City of Hope, Duarte, CA, USA

Raffaella Santoro Institute of Veterinary Biochemistry and Molecular Biology, University of Zürich, Zurich, Switzerland

Peter J. Shaw Department of Cell and Developmental Biology, John Innes Centre, Norwich, UK

Parco M. Siu Department of Health Technology and Informatics, The Hong Kong Polytechnic University, Hung Hom, Kowloon, Hong Kong SAR, China

Laura A. Tollini Curriculum in Genetics and Molecular Biology, University of North Carolina at Chapel Hill, Chapel Hill, NC, USA

Robert Y.L. Tsai Center for Cancer and Stem Cell Biology, Alkek Institute of Biosciences and Technology, Texas A&M Health Science Center, Houston, TX, USA

Renate Voit Division of Molecular Biology of the Cell II, German Cancer Research Center, Heidelberg, Germany

Shea Ping Yip Department of Health Technology and Informatics, The Hong Kong Polytechnic University, Hung Hom, Kowloon, Hong Kong SAR, China

Benjamin Y.M. Yung Department of Health Technology and Informatics, The Hong Kong Polytechnic University, Hung Hom, Kowloon, Hong Kong SAR, China

Yanping Zhang Departments of Radiation Oncology and Pharmacology, University of North Carolina at Chapel Hill, Chapel Hill, NC, USA

Lineberger Comprehensive Cancer Center, University of North Carolina at Chapel Hill, Chapel Hill, NC, USA

Yanxiang Zhao Department of Applied Biology and Chemical Technology, The Hong Kong Polytechnic University, Hung Hom, Kowloon, Hong Kong SAR, China

Part I
Nucleolar Structure and Organization

Chapter 1
Structural Organization of the Nucleolus as a Consequence of the Dynamics of Ribosome Biogenesis

Danièle Hernandez-Verdun

Abbreviations

CDKs	Cyclin dependent kinases
CK2	Casein kinase 2
DFC	Dense fibrillar component
DRB	5,6-Dichloro-1-ribo-furanosylbenzimidazole
EM	Electron microscopy
FC	Fibrillar center
FRET	Fluorescence resonance energy transfer
GC	Granular component
GFC	Giant FC
NADs	Nucleolus-associated chromatin domains
NDF	Nucleolar derived foci
NOR	Nucleolar organizing region
NPM	Nucleophosmin
NS	Nucleostemin
PAGFP	Photoactivatable GFP
PNB	Prenucleolar body
PNC	Perinucleolar compartment
PtK1	Potorous tridactylis kidney
Pol I	RNA polymerase I
rDNA	Ribosomal gene
rRNA	Ribosomal RNA
rProtein	Ribosomal protein
RNP	Ribonucleoprotein

D. Hernandez-Verdun (✉)
Institut Jacques Monod-UMR 7592, CNRS, Université Paris Diderot,
15 rue Hélène Brion, 75205 Paris Cedex 13, France
e-mail: daniele.hernandez-verdun@univ-paris-diderot.fr

M.O.J. Olson (ed.), *The Nucleolus*, Protein Reviews 15,
DOI 10.1007/978-1-4614-0514-6_1, © Springer Science+Business Media, LLC 2011

snoRNP	Small nucleolar RNP
SUMO	Small ubiquitin-like modifier
TIP5	TTF1-interacting protein-5
UBF	Upstream binding factor

1.1 Introduction

The nucleolus is the most prominent visible structure in the nucleus of all eukaryotic cells. Cytologists consequently described it even before it was known to be the site of ribosome biogenesis in eukaryotic cells (Montgomery 1898). Each cell possesses at least one nucleolus that reflects the state of activity or differentiation of that particular cell. It was proposed that the nucleolus is "an organelle formed by the act of building a ribosome" (Mélèse and Xue 1995). Indeed, the size of the nucleolus depends on the level of ribosome production (Smetana and Busch 1974) and the molecular processes occurring in this organelle determine the structural organization of the nucleolus (Hadjiolov 1985). By electron microscopy (EM), this membrane-less organelle presents a structural compartmentation corresponding to the major steps of ribosome biogenesis. This compartmentation has been described in higher eukaryotes as composed of three basic "building blocks": the fibrillar center (FC), the dense fibrillar component (DFC) and the granular component (GC) corresponding to the major steps of the biogenesis of the two ribosome subunits (see below for details) (Goessens 1984; Hernandez-Verdun 1986; Jordan 1991; Mosgoeller 2004). The nucleolus constitutes a model to understand the principles of the organization of the nuclear domains; the dynamics of assembly of these domains after mitosis; and the relationship between nuclear bodies dedicated to related functions, in particular the Cajal body, the PML (promyelocytic leukaemia) body, and the nuclear speckles (for definitions see (Spector 2001)).

The region of the chromosome that carries the ribosomal genes (rDNA) was designated the nucleolar-organizing region (NOR) because the formation of the nucleolus was associated with a particular chromosome translocation in *Zea mays* (McClintock 1934). This hypothesis was verified in an anucleolated *Xenopus laevis* mutant in which rRNAs are not synthesized (Brown and Gurdon 1964). The number of NOR-bearing chromosomes varies in different species, from one chromosome to several chromosome pairs. In humans, the acrocentric chromosomes 13, 14, 15, 21, and 22 are the NOR-bearing chromosomes (Henderson et al. 1972). In haploid budding yeast (*Saccharomyces cerevisiae*), the NOR-bearing chromosome is chromosome XII. In dividing eukaryotes, the nucleoli assemble at the exit from mitosis; they remain functionally active throughout interphase, and disassemble at the beginning of mitosis. Ribosome production varies between G1/S/G2 interphase periods, being maximal in G2 (Sirri et al. 1997, 2000b). Ribosome biogenesis is a multistep process including ribosomal RNA (rRNA) synthesis, modification, and processing, and rRNA assembly with most of the ribosomal proteins (rproteins). The rDNA are repeated genes organized in tandem (Miller and Beatty 1969).

The active rDNA is transcribed by the RNA polymerase I (Pol I) machinery; the pre-rRNAs are processed by specific processing nucleolar proteins and modified by snoRNPs, and during the course of these processes are associated with most of the rproteins corresponding to the small (40S) and large (60S) ribosomal subunits. During the past few decades, the complexity of the nucleolus was deciphered thanks to multiple approaches developed for the in situ localization of proteins, RNAs, and DNAs at photonic and EM resolution and for the third dimension with 3D reconstructions (Dupuy-Coin et al. 1986a; Hozàk et al. 1994, 1989; Le Panse et al. 1999). In parallel, the ability to prepare isolated nucleoli has made biochemical and molecular biology methods useful for the identification of proteins, various RNAs, and chromatin associated with the nucleolus (Hadjiolov 1985; Ochs 1998). It was proposed that the nucleolus is a plurifunctional domain because proteins and RNAs not involved in ribosome biogenesis were observed in the nucleolus, were also recovered in proteomic analyses, and were linked to various pathologies (Andersen et al. 2002, 2005; Pendle 2005; Chamousset et al. 2010; Scherl et al. 2002; Westman et al. 2010). Today new techniques make it possible to measure molecular mobility in living cells and to analyze the nucleolar proteome and interactome (see below).

The organization of the nucleolus as a consequence of ribosome biogenesis has been reviewed in animal and plant cells, during the cell cycle, development, or pathology (for the most recent reviews see (Derenzini et al. 2009; Gébrane-Younès et al. 2005; Hernandez-Verdun et al. 2010; Mosgoeller 2004; Raska et al. 2006; Sirri et al. 2008)) as well as related to the extra ribosomal functions of the nucleolus (Boisvert et al. 2007; Hiscox 2002; Olson 2004; Pederson 1998).

The objective of the present review is to identify common features among the diversity of nucleolar structures and attempts to cover three topics: (1) the nucleolar structures resulting from the dynamics of the ribosome biogenesis, (2) nucleolar assembly, and (3) the nucleoli in the nuclear environment. In conclusion, some open questions that could be important in the future are proposed.

1.2 Nucleoli and Ribosome Biogenesis

The nucleolus is present in all eukaryotic cells during interphase and the size of the nucleolus reflects the dynamics of ribosome production. In dividing cells, ribosome production is high and the size of the nucleolus varies from 0.5 to 7 μm in diameter (Fig 1.1a). In most cancer cells, an increased nucleolar volume is characteristic when compared with the tissue of origin (for reviews see (Busch and Smetana 1974; Hadjiolov 1985, Montanaro et al. 2008)). In aggressive cell lines of breast cancer, the size of the nucleolus increases by 30% during tumor progression and in addition, the translational capacity of the ribosomes is modified (Belin et al. 2009). In differentiated cells, ribosome production decreases and is stopped at final steps of differentiation, for example, in lymphocytes or nucleated erythrocytes. The size of these remnant nucleoli is reduced to 0.1–0.3 μm and the structure is nearly exclusively fibrillar (Fig 1.1b). In the remnant nucleolus of X. laevis erythrocytes, modified

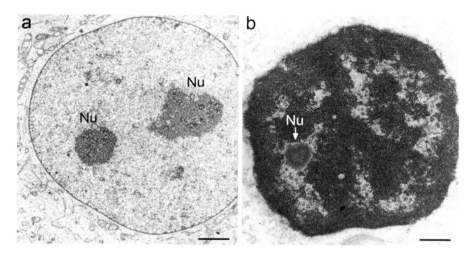

Fig. 1.1 Nuclear organization of a human HeLa cell and a *X. laevis* erythrocyte prepared by conventional methods for EM. The cells were fixed by glutaraldehyde and osmic acid. The sections were contrasted by uranyl acetate and lead citrate. Two nucleoli (Nu) of large size, respectively 3 and 7 μm in diameter, are visible in the HeLa nucleus (**a**). One small nucleolus (Nu) of 0.3 μm in diameter is visible in the erythrocyte nucleus (**b**). Scale bar: $a = 3$ μm, $b = 0.3$ μm

UBF, processing nucleolar proteins, small nucleolar RNAs (snoRNAs), and partially processed pre-rRNAs have been detected (Verheggen et al. 2001). After 40 years of investigation based on multiple approaches, it has become clear that the different steps of ribosome biogenesis have created the basic structure and the general organization of the nucleolus.

1.2.1 Active Ribosome Biogenesis Generates Nucleolar Organization

The basic structures of nucleolar organization were first described using transmission EM. The conventional preparations for transmission EM correspond to thin sections of the biological material contrasted with metals. The cells or tissues are fixed by aldehyde and osmium, embedded in Epon, and the thin sections (50–150 nm thick) are contrasted with uranyl acetate and lead citrate (Puvion et al. 1994). Consequently, the structures containing nucleic acids are highly contrasted by uranyl and lead salts; the resolution is very good but the 3D organization of the nucleolus is more difficult to appreciate without serial sections. Thin serial sections of nucleoli should be numerous and oriented to compare the organization in different cells. For example, 30 serial thin sections in an oriented cell monolayer were necessary for the 3D reconstruction in EM of nucleoli of medium size corresponding to one NOR (Junéra et al. 1995). Alternatively, the observation of large numbers of random sections

Fig. 1.2 Nucleolar organization in a HeLa nucleus. In (**a**) the three nucleolar components are visible: the fibrillar centers (FCs) of different sizes, the dense fibrillar component (DFC), and the granular component (GC). In (**b**) the relationship of a centrally located nucleolus (Nu) and the folded nucleolar envelope (NE) is illustrated. Scale bar: $a = 1$ μm, $b = 5$ μm

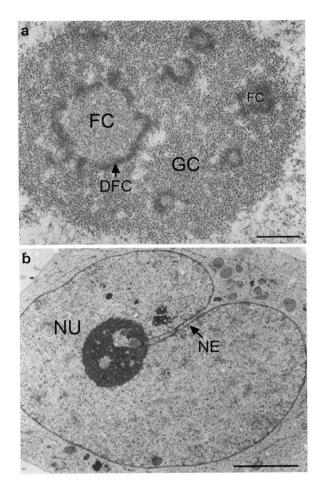

provides a statistical view of this organization but a biased interpretation is not excluded on the basis of the initial hypothesis. In conventional preparations, the nucleoli appear mainly composed of fibrils and granules of 15–20 nm in diameter (Fig 1.2a). The fibrils (about 5 nm in diameter) in low contrast areas form the FCs, first named by Recher et al. (1969), described by Goessens and collaborators (1984), and the object of large debates in the community of nucleologists (see reviews (Jordan 1991; Raska et al. 2006; Scheer and Hock 1999)). The FCs are partly surrounded by the contrasted and highly packed fibrils that form the DFC; the FC/DFCs are included in the GC (Fig 1.2a) (Hernandez-Verdun 1986). Stereological studies on human diploid fibroblasts or Ehrlich tumor cells show that 90% of the nucleolar volume is accounted for by the DFC (about 15%) and the GC (75%) (Hadjiolov 1985). In the nucleoli of higher plants, the proportion of DFC is much higher (Raska et al. 2006). The discrimination between clear (FC) or contrasted

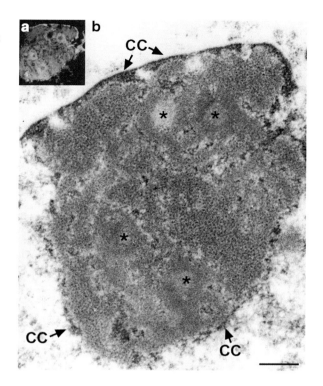

Fig. 1.3 DNA and RNA molecules in a nucleolus. In a PtK1 nucleolus on an EM section (**a**) general view of inverted contrast and (**b**) the nucleic acids are black contrasted with uranyl salt after methylation and acetylation of the amino and carboxyl groups. The *asterisks* indicate four FCs surrounded by DFC, and the RNA granules in the GC are visible in large masses. The condensed chromatin (CC) is visible around the nucleolus as indicated by *arrows* and also inside the nucleolus as chromatin fibers. Scale bar: $b = 0.5$ μm

(DFC) fibrils is not easy in tangential sections without a specific contrast. It is possible to contrast DNA or both DNA and RNA by specific procedures in samples fixed only by aldehyde; in these cases discrimination is easy (Figs. 1.3 and 1.5a) (Derenzini et al. 1993; Gébrane-Younès et al. 1997, Guetg et al. 2010; Junéra et al. 1995; Testillano et al. 1995). The characterization of the nucleolar structure has largely benefited from the EM localization of proteins using specific antibodies, from RNA transcripts using labeled precursors and from specific DNA or RNA sequences using in situ hybridization (Ochs 1998). Ribosome production being a vectorial process, the results collected using all these approaches demonstrate that the different steps of ribosome production generate specific structures. An application of this conclusion is that particular steps of ribosome production can be identified by specific markers of a particular step and consequently localized by light microscopy in the whole cell volume, another way of revealing the organization of the nucleolus (Fig. 1.4). The advantage of light microscopy is easy access to 3D information although with a limited resolution, and more importantly the possibility to analyze and measure the dynamics of molecules in living cells. Presently, the correlative examinations that combine fluorescence microscope images and EM images of the same region open up new possibilities (Robinson et al. 2001; Spiegelhalter et al. 2010). These sophisticated approaches will be of great importance for understanding the nucleolar architecture.

1 Structural Organization of the Nucleolus...

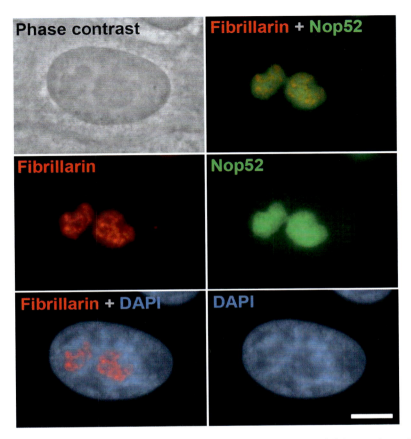

Fig. 1.4 Identification of nucleoli in a human HeLa nucleus. In the upper left image, the nucleoli are visible in contrasted structures when observed by phase contrast light microscopy. In the same nucleus, these contrast structures are decorated in red by fibrillarin (DFC marker) and in *green* by Nop52 (GC marker). In the *upper right* image, the superimposition of the *red* and the *green* labeling shows the distinct distribution of both proteins in the nucleolus. The two lower images show that the DNA in blue is mostly at the nucleolar periphery and that the fibrillarin in red forms a network in the nucleolus. Scale bar: 5 μm

1.2.2 The Building Blocks of the Nucleolar Compartmentation

In higher eukaryotes, all the present data indicate that the nucleolus engaged in active ribosome production is composed of three major building blocks, the FC, DFC, and GC (see details below). In budding yeast (*S. cerevisiae*), only DFC and GC have been described. This is also the case in lower eukaryotes and in insects (Knibiehler et al. 1982, 1984). It has been proposed that the three building blocks emerged during evolution from a bipartite organization in which the FC/DFC components are mixed (Thiry and Lafontaine 2005).

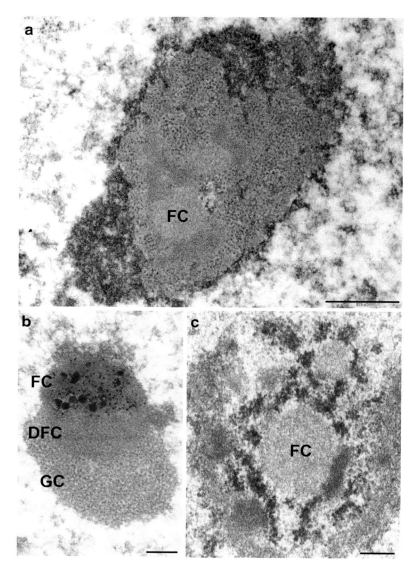

Fig. 1.5 Variability of the nucleolar organization (**a**) in mouse cells, (**b**) after Pol I inhibition, and (**c**) after Roscovitine treatment of HeLa cells. (**a**) Perinucleolar heterochromatin in a mouse NIH3T3 nucleus observed in EM after preferential contrast of the nucleic acids. The DNA and RNA were contrasted with uranyl salt after methylation and acetylation of the amino and carboxyl groups (Testillano et al. 1995). Around the nucleolus, two large clumps of chromatin are visible as well as the perinucleolar chromatin. Two FC are visible in the nucleolus. (**b**) The segregation of the three nucleolar components observed by EM. The HeLa cell was treated with a low concentration of actinomycin D and the Ag-NOR staining (*black dots*) revealed the Ag-NOR proteins in FC. (**c**) The three nucleolar components are not intermingled following treatment with Roscovitine in HeLa cells. Fibrillar center (FC), dense fibrillar component (DFC), and granular component (GC). Scale bar = 1 μm

1 Structural Organization of the Nucleolus...

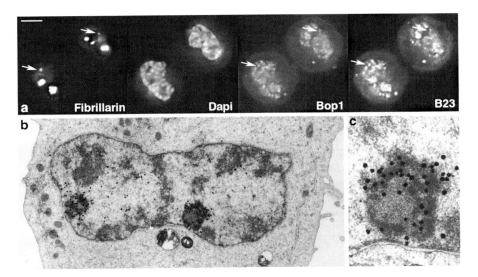

Fig. 1.6 Nucleolar assembly in HeLa cells (**a**) in light microscopy and (**b**) in EM. (**a**) The two daughter cells are in early G1, a few minutes after telophase. The DAPI DNA staining shows that the chromosomes are still partly condensed. At this stage (*left panel*), the fibrillarin is almost completely regrouped around the NORs in the assembling DFC with the exception of a few PNBs indicated by *arrows*; in these PNBs, the markers of the GC, Bop1, and B23, are also detected (*arrows*), indicating that both DFC and GC processing proteins are initially colocalized in the same PNBs. As visible for Bop1 and B23, the CG markers are mostly in PNBs and only a part is localized in the new GC under construction. (**b, c**) Building of the nucleolus observed in EM in one daughter cell in early G1. The labeling corresponds to the sites of RNA synthesis revealed by the autoradiography of tritiated uridine incorporation. Transcription is mostly in the two new nucleoli (**b**) visible close to the nuclear envelope. The high magnification of the right nucleolus shows that transcription is in the DFC around the FC and close to the nuclear envelope. Scale bar = 1 μm

FCs are characterized by a specific EM topology (Figs. 1.2a, 1.3 and 1.6c). They are also characterized by their components, that is, the presence of rDNA as well as Pol I subunits, DNA topoisomerase I, and the transcription upstream binding factor (UBF) (Goessens 1984; Jordan 1991; Puvion-Dutilleul et al. 1991; Scheer and Rose 1984). rDNA transcription is not detected inside, but at the periphery of the FCs (Fig. 1.6c), and this correlates well with the high contrast of the RNA molecules when using the standard EM procedures (Fig. 1.2a) (Hernandez-Verdun and Bouteille 1979; Hozák et al. 1994). The nucleoli of different cell types exhibit a variable number of FCs of different sizes, with an inverse proportion between size and number (Hozák et al. 1989). In the nucleoli of differentiated human lymphocytes, a single FC is visible. If these lymphocytes are stimulated to enter the cell cycle, ribosome production is activated and the FCs unfold because a fraction of the rDNA present in the single FC is transcribed, resulting in the generation of the DFC (Ochs 1998). The same conclusion comes from the 3D organization of the rDNA in the PtK1 cell nucleolus corresponding to one rDNA tandem repeat; that is, one NOR-bearing chromosome per nucleolus with active rDNA alternating with repressed rDNA (Junéra et al. 1995). In these nucleoli, the 3D distribution of the FCs and DFC demonstrates that the DFC constitutes a link between the FCs and this link is superimposable on that of the rDNA distribution revealed by in situ hybridization (Junéra et al. 1997). Therefore, unengaged rDNA transcription machinery is localized in the FCs. However, the composition of the FCs could be more complex and/or have additional and as yet uncharacterized functions. These hypotheses are supported by two different examples. In 3T3 mouse cells, with the depletion of TIP5 (TTF1-interacting protein-5) inducing loss of rDNA silencing and enhancing rRNA production, the FCs are still visible (Guetg et al. 2010). In these conditions, the presence of inactive rDNA is not required for FC formation. In this biological situation, it would be interesting to know if the rDNA spacer sequences are in the FCs, and the rDNA transcribed sequences in the DFC. Another example is provided by the nucleoli of stimulated rat neurons in which the volume of only one FC is 10 times larger than that of the others (Pebusque et al. 1985). The reasons of this cyclic variability are still unknown, but are not related to rDNA transcription in this FC (Dupuy-Coin et al. 1986b). More recently, the group of M. Lafarga demonstrated the presence of one giant FC (GFC) in 58% of the nucleoli in the more active rat neurons (Casafont et al. 2007). The GFC has a diameter of 1–2 µm; in this large GFC, nascent RNAs were not observed but they were observed in the DFC identified by EM (Casafont et al. 2007). In these neurons, ribosome biogenesis is very active and the size of the nucleolus is similar to that of a glial nucleus (Hernandez-Verdun et al. 2010). In the GFC, accumulations of UBF, Pol I, and topoisomerase I were observed as well as the presence of the SUMO-1 conjugation pathway but not of the ubiquitin–proteasome system. A recent proteomic screen for nucleolar SUMO targets showed that SUMOylation modulates the function of the Nop58 snoRNP (Westman et al. 2010). The most intriguing observation is the presence of a unique GFC in these active nucleoli that reflects a kind of specialization compared to other FCs of the same nucleolus or that reflects the association of this GFC with a unique nuclear domain implicated in regulation of ribosome production by modulation of rRNA processing. This is an important open question.

The DFC is characterized by densely packed fibrils. In this nucleolar component, nascent rRNA transcripts were detected as well as early events of rRNA processing (Casafont et al. 2007; Cmarko et al. 2000). The rRNAs were identified using EM hybridization (Puvion-Dutilleul et al. 1991, 1997) and the nascent transcripts were detected with 5′-fluorouridine. Similarly in EM, labeling was found in the DFC after a short incubation with Br-uridine revealed with gold particles, and a working model was proposed (Hozàk et al. 1994). In addition, the nucleolar proteins that participate in the early stages of rRNA processing localize in the DFC, such as fibrillarin and Nopp140 along with snoRNPs (Azum-Gélade et al. 1994; Boulon et al. 2004; Dragon et al. 2002; Grandi et al. 2002; Puvion-Dutilleul et al. 1997; Qiu et al. 2008). Fibrillarin is a good marker of the DFC functioning as methyltransferase and as one of the four core proteins of the Box C/D snoRNPs (Colau et al. 2004).

The GC is characterized in EM by the presence of granules (15–20 nm) corresponding most probably to pre-60S ribosome subunits. These granules are densely packed in compact nucleoli (HeLa cell nucleolus) or organized in networks in reticulated nucleoli (PtK1 cell nucleolus). Assembly of the large ribosome subunits occurs in the GC as demonstrated by the presence of processing proteins involved in processing of the 5.8S and 28S rRNAs. The protein markers of the GC are abundant in cycling cells such as nucleophosmin (NPM, first designated B23 nucleolar protein (Orrick et al. 1973) and now known as NPM/B23), Bop1 (Pestov et al. 2001), Nop52 (Savino et al. 1999), and RRP1B (Chamousset et al. 2010) the two mammalian orthologues of the yeast Rrp1p, nucleostemin (Tsai and McKay 2002) and the PM-Scl 100 subunit of the exosome complex (Allmang et al. 1999). Proteomic analysis of complexes containing RRP1B revealed enrichment of 60S ribosomal proteins and nucleolar proteins involved in mid-late 60S processing (Chamousset et al. 2010). This association most likely contributes to modulate the dynamics of ribosome production and form a large network of interactions because 49 partners were identified by immunoprecipitation of RRP1B by PP1γ phosphatase (Chamousset et al. 2010).

In conclusion, the structure of the nucleolus reveals that the different steps of ribosome subunit production are associated with a topological compartmentation of these processes; different markers of these steps have been identified. The relative quantity of the different compartments reflects the dynamics and complexity of these steps.

1.2.3 Information Coming from Disrupted Ribosome Production

The principal force driving nucleolar organization can also be analyzed after disruption of specific steps in ribosome production. Modifications of the nucleolar structures after inhibition of rDNA transcription or prevention of rRNA processing are well characterized (Gébrane-Younès et al. 2005; Hadjiolov 1985; Hernandez-Verdun 2006; Hernandez-Verdun et al. 2010). In addition, proteins that are not normally associated with ribosome biogenesis have been identified by proteomic analysis in human nucleoli after inhibition of rDNA transcription, indicating changes in the nucleolar proteome (Andersen et al. 2002).

In mammalian nucleoli actively engaged in ribosome production, the arrest of rDNA transcription by low doses of actinomycin D triggers segregation of the FC, DFC, and GC in 1–3 h. This is a typical feature in which the nucleolar components are disengaged and form three juxtaposed structures (Fig. 1.5b). These structural modifications indicate that the synthesis of 47S pre-rRNAs produces a flux necessary to generate intermingling of nucleolar substructures. Following segregation, the FC and DFC markers (UBF and fibrillarin, respectively) appear as an individual "crescent structure" named cap at the periphery of a central body corresponding to the GC marker NPM/B23 or Nop52 (Haaf and Ward 1996). Most probably, the binding affinity of the processing proteins interacting with the highly concentrated pre-rRNAs and snoRNPs still plays an important role in the organization of the three domains. Yet, why they maintain such an association and form caps is presently not clarified. A similar structural reorganization is also seen with the delocalization of molecules from Cajal bodies such as the p80 coilin, Cajal body-specific RNAs (scaRNAs), and nucleoplasmic proteins in cells treated with high doses of actinomycin D that inhibit both Pol I and Pol II transcription (Andersen et al. 2002; Shav-Tal et al. 2005). Interestingly the nucleolar caps are dynamic structures as determined using photobleaching and require energy for their formation (Shav-Tal et al. 2005). It would be important to determine what maintains the association of these molecules as distinct domains without interplay.

In active nucleoli, the inhibition of the CK2 casein kinase by DRB (5,6-dichloro-1-ribo-furanosylbenzimidazole) induces the reversible disconnection between the FC/DFC and the GC. In addition, the FC/DFC unravels, forming a necklace structure (Granick 1975; Haaf and Ward 1996). In EM, the beads of the necklace of PtK1 cells are composed of one FC of mean size 60–65 nm surrounded by the DFC; the presence of UBF, fibrillarin, and nascent RNA transcripts was demonstrated in these beads (Le Panse et al. 1999, Louvet et al. 2005). However, the NPM/B23 and Nop52 proteins are mislocalized in large bodies derived from the GC. When DRB is removed, FC/DFC compaction and contact with GC-derived bodies occurs in 15–20 min and reassembly of the compact nucleolus is achieved in 1 h (Louvet et al. 2005). It was demonstrated that this process is CK2-driven and is ATP/GTP-dependent in permeabilized cell assays capable of promoting this nucleolar reorganization (Louvet et al. 2006). Mutation of the major CK2 site on NPM/B23 reproduced the separation of the GC from the FC/DFC, indicating a functional control by phosphorylation (Louvet et al. 2006). Similarly, the inhibition of the cyclin-dependent kinase (CDK) by specific inhibitors such as Roscovitine impairs DFC and GC interactions (Fig. 1.5c) (Sirri et al. 2002). It has been proposed, at least in yeast cells, that the small ribosome subunits are exported first from the nucleolus, and the building of the large ribosome subunits then continues in the GC (Fromont-Racine et al. 2003; Grandi et al. 2002; Poll et al. 2009). If the transition between the assembly of the small (pre-40S) and large (pre-60S) ribosome subunits occurs at the interface between DFC and GC, this transition could be controlled by phosphorylation of key players. However in higher eukaryotes, these steps are not precisely localized in the nucleolar architecture. In addition, the transition between DFC and GC could depend on the integrity of the snoRNPs under the control of nucleostemin (NS). NS is involved in the processing of the pre-60S ribosomal subunit and its knockdown delays the processing of 32S pre-rRNA into 28S rRNA (Romanova et al. 2009a). The depletion of NS disorganizes the structure of the DFC by redistribution of the snoRNPs (Romanova et al. 2009b).

Taking into account these two examples (segregation and disconnection), we speculate that nucleolar organization reveals, at least in part, the equilibrium between the flux of pre-rRNAs and the sequential sorting of the two types of ribosomal subunits.

1.3 Nucleolar Assembly/Disassembly

The assembly of the nucleoli in higher eukaryotes directly depends on pre-existing machineries and complexes inherited through mitosis from the previous interphase. Nucleolar disassembly occurs during prophase and nucleolar assembly starts in telophase (Hernandez-Verdun 2004; Hernandez-Verdun et al. 2002; Sirri et al. 2008). The processing machineries derived from nucleolar disassembly transit through mitosis and are used to build the new nucleoli. The morphological features of nucleolar disassembly or assembly have been extensively described in animal and plant cells. At the turn of the century, the mechanisms controlling these processes began to be better characterized as reported in a large overview by Dimario (2004). It appears that the networks of the CDKs (Clute and Pines 1999) and phosphates (Trinkle-Mulcahy and Lamond 2006) are involved in these processes, as well as the dynamic localization of the nucleolar building blocks during mitosis (Hernandez-Verdun et al. 2002).

The nucleoli disassemble during early prophase (Gautier et al. 1992; Gébrane-Younès et al. 1997). The final step of nucleolar disruption is obviously the arrest of rDNA transcription. However, the nucleolar processing proteins and snoRNPs are released first; that is, before the arrest of Pol I transcription and nuclear envelope breakdown (Gautier et al. 1994). The proteins from the GC (NPM/B23, PM-Scl 100, Nop52, etc.), and from the DFC (fibrillarin) as well as the snoRNPs (Van Hooser et al. 2005), become distributed over the surface of the chromosomes in such a way that the chromosomes can be isolated with these nucleolar complexes still attached to their surface (Gautier et al. 1992). How the 34 different proteins and snoRNPs are maintained on the chromosome is still unknown. In living cells, these proteins tagged with GFP, concentrate around the chromosomes and migrate with them until telophase (Dundr et al. 2000; Savino et al. 2001). During interphase, NPM/B23 is phosphorylated by CK2 kinase while in early prophase it is phosphorylated by CDK1-cyclinB on T199 (Negi and Olson 2006). CDK1-cyclinB phosphorylation alters the RNA binding affinity of NPM/B23, a modification that explains the nucleolar release of NPM/B23 at this period of the cell cycle (Okuwaki et al. 2002). The consequence of the disengagement of the processing complexes in early prophase while rDNA transcription is not completely repressed is the production of partially processed 45S rRNAs (Dousset et al. 2000). These 45S rRNAs are stabilized during mitosis and will participate in nucleolar assembly (Dousset et al. 2000). The different timings for disruption of the transcription and processing machineries might be due to different control pathways or to modulation of the CDK1-cyclin B pathway. Recently, a FRET (fluorescence resonance energy transfer) biosensor for CDK1-cyclin B activity with high temporal precision in living cells demonstrated the multiple roles of CDK1-cyclin B during prophase (Gavet and Pines 2010). CDK1 is initially activated in HeLa cells 27 ± 7 min before nuclear envelope breakdown. Depending on the localization and concentration of the cyclin and its substrates, the CDK1-cyclinB activity can successively regulate the sequential prophase events (Gavet and Pines 2010).

Nucleolar assembly starts in telophase with restoration of rDNA transcription in competent NORs in cells with multiple NORs (6 out of 10 NORs in HeLa cells) (Roussel et al. 1996) or in two NORs in cells with a pair of NOR-bearing chromosomes in PtK1 cells (Gébrane-Younès et al. 1997). The signature of this assembly is based on detectable resumption of rDNA transcription (Fig. 1.6b). However, it was demonstrated that active rDNA transcription does not possess the ability to organize a complete nucleolus. Nucleolar assembly also depends on the proteins and snoRNAs of the processing complexes (Dousset et al. 2000; Sirri et al. 2002, 2000a). In telophase, processing proteins close to chromosomes assemble in foci (Fig. 1.6a) designated prenucleolar bodies (PNBs) (Ochs et al. 1985). In addition, in some cells containing abundant pre-rRNA processing machineries, the formation of foci designated nucleolus derived foci (NDFs) are observed in the cytoplasm during anaphase and telophase (Dundr et al. 1997, 2000). PNBs and NDFs have a similar composition including early and late processing proteins (Jiménez-Garcia et al. 1994), snoRNAs (Verheggen et al. 2000), and unprocessed rRNAs (Dousset et al. 2000; Dundr and Olson 1998); NDFs finally enter into the nuclei in early G1 (Dundr et al. 2000). It is possible that NDFs are formed in cells when abundant nucleolar processing machinery is present in the cytoplasm during the period of nuclear envelope formation.

The dynamics of the processing nucleolar proteins was analyzed in living cells at the transition mitosis/interphase. The first detectable concentration of proteins in foci occurred on the surface of the chromosomes during telophase (Savino et al. 2001) and in some NDFs visible as mobile bodies in the cytoplasm (Angelier et al. 2005; Dundr et al. 2000). Time-lapse microscopy and FRET, used to analyze the dynamics and interactions of nucleolar proteins in living cells (Louvet et al. 2008), demonstrated the interaction between NPM/B23 and Nop52 in nucleoli and PNBs (Angelier et al. 2005). Interestingly, no FRET was detected during anaphase at the periphery of the chromosomes, whereas it was registered in about 20% of the PNBs at the beginning of telophase, 40% at the end of telophase, and 55% in early G1 nuclei (Angelier et al. 2005). Therefore, interaction occurs between these proteins in PNBs as well as in nucleoli. It is presently unknown if this interaction leads to processing of the 45S rRNA present in PNBs. The flux of proteins between nucleoli and PNBs was measured in living cells at different periods of nucleolar assembly using photoactivation (PA). The PAGFP tagged proteins were photoactivated (for technical details see (Patterson and Lippincott-Schwartz 2002)) in one NOR and the flux of these proteins was analyzed in 3D. The recruitment of the processing complexes, first by the DFC and then by the GC during nucleolar assembly is due to the dynamics of release from PNBs (Muro et al. 2010). PAGFP-fibrillarin migrated from one NOR to every NORs and was excluded from the PNBs, suggesting the absence of its pre-rRNAs targets in PNBs and consequently the processing of the 45S rRNAs. On the contrary, the dynamics of PAGFP-Nop52 between NORs and PNBs controlled their recruitment into the GC (Muro et al. 2010).

During the cell cycle, networks of regulation are necessary to coordinate the different steps of nucleolar assembly that also depend on nucleolar disassembly at the beginning of mitosis (Hernandez-Verdun et al. 2002). In this regulation, the role of the kinases/phosphatases is important and largely coordinated by cell cycle progression (Trinkle-Mulcahy and Lamond 2006). At each crucial point it appears that the dynamics play a major role in determining the functions of the kinases (Gavet and Pines 2010).

1.4 The Nucleolus in the Nucleus

The data reported above concerns only a limited part of the players involved in the organization of the nucleolus. These data reveal the main chronological interest of the researchers for rDNA transcription and rRNA processing with the characterization of the main steps for pre-40S and pre-60S assembly in yeast and in higher eukaryotes (Fatica and Tollervey 2002; Fromont-Racine et al. 2003). To complete this model, the role of r-proteins in nucleolar organization is now in the front line (Choesmel et al. 2007, Choesmel 2008; Lam et al. 2007, Granneman and Tollervey 2007; Poll et al. 2009) and will certainly be a promising field of research in the near future. To understand the nucleolar architecture, it is also important to consider the dynamics of the nucleolar structures, the intra- and peri-nucleolar chromatin, and the particular relationship of the nucleolus with the nuclear envelope.

1.4.1 Dynamics of the Nucleolar Structures

The analysis in living cells of intranuclear dynamics has demonstrated that during interphase even after nucleolar assembly, nucleolar proteins rapidly associate with and dissociate from nucleolar components in continuous exchanges with the nucleoplasm (Phair and Misteli 2000). The flux between nucleoli in the same nucleus is impressive as a pool of GC proteins (NPM/B23 or Nop52) activated in one nucleolus is homogeneously redistributed in the whole volume of nucleoli in 2 min (Muro et al. 2008).

The diffusion coefficient of fibrillarin (estimated between 0.02 and 0.046 $\mu m^2\ s^{-1}$) was ten times lower in the nucleolus than in the nucleoplasm (Chen and Huang 2001, Phair and Misteli 2000; Snaar et al. 2000). This value is believed to reflect the time of residency of fibrillarin engaged in nucleolar activity or binding. The nucleolar proteins engaged in rRNA transcription and processing (e.g., UBF, and NPM/B23, Nop52, nucleolin, and Rpp29) also move with rapid recovery rates in the nucleolus as does fibrillarin (Chen and Huang 2001; Louvet et al. 2005). Conversely, the recovery rates of r-proteins are low (~3 times lower than that of nucleolar proteins); this was proposed to reflect a slower process for the assembly of ribosomes compared to transcription and processing (Chen and Huang 2001). Alternatively, this could be due to more stable associations of r-proteins with pre-rRNAs. In contrast to the well-defined nucleolar structures visible by EM, all the nucleolar proteins involved in ribosome biogenesis presently examined cycle between the nucleolus and the nucleoplasm in interphase cells. NPM/B23 undergoes different phosphorylation events during the cell cycle. It was recently demonstrated by FRAP that the kinetics of NPM/B23 depends on its phosphorylation status (Negi and Olson 2006). During interphase, the half-time ($t_{1/2}$) of recovery of NPM/B23 is 22 s in nucleoli but when the CK2 phosphorylation site is mutated (S125A) the $t_{1/2}$ increases to 44 s, and when a mutant mimicking the phosphorylation charges of the four sites of mitotic

CDK1 phosphorylation is examined, the $t_{1/2}$ decreases to 12 s. This could indicate that the S125A-B23 protein has a higher affinity for the nucleolar components (Negi and Olson 2006). Alternatively, this could correspond to a decreased turnover in the nucleolar complexes in correlation with the uncoupling of the DFC and GC occurring by overexpression of S125A-B23 (Louvet et al. 2006). Because overexpression, during interphase of NPM/B23 mimicking four sites of mitotic phosphorylation, increased the mobility of these proteins, it is tempting to propose that this results from a defect in affinity for rRNAs caused by these mutant NPM/B23s as demonstrated for mitotic phosphorylation of NPM/B23 (Okuwaki et al. 2002).

Inhibition of Pol I transcription by actinomycin D does not prevent traffic of nucleolar proteins. However, even if the diffusion coefficients of the nucleolar proteins in the nucleoplasm are similar for active and repressed Pol I transcription, the traffic in segregated nucleoli appears to change differently for different nucleolar components. For example, the traffic of UBF in the nucleolus is decreased by actinomycin D, whereas it is unaltered for nucleolin and increased for r-proteins (Chen and Huang 2001). In addition, many RNA binding proteins relocalize from the nucleoplasm to a specific nucleolar cap during transcriptional inhibition of Pol I and II transcription (Shav-Tal et al. 2005). In conclusion, rapid diffusion of nucleolar proteins occurs in the nucleoplasm and their renewal in the nucleolus is permanent.

1.4.2 Nucleolus and Chromatin

In light microscopy, DNA staining of nuclei reveals empty black areas surrounded by DNA. In phase contrast, these black areas correspond to highly contrasted areas, and they are the nucleoli (Fig. 1.4). This indicates that the amount of chromatin is relatively low inside the nucleolus compared to that of rRNAs, rRNPs, and snoRNPs. Standard and specific EM contrast demonstrates the presence of condensed chromatin at the nucleolar periphery and inside the nucleolus in the GC, and noncondensed chromatin in FC/DFC (Derenzini et al. 2006; Gébrane-Younès et al. 2005). The osmium ammine DNA tracer shows three levels of chromatin organization in the nucleolus: clumps of nucleosomes, chromatin fibers, and DNA filaments, the third only found in the FC. It was proposed that the DNA filaments in the FC correspond to rDNA either transcribed or silent (Derenzini et al. 2006). This could be a characteristic feature of competent NORs compared to incompetent NORs. This is in accordance with the fact that when rDNA transcription is arrested during mitosis, UBF is still associated with noncondensed DNA in the two NORs of PtK1 cells (Gébrane-Younès et al. 1997). In human cells, both competent and repressed NORs are present (Roussel et al. 1996). It was demonstrated in HeLa cells that repressed NORs associate with nucleoli (Sullivan et al. 2001); most of the repressed NORs are included in the nucleoli and some are on the loop of condensed chromatin connecting nucleoli to NOR-bearing chromosomes (Kalmarova et al. 2007).

At the nucleolar periphery, the association of chromatin domains is variable depending on the activity of the cell and of the species (Figs. 1.3 and 1.5a). Around the nucleolus, the chromosomal motion is constrained just at the nuclear periphery and this seems to reflect the inactivity of the perinucleolar chromatin (Chubb et al. 2002). In rat hepatocytes, stimulation of the cell cycle by hepatectomy increases the amount of chromatin fibers in the GC while the nucleolar volume increases, but the sequences of these DNAs have not been identified (Derenzini et al. 1982). In mouse cells, the centromeric heterochromatin regroups around the nucleolus during the first cell cycle of development (Martin et al. 2006). It was recently demonstrated that the complex that maintains the silencing of half of the rRNA genes (Guetg et al. 2010) mediates heterochromatin formation of the centromeric repeats in mouse cells. It would be interesting to know if this is a general process in every species or if it is unique to mouse cells. In human cancer cells, the perinucleolar compartment (PNC) (Matera et al. 1995) forms a reticulated mesh of 0.25–4 μm on the surface of the nucleolus (Huang et al. 1998). The PNC selectively forms in malignant cells derived from solid tumor tissues (Norton et al. 2008). The structural integrity of the PNC depends on Pol III transcription and is directly associated with an as yet unidentified specific DNA locus (Norton et al. 2009; Pollock and Huang 2010).

Recently the initial genomics of the nucleolus-associated chromosomal domains (NADs) demonstrated that 4% of the entire genome sequences interact with nucleoli in HeLa cells. These sequences correspond to rDNA, pericentromeric and centromeric repetitive sequences, or are involved in specific biological processes (Németh et al. 2010). This elevated number of sequences is not completely surprising considering that the analyses included DNAs inside and around the nucleoli. However, there is a specific enrichment in NADs corresponding to the high density of AT-rich sequences, low gene density, and significant enrichment in transcriptionally repressed genes (van Koningsbruggen et al. 2010). This is in agreement with the hypothesis that the organization of repetitive DNA of the short arms of the acrocentric chromosome (the NOR-bearing chromosome) is reflected in the topographic organization of the human nucleolus (Kaplan et al. 1993).

1.4.3 Relationship Between the Nucleolus and the Nuclear Envelope

Nucleoli show extensive nuclear envelope contact in the yeast *S. cerevisiae* (Berger et al. 2008; Taddei et al. 2010). Similarly, the nucleoli are located at or near the nuclear envelope in higher eukaryotes (Bourgeois and Hubert 1988). This association was demonstrated in 3D reconstructions of serial EM sections (Dupuy-Coin et al. 1986a). In HeLa cell nuclei, the folding of the nuclear envelope forms several nucleolar canals that are in direct contact with centrally located nucleoli (Fig. 1.2b). Micronuclei containing only one chromosome can be induced in PtK1 cells (Labidi et al. 1990). In these micronucleated cells, the nuclear organization was observed in serial EM sections. The 3D reconstitution of the nucleolar envelope demonstrated a folded nuclear envelope forming a canal only in micronuclei containing a nucleolus, and a spherical nuclear envelope in micronuclei without a nucleolus (Géraud et al. 1989).

We propose that the canal formed by the folding of the nuclear envelope is induced by the presence of active rDNA. The role of this canal is presently unknown. In particular, there is no definite proof that this canal is involved in nuclear export. However, specific structures, called GLFG-body containing hNup98 nucleoporin, have been described close to the nucleolus (Griffis et al. 2002). It was also demonstrated that the GLFG repeat domain of hNup98 interacts with the nuclear export protein of Influenza A virus (Chen et al. 2010).

Nuclear lamins are known to be associated with the nucleolus and participate in chromatin organization (Benavente 1991). Recently, it was demonstrated that lamin B1 maintains the functional plasticity of nucleoli (Martin et al. 2009) and participates in the post-mitotic structural reorganization of the nucleus and nucleoli (Martin et al. 2010). In correlation with these conclusions, DNA sequencing of the NADs demonstrated that some chromatin loci specifically associate with either the nucleolus or the nuclear envelope (van Koningsbruggen et al. 2010).

1.5 Conclusions

The nucleolus is a model of coordination between nuclear functions because several complex networks must cooperate to generate the rRNAs, to process and modify these RNAs, and to assemble the rRNAs with r-proteins. In the last ten years, important technical progress has considerably modified our vision of the nucleolus, in particular the analysis of its dynamics in living cells. In contrast to the well-defined nucleolar structures visible by EM, all the nucleolar proteins involved in ribosome biogenesis that have been examined cycle between the nucleolus and the nucleoplasm in interphase cells. The flux is rapid and future studies would certainly benefit from recent progress that makes it possible to analyze the dynamics at the scale of the msec (millisecond) and in 3D, as well as the new spatial resolution and sensitivity of correlative light and EM tomography (Kukulski et al. 2011). Atomic force microscopy could also be useful to examine how compact is the structure of the nucleolus (Pederson 2010) and how the nucleolar surface is modified during nucleolar segregation.

When Pol I transcription is inhibited, segregation of the nucleolar components occurs. Why these nucleolar components remain assembled in caps around a central body when the flux of pre-rRNA is stopped is presently unknown. In the caps and central body, the traffic of proteins from segregated nucleoli to nucleoplasm is maintained. This indicates a binding affinity with partners no longer engaged in ribosome biogenesis. In this case, the interaction does not depend on the flux of pre-rRNA. It would be important to know if the natural nucleolar segregation occurring in cells at the terminal differentiation stage is similar; that is, close to or different from that of the segregation induced by actinomycin D.

During the cell cycle of mammalian cells, nucleolar assembly involves the formation of PNBs that form around inherited 45S rRNAs synthesized during prophase. How these pre-rRNAs are stabilized during mitosis is unknown.

The role of these 45S rRNAs in nucleolar assembly could be the formation of reservoirs of processing proteins to equilibrate the recruitment of processing complexes on transcription sites. The dynamics of flux between NORs and PNBs is in favor of this hypothesis (Muro et al. 2010). During *X. laevis* development, maternally inherited 40S rRNAs enter the embryo nuclei before Pol I transcription and form PNBs (Verheggen et al. 1998, 2000). The formation of the PNBs is observed during the assembly of nucleoli in all cycling eukaryotes that undergo mitosis (at the opposite of budding one). Why this step has been conserved throughout evolution is still to be established.

Another challenge will be to identify the role of the FCs, structures that emerged during the evolution of eukaryotes. In mammalian cells, FCs are characterized by their structure and contrast in EM as well by the presence of the rDNA and rDNA machinery; however, rDNA transcription activity is at its periphery. It is proposed that non-transcribed rDNA localizes in FCs. However, in mouse cells with no repressed rDNA sequences, the FC was still observed (Guetg et al. 2010). This indicates that other actors are essential for the formation of FCs. The giant FC in neurons demonstrated the variability of the FC, only one GFC per nucleolus and again without indication of its role (Casafont et al. 2007). It would also be important to know if only one inserted rDNA sequence can generate an FC.

The nucleolus occupies one-third of the nuclear volume in *S. cerevisiae* and the ratio is of the same order as in human cancer nuclei. This large nucleolar volume corresponds to "hot" spots of concentration of pre-rRNAs and snoRNAs. This concentration creates a space different from that of the nucleoplasm with different rules. During the evolution from prokaryotes to eukaryotes, compartmentalization of the cell functions emerged. In eukaryotes, the nucleolus is the consequence of the compartmentation of the functions within the nuclei and probably linked to the creation of multiple copies of rDNA on specific domains of the chromosomes, the NORs.

Acknowledgments We are grateful to A.-L. Haenni for critical reading of the manuscript.

References

Allmang C, Petfalski E, Podtelejnikov A, Mann M, Tollervey D, Mitchell P (1999) The yeast exosome and human PM-Scl are related complexes of 3'-5' exonucleases. Genes Dev 13:2148–2158

Andersen JS, Lyon CE, Fox AH, Leung AKL, Lam YW, Steen H, Mann M, Lamond AI (2002) Directed proteomic analysis of the human nucleolus. Cur Biol 12:1–11

Andersen JS, Lam YW, Leung AK, Ong SE, Lyon CE, Lamond AI, Mann M (2005) Nucleolar proteome dynamics. Nature 433:77–83

Angelier N, Tramier M, Louvet E, Coppey-Moisan STM, De Mey JR, Hernandez-Verdun D (2005) Tracking the Interactions of rRNA Processing Proteins during Nucleolar Assembly in Living Cells. Mol Biol Cell 16:2862–2871

Azum-Gélade MC, Noaillac-Depeyre J, Caizergues-Ferrer M, Gas N (1994) Cell cycle redistribution of U3 snRNA and fibrillarin. Presence in the cytoplasmic nucleolus remnant and in the prenucleolar bodies at telophase. J Cell Sci 107:463–475

Belin S, Beghin A, Solano-Gonzalez E, Bezin L, Brunet-Manquat S, Textoris J, Prats AC, Mertani HC, Dumontet C, Diaz JJ (2009) Dysregulation of ribosome biogenesis and translational capacity is associated with tumor progression of human breast cancer cells. PLoS One 4:e7147

Benavente R (1991) Postmitotic nuclear reorganization events analyzed in living cells. Chromosoma 100:215–220

Berger AB, Cabal GG, Fabre E, Duong T, Buc H, Nehrbass U, Olivo-Marin JC, Gadal O, Zimmer C (2008) High-resolution statistical mapping reveals gene territories in live yeast. Nat Methods 5:1031–7

Boisvert FM, van Koningsbruggen S, Navascues J, Lamond AI (2007) The multifunctional nucleolus. Nat Rev Mol Cell Biol 8:574–85

Boulon S, Verheggen C, Jady BE, Girard C, Pescia C, Paul C, Ospina JK, Kiss T, Matera AG, Bordonne R, Bertrand E (2004) PHAX and CRM1 are required sequentially to transport U3 snoRNA to nucleoli. Mol Cell 16:777–87

Bourgeois CA, Hubert J (1988) Spatial relationship between the nucleolus and the nuclear envelope: structural aspects and functional significance. Int Rev Cytol 111:1–52

Brown DD, Gurdon JB (1964) Absence of Ribosomal Rna Synthesis in the Anucleolate Mutant of Xenopus Laevis. Proc Natl Acad Sci U S A 51:139–46

Busch H, Smetana K (1974) The nucleus of the cancer cells. Academic Press, New York, pp 41–80

Casafont I, Bengoechea R, Navascues J, Pena E, Berciano MT, Lafarga M (2007) The giant fibrillar center: a nucleolar structure enriched in upstream binding factor (UBF) that appears in transcriptionally more active sensory ganglia neurons. J Struct Biol 159:451–61

Chamousset D, De Wever V, Moorhead GB, Chen Y, Boisvert FM, Lamond AI, Trinkle-Mulcahy L (2010) RRP1B Targets PP1 to Mammalian Cell Nucleoli and Is Associated with Pre-60S Ribosomal Subunits. Mol Biol Cell 21:4212–26

Chen D, Huang S (2001) Nucleolar components involved in ribosome biogenesis cycle between the nucleolus and nucleoplasm in interphase cells. J Cell Biol 153:169–176

Chen J, Huang S, Chen Z (2010) Human cellular protein nucleoporin hNup98 interacts with influenza A virus NS2/nuclear export protein and overexpression of its GLFG repeat domain can inhibit virus propagation. J Gen Virol 91:2474–84

Choesmel V, Bacqueville D, Rouquette J, Noaillac-Depeyre J, Fribourg S, Cretien A, Leblanc T, Tchernia G, Da Costa L, Gleizes PE (2007) Impaired ribosome biogenesis in Diamond-Blackfan anemia. Blood 109:1275–83

Choesmel V, Fribourg S, Aguissa-Toure AH, Pinaud N, Legrand P, Gazda HT, Gleizes PE (2008) Mutation of ribosomal protein RPS24 in Diamond-Blackfan anemia results in a ribosome biogenesis disorder. Hum Mol Genet 17:1253–63

Chubb JR, Boyle S, Perry P, Bickmore WA (2002) Chromatin motion is constrained by association with nuclear compartments in human cells. Curr Biol 12:439–445

Clute P, Pines J (1999) Temporal and spatial control of cyclin B1 destruction in metaphase. Nature cell Biol 1:82–87

Cmarko D, Verschure PJ, Rothblum LI, Hernandez-Verdun D, Amalric F, van Driel R, Fakan S (2000) Ultrastructural analysis of nucleolar transcription in cells microinjected with 5-bromo-UTP. Histochem Cell Biol 113:181–187

Colau G, Thiry M, Leduc V, Bordonne R, Lafontaine DL (2004) The small nucle(ol)ar RNA cap trimethyltransferase is required for ribosome synthesis and intact nucleolar morphology. Mol Cell Biol 24:7976–86

Derenzini M, Hernandez-Verdun D, Bouteille M (1982) Visualization in situ of extended DNA filaments in nucleolar chromatin of rat hepatocytes. Exp Cell Res 141:463–469

Derenzini M, Farabegoli F, Treré D (1993) Localization of the DNA in the fibrillar components of the nucleolus: a cytochemical and morphometrical study. J Histochem Cytochem 41:829–836

Derenzini M, Pasquinelli G, O'Donohue MF, Ploton D, Thiry M (2006) Structural and functional organization of ribosomal genes within the mammalian cell nucleolus. J Histochem Cytochem 54:131–45

Derenzini M, Montanaro L, Trere D (2009) What the nucleolus says to a tumour pathologist. Histopathology 54:753–62

Dimario PJ (2004) Cell and molecular biology of nucleolar assembly and disassembly. Int Rev Cytol 239:99–178

Dousset T, Wang C, Verheggen C, Chen D, Hernandez-Verdun D, Huang S (2000) Initiation of nucleolar assembly is independent of RNA polmerase I transcription. Mol Biol Cell 11:2705–2717

Dragon F, Gallagher JE, Compagnone-Post PA, Mitchell BM, Porwancher KA, Wehner KA, Wormsley S, Settlage RE, Shabanowitz J, Osheim Y, Beyer AL, Hunt DF, Baserga SJ (2002) A large nucleolar U3 ribonucleoprotein required for 18S ribosomal RNA biogenesis. Nature 417:967–70

Dundr M, Olson MOJ (1998) Partially processed pre-rRNA is preserved in association with processing components in nucleolus derived foci during mitosis. Mol Biol Cell 9:2407–2422

Dundr M, Meier UT, Lewis N, Rekosh D, Hammarskjöld ML, Olson MOJ (1997) A class of non-ribosomal nucleolar components is located in chromosome periphery and in nucleolus-derived foci during anaphase and telophase. Chromosoma 105:407–417

Dundr M, Misteli T, Olson MOJ (2000) The dynamics of postmitotic reassembly of the nucleolus. J Cell Biol 150:433–446

Dupuy-Coin AM, Moens P, Bouteille M (1986a) Three-dimensional analysis of given cell structures: nucleolus, nucleoskeleton and nuclear inclusions. Methods Achiev Exp Pathol 12:1–25

Dupuy-Coin AM, Pebusque MJ, Seite R, Bouteille M (1986b) Localization of transcription in nucleoli of rat sympathetic neurons. A quantitative ultrastructural autoradiography study. J Submicrosc Cytol 18:21–7

Fatica A, Tollervey D (2002) Making ribosomes. Curr opin Cell Biol 14:313–318

Fromont-Racine M, Senger B, Saveanu C, Fasiolo F (2003) Ribosome assembly in eukaryotes. Gene 313:17–42

Gautier T, Robert-Nicoud M, Guilly MN, Hernandez-Verdun D (1992) Relocation of nucleolar proteins around chromosomes at mitosis- A study by confocal laser scanning microscopy. J Cell Sci 102:729–737

Gautier T, Fomproix N, Masson C, Azum-Gélade MC, Gas N, Hernandez-Verdun D (1994) Fate of specific nucleolar perichromosomal proteins during mitosis: Cellular distribution and association with U3 snoRNA. Biol Cell 82:81–93

Gavet O, Pines J (2010) Progressive activation of CyclinB1-Cdk1 coordinates entry to mitosis. Dev Cell 18:533–43

Gébrane-Younès J, Fomproix N, Hernandez-Verdun D (1997) When rDNA transcription is arrested during mitosis, UBF is still associated with non-condensed rDNA. J Cell Sci 110:2429–2440

Gébrane-Younès J, Sirri V, Junéra HR, Roussel P, Hernandez-Verdun D (2005) Nucleolus: an essential nuclear domain. In: Hemmrich P, Diekmann S (eds) Visions of the cell nucleus. American Scientific, Stevenson Ranch, CA, pp 120–135

Géraud G, Laquerriere F, Masson C, Arnoult J, Hernandez-Verdun D (1989) Three-dimensional organization of micronuclei induced by colchicine in PtK_1 cells. Exp Cell Res 181:27–39

Goessens G (1984) Nucleolar structure. Int Rev Cytol 87:107–58

Grandi P, Rybin V, Bassler J, Petfalski E, Strauss D, Marzioch M, Schafer T, Kuster B, Tschochner H, Tollervey D, Gavin AC, Hurt E (2002) 90S pre-ribosomes include the 35S pre-rRNA, the U3 snoRNP, and 40S subunit processing factors but predominantly lack 60S synthesis factors. Mol Cell 10:105–15

Granick D (1975) Nucleolar necklaces in chick embryo fibroblast cells. I. Formation of necklaces by dichlororibobenzimidazole and other adenosine analogues that decrease RNA synthesis and degrade preribosomes. J Cell Biol 65:398–417

Granneman D, Tollervey D (2007) Building ribosomes: even more expensive than expected? Curr Biol 17:R415–417

Griffis ER, Altan N, Lippincott-Schwartz J, Powers MA (2002) Nup98 is a mobile nucleoporin with transcription-dependent dynamics. Mol Biol Cell 13:1282–97

Guetg C, Lienemann P, Sirri V, Grummt I, Hernandez-Verdun D, Hottiger MO, Fussenegger M, Santoro R (2010) The NoRC complex mediates the heterochromatin formation and stability of silent rRNA genes and centromeric repeats. EMBO J 29:2135–46

Haaf T, Ward DC (1996) Inhibition of RNA polymerase II transcription causes chromatin decondensation, loss of nucleolar structure, and dispersion of chromosomal domains. Exp Cell Res 224:163–173

Hadjiolov AA (1985) The nucleolus and ribosome biogenesis. Springer-Verlag, Wien, New-York, pp 1–268

Henderson AS, Warburton D, Atwood KC (1972) Location of ribosomal DNA in the human chromosome complement. Proc Natl Acad Sci USA 69:3394–8

Hernandez-Verdun D (1986) Structural organization of the nucleolus in mammalian cells. Karger, Basel, pp 26–62

Hernandez-Verdun D (2004). Behavior of the nucleolus during mitosis. In: Olson MOJ (ed) The nucleolus. Landes Biosciences, Austin, pp 41–57

Hernandez-Verdun D (2006) The nucleolus: a model for the organization of nuclear functions. Histochem Cell Biol 126:135–48

Hernandez-Verdun D, Bouteille M (1979) Nucleologenesis in chick erythrocyte nuclei reactivated by cell fusion. J Ultrastruct Res 69:164–79

Hernandez-Verdun D, Roussel P, Gébrane-Younès J (2002) Emerging concepts of nucleolar assembly. J Cell Sci 115:2265–2270

Hernandez-Verdun D, Roussel P, Thiry M, Sirri V, Lafontaine DLJ (2010) The nucleolus: structure/function relationship in RNA metabolism. John Wiley & Sons, Ltd, pp 415–431

Hiscox JA (2002) The nucleolus-a gateway to viral infection? Arch Virol 147:1077–89

Hozàk P, Novak JT, Smetana K (1989) Three-dimensional reconstructions of nucleolus-organizing regions in PHA-stimulated human lymphocytes. Biol Cell 66:225–233

Hozàk P, Cook PR, Schöfer C, Mosgöller W, Wachtler F (1994) Site of transcription of ribosomal RNA and intranucleolar structure in HeLa cells. J Cell Sci 107:639–648

Huang S, Deerinck TJ, Ellisman MH, Spector DL (1998) The perinucleolar compartment and transcription. J Cell Biol 143:35–47

Jiménez-Garcia LF, MdL S-V, Ochs RL, Rothblum LI, Hannan R, Spector DL (1994) Nucleologenesis: U3 snRNA-containing prenucleolar bodies move to sites of active Pre-rRNA transcription after mitosis. Mol Biol Cell 5:955–966

Jordan EG (1991) Interpreting nucleolar structure; where are the transcribing genes? J Cell Sci 98:437–449

Junéra HR, Masson C, Géraud G, Hernandez-Verdun D (1995) The three-dimensional organization of ribosomal genes and the architecture of the nucleoli vary with G1, S and G2 phases. J Cell Sci 108:3427–3441

Junéra HR, Masson C, Géraud G, Suja J, Hernandez-Verdun D (1997) Involvement of in situ conformation of ribosomal genes and selective distribution of UBF in rRNA transcription. Mol Biol Cell 8:145–156

Kalmarova M, Smirnov E, Masata M, Koberna K, Ligasova A, Popov A, Raska I (2007) Positioning of NORs and NOR-bearing chromosomes in relation to nucleoli. J Struct Biol 160:49–56

Kaplan FS, Murray J, Sylvester JE, Gonzalez IL, O'Connor JP, Doering JL, Muenke M, Emanuel BS, Zasloff MA (1993) The topographic organization of repetitive DNA in the human nucleolus. Genomics 15:123–32

Knibiehler B, Mirre C, Rosset R (1982) Nucleolar organizer structure and activity in a nucleolus without fibrillar centres:the nucleolus in an established drosophila cell line. J Cell Sci 57:351–364

Knibiehler B, Mirre C, Navarro A, Rosset R (1984) Studies on chromatin organization in a nucleolus without fibrillar centres. Presence of sub-nucleolar structure in KCo cells of Drosophila. Cell Tissue Res 236:279–288

Kukulski W, Schorb M, Welsch S, Picco A, Kaksonen M, Briggs JAG (2011) Correlated fluorescence and 3D electron microscopy with high sensitivity and spatial precision. J Cell Biol 192(1):111–9

Labidi B, Broders F, Meyer JL, Hernandez-Verdun D (1990) Distribution of rDNA and 28S, 18S, and 5S rRNA in micronuclei containing a single chromosome. Biochem Cell Biol 68: 957–64

Lam YW, Lamond AI, Mann M, Andersen JS (2007) Analysis of nucleolar protein dynamics reveals the nuclear degradation of ribosomal proteins. Curr Biol 17:749–60

Le Panse S, Masson C, Héliot L, Chassery J-M, Junéra HR, Hernandez-Verdun D (1999) 3-D Organization of single ribosomal transcription units after DRB inhibition of RNA polymerase II transcription. J Cell Sci 112:2145–2154

Louvet E, Junera HR, Le Panse S, Hernandez-Verdun D (2005) Dynamics and compartmentation of the nucleolar processing machinery. Exp Cell Res 304:457–470

Louvet E, Junera HR, Berthuy I, Hernandez-Verdun D (2006) Compartmentation of the nucleolar processing proteins in the granular component is a CK2-driven process. Mol Biol Cell 17:2537–46

Louvet E, Tramier M, Angelier N, Hernandez-Verdun D (2008) Time-lapse microscopy and fluorescence resonance energy transfer to analyze the dynamics and interactions of nucleolar proteins in living cells. Humana, Totowa, NJ

Martin C, Beaujean N, Brochard V, Audouard C, Zink D, Debey P (2006) Genome restructuring in mouse embryos during reprogramming and early development. Dev Biol 292:317–32

Martin C, Chen S, Maya-Mendoza A, Lovric J, Sims PF, Jackson DA (2009) Lamin B1 maintains the functional plasticity of nucleoli. J Cell Sci 122:1551–62

Martin C, Chen S, Jackson DA (2010) Inheriting nuclear organization: can nuclear lamins impart spatial memory during post-mitotic nuclear assembly? Chromosome Res 18:525–41

Matera AG, Frey MR, Margelot K, Wolin SL (1995) A perinucleolar compartment contains several RNA polymerase III transcripts as well as the polypyrimidine tract-binding protein, hnRNP I. J Cell Biol 129:1181–93

McClintock B (1934) The relation of a particular chromosomal element to the development of the nucleoli in *Zea mays*. Z Zellforsch Mikrosk Anat 21:294–328

Mélèse T, Xue Z (1995) The nucleolus: an organelle formed by the act of building a ribosome. Curr Opin Cell Biol 7:319–324

Miller OL Jr, Beatty BR (1969) Visualization of nucleolar genes. Science 164:955–7

Montanaro L, Trere D, Derenzini M (2008) Nucleolus, ribosomes, and cancer. Am J Pathol 173:301–310

Montgomery T (1898) Comparative cytological studies, with especial regard to the morphology of the nucleolus. J Morphol 15:265–582

Mosgoeller W (2004) Nucleolar ultrastructure in vertebrates. In: Olson MOJ (ed) The nucleolus. Landes Biosciences, Austin, pp 10–20

Muro E, Hoang TQ, Jobart-Malfait A, Hernandez-Verdun D (2008) In nucleoli, the steady state of nucleolar proteins is leptomycin B-sensitive. Biol Cell 100:303–13

Muro E, Gébrane-Younès J, Jobart-Malfait A, Louvet E, Roussel P, Hernandez-Verdun D (2010) The traffic of proteins between nucleolar organizer regions and prenucleolar bodies governs the assembly of the nucleolus at exit of mitosis. Nucleus 1:202–211

Negi SS, Olson MO (2006) Effects of interphase and mitotic phosphorylation on the mobility and location of nucleolar protein B23. J Cell Sci 119:3676–85

Németh A, Conesa A, Santoyo-Lopez J, Medina I, Montaner D, Péterfia B, Solovei I, Cremer T, Dopazo J, Längst G (2010) Initial genomics of the human nucleolus. PLoS Genet 6(3):e1000889

Norton JT, Pollock CB, Wang C, Schink JC, Kim JJ, Huang S (2008) Perinucleolar compartment prevalence is a phenotypic pancancer marker of malignancy. Cancer 113:861–9

Norton JT, Wang C, Gjidoda A, Henry RW, Huang S (2009) The perinucleolar compartment is directly associated with DNA. J Biol Chem 284:4090–101

Ochs RL (1998) Methods used to study structure and function of the nucleolus. Methods Cell Biol 53:303–21

Ochs RL, Lischwe MA, Shen E, Caroll RE, Busch H (1985) Nucleologenesis: composition and fate of prenucleolar bodies. Chromosoma 92:330–336

Okuwaki M, Tsujimoto M, Nagata K (2002) The RNA binding activity of a ribosome biogenesis factor, nucleophosmin/B23, is modulated by phosphorylation with a cell cycle-dependent kinase and by association with its subtype. Molec Biol Cell 13:2016–2030

Olson MOJ (2004) Nontraditional roles of the nucleolus. In: Olson MOJ (ed) The nucleolus. Landes Biosciences, Austin, pp 329–342

Orrick LR, Olson MOJ, Busch H (1973) Comparison of nucleolar proteins of normal rat liver and Novikoff hepatoma ascites cells by two-dimensional polyacrylamide gel electrophoresis. Proc Natl Acad Sci U S A 70:1316–1320

Patterson GH, Lippincott-Schwartz J (2002) A photoactivatable GFP for selective photolabeling of proteins and cells. Science 297:1873–7

Pebusque MJ, Vio-Cigna M, Aldebert B, Seite R (1985) Circadian rhythm of nucleoli in rat superior cervical ganglion neurons: the two types of fibrillar centres and their quantitative relationship with the nucleolar organizing regions. J Cell Sci 74:65–74

Pederson T (1998) The plurifunctional nucleolus. Nucleic Acids Res 26:3871–3876

Pederson T (2010) "Compact" nuclear domains: reconsidering the nucleolus. Nucleus 1:444–445

Pendle AF, Clark GP, Boon R, Lewandowska D, Lam YW, Andersen J, Mann M, Lamond AI, Brown JW, Shaw PJ (2005) Proteomic analysis of the Arabidopsis nucleolus suggests novel nucleolar functions. Mol Biol Cell 16:260–9

Pestov DG, Strezoska Z, Lau LF (2001) Evidence of p53-dependent cross-talk between ribosome biogenesis and the cell cycle: effects of nucleolar protein Bop1 on G(1)/S transition. Mol Cell Biol 21:4246–55

Phair RD, Misteli T (2000) High mobility of proteins in the mammalian cell nucleus. Nature 404:604–609

Poll G, Braun T, Jakovljevic J, Neueder A, Jakob S, Woolford JL Jr, Tschochner H, Milkereit P (2009) rRNA maturation in yeast cells depleted of large ribosomal subunit proteins. PLoS One 4:e8249

Pollock C, Huang S (2010) The perinucleolar compartment. Cold Spring Harb Perspect Biol 2:a000679

Puvion E, Hernandez-Verdun D, Haguenau F (1994) The nucleus and the nucleolus. The contribution of French electron microscopists. Biol Cell 80:91–5

Puvion-Dutilleul F, Bachellerie J-P, Puvion E (1991) Nucleolar organization of HeLa cells as studied by in situ hybridization. Chromosoma 100:395–409

Puvion-Dutilleul F, Puvion E, Bachellerie J-P (1997) Early stages of pre-rRNA formation within the nucleolar ultrastructure of mouse cells studied by in situ hybridization with 5'ETS leader probe. Chromosoma 105:496–505

Qiu H, Eifert J, Wacheul L, Thiry M, Berger AC, Jakovljevic J, Woolford JL Jr, Corbett AH, Lafontaine DL, Terns RM, Terns MP (2008) Identification of genes that function in the biogenesis and localization of small nucleolar RNAs in Saccharomyces cerevisiae. Mol Cell Biol 28:3686–99

Raska I, Shaw PJ, Cmarko D (2006) New insights into nucleolar architecture and activity. Int Rev Cytol 255:177–235

Recher L, Whitescarver J, Briggs L (1969) The fine structure of a nucleolar constituent. J Ultrastruct Res 29:1–14

Robinson JM, Takizawa T, Pombo A, Cock PR (2001) Correlative fluorescence and electron microscopy on ultrathin cryosections: bridging the resolution gap. J Histochem Cytochem 49:803–8

Romanova L, Grand A, Zhang L, Rayner S, Katoku-Kikyo N, Kellner S, Kikyo N (2009a) Critical role of nucleostemin in pre-rRNA processing. J Biol Chem 284:4968–4977

Romanova L, Kellner S, Katoku-Kikyo N, Kikyo N (2009b) Novel role of nucleostemin in the maintenance of nucleolar architecture and integrity of small nucleolar ribonucleoproteins and the telomerase complex. J Biol Chem 284:26685–26694

Roussel P, André C, Comai L, Hernandez-Verdun D (1996) The rDNA transcription machinery is assembled during mitosis in active NORs and absent in inactive NORs. J Cell Biol 133:235–246

Savino TM, Bastos R, Jansen E, Hernandez-Verdun D (1999) The nucleolar antigen Nop52, the human homologue of the yeast ribosomal RNA processing RRP1, is recruited at late stages of nucleologenesis. J Cell Sci 112:1889–1900

Savino TM, Gébrane-Younès J, De Mey J, Sibarita J-B, Hernandez-Verdun D (2001) Nucleolar assembly of the rRNA processing machinery in living cells. J Cell Biol 153:1097–1110

Scheer U, Hock R (1999) Structure and function of the nucleolus. Curr Opin Cell Biol 11:385–390

Scheer U, Rose KM (1984) Localisation of RNA polymerase I in interphase cells and mitotic chromosomes by light and electron microscopic immunocytochemistry. Proc Natl Acad Sci U S A 81:1431–1435

Scherl A, Couté Y, Déon C, Callé A, Kindbeiter K, Sanchez J-C, Greco A, Hochstrasser D, Diaz J-J (2002) Functional proteomic analysis of human nucleolus. Mol Biol Cell 13: 4100–4109

Shav-Tal Y, Blechman J, Darzacq X, Montagna C, Dye BT, Patton JG, Singer RH, Zipori D (2005) Dynamic sorting of nuclear components into distinct nucleolar caps during transcriptional inhibition. Mol Biol Cell 16:2395–2413

Sirri V, Roussel P, Gendron MC, Hernandez-Verdun D (1997) Amount of the two major Ag-NOR proteins, nucleolin, and protein B23 is cell-cycle dependent. Cytometry 28:147–56

Sirri V, Roussel P, Hernandez-Verdun D (2000a) In vivo release of mitotic silencing of ribosomal gene transcription does not give rise to precursor ribosomal RNA processing. J Cell Biol 148:259–270

Sirri V, Roussel P, Hernandez-Verdun D (2000b) The AgNOR proteins: qualitative and quantitative changes during the cell cycle. Micron 31:121–6

Sirri V, Hernandez-Verdun D, Roussel P (2002) Cyclin-dependent kinases govern formation and maintenance of the nucleolus. J Cell Biol 156:969–981

Sirri V, Urcuqui-Inchima S, Roussel P, Hernandez-Verdun D (2008) Nucleolus: the fascinating nuclear body. Histochem Cell Biol 129:13–31

Smetana K, Busch H (1974) The nucleolus and nucleolar DNA. Academic press, New York, pp 73–147

Snaar S, Wiesmeijer K, Jochemsen AG, Tanke HJ, Dirks RW (2000) Mutational analysis of fibrillarin and its mobility in living human cells. J Cell Biol 151:653–662

Spector DL (2001) Nuclear domains. J Cell Sci 114:2891–2893

Spiegelhalter C, Tosch V, Hentsch D, Koch M, Kessler P, Schwab Y, Laporte J (2010) From dynamic live cell imaging to 3D ultrastructure: novel integrated methods for high pressure freezing and correlative light-electron microscopy. PLoS One 5:e9014

Sullivan GJ, Bridger JM, Cuthbert AP, Newbold RF, Bickmore WA, McStay B (2001) Human acrocentric chromosomes with transcriptionally silent nucleolar organizer regions associated with nucleoli. EMBO J 20:2867–2877

Taddei A, Schober H, Gasser SM (2010) The budding yeast nucleus. Cold Spring Harb Perspect Biol 2:a000612

Testillano PS, Gonzalez-Melendi P, Ahmadian P, Risueno MC (1995) The methylation-acetylation method: an ultrastructural cytochemistry for nucleic acids compatible with immunogold studies. J Struct Biol 114:123–39

Thiry M, Lafontaine DL (2005) Birth of a nucleolus: the evolution of nucleolar compartments. Trends Cell Biol 15:194–9

Trinkle-Mulcahy L, Lamond AI (2006) Mitotic phosphatases: no longer silent partners. Curr Opin Cell Biol 18:623–631

Tsai RYL, McKay RDG (2002) A nucleolar mechanism controlling cell proliferation in stem cells and cancer cells. Genes Dev 16:2991–3003

Van Hooser AA, Yuh P, Heald R (2005) The perichromosomal layer. Chromosoma 114:377–88

van Koningsbruggen S, Gierlinski M, Schofield P, Martin D, Barton GJ, Ariyurek Y, den Dunnen JT, Lamond AI (2010) High-resolution whole-genome sequencing reveals that specific chromatin domains from most human chromosomes associate with nucleoli. Mol Biol Cell 21:3735–48

Verheggen C, Le Panse S, Almouzni G, Hernandez-Verdun D (1998) Presence of pre-rRNAs before activation of polymerase I transcription in the building process of nucleoli during early development of *Xenopus laevis*. J Cell Biol 142:1167–1180

Verheggen C, Almouzni G, Hernandez-Verdun D (2000) The ribosomal RNA processing machinery is recruited to the nucleolar domain before RNA polymerase I during *Xenopus laevis* development. J Cell Biol 149:293–305

Verheggen C, Le Panse S, Almouzni G, Hernandez-Verdun D (2001) Maintenance of nucleolar machineries and pre-rRNAs in remnant nucleolus of erythrocyte nuclei and remodeling in *Xenopus* egg extracts. Exp Cell Res 269:23–34

Westman BJ, Verheggen C, Hutten S, Lam YW, Bertrand E, Lamond AI (2010) A proteomic screen for nucleolar SUMO targets shows SUMOylation modulates the function of Nop5/Nop58. Mol Cell 39:618–31

Chapter 2
The Dynamic Proteome of the Nucleolus

François-Michel Boisvert, Yasmeen Ahmad, and Angus I. Lamond

2.1 Introduction

The primary function of the nucleolus is as the site of ribosome subunit biogenesis in eukaryotic cells. Nucleoli reassemble at the end of mitosis around the tandemly repeated clusters of rDNA genes forming a subnuclear compartment that locally concentrates the dedicated transcription and processing machineries that are responsible for generating ribosome subunits. The process of assembling a ribosome subunit requires the initial transcription of the ribosomal DNA (rDNA) genes by a specialized RNA polymerase – RNA pol I. These rDNA genes are arranged in arrays of head-to-tail tandem repeats, termed nucleolar organizer regions (NORs). In humans, approximately 400 copies of 43-kb repeat units are distributed along all acrocentric chromosomes (chromosomes 13, 14, 15, 21 and 22) to form NORs. In many cell types, only a subset of rDNA genes are transcriptionally active, even though inactive rDNAs are still assembled into nucleoli. The initial 47S ribosomal RNA (rRNA) precursor transcript transcribed by RNA pol I is subsequently cleaved to form the mature 28S, 18S and 5.8S rRNAs, post-transcriptionally modified through interaction with small nucleolar ribonucleoproteins (snoRNPs) and additional protein processing factors. Finally, the processed and modified rRNAs are assembled with the many ribosomal proteins, prior to interaction with the export machinery and transport to the cytoplasm.

The isolation and characterization of organelles by subcellular fractionation is a well-established technique in cell biology. Many organelles have been isolated and analysed in the past century (see, e.g. Spector et al. (1997) for reviews and protocols). These studies have provided invaluable information on the functions and properties of individual organelles. With recent advances in mass spectrometry based proteomic

---

A.I. Lamond (✉)
Wellcome Trust Centre for Gene Regulation and Expression,
College of Life Sciences, University of Dundee, Dundee DD15EH, UK
e-mail: angus@lifesci.dundee.ac.uk

M.O.J. Olson (ed.), *The Nucleolus*, Protein Reviews 15,
DOI 10.1007/978-1-4614-0514-6_2, © Springer Science+Business Media, LLC 2011

technology, it has been possible to determine the major protein composition of various cytoplasmic organelles, for example the mitochondria (Pflieger et al. 2002), the Golgi apparatus (Bell et al. 2001; Taylor et al. 2000) and the chloroplast thylakoid membrane (Gomez et al. 2002). The isolation of subnuclear structures, in contrast with these cytoplasmic organelles, is made more difficult because they are not surrounded by membrane. Despite this limitation, isolation of several nuclear compartments, such as the nuclear envelope (Dreger et al. 2001), nuclear pore complexes (Cronshaw et al. 2002), interchromatin granule clusters (Mintz et al. 1999) and Cajal bodies (Lam et al. 2002), has been reported. The most well-studied nuclear organelle, the nucleolus, whose high density and structural stability allow effective purification using a straightforward procedure is an ideal structure for proteomic characterization. The ability to isolate nucleoli in large scale provided an excellent starting material for identifying, purifying and studying proteins in this nuclear compartment (Andersen et al. 2005; Andersen et al. 2002; Scherl et al. 2002).

Until recently, our knowledge of the protein content of nucleoli was quite limited (Fig. 2.1). However, the ability to purify nucleoli in large scale (Fig. 2.1a), combined with the major advances in the identification and analysis of proteins using mass spectrometry, has provided a wealth of information regarding the nucleolar proteome. Knowledge of nucleolar protein content has grown during the past 10 years from less than 100 proteins to well over 6,000 nucleolar proteins (Fig. 2.1b). Proteomic analyses have characterized the nucleolar proteome in both human and plant cells, identifying more than 200 plant and over 6,000 human proteins that stably co-purify with isolated nucleoli (Ahmad et al. 2009b; Andersen et al. 2002, 2005; Boisvert et al. 2010; Lam et al. 2010; Pendle et al. 2005; Scherl et al. 2002). A comparison of human and budding yeast data showed that ~90% of the nucleolus-related yeast proteins that have a clear human homologue are detected in the human nucleolar proteome (Andersen et al. 2005). This demonstrates that the nucleolar proteome is highly conserved through evolution.

Bibliographic and bioinformatic analyses of the proteomic data have allowed the classification of nucleolar proteins into functional groups and suggested potential functions for ~150 previously uncharacterized human proteins (Ahmad et al. 2009b; Coute et al. 2006; Hinsby et al. 2006; Leung et al. 2003). A classification of the molecular functions of the nucleolar proteins shows that only approximately 30% have a function obviously related to the production of ribosome subunits (Boisvert et al. 2007). However, the diverse identities and functions of many of the other nucleolar proteins are consistent with additional processes occurring within the nucleolus. This includes many pre-mRNA processing factors and proteins that are involved in cell-cycle control as well as DNA replication and repair (reviewed in Boisvert et al. 2007). An additional dimension has been added to the analysis of the nucleolar proteome by studies characterizing the dynamic changes in the proteome of the nucleolus under different metabolic conditions, such as inhibition of transcription following treatment of cells with actinomycin D (Andersen et al. 2005), in response to DNA damage (Boisvert et al. 2010; Boisvert and Lamond 2010) or following viral infection (Lam et al. 2010). The ability to analyse quantitatively and with high throughput the parallel increases and decreases in levels of many protein components has highlighted just how dynamic the nucleolar proteome can be.

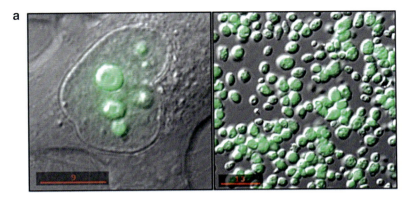

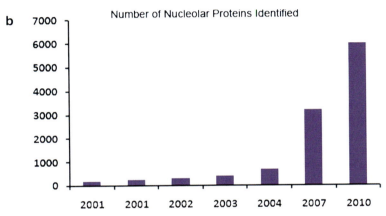

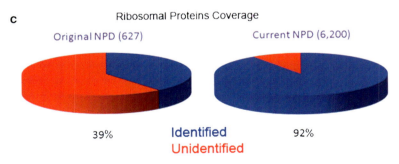

Fig. 2.1 Nucleolar proteome. (a) Overlay of a GFP-tagged nucleolar protein as shown over the DIC image in the whole cell (*left*) or following isolation of nucleoli showing intact structure and the absence of any visible contaminants from other cellular compartments. (b) Number of proteins that have been identified by mass spectrometry on purified nucleoli over the years following the improvement of methods and instruments. (c) Comparison of the number of ribosomal proteins that have been identified from the original published Nucleolar Protein Database to the present database

2.2 Isolation of Nucleoli

The starting point for the proteomic study of a cellular organelle or complex is the ability to isolate it intact, in high purity and in ideally large quantities. The relatively high density and structural stability of the nucleolus, as compared with other cellular structures, facilitates its efficient isolation even though it is not enclosed by a lipid membrane. Since the initial purification of nucleoli from human tumour cells and rodent liver cells in the early 1960s (Busch et al. 1963; Maggio 1966), several studies have been reported on the characterization of isolated nucleoli. Nucleoli have been purified from a large variety of mammalian tissues, including liver, thyroid (Voets et al. 1979) and brain (Banks and Johnson 1973) and from cells of non-mammalian species such as *Xenopus* (Saiga and Higashinakagawa 1979) and Tetrahymena (Matsuura and Higashinakagawa 1992).

The nucleolar isolation procedure is robust, and therefore the general strategy has been essentially unchanged over almost 40 years. Isolated cell nuclei are subjected to sonication, adjusting the power so that the nucleoli remain intact while the rest of the nuclei are fragmented as judged by microscopy. Then the nucleoli are isolated by centrifugation through a density gradient, on the basis of their high density compared with other nuclear components. Modifications of the basic procedure cater to the isolation of nucleoli from different cell types and organisms. For example, the procedure for isolating nucleoli from adherent HeLa cells was not suitable for suspension cultured HeLa S3 cells (our unpublished results). A critical factor is the salt concentration, especially, magnesium ion concentration used in the buffer during sonication, because the structural intactness of nucleoli decreases if salt concentration is too low (Vandelaer et al. 1996). However, if magnesium concentration is too high, nuclei cannot be efficiently disrupted by sonication (Lam et al. 2002) and hence the purity of the isolated nucleoli is compromised. In our experience, nuclei from adherent HeLa cells can be effectively sonicated in 0.35 M sucrose containing 0.5 mM $MgCl_2$, which lies between the large range of magnesium concentrations reported in other studies involving nucleolar isolation (Cheutin et al. 2002; Scherl et al. 2002).

It is essential to assess the purity and intactness of the isolated nucleoli before MS analysis. The quality of isolated HeLa nucleoli can be assessed using several criteria. First, the fraction isolated contains round or ovoid particles of uniform size, more than 95% of which can be labelled by the RNA dye Pyronin Y and by anti-nucleolar antibodies. These particles are morphologically similar to nucleoli detected in intact HeLa cells, as judged by both light and electron microscopy (Fig. 2.2). The ultrastructure of the isolated nucleoli shows that the internal nucleolar substructures (FC, DFC and GC) remain intact (Fig. 2.2). The purity of the isolated nucleoli can be further confirmed by western blotting, which should show that proteins known to be largely excluded from the nucleolus are virtually undetectable in the isolated nucleoli, while known nucleolar proteins (e.g. nucleolin and fibrillarin) are highly enriched. This was confirmed in an initial MS analysis, conducted to estimate the purity of the isolated nucleoli (Andersen et al. 2005). Of the 80 proteins found in

2 The Dynamic Proteome of the Nucleolus 33

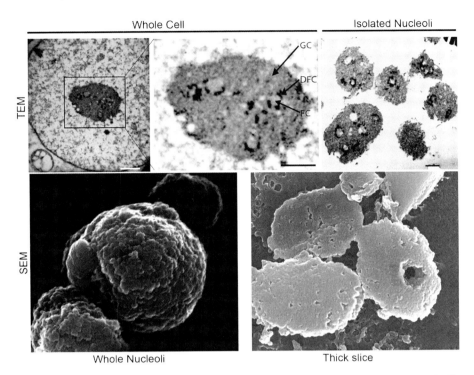

Fig. 2.2 Electron microscopy images of isolated nucleoli. Purified nucleoli are morphologically similar to nucleoli in intact cells. Transmission electron microscopy (TEM – *upper panels*) and scanning electron microscopy (SEM – *lower panels*) of isolated nucleoli were used to image nucleoli within intact HeLa cells (*left*) and nucleoli purified from HeLa cells (*right*). The central panels are enlargements of the indicated nucleoli in the intact cells. The isolated nucleoli are morphologically intact, retaining a clearly defined granular compartment (GC), dense fibrillar centre (DFC) and fibrillar centre (FC). *Scale bars* are 1 mm

the initial screen, many were known nucleolar proteins, while obvious protein contaminants were absent. The detection of fibrillarin and nucleolin in the non-nucleolar fraction by Western blotting likely reflects the physiological presence of these proteins in a diffuse pool in the nucleoplasm (Chen and Huang 2001; Phair and Misteli 2000). It is, however, also possible that there may be some "leakage" of nucleolar proteins during the isolation procedure.

One indication of the intactness of the isolated HeLa nucleoli is that these nucleoli can incorporate BrUTP in vitro (Andersen et al. 2005). The incorporated BrUTP is located in distinct foci inside the nucleoli, similar to the published nascent RNA pattern in nucleoli in vivo (Masson et al. 1996). The in vitro incorporation of BrUTP in isolated nucleoli was inhibited by actinomycin D, but not by α-amanitin, which selectively inhibits RNA polymerase II (Andersen et al. 2005). This confirms that nucleolar BrUTP incorporation was due to the action of RNA polymerase I, and that the isolated nucleoli, in agreement with previous reports, were transcriptionally active, at least for elongation (Cheutin et al. 2002). The properties of isolated nucleoli may vary greatly according to cell types, as Vandalaer et al. (Vandelaer et al. 1996) reported that nucleoli isolated from ELT cells using this method showed damage, especially in fibrillar centres, and were inefficient in transcription. In our experience, this nucleolar isolation (Vandelaer et al. 1996) method gave an extremely low yield and therefore was unsuitable for HeLa cells. Altogether, the combination of morphological, biochemical and functional studies demonstrate it is possible to isolate nucleoli that are both structurally and functionally intact.

2.3 Proteomic Analysis of Nucleolar Proteins

While nucleoli have been studied for over two centuries, it is only recently that an extensive list of proteins present within nucleoli at different times has emerged, thanks to the advance in the techniques of mass spectrometry based proteomics (reviewed by Coute et al. 2006) as well as better purification procedures. Adaptation of nucleolar purification procedures by sedimentation over sucrose cushion of sonicated nuclei led to the isolation of relatively intact and pure nucleoli, which were then used in high throughput proteomic-based experiments (Andersen et al. 2002). From the ~300 proteins identified in the first experiments (Andersen et al. 2002; Scherl et al. 2002), improvements in mass spectrometers have now identified over 50,000 peptides from more than 6,000 human proteins that co-purify with isolated nucleoli, providing significantly enhanced coverage of the nucleolar proteome (Ahmad et al. 2009b) (http://www.lamondlab.com/NOPdb3.0/). Interestingly, however, despite such an increase in the number of identified nucleolar proteins, when the proteins are categorized in terms of their functions, the distribution of functional categories is not altered significantly (Boisvert et al. 2007). This suggests that while the nucleolar proteome identified may still not be complete, it nonetheless fairly reflects the distribution of protein categories in the nucleolus, which is unlikely to change dramatically, even if more nucleolar proteins are discovered in the future. The functional categories indicated by proteomic studies are therefore likely to be a realistic reflection of nucleolar functions.

Interpretation of protein inventories derived using proteomics to identify proteins in purified organelles is complicated by the fact that many proteins are not exclusive to one compartment but instead partitioned between separate subcellular locations (Gauthier and Lazure 2008; Hall et al. 2009). Recent developments in quantitative proteomics allow the subcellular spatial distribution of proteins to be mapped and thus have led to a better definition of a nucleolar protein (Boisvert et al. 2010; Boisvert and Lamond 2010). The measurements using the spatial proteomics method allow classification of proteins according to whether they are enriched in the nucleolus compared to other compartments, or whether they are less abundant in that organelle. It is also important to recognize that these values are not fixed and can change over time. This highlights the importance of not only identifying the presence of a protein in any specific cellular organelle or structure, but also measuring its relative abundance in different locations and assessing how this subcellular localization can change between different compartments under different cell growth and physiological conditions.

Global proteomic analyses of the different proteins have identified components that have been associated with functions unrelated to ribosome subunit biogenesis. Several proteomic analyses have also been undertaken to characterize the nucleolar proteome in non-mammalian species such as trypanosomes (Degrasse et al. 2008), *Arabidopsis* (Brown et al. 2005) and budding yeast (Huh et al. 2003). A comparison of human and budding yeast nucleolar data shows that over 90% of yeast proteins with clear human homologues can be detected in the human nucleolar proteome. This demonstrates that the core nucleolar proteome is largely conserved through evolution. Bibliographic and bioinformatic analyses of the proteomic data have allowed the classification of nucleolar proteins into functional groups and suggested potential functions for several previously uncharacterized human proteins. This shows that approximately 30% have a function related to the production of ribosome subunits (Boisvert et al. 2007). However, the diverse identities and functions of many of the other nucleolar proteins are consistent with additional processes occurring within the nucleolus. This includes many pre-mRNA processing factors and proteins that are involved in cell-cycle control as well as DNA replication and repair. The most striking feature of the functional distribution of the nucleolar proteome is the high proportion of novel and previously uncharacterized factors, a surprising fact for an organelle intensively investigated for over two centuries. Of the known proteins, the most common functional motifs found in approximately 20% of these proteins are nucleic acid and nucleotide binding domains. The DEAD-box helicase motifs in particular, characteristic of the superfamily of RNA dependent ATPases, were also represented highly in the nucleolar proteome, consistent with the control of RNA interactions being an important feature of nucleolar function. Consistent with its major role in transcription and processing of rRNAs and their subsequent assembly into ribosomal subunits, the nucleolar proteome includes many ribosomal proteins, processing factors and components required for transcription of the rRNA gene clusters, as well as human homologues of genes known to be involved in these processes in other organisms.

An additional dimension to the analysis of the nucleolar proteome involves characterizing the dynamic changes in the proteome of the nucleolus under different metabolic conditions, such as inhibition of transcription following treatment of cells with actinomycin D (Andersen et al. 2005), following viral infection (Cawood et al. 2007; Hirano et al. 2009; Hiscox 2007; Hiscox et al. 2010; Lam et al. 2010), following DNA damage (Boisvert et al. 2010; Boisvert and Lamond 2010) or through studies of protein turnover (Lam et al. 2007). The ability to analyse quantitatively and with high throughput the parallel increases and decreases in levels of many protein components has highlighted just how dynamic the nucleolar proteome can be. It will be interesting in the future to compare in detail how the nucleolar proteome varies between transformed human cell lines and primary cells.

The tumour suppressor p53 plays an important role involving the nucleolus in regulating aspects of stress responses and control of cell cycle progression. Under normal conditions, p53 is a short-lived protein that is present in cells at a barely detectable level. Exposure of cells to various form of exogenous stress, such as DNA damage, heat shock, hypoxia, etc., triggers the stabilization of p53, which is then responsible for an ensuing cascade of events, resulting in either cell cycle arrest, or in apoptosis. Accumulation of p53 induces the p21-mediated inhibition of cyclin D/cdk4 and cyclinE/cdk2, resulting in cell cycle arrest in G1. The stability of the p53 protein in mammals is primarily regulated in non-transformed cells by the interplay of two proteins, hdm2 and p14Arf in humans (the equivalent mouse proteins are mdm2 and p19Arf) (Prives 1998). Hdm2 functions as a specific E3 ubiquitin ligase for p53, resulting in a low level of p53 due to proteasome-mediated degradation of ubiquitin-conjugated p53 in the cytoplasm. A variety of stimuli, including stress pathways and oncogenic signals, increase expression of Arf, which then associates with hdm2 to inhibit the ubiquitination, nuclear export and subsequent degradation of p53. It has been proposed that Arf physically sequesters hdm2 in nucleoli, thereby relieving nucleoplasmic p53 from hdm2-mediated degradation (Weserska-Gadek and Horky 2003). Arf is predominantly a nucleolar protein and might also regulate ribosome biogenesis by retarding the processing of early 47S/45S and 32S rRNA precursors, perhaps through interaction with B23 (Bertwistle et al. 2004). Exposure of cells to various forms of stress, such as DNA damage, heat shock and aberrant ribosome subunit biogenesis results in an increase in p53 level and hence cell cycle arrest. Thus, the nucleolus acts as a sensor for cellular stress signals through p53 stabilization.

In p53 wild-type cells, p53 appears to cause a shut-down of nucleolar activity, which results in a specific segregation of nucleolar proteins within the nucleolus (Boisvert and Lamond 2010). However, this seems to be dependent on p53, because the effect is reduced in p53 knock-out cells (Boisvert and Lamond 2010). One consequence is that ribosomal proteins no longer accumulate in the nucleolus following DNA damage. This suggests a possible early role for p53 in shutting down the rDNA transcription machinery, as well as stopping the nucleolar recruitment, and/or retention of ribosomal proteins in the nucleolus, indicating that cells rapidly stop ribosome subunit production following DNA damage. Several recent reports showed that p53 becomes activated after silencing of ribosomal proteins such as RPL23 (Zhang et al. 2010), RPL11 (Lohrum et al. 2003), RPS6 (Volarevic et al. 2000) and TIF1A (Yuan et al. 2005). Other evidence emerging from a number of mouse models supports the existence of this ribosomal dependent p53 checkpoint in vivo (Fumagalli et al. 2009). During normal cellular growth, ribosomal proteins are assembled into ribosome subunits, but several ribosomal proteins including RPL11, RPL5, RPL23, RPS7 and RPS9 have now been shown to be released from the nucleolus following stress and to bind HDM2, resulting in stabilization of p53 (Fumagalli et al. 2009; Lindstrom and Nister 2010; Ohashi et al. 2010; Zhang et al. 2006). However, proteomic analysis suggests that p53 is actually necessary for the initial release of ribosomal proteins from the nucleolus following stress, and that this release probably results in an amplification of the p53 response by preventing HDM2-mediated degradation of p53 (Boisvert and Lamond 2010).

2 The Dynamic Proteome of the Nucleolus

While many of the proteins identified in nucleolar proteomic studies are either known nucleolar proteins or else homologues of nucleolar proteins in other species, there are still a large number of proteins that are either previously unidentified, or else have not been shown previously to be localized in the nucleolus. To confirm whether these proteins are indeed localized in nucleoli, and not contaminants, systematic tagging of putative nucleolar proteins with fluorescent proteins and subcellular localization in cells following transient transfection have been analyzed by fluorescence microscopy. The relatively small number of FP-tagged proteins that did not localize in the nucleolus are not necessarily contaminants, however, because a protein may only be accumulated in the nucleolus during a particular phase of the cell cycle or under specific metabolic conditions. For example, microscopy analysis has showed that many proteins rapidly cycle between the nucleolus and nucleoplasm (e.g. Chen and Huang 2001). It is also possible that the fluorescent protein attached to the nucleolar protein interferes with the correct localization. The isolated nucleoli may therefore contain factors that are predominately localized in the nucleoplasm but transiently cycle through the nucleolus. Mass spectrometry is sufficiently sensitive to detect these low abundant proteins. For example, PSP1, a protein first identified in the proteome of the nucleolus, was present in a previously unknown nuclear domain "paraspeckles" and was apparently not nucleolar. However, drug treatment and fluorescence loss in photobleaching (FLIP) experiments confirmed that this protein interacted dynamically with the nucleoplasm and nucleolus in a transcriptional-dependent manner (Fox et al. 2005). To fully confirm the presence of a protein in the nucleolus, it is therefore necessary to take this into consideration and perform FLIP experiments on the transfected cells. This demonstrates that most of the identified proteins were nucleolar of steady state in interphase cells, and that confirms the proteomic approach is highly reliable in discovering nucleolar proteins.

2.4 Presentation and Publication of Data

The large amount of data acquired from proteomics studies requires a systematic way to analyse and integrate them with the information already deposited in publicly available databases. To facilitate this, a Nucleolar Online Proteomics Database (NOPdb) was created and published in 2006 (Leung et al. 2006). More recently, this database has been updated and revamped to version 3.0 (Ahmad et al. 2009a). The NOPdb consists of a backend database and a frontend interface to allow researchers to search for nucleolar proteomics data (Fig. 2.3).

The NOPdb archives all human nucleolar proteins identified to date by the Lamond group and their collaborators using MS analyses performed on purified preparations of human nucleoli (Andersen et al. 2005; Boisvert et al. 2010; Boisvert and Lamond 2010; Lam et al. 2007; Leung et al. 2006). The current version 3.0 of the NOPdb includes over 50,000 peptides contained in over 6,200 human proteins identified in different human cells lines. The coverage of the human nucleolar

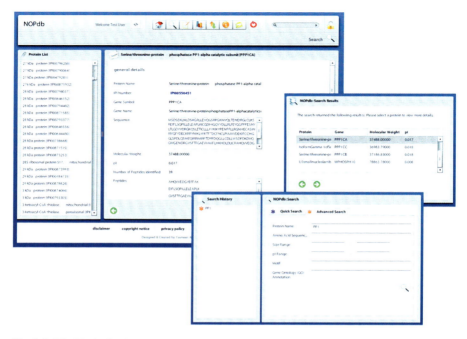

Fig. 2.3 The Nucleolar Protein Database (NOPdb). The NOPdb3.0 is an online resource available at http://www.lamondlab.com/NOPdb3.0/. It is searchable by protein names, gene names, amino acid or nucleotide sequences, sequence motifs or by limiting the range for isoelectric points and/or molecular weights. This web-based database displays interactive entries for each nucleolar protein identified in our studies. Information for each protein includes a summary of the known features, genomic location, unigene entry and proteome. The database is continually updated to include newly identified nucleolar proteins

proteome has increased over the years, as demonstrated by coverage of ribosomal proteins increasing from ~28% in the earlier versions of the NOPdb (version 2.0) to incorporate over 80% in NOPdb3.0. It is estimated that NOPdb3.0 contains over 80% of the main human nucleolar proteins. The proteins in the database are regularly updated as more experiments are performed in the Lamond laboratory.

The NOPdb3.0 is an online resource available at http://www.lamondlab.com/NOPdb3.0/. It is searchable either by protein names, gene names, amino acid or nucleotide sequences, sequence motifs or by limiting the range for isoelectric points and/or molecular weights. The database is also searchable by Interpro motif numbers (database of protein families, domains and functional sites) (Bateman et al. 2004; Letunic et al. 2004; Mulder et al. 2003) and by gene onotology (GO) terms (describe gene products in terms of their associated biological processes, cellular components and molecular functions in a species-independent manner) (Ashburner et al. 2000).

The NOPdb3.0 provides a range of information on proteins, including protein name, accession number, gene symbol, gene name, sequence, molecular weight, isoelectric point (PI), peptides identified, experiments in which the protein was identified, motifs and GO annotation.

The NOPDB application facilitates mining of stored data thanks to the data being stored in a relational structure that is well documented. Thus, tools can be built to search, analyse, read and interpret the data. This mining capability is evident within the search feature of the application. Furthermore, the NOPdb3.0 uses application programming interface (API) to create dynamically generated graphs, allowing researchers to visualize the data produced from experiments and enabling cross analysis between experiments.

2.5 Perspectives

The nucleolus can be isolated intact from mammalian cells using a simple and straightforward procedure. This makes the nucleolus a model nuclear organelle for proteomic studies. The continuing advances in mass spectrometry techniques toward high sensitivity and automation enable identification of most of the proteins present in isolated organelles. The basic map of HeLa nucleolar proteins is therefore now largely charted. Future analyses of cell nucleoli will identify also some cell type-specific nucleolar proteins, through the analysis of nucleoli purified from a variety of sources, including primary cells and cell lines derived from different tissues. Some proteins may interact with nucleoli under only specific metabolic conditions and therefore have not been detected in current studies. For example, it will be important to isolate and analyse nucleoli from cells at specific cell cycle stages, during different cell differentiation states, following various forms of cell transformation and during senescence. One challenge of these experiments is the need not only to detect the identity of proteins but also to quantitate the changes in their abundance under different conditions. More quantitative methods have now allowed measurement of the relative protein enrichment in nucleoli, which should provide a standard for annotating nucleolar proteins (Boisvert et al. 2010).

In conclusion, although more work remains to be done, we believe that the human nucleolar proteome detailed so far represents a significant advance toward defining a comprehensive inventory of nucleolar proteins. These data should be of value for future studies on the range of biological roles performed by the nucleolus, including, for example, stress responses as well as ribosome subunit biogenesis, and the mechanisms involved in its assembly and function. Future studies will expand our knowledge of the nucleolar proteomics in other model organisms and will provide a more detailed quantitative picture of the levels of each protein and how this changes under a range of metabolic conditions.

References

Ahmad Y, Boisvert FM, Gregor P, Cobley A, Lamond AI (2009) NOPdb: nucleolar proteome database-2008 update. Nucleic Acids Res 37:D181–D184

Andersen JS, Lyon CE, Fox AH, Leung AK, Lam YW, Steen H, Mann M, Lamond AI (2002) Directed proteomic analysis of the human nucleolus. Curr Biol 12:1–11

Andersen JS, Lam YW, Leung AK, Ong SE, Lyon CE, Lamond AI, Mann M (2005) Nucleolar proteome dynamics. Nature 433:77–83

Ashburner M, Ball CA, Blake JA, Botstein D, Butler H, Cherry JM, Davis AP, Dolinski K, Dwight SS, Eppig JT, Harris MA, Hill DP, Issel-Tarver L, Kasarskis A, Lewis S, Matese JC, Richardson JE, Ringwald M, Rubin GM, Sherlock G (2000) Gene ontology: tool for the unification of biology. The Gene Ontology Consortium. Nat Genet 25:25–29

Banks SP, Johnson TC (1973) Developmental alterations in RNA synthesis in isolated mouse brain nucleoli. Biochim Biophys Acta 294:450–460

Bateman A, Coin L, Durbin R, Finn RD, Hollich V, Griffiths-Jones S, Khanna A, Marshall M, Moxon S, Sonnhammer EL, Studholme DJ, Yeats C, Eddy SR (2004) The Pfam protein families database. Nucleic Acids Res 32:D138–D141

Bell AW, Ward MA, Blackstock WP, Freeman HN, Choudhary JS, Lewis AP, Chotai D, Fazel A, Gushue JN, Paiement J, Palcy S, Chevet E, Lafreniere-Roula M, Solari R, Thomas DY, Rowley A, Bergeron JJ (2001) Proteomics characterization of abundant Golgi membrane proteins. J Biol Chem 276:5152–5165

Bertwistle D, Sugimoto M, Sherr CJ (2004) Physical and functional interactions of the Arf tumor suppressor protein with nucleophosmin/B23. Mol Cell Biol 24:985–996

Boisvert FM, Lamond AI (2010) p53-Dependent subcellular proteome localization following DNA damage. Proteomics 10:4087–4097

Boisvert FM, van Koningsbruggen S, Navascues J, Lamond AI (2007) The multifunctional nucleolus. Nat Rev Mol Cell Biol 8:574–585

Boisvert FM, Lam YW, Lamont D, Lamond AI (2010) A quantitative proteomics analysis of subcellular proteome localization and changes induced by DNA damage. Mol Cell Proteomics 9:457–470

Brown JW, Shaw PJ, Shaw P, Marshall DF (2005) Arabidopsis nucleolar protein database (AtNoPDB). Nucleic Acids Res 33:D633–D636

Busch H, Muramatsu M, Adams H, Steele WJ, Liau MC, Smetana K (1963) Isolation of nucleoli. Exp Cell Res 24(Suppl 9):150–163

Cawood R, Harrison SM, Dove BK, Reed ML, Hiscox JA (2007) Cell cycle dependent nucleolar localization of the coronavirus nucleocapsid protein. Cell Cycle 6:863–867

Chen D, Huang S (2001) Nucleolar components involved in ribosome biogenesis cycle between the nucleolus and nucleoplasm in interphase cells. J Cell Biol 153:169–176

Cheutin T, O'Donohue MF, Beorchia A, Vandelaer M, Kaplan H, Defever B, Ploton D, Thiry M (2002) Three-dimensional organization of active rRNA genes within the nucleolus. J Cell Sci 115:3297–3307

Coute Y, Burgess JA, Diaz JJ, Chichester C, Lisacek F, Greco A, Sanchez JC (2006) Deciphering the human nucleolar proteome. Mass Spectrom Rev 25:215–234

Cronshaw JM, Krutchinsky AN, Zhang W, Chait BT, Matunis MJ (2002) Proteomic analysis of the mammalian nuclear pore complex. J Cell Biol 158:915–927

Degrasse JA, Chait BT, Field MC, Rout MP (2008) High-yield isolation and subcellular proteomic characterization of nuclear and subnuclear structures from trypanosomes. Methods Mol Biol 463:77–92

Dreger M, Bengtsson L, Schoneberg T, Otto H, Hucho F (2001) Nuclear envelope proteomics: novel integral membrane proteins of the inner nuclear membrane. Proc Natl Acad Sci USA 98:11943–11948

Fox AH, Bond CS, Lamond AI (2005) P54nrb forms a heterodimer with PSP1 that localizes to paraspeckles in an RNA-dependent manner. Mol Biol Cell 16:5304–5315

2 The Dynamic Proteome of the Nucleolus

Fumagalli S, Di Cara A, Neb-Gulati A, Natt F, Schwemberger S, Hall J, Babcock GF, Bernardi R, Pandolfi PP, Thomas G (2009) Absence of nucleolar disruption after impairment of 40S ribosome biogenesis reveals an rpL11-translation-dependent mechanism of p53 induction. Nat Cell Biol 11:501–508

Gauthier DJ, Lazure C (2008) Complementary methods to assist subcellular fractionation in organellar proteomics. Expert Rev Proteomics 5:603–617

Gomez SM, Nishio JN, Faull KF, Whitelegge JP (2002) The chloroplast grana proteome defined by intact mass measurements from liquid chromatography mass spectrometry. Mol Cell Proteomics 1:46–59

Hall SL, Hester S, Griffin JL, Lilley KS, Jackson AP (2009) The organelle proteome of the DT40 Lymphocyte cell line. Mol Cell Proteomics 8(6):1295–1305

Hinsby AM, Kiemer L, Karlberg EO, Lage K, Fausboll A, Juncker AS, Andersen JS, Mann M, Brunak S (2006) A wiring of the human nucleolus. Mol Cell 22:285–295

Hirano Y, Ishii K, Kumeta M, Furukawa K, Takeyasu K, Horigome T (2009) Proteomic and targeted analytical identification of BXDC1 and EBNA1BP2 as dynamic scaffold proteins in the nucleolus. Genes Cells 14:155–166

Hiscox JA (2007) RNA viruses: hijacking the dynamic nucleolus. Nat Rev Microbiol 5:119–127

Hiscox JA, Whitehouse A, Matthews DA (2010) Nucleolar proteomics and viral infection. Proteomics 10:4077–4086

Huh WK, Falvo JV, Gerke LC, Carroll AS, Howson RW, Weissman JS, O'Shea EK (2003) Global analysis of protein localization in budding yeast. Nature 425:686–691

Lam YW, Lyon CE, Lamond AI (2002) Large-scale isolation of Cajal bodies from HeLa cells. Mol Biol Cell 13:2461–2473

Lam YW, Lamond AI, Mann M, Andersen JS (2007) Analysis of nucleolar protein dynamics reveals the nuclear degradation of ribosomal proteins. Curr Biol 17:749–760

Lam YW, Evans VC, Heesom KJ, Lamond AI, Matthews DA (2010) Proteomics analysis of the nucleolus in adenovirus-infected cells. Mol Cell Proteomics 9:117–130

Letunic I, Copley RR, Schmidt S, Ciccarelli FD, Doerks T, Schultz J, Ponting CP, Bork P (2004) SMART 4.0: towards genomic data integration. Nucleic Acids Res 32:D142–D144

Leung AK, Andersen JS, Mann M, Lamond AI (2003) Bioinformatic analysis of the nucleolus. Biochem J 376:553–569

Leung AKL, Trinkle-Mulcahy L, Lam YW, Andersen JS, Mann M, Lamond AI (2006) NOPdb: nucleolar proteome database. Nucleic Acids Res 34:D218–D220

Lindstrom MS, Nister M (2010) Silencing of ribosomal protein S9 elicits a multitude of cellular responses inhibiting the growth of cancer cells subsequent to p53 activation. PLoS One 5:e9578

Lohrum MA, Ludwig RL, Kubbutat MH, Hanlon M, Vousden KH (2003) Regulation of HDM2 activity by the ribosomal protein L11. Cancer Cell 3:577–587

Maggio R (1966) Some properties of isolated nucleoli from guinea-pig liver. Biochim Biophys Acta 119:641–644

Masson C, Bouniol C, Fomproix N, Szollosi MS, Debey P, Hernandez-Verdun D (1996) Conditions favoring RNA polymerase I transcription in permeabilized cells. Exp Cell Res 226:114–125

Matsuura T, Higashinakagawa T (1992) In vitro transcription in isolated nucleoli of *Tetrahymena pyriformis*. Dev Genet 13:143–150

Mintz PJ, Patterson SD, Neuwald AF, Spahr CS, Spector DL (1999) Purification and biochemical characterization of interchromatin granule clusters. EMBO J 18:4308–4320

Mulder NJ, Apweiler R, Attwood TK, Bairoch A, Barrell D, Bateman A, Binns D, Biswas M, Bradley P, Bork P, Bucher P, Copley RR, Courcelle E, Das U, Durbin R, Falquet L, Fleischmann W, Griffiths-Jones S, Haft D, Harte N, Hulo N, Kahn D, Kanapin A, Krestyaninova M, Lopez R, Letunic I, Lonsdale D, Silventoinen V, Orchard SE, Pagni M, Peyruc D, Ponting CP, Selengut JD, Servant F, Sigrist CJA, Vaughan R, Zdobnov EM (2003) The InterPro Database, 2003 brings increased coverage and new features. Nucleic Acids Res 31:315–318

Ohashi S, Natsuizaka M, Wong GS, Michaylira CZ, Grugan KD, Stairs DB, Kalabis J, Vega ME, Kalman RA, Nakagawa M, Klein-Szanto AJ, Herlyn M, Diehl JA, Rustgi AK, Nakagawa H (2010)

Epidermal growth factor receptor and mutant p53 expand an esophageal cellular subpopulation capable of epithelial-to-mesenchymal transition through ZEB transcription factors. Cancer Res 70(10):4174–4184

Pendle AF, Clark GP, Boon R, Lewandowska D, Lam YW, Andersen J, Mann M, Lamond AI, Brown JW, Shaw PJ (2005) Proteomic analysis of the Arabidopsis nucleolus suggests novel nucleolar functions. Mol Biol Cell 16:260–269

Pflieger D, Le Caer JP, Lemaire C, Bernard BA, Dujardin G, Rossier J (2002) Systematic identification of mitochondrial proteins by LC-MS/MS. Anal Chem 74:2400–2406

Phair RD, Misteli T (2000) High mobility of proteins in the mammalian cell nucleus. Nature 404:604–609

Prives C (1998) Signaling to p53: breaking the MDM2-p53 circuit. Cell 95:5–8

Saiga H, Higashinakagawa T (1979) Properties of in vitro transcription by isolated *Xenopus* oocyte nucleoli. Nucleic Acids Res 6:1929–1940

Scherl A, Coute Y, Deon C, Calle A, Kindbeiter K, Sanchez JC, Greco A, Hochstrasser D, Diaz JJ (2002) Functional proteomic analysis of human nucleolus. Mol Biol Cell 13:4100–4109

Spector DL, Goldman RD, and Leinw LA (1997) Cells: a laboratory manual, pp. 41.1–41.7, Cold Spring Harbor Laboratory Press, Cold Spring Harbor

Taylor RS, Wu CC, Hays LG, Eng JK, Yates JR III, Howell KE (2000) Proteomics of rat liver Golgi complex: minor proteins are identified through sequential fractionation. Electrophoresis 21:3441–3459

Vandelaer M, Thiry M, Goessens G (1996) Isolation of nucleoli from ELT cells: a quick new method that preserves morphological integrity and high transcriptional activity. Exp Cell Res 228:125–131

Voets R, Lagrou A, Hilderson H, Van Dessel G, Dierick W (1979) RNA synthesis in isolated bovine thyroid nuclei and nucleoli. alpha-Amanitin effect, a hint to the existence of a specific regulatory system. Hoppe Seylers Z Physiol Chem 360:1271–1283

Volarevic S, Stewart MJ, Ledermann B, Zilberman F, Terracciano L, Montini E, Grompe M, Kozma SC, Thomas G (2000) Proliferation, but not growth, blocked by conditional deletion of 40S ribosomal protein S6. Science 288:2045–2047

Weserska-Gadek J, Horky M (2003) How the nucleolar sequestration of p53 protein or its interplayers contributes to its (re)-activation. Ann N Y Acad Sci 1010:266–272

Yuan X, Zhou Y, Casanova E, Chai M, Kiss E, Grone HJ, Schutz G, Grummt I (2005) Genetic inactivation of the transcription factor TIF-IA leads to nucleolar disruption, cell cycle arrest, and p53-mediated apoptosis. Mol Cell 19:77–87

Zhang F, Hamanaka RB, Bobrovnikova-Marjon E, Gordan JD, Dai MS, Lu H, Simon MC, Diehl JA (2006) Ribosomal stress couples the unfolded protein response to p53-dependent cell cycle arrest. J Biol Chem 281:30036–30045

Zhang Y, Shi Y, Li X, Du W, Luo G, Gou Y, Wang X, Guo X, Liu J, Ding J, Wu K, Fan D (2010) Inhibition of the p53-MDM2 interaction by adenovirus delivery of ribosomal protein L23 stabilizes p53 and induces cell cycle arrest and apoptosis in gastric cancer. J Gene Med 12:147–156

Chapter 3
The Structure of rDNA Chromatin

Peter J. Shaw and Peter C. McKeown

3.1 Introduction

In all eukaryotic organisms so far studied, the genes for three of the ribosomal RNAs are carried in single transcription units – in the order, the S-rRNA (18S), 5.8S rRNA and L-rRNA (25S/28S). They are transcribed by RNA polymerase I as a single precursor RNA molecule, which is subsequently processed in a number of steps, removing first a leader sequence (external 5′ transcribed spacer or ETS), then two internal transcribed spacers (ITS1 and ITS2) that flank the 5.8S rRNA and finally short 3′ external transcribed spacer. The rDNA repeats are separated by intergenic spacer regions, often called the non-transcribed spacer (NTS). The rRNAs are highly conserved between species, whereas the various spacer regions are highly divergent in sequence. Plants have fairly short transcribed spacers; among vertebrates, birds have long internal transcribed spacers, whereas mammals have very long 5′ external transcribed spacers. The intergenic spacers are also highly divergent in length and sequence, with plants and *Saccharomyces cerevisiae* having NTS length of 2–3 kb and other species such as vertebrates having much longer sequences (20–30 kb) (Hadjiolov 1985; Busch and Smetana 1970; Raska et al. 2006a, b; Shaw 2010; Shaw and Jordan 1995).

It is clear that a single copy of the rDNA repeat could not be transcribed at a sufficient rate for the cell's requirement for ribosomes, and thus all organisms carry multiple rDNA copies. These are generally arranged in one or more tandem repeat arrays; it is tempting to speculate that the arrangement of rDNA in tandem repeats is likely to produce high local concentrations of the various factors required for ribosome biogenesis and thus increase efficiency of ribosome production, but this has yet to be proved. In almost all eukaryotes, the fourth ribosomal RNA, 5S, is

---

P.J. Shaw (✉)
Department of Cell and Developmental Biology, John Innes Centre, Colney, Norwich, UK
e-mail: peter.shaw@jic.ac.uk

M.O.J. Olson (ed.), *The Nucleolus*, Protein Reviews 15,
DOI 10.1007/978-1-4614-0514-6_3, © Springer Science+Business Media, LLC 2011

located as tandem arrays elsewhere in the genome. Even in *S. cerevisiae*, in which the 5S gene is part of the same repeating unit as the other rRNA genes, the 5S is transcribed in the opposite direction from the rDNA unit, and by a different RNA polymerase (RNA polymerase III) (Hadjiolov 1985; Shaw and Jordan 1995). The reason for this separation is unknown.

The number of rDNA repeats is very variable throughout phylogeny. In birds and mammals there are typically 100–300 per haploid set, whereas amphibians may have thousands of copies. Higher plants also have thousands of rDNA copies, although some algae have only hundreds. The rDNA can also be differentially amplified, either extrachromosomally as in amphibian oocytes, or differentially in polytene chromosomes as in Drosophila salivary gland nuclei (Hadjiolov 1985). The reason that some organisms have thousands of repeats is unknown. It is clear that a couple of hundred copies is enough to service most cells' ribosome requirement, at least in mammals and many other species. It is possible that there are cell types or developmental stages in some organisms that require many more rDNA copies than are normally transcribed, but even in the yeast *S. cerevisiae* only about half the (~150) copies are transcribed, and in a number of organisms, including yeast, viable mutants have been made with only a fraction of the normal number of rDNA copies (Takeuchi et al. 2003). There is a broad correlation between genome size and number of rDNA copies (Prokopowich et al. 2003), and this has led to a hypothesis that the rDNA may be acting as a sensor for DNA damage, protecting the rest of the genome by inducing DNA repair mechanisms or apoptosis. The extra copies present in the rDNA repeats would initially presumably buffer such damage by ensuring that sufficient undamaged copies were available for ribosome biosynthesis (Kobayashi 2008). There is certainly clear evidence that disrupting the nucleolus activates p53 DNA damage pathways in human culture cells (Rubbi and Milner 2003).

3.2 Genomic Organization of rDNA

The rDNA tandem repeats are carried on one or more chromosomal locations. It was recognized early on that secondary constrictions in the metaphase chromosomes were the sites at which nucleoli were formed in early G1, and these sites were termed nucleolar organizers regions (NORs) (Heitz 1931; McClintock 1934). With the advent of in situ hybridization techniques, it was confirmed that these sites did indeed contain rDNA repeats (Gall and Pardue 1969), but in some species or hybrids there were additional arrays of rDNA repeats which did not give rise to nucleoli. These rDNA arrays are now known to be silent; it is only those arrays that have been transcriptionally active during the preceding interphase that retain a partially decondensed state at metaphase, visualized with stains as a secondary constriction (primarily due to the binding of UBF, at least in animal nucleoli – see Chap. 5).

The NORs may be carried on any chromosome, usually on autosomes, but also on sex chromosomes as in many insects, such as Drosophila, and certain mammals.

The number of chromosomes that carry NORs varies from one to six or more, with no apparent pattern. In general, NORs are predominantly found close to the telomere of the shorter arm of chromosomes on which they are carried (Lima-de-Faria 1976). rDNA copies are multiplied by repeated recombination, and their homogeneity is maintained by gene conversion events between the tandem repeats (Kobayashi 2008). In *S. cerevisiae*, at least, recombination is induced by Fob1, which causes double strand break formation; in a *fob1* mutant rDNA repeat fluctuation is reduced or eliminated. The histone deacetylase Sir2p also has a role in rDNA copy number regulation – in a *sir2* mutant, copy number fluctuates wildly (Kobayashi 2008). In yeast, *SIR1* and *FOB1* affect cellular aging, and a connection with rDNA was originally shown by the generation of extrachromosomal rDNA circles from the rDNA repeats, which accumulate preferentially in the mother cells of budding yeast (Sinclair and Guarente 1997).

A full sequence analysis of the rDNA repeat arrays has not yet been carried out in any higher eukaryotic genome. The accurate genomic sequencing of multiple repeated sequences of such length is extremely difficult and may still be beyond current sequencing technology. All current genome sequence assemblies contain consensus sequences or sequences derived from the ends of the arrays, and we still have no real idea of what proportion of rDNA repeats are functional and whether any are pseudogenes. It is also possible that the intergenic rDNA sequences have different variants present in a single tandem array, or contain other interspersed sequences or repeating elements. Fibre FISH of rDNA in humans has suggested that some rDNA repeats are inverted, and may not be functional (Caburet et al. 2005), but this has yet to be fully confirmed. Fibre FISH of rice NORs did not find any evidence for rearrangements or inversions in the repeats (Mizuno et al. 2008).

3.3 Structural Organization of rDNA Chromatin

DNA specific dyes such as DAPI generally show the nucleolus as a dark region, suggesting a relatively low concentration of DNA (e.g. see Fig. 3.1). In fact in most cells the nucleolus is the most transcriptionally active region of the nucleus, and the relatively low level of DNA staining shows that these active genes are highly decondensed. With sensitive imaging, faint internal nucleolar structures and foci can often be seen with DNA stains. These most likely correspond to fibrillar centres (see Chap. 1), which contain DNA, probably at an intermediate level of compaction. On the basis of transmission EM images, the fibrillar centres were originally assumed to be the interphase counterparts of the NORs, and to contain DNA at a level of compaction comparable to that seen in the NORs at mitosis (Hadjiolov 1985; Shaw et al. 1995; Jordan 1984).

Fluorescence in situ hybridization shows the rDNA more clearly. Occasionally, particularly in plants, more condensed regions of chromatin are seen within FCs; these have been termed heterogeneous fibrillar centres (Jordan 1984). In addition, rDNA repeats are frequently seen as knobs at the periphery of the nucleoli or even

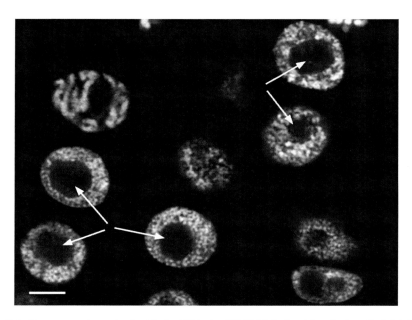

Fig. 3.1 Pea root nuclei stained with the DNA dye DAPI (4′,6-diamidino-2-phenylindole). The nucleoli, indicated by *arrows*, contain a low level of DNA and are therefore seen as *dark* voids within the nuclei. Bar = 5 μm

inside, condensed into a heterochromatic state. This is particularly common in plants, probably because a large number of rDNA repeats is present, of which many copies are silent (e.g. Gonzalez-Melendi et al. 2001). The condensed heterochromatic rDNA corresponds to inactive genes as shown by lack of incorporation of BrU into nascent RNA in labelling experiments (Thompson et al. 1997), while the active copies are decondensed within the body of the nucleolus (see Fig. 3.2). In some species, such as the diploid species rye, the internal path of the decondensed rDNA can be clearly seen within the nucleolus, whereas in the closely related species hexaploid wheat, the internal labelling is more complex and may contain small condensed regions of rDNA, while some NORs remain inactive and unassociated with a nucleolus (Leitch et al. 1992). In pea, the four NORs all contribute to the nucleolus, and the size of the knobs varies inversely with the size and presumed activity of the nucleolus, showing that increased nucleolar activity causes more of the rDNA copies to decondense and become active (Highett et al. 1993).

The transcriptionally active rDNA was imaged in spread preparations by Miller and Beatty (1969) as elongated threads with many (50–100) RNA polymerase complexes on to each gene, and increasing lengths of nascent RNA attached along the length of the gene. It has been very difficult to determine the exact conformation of the transcription units in intact nucleoli. The clearest results to date have been obtained by labelling nascent RNA in the nucleolus with BrU (Dundr and Raska 1993; Wansink et al. 1993; Hozak et al. 1994). This has shown foci within the DFC region of the nucleolus of mammalian culture cells, sometimes in contact with the

3 The Structure of rDNA Chromatin

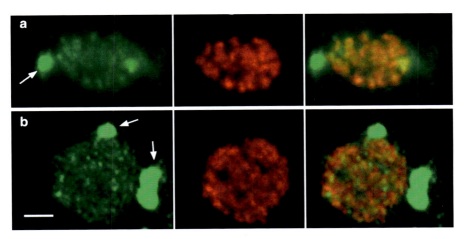

Fig. 3.2 Pea root nucleoli double labelled by in situ hybridization (*green*) to show rDNA and BrU incorporation (*red*) to show nascent rRNA transcripts. The transcript labelling is restricted to the inside of the nucleolus, and the condensed peripheral knobs of rDNA (*arrows*) have no transcript labelling, showing that they are inactive. (**a**) An in situ probe to S-rRNA (18S) portion of the rDNA shows good colocalization with the transcripts. (**b**) An in situ probe to the NTS intergenic spacer region of the rDNA shows poorer localization with the transcripts. Bar = 2 μm

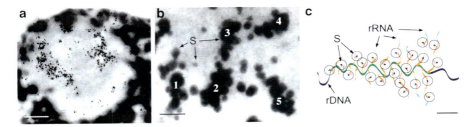

Fig. 3.3 Pea root nucleoli labelled with BrU to mark nascent rRNA transcripts, detected with silver-enhanced 1 nm gold by transmission electron microscopy. (**a**) Low magnification image shows the nascent RNA is concentrated in specific regions of the nucleolus corresponding to parts of the dense fibrillar component. Bar = 1 μm. (**b**) Higher magnification image shows that in many places the gold label is arranged as elongated clusters about 200–300 nm in length, each containing 20–30 particles. Bar = 100 nm. (**c**) Diagram drawn to scale showing the interpretation of the gold clusters in (**b**) as individual transcription units. S points to individual silver-enhanced gold particles in (**b**) and (**c**). Bar = 50 nm

periphery of the FCs (Koberna et al. 2002). This agrees with the results in plant cells, where the more extensive DFC made the interpretation even clearer. Gonzalez-Melendi et al. (2001) analyzed the labelled BrU at the electron microscope level with silver-enhanced 1 nm gold and showed that the labelling of the nascent RNA consisted of discrete, elongated clusters of gold particles about 200–300 nm in length. The clusters were often approximately conical in shape, and the authors suggested that these corresponded to individual transcription units, compacted by a factor of 5–8 compared to Miller spreads (see Fig. 3.3).

There is now general agreement that chromatin carrying transcribable genes can adopt at least three different states: repressed or silenced by mechanisms often described as epigenetic; untranscribed, but potentially transcribable (potentiated or poised for transcription); and transcriptionally active. In the case of rDNA, three different states of structural compaction have been described: heterochromatin-like rDNA present in the perinucleolar knobs and in smaller structures within the nucleolus; inactive, but less condensed rDNA within the mitotic NORs and at least in the centre of fibrillar centres; and fully decondensed, actively transcribed genes. Broadly, the different rDNA conformations seen are likely to be equivalent to the three chromatin states; there is little doubt about the repressive state of the heterochromatic knobs or rDNA, or about the fully extended state of transcribed genes. What exactly the intermediate, potentiated state corresponds to in structural terms is still a matter for debate.

3.4 rDNA Chromatin Composition: Proteins and Modifications

In eukaryotes DNA associates with histones to form chromatin, a proteinaceous super-structure. The different forms of rDNA chromatin described above modulate the chromatin for optimal transcription or for repression, as is the case with regions of the genome transcribed by other polymerases. rDNA chromatin has also additional specialisms to prevent recombination of the rDNA repeats, to block aberrant transcription by other polymerases and to organize the nucleolar structure. So the proteins which bind rDNA to form the nucleolar and perinucleolar chromatin include those also associated with other parts of the genome, or their Pol I-associated equivalents, and other proteins involved with the specialized functions of rDNA. Typically, the most abundant chromatin proteins are the histones, which assemble into nucleosomes containing histone octamers bound to DNA. In most instances, the histones which associate with rDNA are identical to those which bind other regions of the genome, although their abundance may differ between nucleolus and nucleoplasm. This is especially the case for the linker histones, for which nucleolar forms are known in several systems (Stoldt et al. 2007; McKeown and Shaw 2009). However, as the nucleolus is not separated from the nucleoplasm by any form of membrane, other non-rDNA specific histones will not be completely excluded. Histone deposition may also regulate transcription through ATP-dependent positioning of nucleosomes, as shown at human rDNA promoters (Felle et al. 2010), or by the activity of histone chaperones such as *Arabidopsis* FKBP53 which represses rRNA transcription (Li and Luan 2010).

Non-histone proteins which are likely to associate with chromatin can also be significantly enriched in nucleoli (Andersen et al. 2002; Pendle et al. 2005) (see Chap. 2). rDNA-specific functions of nucleolar chromatin are probably mediated by such non-histone components, which include abundant structural proteins such as UBF, nucleolin and nucleophosmin/B23 (see Chaps. 5, 9 and 10). These non-histone proteins may bind rDNA via chromodomains or bromodomains, as is the case for the mammalian transcriptional regulator CHD7, which binds open rDNA to promote transcription (Zentner et al. 2010). As with most histones, this chromodomain protein acts as part of rDNA chromatin, but is also located in other parts of the nucleoplasm. In some instances, non-histone proteins are actually able to supplant histones as the primary DNA-associated protein component of chromatin. This is the case for UBF (McStay and Grummt 2008; see Chap. 5). Although UBF is specific to metazoans, in yeast, the HMG protein Hmo1 replaces histones at active rDNA in an apparently similar way (Merz et al. 2008). Nucleolus-specific HMG proteins have also been reported from maize (ZmHMG), and have been shown to bind rDNA through C-terminal hooks with an affinity for AT-rich DNA (Zhao et al. 2009).

Interestingly, the hook domains that allow ZmHMG-rDNA binding are phosphorylated *in vivo*, which reduces the protein's affinity for DNA (Zhao et al. 2009). This is a parallel to the role of some histone acetylations in reducing the physical affinity of basic histones with the acidic DNA backbone. In fact, many rDNA-binding proteins are covalently modified and enzymes catalyzing these modifications form a part of the nucleolar network of chromatin regulation (Grummt 2007). As with the protein components, rDNA-binding protein covalent modifications often combine features found genome-wide with rDNA-specific modifications. This is the case for histone modifications which are part of the common regulatory network of the genome (Bartova et al. 2010) but histone modifications specific for rDNA have also been described (McKeown and Shaw 2009). Examples include H4 acetylation in replicating *Arabidopsis* nucleoli (Jasencakova et al. 2000) and the association of active rDNA promoters in HeLa cells with two hitherto-uncharacterized dimethyl arginine histone modifications, H3R8me2 and H4R3me2 (Majumder et al. 2010). However, many of the regulatory functions of rDNA-associated histones are still mediated by the same covalent modifications associated with other parts of the genome (e.g., H3K14ac and H3S10 phosphorylations that are associated with active rDNA in Drosophila polytene chromosomes; Plata et al. 2009), although these may also have novel nucleolar functions. For example, the establishment of nucleolar dominance in *Arabidopsis* hybrids involves H3K9me, H3K9ac and, again, H3K14ac (Earley et al. 2006). The release of sequestered Cdc14 in yeast is another example of familiar modifications playing unusual roles: it requires a pathway involving nucleolar H2B ubiquitination that triggers subsequent H3K4 and H3K79 methylation, and a second H3K36me-dependent pathway (Hwang and Madhani 2009). Whether rDNA-associated histone modifications occur at the same nucleosomes as those also found in other parts of the genome remains to be established, but it is clear that such combinations could increase the combinatorial complexity of histone regulation of rDNA and potentially explain the many functions played by rDNA histones.

Additional covalent modifications result from nucleolus-specific chromatin-modifying enzymes, which can target both histones and non-histone proteins. Mammalian histone arginine dimethylation, for example, occurs through the action of PRMT5, and knocking down this enzyme increases rRNA synthesis (Majumder et al. 2010). Two of the growing number of characterized JmjC histone demethylases, PHF8 and KDM2A, also regulate human rRNA synthesis and, as a good example of the complexity of the chromatin code, do so in opposite directions. PHF8 increases rRNA synthesis by demethylating the heterochromatic marks H3K9me1 and H3K9me2 *in vivo* (Feng et al. 2010; Zhu et al. 2010). As PHF8 protein co-localized with fibrillarin and nucleophosmin *in vivo*, its rDNA-specific activity is mirrored by its nucleolar localization (Zhu et al. 2010). On the other hand, KDM2A reduced rDNA transcription levels in response to starvation by demethylating the "active" histone marks, H3K36me1 and H3K36me2 (Tanaka et al. 2010). Again, KDM2A acts only on rDNA and is localized specifically to nucleoli (Tanaka et al. 2010). Proteins that modulate the degree of methylation of DNA itself may also be targeted to the nucleolus and preferentially target rDNA (e.g. the human MBD2; Ghoshal et al. 2004). A particular function of non-histone covalent modification is to couple rRNA synthesis, which is also regulated by covalent modification of chromatin-binding complexes, to cellular energy levels. The key sensor of this in human cells is the eNoSC complex that contains SIRT1 (Murayama et al. 2008). SIRT1 deacetylates H3 in response to a falling NAD(+)/NADH ratio while SUB39H1, which is also part of eNoSC, promotes H3K9 dimethylation. The well-described rDNA-binding chromatin complex of yeast, NoRC (Santoro et al. 2002), also responds to cellular energy, this time through the covalent modification of TIP5 by K633 acetylation (Zhou et al. 2009). This occurs in response to reduced cellular energy, enhances the binding of NoRC to rDNA promoters and leads to the propagation of heterochromatic histone marks (Zhou et al. 2009). NoRC is itself tethered to the rDNA chromatin by interaction with H4K16ac, a characteristic of active genes, but it leads to subsequent heterochromatinization (Zhou and Grummt 2005). Recent evidence shows that NoRC is recruited to a subset of rDNA repeats by a 200–300 nt RNA species derived from the intergenic regions of rDNA (Mayer et al. 2008; Santoro et al. 2010).

rDNA chromatin not only ensures that Pol I transcription occurs at the optimal rate for the cell but also acts to prevent aberrant transcriptional activity. This happens as part of the cellular DNA damage response in mice and is mediated by a signalling cascade which leads to modification and inactivation of the Pol I complex (Kruhlak et al. 2007). In this way, transcription at double-stranded DNA breaks is averted. Transcription of rDNA sequences by Pol II may also be deleterious for the cell and is blocked by the interaction of multiple rDNA-associated chromodomain proteins in *S. cerevisiae* and *Schizosaccharomyces pombe* (Thon and Verhein-Hansen 2000). The NTS regions of the rDNA can also be aberrantly transcribed by Pol II, potentially producing genotoxic small RNA molecules, and it has been suggested that the mechanism that prevents this requires Pol I transcription and mechanisms that restrict histone acetylation to the rRNA genes (Cesarini et al. 2010). It has been hypothesized that the maintenance of a stable genome requires correct organization of rDNA chromatin (Kobayashi 2008). In support of this, rDNA chromatin seems to be able to affect the stability of heterochromatic repeats in *trans*, as loss of the NoRC component TIP5 leads to instability of microsatellite repeats (Guetg et al. 2010) and reduction in the number of Drosophila rDNA repeats themselves leads to a general release of heterochromatin silencing

3 The Structure of rDNA Chromatin   51

throughout the nucleus (Paredes and Maggert 2009). As rDNA repeats are the most common genes in the genome, these effects may occur because of disrupted balance between euchromatin and heterochromatin (Kobayashi 2008; Paredes and Maggert 2009). However, at least in yeast, there is evidence that rDNA organization can affect the wider genome by regulating the global distribution of condensin (Wang and Strunnikov 2008), so this could again be a process that has the potential for regulation.

As well as helping to determine genome stability, rDNA chromatin also mediates correct nuclear architecture. rDNA transcription leads to the formation of the nucleolus, which is the largest nuclear structure and with which many other structures are associated, but the two processes are not automatically correlated. Instead, the proteins which bind to rDNA chromatin contribute to the assembly of a correctly-formed nucleolus in response to Pol I transcription. In the reverse case, loss of Pol I activity leads to a regulated disassembly of the nucleolus. In humans, this can occur in response to starvation because of the action of TOR, which acts by tethering HDACs to rDNA chromatin and triggering H4 hypoacetylation (Tsang et al. 2003). The down-regulation of rDNA transcription on nucleolar organization is reversible and does not lead to permanent disassembly of the nucleolus. However, this flexibility requires lamin B1, a member of the lamin family with roles in several diseases (e.g. Scaffidi and Misteli 2006); in cells in which lamin B1 is reduced, reduced rDNA transcription leads to irreversible nucleolar disassembly (Martin et al. 2009). Cohesin mutants of *S. cerevisiae* also have defects in nuclear reorganization programmes, which lead to aberrant nucleolar morphology (Gard et al. 2009), suggesting that associations between rDNA chromatin and structural proteins in the nucleoplasm play key roles in nucleolar morphology. In yeast, rDNA stability also requires association of silenced rDNA chromatin with the nuclear envelope (Mekhail et al. 2008).

3.5   Conclusions

In this chapter, we have discussed the chromatin-associated proteins which bind rDNA and the nucleolar structures into which it is organized during transcription. In agreement with the increasing number of roles ascribed to the nucleolus (Boisvert et al. 2007), these include core transcriptional functions as well as roles in nuclear organization and genome stability. However, many more chromatin proteins remain to be characterized, a point emphasized by an elegant mutagenesis screen performed by Hontz et al. (2009), which identified 68 *Saccharomyces* genes as regulators of Pol I transcription, including kinases and phosphatases, and 14 entirely uncharacterized genes. Further research into regulators of rDNA transcription and novel protein components will certainly reveal further complexity in the subtle roles of rDNA chromatin.

References

Andersen JS, Lyon CE, Fox AH, Leung AKL, Lam YW, Steen H, Mann M, Lamond AI (2002) Directed proteomic analysis of the human nucleolus. Curr Biol 12(1):1–11

Bartova E, Horakova AH, Uhlirova R, Raska I, Galiova G, Orlova D, Kozubek S (2010) Structure and epigenetics of nucleoli in comparison with non-nucleolar compartments. J Histochem Cytochem 58(5):391–403. doi:10.1369/jhc.2009.955435

Boisvert F-M, van Koningsbruggen S, Navascues J, Lamond AI (2007) The multifunctional nucleolus. Nat Rev Mol Cell Biol 8(7):574–585

Busch H, Smetana K (1970) The Nucleolus. Academic Press, New York

Caburet S, Conti C, Schurra C, Lebofsky R, Edelstein SJ, Bensimon A (2005) Human ribosomal RNA gene arrays display a broad range of palindromic structures. Genome Res 15(8):1079–1085. doi:10.1101/gr.3970105

Cesarini E, Mariotti FR, Cioci F, Camillori G (2010) RNA polymerase I transcription silences noncoding RNAs at the ribosomal DNA locus in *Saccharomyces cerevisiae*. Eukaryot Cell 9(2):325–335. doi:10.1128/EC.00280-09

Dundr M, Raska I (1993) Nonisotopic ultrastructural mapping of transcription sites within the nucleolus. Exp Cell Res 208(1):275–281. doi:10.1006/excr.1993.1247

Earley K, Lawrence RJ, Pontes O, Reuther R, Enciso AJ, Silva M, Neves N, Gross M, Viegas W, Pikaard CS (2006) Erasure of histone acetylation by *Arabidopsis* HDA6 mediates large-scale gene silencing in nucleolar dominance. Genes Dev 20(10):1283–1293. doi:10.1101/gad.1417706

Felle M, Exler JH, Merkl R, Dachauer K, Brehm A, Grummt I, Längst G (2010) DNA sequence encoded repression of rRNA gene transcription in chromatin. Nucl Acids Res 38(16): 5304–5314. doi:10.1093/nar/gkq263

Feng W, Yonezawa M, Ye J, Jenuwein T, Grummt I (2010) PHF8 activates transcription of rRNA genes through H3K4me3 binding and H3K9me1/2 demethylation. Nat Struct Mol Biol 17(4): 445–450. doi:10.1038/nsmb.1778

Gall JG, Pardue ML (1969) Formation and detection of RNA-DNA hybrid molecules in cytological preparations. Proc Natl Acad Sci USA 63:378–383

Gard S, Light W, Xiong B, Bose T, McNairn AJ, Harris B, Fleharty B, Seidel C, Brickner JH, Gerton JL (2009) Cohesinopathy mutations disrupt the subnuclear organization of chromatin. J Cell Biol 187(4):455–462. doi:10.1083/jcb.200906075

Ghoshal K, Majumder S, Datta J, Motiwala T, Bai S, Sharma SM, Frankel W, Jacob ST (2004) Role of human ribosomal RNA (rRNA) promoter methylation and of Methyl-CpG-binding protein MBD2 in the suppression of rRNA gene expression. J Biol Chem 279(8):6783–6793. doi:10.1074/jbc.M309393200

Gonzalez-Melendi P, Wells B, Beven AF, Shaw PJ (2001) Single ribosomal transcription units are linear, compacted Christmas trees in plant nucleoli. Plant J 27(3):223–233

Grummt I (2007) Different epigenetic layers engage in complex crosstalk to define the epigenetic state of mammalian rRNA genes. Hum Mol Genet 16(R1):R21–R27. doi:10.1093/hmg/ddm020

Guetg C, Lienemann P, Sirri V, Grummt I, Hernandez-Verdun D, Hottiger MO, Fussenegger M, Santoro R (2010) The NORC complex mediates the heterochromatin formation and stability of silent rRNA genes and centromeric repeats. EMBO J 29(13):2135–2146. doi:10.1038/emboj.2010.17

Hadjiolov AA (1985) The nucleolus and ribosome biogenesis, vol 12, Cell Biol Monogr. Springer, Wien

Heitz E (1931) Die ursache der gesetzmässigen zahl, lage, form und grösse pflanzlicher nukleolen. Planta 12:775–844

Highett MI, Rawlins DJ, Shaw PJ (1993) Different patterns of rDNA distribution in *Pisum sativum* nucleoli correlate with different levels of nucleolar activity. J Cell Sci 104(Pt3):843–852

Hontz RD, Niederer RO, Johnson JM, Smith JS (2009) Genetic identification of factors that modulate ribosomal DNA transcription in *Saccharomyces cerevisiae*. Genetics 182(1):105–119. doi:10.1534/genetics.108.100313

3 The Structure of rDNA Chromatin

Hozak P, Cook PR, Schofer C, Mosgoller W, Wachtler F (1994) Site of transcription of ribosomal-RNA and intranucleolar structure in HeLa cells. J Cell Sci 107(Pt2):639–648

Hwang WW, Madhani HD (2009) Nonredundant requirement for multiple histone modifications for the early anaphase release of the mitotic exit regulator CDC14 from nucleolar chromatin. PLoS Genet 5(8):e1000588

Jasencakova Z, Meister A, Walter J, Turner BM, Schubert I (2000) Histone H4 acetylation of euchromatin and heterochromatin is cell cycle dependent and correlated with replication rather than with transcription. Plant Cell 12(11):2087–2100

Jordan EG (1984) Nucleolar nomenclature. J Cell Sci 67:217–220

Kobayashi T (2008) A new role of the rDNA and nucleolus in the nucleus – rDNA instability maintains genome integrity. Bioessays 30(3):267–272. doi:10.1002/bies.20723

Koberna K, Malinsky J, Pliss A, Masata M, Vecerova J, Fialova M, Bednar J, Raska I (2002) Ribosomal genes in focus: New transcripts label the dense fibrillar components and form clusters indicative of "Christmas trees" in situ. J Cell Biol 157(5):743–748

Kruhlak M, Crouch EE, Orlov M, Montano C, Gorski SA, Nussenzweig A, Misteli T, Phair RD, Casellas R (2007) The ATM repair pathway inhibits RNA polymerase I transcription in response to chromosome breaks. Nature 447(7145):730–734. doi:10.1038/nature05842

Leitch AR, Mosgoller W, Shi M, Heslop-Harrison JS (1992) Different patterns of rDNA organization at interphase in nuclei of wheat and rye. J Cell Sci 101(Pt 4):751–757

Li H, Luan S (2010) ATFKBP53 is a histone chaperone required for repression of ribosomal RNA gene expression in *Arabidopsis*. Cell Res 20(3):357–366

Lima-de-Faria A (1976) The chromosome field. I. Prediction of the location of ribosomal cistrons. Hereditas 83:1–22

Majumder S, Alinari L, Roy S, Miller T, Datta J, Sif S, Baiocchi R, Jacob ST (2010) Methylation of histone H3 and H4 by PRMT5 regulates ribosomal RNA gene transcription. J Cell Biochem 109(3):553–563. doi:10.1002/jcb.22432

Martin C, Chen S, Maya-Mendoza A, Lovric J, Sims PFG, Jackson DA (2009) Lamin B1 maintains the functional plasticity of nucleoli. J Cell Sci 122(10):1551–1562. doi:10.1242/jcs.046284

Mayer C, Neubert M, Grummt I (2008) The structure of NORC-associated RNA is crucial for targeting the chromatin remodelling complex NORC to the nucleolus. EMBO Rep 9(8):774–780. doi:10.1038/embor.2008.109

McClintock B (1934) The relation of a particular chromosomal element to the development of the nucleoli in *Zea mays*. Z Zellforsch Mikrosk Anat 21:294–328

McKeown P, Shaw P (2009) Chromatin: Linking structure and function in the nucleolus. Chromosoma 118(1):11–23. doi:10.1007/s00412-008-0184-2

McStay B, Grummt I (2008) The epigenetics of rRNA genes: From molecular to chromosome biology. Annu Rev Cell Dev Biol 24(1):131–157. doi:10.1146/annurev.cellbio.24.110707.175259

Mekhail K, Seebacher J, Gygi SP, Moazed D (2008) Role for perinuclear chromosome tethering in maintenance of genome stability. Nature 456(7222):667–670. doi:10.1038/nature07460

Merz K, Hondele M, Goetze H, Gmelch K, Stoeckl U, Griesenbeck J (2008) Actively transcribed rRNA genes in *S. cerevisiae* are organized in a specialized chromatin associated with the high-mobility group protein Hmo1 and are largely devoid of histone molecules. Genes Dev 22(9):1190–1204. doi:10.1101/gad.466908

Miller OLJ, Beatty RR (1969) Visualization of nucleolar genes. Science 164:955–957

Mizuno H, Sasaki T, Matsumoto T (2008) Characterization of internal structure of the nucleolar organizing region in rice (*Oryza sativa* l.). Cytogenet Genome Res 121(3–4):282–285. doi:10.1159/000138898

Murayama A, Ohmori K, Fujimura A, Minami H, Yasuzawa-Tanaka K, Kuroda T, Oie S, Daitoku H, Okuwaki M, Nagata K, Fukamizu A, Kimura K, Shimizu T, Yanagisawa J (2008) Epigenetic control of rDNA loci in response to intracellular energy status. Cell 133(4):627–639

Paredes S, Maggert KA (2009) Ribosomal DNA contributes to global chromatin regulation. Proc Natl Acad Sci USA 106(42):17829–17834. doi:10.1073/pnas.0906811106

Pendle AF, Clark GP, Boon R, Lewandowska D, Lam YW, Andersen J, Mann M, Lamond AI, Brown JWS, Shaw PJ (2005) Proteomic analysis of the *Arabidopsis* nucleolus suggests novel nucleolar functions. Mol Biol Cell 16(1):260–269. doi:10.1091/mbc.E04-09-0791

Plata M, Kang H, Zhang S, Kuruganti S, Hsu S-J, Labrador M (2009) Changes in chromatin structure correlate with transcriptional activity of nucleolar rDNA in polytene chromosomes. Chromosoma 118(3):303–322. doi:10.1007/s00412-008-0198-9

Prokopowich CD, Gregory TR, Crease TJ (2003) The correlation between rDNA copy number and genome size in eukaryotes. Genome 46(1):48–50. doi:10.1139/g02-103

Raska I, Shaw PJ, Cmarko D (2006a) New insights into nucleolar architecture and activity. Int Rev Cytol 255:177–235

Raska I, Shaw PJ, Cmarko D (2006b) Structure and function of the nucleolus in the spotlight. Curr Opin Cell Biol 18(3):325–334

Rubbi CP, Milner J (2003) Disruption of the nucleolus mediates stabilization of p53 in response to DNA damage and other stresses. EMBO J 22(22):6068–6077

Santoro R, Li J, Grummt I (2002) The nucleolar remodeling complex NORC mediates heterochromatin formation and silencing of ribosomal gene transcription. Nat Genet 32(3):393–396

Santoro R, Schmitz KM, Sandoval J, Grummt I (2010) Intergenic transcripts originating from a subclass of ribosomal DNA repeats silence ribosomal RNA genes in trans. EMBO Rep 11(1):52–58. doi:10.1038/embor.2009.254

Scaffidi P, Misteli T (2006) Lamin A-dependent nuclear defects in human aging. Science 312(5776):1059–1063. doi:10.1126/science.1127168

Shaw PJ (2010) Nucleolus. In: Encyclopaedia of life sciences. John Wiley & Sons, Ltd, Chichester, UK. doi: 10.1002/9780470015902.a0001352.pub3, http://www.els.net/

Shaw PJ, Jordan EG (1995) The nucleolus. Annu Rev Cell Dev Biol 11:93–121

Shaw PJ, Highett MI, Beven AF, Jordan EG (1995) The nucleolar architecture of polymerase I transcription and processing. EMBO J 14(12):2896–2906

Sinclair DA, Guarente L (1997) Extrachromosomal rDNA circles–a cause of aging in yeast. Cell 91(7):1033–1042. doi:S0092-8674(00)80493-6

Stoldt S, Wenzel D, Schulze E, Doenecke D, Happel N (2007) G1 phase-dependent nucleolar accumulation of human histone H1x. Biol Cell 99:541–552

Takeuchi Y, Horiuchi T, Kobayashi T (2003) Transcription-dependent recombination and the role of fork collision in yeast rDNA. Genes Dev 17(12):1497–1506. doi:10.1101/gad.1085403

Tanaka Y, Okamoto K, Teye K, Umata T, Yamagiwa N, Suto Y, Zhang Y, Tsuneoka M (2010) Jmjc enzyme KDM2A is a regulator of rRNA transcription in response to starvation. EMBO J 29(9):1510–1522. doi:10.1038/emboj.2010.56

Thompson WF, Beven AF, Wells B, Shaw PJ (1997) Sites of rDNA transcription are widely dispersed through the nucleolus in *Pisum sativum* and can comprise single genes. Plant J 12(3):571–581

Thon G, Verhein-Hansen J (2000) Four chromo-domain proteins of *Schizosaccharomyces pombe* differentially repress transcription at various chromosomal locations. Genetics 155(2):551–568

Tsang CK, Bertram PG, Ai W, Drenan R, Zheng XFS (2003) Chromatin-mediated regulation of nucleolar structure and RNA pol I localization by TOR. EMBO J 22(22):6045–6056

Wang B-D, Strunnikov A (2008) Transcriptional homogenization of rDNA repeats in the episome-based nucleolus induces genome-wide changes in the chromosomal distribution of condensin. Plasmid 59(1):45–53

Wansink DG, Schul W, van der Kraan I, van Steensel B, van Driel R, de Jong L (1993) Fluorescent labeling of nascent RNA reveals transcription by RNA polymerase II in domains scattered throughout the nucleus. J Cell Biol 122(2):283–293

Zentner GE, Hurd EA, Schnetz MP, Handoko L, Wang C, Wang Z, Wei C, Tesar PJ, Hatzoglou M, Martin DM, Scacheri PC (2010) CHD7 functions in the nucleolus as a positive regulator of ribosomal RNA biogenesis. Hum Mol Genet 19(18):3491–3501. doi:10.1093/hmg/ddq265

Zhao J, Paul LK, Grafi G (2009) The maize HMGA protein is localized to the nucleolus and can be acetylated in vitro at its globular domain, and phosphorylation by CDK reduces its binding activity to AT-rich DNA. Biochim Biophys Acta 1789(11–12):751–757

Zhou Y, Grummt I (2005) The PHD finger/bromodomain of NORC interacts with acetylated histone H4K16 and is sufficient for rDNA silencing. Curr Biol 15(15):1434–1438. doi:10.1016/j.cub.2005.06.057

Zhou Y, Schmitz K-M, Mayer C, Yuan X, Akhtar A, Grummt I (2009) Reversible acetylation of the chromatin remodelling complex NORC is required for non-coding RNA-dependent silencing. Nat Cell Biol 11(8):1010–1016. doi:10.1038/ncb1914

Zhu Z, Wang Y, Li X, Wang Y, Xu L, Wang X, Sun T, Dong X, Chen L, Mao H, Yu Y, Li J, Chen PA, Chen CD (2010) PHF8 is a histone H3K9me2 demethylase regulating rRNA synthesis. Cell Res 20(7):794–801

Chapter 4
The Epigenetics of the Nucleolus: Structure and Function of Active and Silent Ribosomal RNA Genes

Raffaella Santoro

4.1 Introduction

The nucleolus is a well-defined nuclear compartment in which synthesis of rRNA and the assembly of ribosomes take place. Transcription of rRNA genes generates the 45S pre-rRNA precursor that is subsequently cleaved and processed into 28S, 18S and 5.8S rRNAs. These rRNAs are then packaged with ribosomal proteins to form the large and small subunits of ribosomes. As an average mammalian cell can produce up to 10,000 ribosomes/min, cells have to invest a very large portion of their own metabolic effort to meet demand from protein synthesis. To limit excessive energy consumption to produce ribosomes that could potentially deplete the cells from nutrients required for other essential processes, cells keep transcriptional activity of rRNA genes under tight surveillance. Changes in this commitment are likely to have extensive repercussions on the cellular economy, limiting proliferation rates and perhaps even cell fate. This is exemplified by the fact that conditions that harm cellular metabolism, downregulate rRNA transcription. Conversely, rRNA transcription is upregulated on reversal of such conditions and by agents that stimulate growth (Moss 2004).

To produce an elevated number of ribosomes, cells have to achieve synthesis of large amounts of rRNA. Indeed, synthesis of rRNA represents the major transcriptional activity of the cell, accounting for 60% of total transcription in rapidly growing yeast cells and 35% in proliferating mammalian cells (Moss et al. 2007). To do this, cells evolved a unique and efficient transcription system by using a specific and efficient RNA polymerase (RNA polymerase I, Pol I) and by amplifying the number

---

R. Santoro (✉)
Institute of Veterinary Biochemistry and Molecular Biology,
University of Zürich, Zurich, Switzerland
e-mail: raffaella.santoro@vetbio.uzh.ch

M.O.J. Olson (ed.), *The Nucleolus*, Protein Reviews 15,
DOI 10.1007/978-1-4614-0514-6_4, © Springer Science+Business Media, LLC 2011

of rRNA genes to hundreds or even thousands of copies per genome. In contrast to RNA Pol II genes that seldom have more than one isolated polymerase, rDNA transcription units are teeming with polymerases and nascent transcript complexes (Fraser 2006; Jackson et al. 1998). Average Pol I density has been measured on rRNA genes from yeast (about 50 polymerases/gene), CHO cells (114 polymerases/gene), and rat liver cells (101 polymerases/gene) which correspond to one polymerase every 132, 123, and 139 nucleotides, respectively (French et al. 2003; Harper and Puvion-Dutilleul 1979; Puvion-Dutilleul and Bachellerie 1979). The presence of several dozen to hundreds of rDNA copies that transcribe at such high rates contributes to generate an elevated number of rRNA moieties.

Although cells possess many rRNA gene copies, not all the rDNA units are used for rRNA production. Electron microscopic visualization of rRNA genes from many different cell types by the Miller spreading method typically shows two classes of rRNA genes: genes covered by elongating polymerases that efficiently synthesize rRNA moieties (active genes) and genes not associated with Pol I and thus not transcribing (silent genes). The coexistence of active and silent rRNA genes in the same cell, led researchers to propose two modes of action that cells can use to modulate rRNA levels: (1) by controlling the transcription rate per gene by acting directly on the Pol I transcription machinery; (2) by regulating the number of genes to be transcribed. Probably, these two strategies are not mutually exclusive.

Here I explore what is known about the composition of these two classes of rRNA genes with a particular focus on chromatin structure and epigenetics. I discuss also the mechanisms that account for establishment and inheritance of active and silent rRNA genes through cell generation. Finally, I discuss emerging themes, highlighting the role of silent rRNA genes, whether they represent a reservoir for the cells to draw on in case of elevated ribosome demand or if their role goes beyond the ribosome factory.

4.2 Chromatin and Epigenetic Features of Active and Silent Genes

Eukaryotic genomes contain many rRNA gene copies, ranging from hundreds to thousands in some plants, organized in tandem arrays and distributed among different chromosomes (Long and Dawid 1980; Santoro 2005). Humans and mice contain about 200 rRNA copies per haploid genome. In humans, rRNA genes are located between the short arm and the satellite body of acrocentric chromosomes 13, 14, 15, 21, and 26. In the mouse, rDNA clusters are placed within the centromeric regions of chromosomes 12, 15, 16, 18, and 19 (Dev et al. 1977; Kurihara et al. 1994). The *Saccharomyces cerevisiae* rRNA genes are located on the right arm of chromosome XII in a tandem array of 150–200 copies, representing almost 10% of the yeast genome.

One of the earliest and yet still highly informative methods of studying eukaryotic gene expression is by direct electron microscopic visualization of the transcribing chromatin (Miller and Beatty 1969). In *S. cerevisiae*, as a result of their ease of identification, rRNA genes are the most amenable to study in Miller spreads. When spread native rDNA chromatin is visualized, two kinds of rRNA gene units can be observed: (1) transcribing rRNA genes (active copies) that have a characteristic tree-like appearance (referred as "Christmas tree"), with a DNA "trunk" from which close-packed ribonucleoprotein "branches" of increasing length extend; (2) genes that do not associate with Pol I and are not transcribed (silent copies). Although the genome complexity of higher eukaryotes does not yet allow visualization of rDNA chromatin by Miller spreads, later biochemical studies assessed that the coexistence of active and silent rRNA genes in each cell is not limited to *S. cerevisiae*. Differences in chromatin composition between mammalian active and silent rRNA genes were initially explored by in vivo crosslinking analyses of Friend cells using psoralen, an intercalating drug that can introduce crosslinks into DNA sites that are not protected by nucleosomes (Conconi et al. 1989; Sogo et al. 1984). Using this method, it was demonstrated that two distinct types of ribosomal chromatin coexist in each cell. The fraction of rRNA genes inaccessible to psoralen (f-band) contains nucleosomes while the rDNA units accessible to psoralen (s-band) display a chromatin structure free of regularly spaced nucleosomes. The demonstration that nascent rRNA is selectively associated with the heavily psoralen-cross-linked s-band led to the conclusion that the nucleosome-free fraction of rDNA is actively transcribed in vivo (active genes) and nucleosomal rDNA fraction corresponds to silent genes (Conconi et al. 1989).

Further studies demonstrated that active and silent rRNA genes are also characterized by different epigenetic marks. CpG methylation, an epigenetic mark associated with heritable gene silencing and heterochromatic structures, was found enriched in the rDNA chromatin fraction inaccessible to psoralen (silent genes) and absent from rDNA units accessible to psoralen (active genes) (Stancheva et al. 1997). Later studies demonstrated a direct role of DNA methylation in repressing rRNA transcription. Treatment of mouse and human cells with 5-azacytidine, an inhibitor of cytosine methylation, increased 45S pre-rRNA levels, suggesting that lack of DNA methylation alleviates transcriptional repression of the corresponding fraction of silent rRNA genes (Santoro and Grummt 2001). Notably, methylation-dependent transcriptional silencing could be reproduced in vitro but only when methylated rDNA templates were assembled into chromatin. Conversely, transcription on naked rDNA templates was not affected, a finding that implies that CpG methylation induces structural changes in rDNA chromatin that are incompatible for transcription. The repressive action of DNA methylation on rRNA transcription was ascribed to a few critical CpGs within the rDNA promoter region. In mouse, methylation of a single CpG within the UCE (upstream control element) of the rDNA promoter located at −133 impairs binding of the Pol I transcription factor UBF (upstream binding factor) to rDNA chromatin, thereby preventing initiation complex formation. In human, CpGs located at −60 and −68 seem to act in a similar manner (R. Santoro, personal communication). Consistent with this, methylation of one single HpaII site (CCGG), located in the rat promoter region of silent rDNA chromatin inaccessible to psoralen crosslinking, showed particularly strong correlation with the repressed transcriptional state (Stancheva et al. 1997). The correlation between rDNA methylation and transcriptional silencing is further supported by studies on tumors where rRNA transcription is usually upregulated. Hypomethylation of the rRNA genes has been observed in lung cancer, Wilms tumor, and hepatocellular carcinomas (Ghoshal et al. 2004; Powell et al. 2002; Qu et al. 1999; Shiraishi et al. 1999). Although all these results strongly indicated that DNA methylation represses rRNA transcription, recent data proposed a positive role of CpG methylation in rRNA synthesis and processing. This study showed that cells derived from human colorectal carcinoma HCT116 cells having somatic knockouts for DNA methyltransferase 1 and 3B (Dnmt1 and Dnmt3B, respectively) lack rDNA methylation and increase the fraction of psoralen accessible (non nucleosomal, active) genes. These cells displayed reduced rRNA synthesis and processing, and accumulated unprocessed 45S rRNA, leading the authors to conclude that the role of rDNA methylation is to repress cryptic RNA polymerase II transcription of rRNA genes (Gagnon-Kugler et al. 2009). This result is consistent with the emerging idea that the presence of DNA methylation in the bodies of transcribed regions, a feature common among plants and animals, can play a role in silencing of cryptic promoters (Inagaki and Kakutani 2010). However, caution must be taken in interpreting the role of rDNA methylation using HCT116 somatic knockouts for Dnmt1 and 3B. First, these are selected cell clones for enzymes whose inactivation leads to fetal (Dnmt1) or embryonic (Dnmt3B) lethality in mouse (Li et al. 1992; Okano et al. 1999). Second, later analyses showed that HCT116 KO cells express an alternatively spliced form of Dnmt1 that lacks exons 3–5 and is yet catalytically active (Spada et al. 2007). Third, these cells display a structurally disorganized nucleolus, which is fragmented into small nuclear masses, and contains prominent nucleolar proteins (i.e., fibrillarin and Ki-67) and rRNA genes that are scattered throughout the nucleus (Espada et al. 2007). Considering also that these cells possess a genome completely demethylated, with obvious consequences at the level of genome-wide transcription and genome stability, this study does not allow us to determine whether rRNA synthesis and processing abnormalities are an indirect (i.e., reduced levels of a Pol I factor) or a direct consequence of demethylation. Thus, this positive effect of DNA methylation on rRNA transcription/processing must be further validated using different experimental approaches.

The finding that the fraction of silent rRNA genes is enriched in CpG methylated sequences made it possible to analyze the composition of silent and active rDNA chromatin in higher eukaryotes and plants. An assay based on chromatin immunoprecipitation (ChIP) coupled to CpG methylation measurement (ChIP-chop) was developed (Santoro et al. 2002), allowing the identification of protein factors, including posttranslationally modified histones that bind either to active (i.e., lack of meCpG) or to silent (i.e., enriched in meCpGs) genes. Using this approach, several studies showed that the promoter of mouse and human active rRNA genes was associated with Pol I transcription factors and with histones modified with active marks (i.e., H4Ac and H3K4me2) (Santoro and Grummt 2005; Santoro et al. 2002). In contrast, silent rRNA genes are associated with the heterochromatin protein 1

4 The Epigenetics of the Nucleolus... 61

(HP1) and with histones modified with silent marks like H3K9me2, H3K27me3, and H4K20me3 (Santoro and Grummt 2001, 2005; Santoro et al. 2002). A similar epigenetic pattern was also described in plants (Lawrence et al. 2004). Thus, active and silent rRNA genes are demarcated both by their pattern of DNA methylation and by specific posttranslationally modified histones.

Although the ChIP-chop method represents a valid assay to analyze and distinguish the epigenetic composition of active and silent rDNA chromatin in organisms that evolved the CpG methylation system, it cannot be applied for cellular systems (i.e., *S. cerevisiae*) that lack the CpG methylation machinery. Thus, in yeast, the psoralen method remains still the unique possibility to analyze biochemically rDNA chromatin composition. Recently, ChEC (chromatin endogenous cleavage) with MNase-fusion proteins, which allows for the precise localization of chromatin-associated factors on genomic DNA (Schmid et al. 2004), was combined with psoralen photo-cross-linking analyses to study the association of histones with rDNA (Merz et al. 2008). This study demonstrated for the first time that, in contrast with the inactive yeast rDNA repeats, the actively transcribed rRNA genes are largely devoid of histones and associate with Pol I and the high-mobility group protein Hmo1, a transcription factor remotely related to animal UBF (Merz et al. 2008; Gadal et al. 2002). Thus, these results confirm the conclusions previously drawn from psoralen experiments that yeast active rRNA genes are free of regularly spaced nucleosomes (Dammann et al. 1993). This would agree with the observation that the transcribing region of active rDNA units are teeming with polymerases and that a nucleosomal array might represent an obstacle to the elongation process, denying access to DNA. However, this does not seem to be the case for higher eukaryotes where the active rRNA genes were shown to be associated with histones (particularly the variant H3.3) modified either with active and/or with a specific set of silent marks (H3K9me2) (Prior et al. 1983; Schwartz and Ahmad 2005; Yuan et al. 2007; Zhou et al. 2002). The unexpected presence of histone H3K9me2, a typical silent histone mark, on active mammal rRNA genes was found to be mediated by G9a, a histone H3K9 methyltransferase, that was found associated with the transcribing region of active genes (Yuan et al. 2007). Knockdown of G9a leads to decreased levels of H3K9me2 and heterochromatin protein 1γ (HP1γ) at the transcribed region and downregulation of pre-rRNA synthesis, suggesting that establishment of silent histone marks are required for an efficient elongation process. This is consistent with other studies showing that H3K9 methylation, as well as HP1γ, is enriched in the coding region of Pol II active genes and that G9a is localized in euchromatin and is a coactivator of nuclear receptors (Lee et al. 2006; Piacentini et al. 2003; Tachibana et al. 2002; Vakoc et al. 2005). The functional link between HP1 and transcription was recently described in *Drosophila*, where it was shown that HP1c guides the recruitment of FACT (facilitates chromatin transcription) to active genes and absence of HP1c partially impairs the recruitment of FACT into heat-shock loci and causes a defect in heat-shock gene expression (Kwon et al. 2010). A recent study suggests that a similar mechanism can likely take place in mammalian rRNA genes. A biochemical analysis demonstrated that RNA Pol I can transcribe through nucleosomal templates and that this requires structural rearrangement of the nucleosomal core particle mediated by two subunits of the histone chaperone FACT, SSRP1 and Spt16 (Birch et al. 2009). This suggests that the nucleosomal barriers can be overcome by Pol I-associated FACT activity, perhaps in conjunction with other FACT-like histone chaperones and chromatin remodellers, to allow for productive elongation of transcription and rRNA synthesis. Consistent with this, several studies indicated that chromatin transcription by mammalian Pol I requires histone chaperone activities (i.e., nucleolin and nucleophosmin (B23)) and chromatin remodeling activities (i.e., CSB (Cockayne syndrome group B) protein, a member of the SWI/SNF family of ATP-dependent chromatin-remodeling activities; a complex containing nuclear myosin 1 (NM1), WSTF (William's syndrome transcription factor), and SNF2h (sucrose non-fermenting protein 2 homologue)) (Angelov et al. 2006; Murano et al. 2008; Percipalle et al. 2006; Rickards et al. 2007; Yuan et al. 2007). The association of histones in mammalian rRNA transcribing regions as well as the requirement of histone modifier and chromatin remodeling activities for transcription suggests that the chromatin of transcribing regions of mammalian active rRNA genes is more similar to Pol II genes than to yeast active rRNA genes. Although the exact nucleosomal arrangement at the transcribed regions of the mammalian rRNA genes is currently unknown, on the basis of psoralen cross-linking experiments (Conconi et al. 1989), chromatin of transcribing active rDNA region should posses a chromatin structure of unphased nucleosomes and not of an intact nucleosome array. Clearly, passage of the polymerase through a chromatin template might be a potential control point for Pol I transcription. One contributory influence on the chromatin status of mammalian rDNA transcribing region is UBF whose association with rDNA is not restricted to the promoter but extends across the entire transcribed portion (O'Sullivan et al. 2002). UBF has the abilities to stimulate promoter escape (Panov et al. 2006) and modulate Pol I transcription elongation rates (Stefanovsky et al. 2006), as well as to decondense rDNA chromatin (Chen et al. 2004; Mais et al. 2005; Wright et al. 2006) by preventing the assembly of transcriptionally inactive higher order chromatin structures catalyzed by linker histone H1 (Sanij et al. 2008). Consistent with this, recent data showed that depletion of UBF led to a switch from active to silent rDNA chromatin, underscoring the role of UBF in organizing chromatin of active rRNA genes (Sanij et al. 2008).

Recently, nucleosome positioning at the rDNA promoter was proposed as an additional feature that characterizes active and silent rRNA genes. In mouse, two distinct nucleosome positions at the promoters of active and silent mouse rRNA genes were identified (Li et al. 2006). In mouse active genes, a nucleosome occupies sequences from −157 to the transcription start site, whereas in silent genes the nucleosome covers sequences from −132 to +22. The positioning of a nucleosome over the promoter region of silent genes was found to be mediated by the nucleolar remodeling complex NoRC, whose function is described in the next paragraph. The position of a nucleosome at the promoter sequence of an active gene is consistent with previous findings showing that binding of the transcription termination factor TTF-I to the promoter-proximal terminator T0 located adjacent to the transcription start site recruits an ATP-dependent nucleosome remodeling activity. This activity locates the nucleosome over the rDNA promoter and allows transcription of mouse rDNA templates reconstituted into chromatin using *Drosophila* embryo extracts (Langst et al. 1997). This specific nucleosomal architecture of active genes would bring the UCE and the core element into close proximity and might facilitate specific interactions between the TBP-containing promoter selectivity factor TIF-IB/SL1 and the HMG box-containing architectural factor UBF. In this scenario, the nucleosome positioned at the rDNA promoter may provide the correct scaffolding for productive interactions between TIF-IB/SL1 and UBF bound at the two recognition sites, which are separated by 120 bp. A similar structure may also be driven by UBF itself that dimerizes and, after binding to DNA, has the ability to induce formation of an "enhancesome," in which ~140 bp of DNA is organized in a 360° turn as a result of six in-phase bends generated by three of the six HMG boxes in each UBF monomer (Bazett-Jones et al. 1994). The requirement of a proper promoter architecture is also suggested by experiments showing that changing either the distance between the two promoter elements or increasing the length between TTF-I binding site and the transcription start site represses transcription (Clos et al. 1986; Langst et al. 1998). Interestingly, sequence-dependent features of the mouse rDNA promoter disfavor the reconstitution of a nucleosome positioned as such on an active rRNA promoter, a further indication that remodeling activities are required to reorganize chromatin for transcription (Felle et al. 2010). Several chromatin-remodeling complexes specific for active rDNA repeats have been identified, including CSB and a complex containing NM1, WSTF, and SNF2h (Percipalle et al. 2006; Yuan et al. 2007). The positive role in rRNA transcription of both chromatin remodelers seems to be dependent on TTF1. However, whether these complexes are able to position the nucleosome over the rDNA promoter at the same location found in active genes has not yet been investigated. Interestingly, both CSB and WSTF were present on a 2–3 MDa complex, termed B-WICH, which contains NM1, WSTF, SNF2h, CSB, and other proteins involved in transcription and processing of rRNA, such as RNA helicase II/Gua, and the myb-binding protein 1a (Cavellan et al. 2006). Whether and how WSTF and CSB might act in the same complex to facilitate rRNA transcription/elongation on chromatin is yet not clear.

In conclusion, chromatin composition does not only distinguish between the two classes of rRNA genes but also represent an important control point to modulate and regulate rRNA transcription.

4.3 Inheritance of rDNA Chromatin Structures

Psoralen crosslinking analysis of rDNA chromatin in mouse cells showed that the levels of active and silent rDNA chromatin are similar both in growing and resting cells as well as during interphase and metaphase, although their run-on activities differ significantly (Conconi et al. 1989). This result suggested that chromatin of active and silent rRNA genes is stably propagated throughout the cell cycle and maintained independently of transcriptional activity. This is also consistent with data showing that in mouse cells, the epigenetic and chromatin state of a given

CpG-methylated silent rRNA gene is propagated to the daughter cells (Li et al. 2005). Moreover, studies in a HeLa cell line showed that Pol I, UBF, and SL1 are always associated with the same nucleolar organizing regions (NORs), the chromosomal regions containing rRNA genes (Roussel et al. 1996). But again, yeast seems to not follow this rule. In *S. cerevisiae*, the replication machinery entering upstream to a transcriptionally active ribosomal rRNA gene generates two newly replicated coding regions regularly packaged into nucleosomes, indicating that the active chromatin structure cannot be directly inherited at the replication fork and that regeneration of the active chromatin structure along the coding region is always a post-replicative process involving disruption of preformed nucleosomes (Lucchini and Sogo 1995). Although in terms of inheritance these results showed that yeast greatly diverges from mammals, these data point out that in the first round of post-replicative transcription yeast might use mechanisms analogous to those described in mammals (see above) to overcome the nucleosome barrier.

The absence of rDNA epigenetic memory in yeast can be explained by the fact that all rRNA genes are clustered on one chromosome and that yeast lacks DNA methylation, a relatively stable mark that provides heritable, long-term silencing (Wu and Zhang 2010). Supporting the idea that yeast and mammals do not share the same mechanisms to inherit the specific rDNA chromatin and epigenetic state during cell division, electron microscopy analysis showed that yeast active and silent rRNA gene copies are randomly distributed (Dammann et al. 1995; French et al. 2003). Conversely, in higher eukaryotes, the rDNA copies are clustered and distributed on active and silent NORs, indicating that there are regulatory mechanisms that act on a scale much larger than a single rRNA gene (Pikaard 2000; Schlesinger et al. 2009).

In eukaryotic organisms, chromosomal DNA replication initiates at multiple sites on the chromosomes at different times following a temporal replication program (Goren and Cedar 2003; Santoro and De Lucia 2005). The presence of a temporal-order replication program in all eukaryotic cells argues that such a program does have functional importance. A large body of evidence indicated that DNA replication timing is a regional epigenetic marking mechanism that is correlated with gene expression. Regions of monoallelic expression have been found to replicate asynchronously, with one allele duplicated earlier than the other (Goldmit and Bergman 2004). For example, the establishment of late replication timing represents one of the earliest events of X-chromosome inactivation in female embryos (Keohane et al. 1998), and for many autosomal regions, asynchronous replication is instrumental in determining the preferred expression of one allele in each cell (Gimelbrant et al. 2005). The "window of opportunity" model provides one of the most interesting suggestions for explaining the need for replication timing (Goren and Cedar 2003). According to this model, an active gene that replicates in early S phase is exposed to factors that are required for the formation of active transcription complexes, whereas a silent gene replicating in late S phase experiences a different nuclear environment, which is more conducive for the generation of repressive structures. In support of this, reporter genes microinjected into nuclei of cells in early S phase are packaged into chromatin containing deacetylated histones and they are better templates for transcription. The opposite was found to be true when the reporter gene was introduced during late S phase (Zhang et al. 2002).

In mouse and human cells, rRNA genes are replicated in a biphasic manner: the active rRNA genes replicate early, whereas silent rDNA arrays replicate late (Berger et al. 1997; Li et al. 2005). rDNA replication timing is controlled allelically, with one allele replicating early and one replicating late in almost every cell (Schlesinger et al. 2009). Although the mechanisms of inheritance of active rDNA chromatin remain still elusive, the identification of the nucleolar remodeling complex NoRC led to important advances in the elucidation of the mechanisms controlling maintenance of silent rDNA chromatin in mammals (Santoro et al. 2002; Strohner et al. 2001). NoRC consists of TIP5 (TTF1-interacting protein 5) and the ATPase SNF2h and is the key determinant in setting heterochromatic and silent features at the rDNA locus during cell division (Li et al. 2005; Santoro et al. 2002; Zhou et al. 2002). NoRC is targeted to the rDNA promoter via TTF1 and represses rRNA transcription through recruitment of histone-modifying and DNA methylating activities (i.e., HDAC1, SETDB1, SIRT1, MOF, Dnmts) (Santoro and Grummt 2005; Santoro et al. 2002; Zhou et al. 2002, 2009). The association of NoRC with rRNA genes was shown to take place immediately after rDNA replication in late S phase (Li et al. 2005), suggesting a role of NoRC in maintaining the epigenetic and chromatin state of newly duplicated silent rRNA genes. NoRC was shown to position a nucleosome over the rDNA promoter of silent genes (from −132 to +22) (Li et al. 2006). Noteworthy, in this "inactive" position, the critical CpG dinucleotide at −133, whose methylation prevents binding of UBF, is placed at the 5′ boundary of the nucleosome (Santoro and Grummt 2001). In this position, not hindered by a nucleosome, the CpG-133 would be exposed to methylation mediated by Dnmts associated with NoRC (Santoro et al. 2002). In support of this, impairment of nucleosome remodeling activity of NoRC abrogates transcriptional repression and CpG methylation of an rDNA reporter gene (Santoro and Grummt 2005). An important event required for NoRC-mediated rDNA silent chromatin formation is the association of TIP5 with a non-coding RNA (pRNA) (Mayer et al. 2006). pRNA was shown to stabilize binding of NoRC to rDNA via formation of a DNA:RNA triplex at the T0 element and to induce a conformational change of TIP5 that probably allows interactions with co-repressors that promote heterochromatin formation and rDNA silencing (Mayer et al. 2006, 2008; Schmitz et al. 2010). In mouse, pRNA is a 150–250-nucleotide rRNA that matches the rDNA promoter sequences. pRNA is made by processing of an intergenic (IGS) rRNA whose synthesis is mediated by Pol I and originates from the spacer promoter, located 2 kb upstream the main gene promoter. IGS transcripts are rare, being 1,000-fold less abundant than pre-rRNA. The spacer promoter and the main gene promoter have some sequence homology, binding of TIF-IB/SL1, and the Pol I-associated factor TIF-IA is slightly decreased at the spacer promoter while Pol I is threefold more abundant at the spacer promoter than the main gene promoter (Santoro et al. 2010). The differences in IGS and pre-rRNA levels can be probably ascribed either to stalled Pol I at the spacer promoter or to rapid degradation, as suggested by data showing that binding to NoRC stabilizes pRNA (Mayer et al. 2006). Recent results indicated that there are two

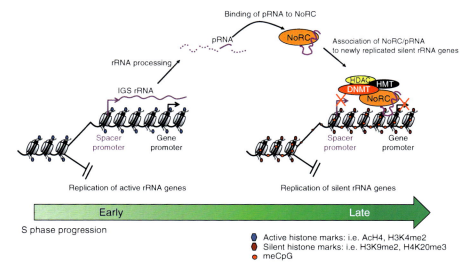

Fig. 4.1 Model showing inheritance of silent chromatin structure of rRNA genes replicating in late S phase. During early S phase, a subset of active genes containing nine enhancer repeats transcribe IGS rRNA. Transcripts originate from the spacer promoter. Immediately after synthesis, IGS rRNA is processed during mid to late S phase to yield pRNA that is indispensable for NoRC-dependent rDNA silencing. NoRC binds to pRNA, associates with new late-replicating genes, and re-establish the silent chromatin structure (including CpG methylation)

additional factors that can lead to low IGS rRNA abundance (Santoro et al. 2010). IGS rRNA synthesis was shown to occur during a restricted time window of S phase (early) and to originate from a specific set of active and hypomethylated rRNA genes that contain nine enhancer repeats located between the spacer and the main gene promoter (Fig. 4.1). These spacer transcripts are then processed during mid to late S phase to yield pRNA that is indispensable for establishment of silent rDNA chromatin mediated by NoRC (Santoro et al. 2010). This mode of action suggests that pRNA acts in *trans* to inherit DNA methylation and transcriptional repression of late-replicating silent rDNA copies. Notably, timing of IGS rRNA transcription (early S phase) and processing (mid-late S phase) into pRNA correlates with the time when NoRC associates with newly replicated silent genes (mid-late S phase) to re-establish silent chromatin (Li et al. 2005). This finding indicates that the cell carefully tunes the timing of IGS rRNA transcription/processing to inherit rDNA silencing during cell division, suggesting that replication timing serves to coordinate synthesis and availability of factors at the time when they have to bind selectively to newly replicated chromatin to propagate their epigenetic state to next cell generation. Recent results revealed another layer of epigenetic control that involves acetylation state of K633 of TIP5 able to modulate pRNA-NoRC association during S phase progression (Zhou et al. 2009). The acetyltransferase MOF (males absent on the first) acetylates TIP5 while the NAD$^+$-dependent deacetylase SIRT1 (sirtuin-1) removes the acetyl group from K633. Acetylation of TIP5 that weakens the NoRC-pRNA association is required for silent nucleosome positioning and fluctuates during S phase. TIP5-K633 is not acetylated in early S phase (the first 2 h after entry into S phase), then is acetylated during early-mid S phase (3–4 h after entry into S phase), and finally deacetylated during mid-late S phase (from 5 h after entry into S phase), the time when NoRC binds to newly replicated silent rRNA genes. The model proposed by the authors is the following: "1- in early S phase, TIP5 is recruited to rDNA by interactions with promoter-bound TTF-I, pRNA and H4K16ac; 2- in mid phase, MOF acetylates chromatin-bound TIP5 at K633, leading to transient dissociation of pRNA and NoRC and shifting of the promoter-bound nucleosome downstream of the transcription start site; 3- in late S phase, this nucleosome positioning allows de novo methylation by NoRC-associated DNMTs. To establish heterochromatic histone modifications and maintain the silent chromatin state, K633 has to be deacetylated by SIRT1, allowing re-association of NoRC with pRNA and the establishment of heterochromatic histone modifications (mediated by HTM and HDAC1) at the rDNA promoter" (Zhou et al. 2009). However, although fascinating, this model does not take into account that chromatin is completely erased during the passage of the replication fork. In other words, the TIP5-acetylation mediated nucleosome repositioning that was proposed, but not demonstrated, to occur before replication of silent rRNA genes will be inevitably erased during the passage of the replication fork. Thus, although deacetylation of K633-TIP5 in mid-late S phase correlates well with the timing of NoRC-pRNA binding to newly replicated silent rRNA genes, the proposed model is weak in offering explanations for the role of acetylated TIP5 in the time window of early-mid S phase that precedes replication of silent rRNA genes. Replication of DNA requires disruption of parental nucleosomes, implying that mechanisms must exist that are able to loosen chromatin compaction and facilitate the disassembly of nucleosomes before passage of replication machinery. Following this line, acetylation of TIP5 and consequent weakening of NoRC-pRNA association and binding to rDNA can be part of these temporally coordinated changes aimed to decompact rDNA silent chromatin structure before passage of the replication fork. Similar chromatin structural changes have been also attributed to phosphorylation of histone H1 in late G1 and S phase that, by decreasing binding to nucleosomal DNA, might lead to a less compacted higher order chromatin structure (Fasy et al. 1979; Flickinger 2001; Gunjan et al. 2001). As a consequence of this, it was proposed that the accessibility of the pre-replication complex to the origin of replication and, probably, the initiation process itself through the chromatin barrier would be facilitated to some extent.

The finding that transcripts from the spacer promoter have an indispensable function in epigenetic silencing of rDNA is in apparent disagreement with previous studies showing that the spacer promoter enhances transcription from the main rDNA promoter (Caudy and Pikaard 2002; De Winter and Moss 1986; Grimaldi and Di Nocera 1988; Paalman et al. 1995; Putnam and Pikaard 1992; Tower et al. 1989). In the "read-through enhancement" model, it was proposed that Pol I molecules, which are directed by the spacer promoter to transcribe through the enhancer, release rDNA transcription factors from the enhancer and make them available to the gene promoter, thereby stimulating gene promoter transcription (De Winter and Moss 1987). A corollary of read-through enhancement models is that the spacer promoter must act to somehow amplify or increase the effect exerted by the enhancer repeats alone and that the level of stimulation is proportional to the transcriptional strength of the spacer promoter. However, replacement of mouse spacer promoter by the much more active Chinese hamster spacer promoter did not change the level of gene promoter stimulation (Paalman et al. 1995). According to these results, enhancement of pre-rRNA synthesis does not depend on transcripts originating from spacer promoter and implies that spacer promoter affects the main gene promoter using alternative mechanisms. One possibility is formation of a loop between the spacer promoter and the main gene promoter similar to that described to occur between rDNA main promoter and terminator regions (Nemeth et al. 2008). The spatial juxtaposition of both promoters might enhance transcription from the main gene promoter by delivering Pol I factors and it would not require IGS rRNA synthesis. Consistent with this, recent results identified binding of CTCF and enrichment of the histone variant H2A.Z at the spacer promoter (van de Nobelen et al. 2010). CTCF is a conserved and ubiquitously expressed protein, which binds DNA and organizes chromatin into loops (Phillips and Corces 2009) while H2A.Z is a mark associated with "poised" promoters (Fan et al. 2002). Loop formation within the IGS rDNA can be also mediated via dimerization of TTF1 bound to terminator elements T0 and T-1, located upstream of the transcription start site of the main gene promoter and downstream of the spacer promoter, respectively (R. Santoro, unpublished data). Involvement of TTF1 in a structure mediating interaction between the main gene promoter and the 3′-rDNA region has also been recently proposed (Nemeth et al. 2008). The involvement of TTF1 in forming the spacer-main gene promoter loop not only suggests that IGS rRNA synthesis is not required but that it might not occur at all. The major obstacle that Pol I would encounter in transcribing IGS rDNA is TTF1 that, if bound to T0 and T-1 elements, might prematurely terminate IGS rRNA transcripts. Thus, when IGS rRNA is synthesized, TTF1 should not be bound to either T0 and/or T-1 elements. As binding of TTF1 to T0 is a prerequisite for 45S pre-rRNA synthesis (Langst et al. 1998), it is unlikely that transcription from spacer promoter enhances the strength of the main gene promoter in the absence of TTF1. Whether and how binding of TTF1 to T0 and T-1 is abrogated during synthesis of IGS rRNA in early S phase remains an issue to be investigated. Taken together, all these observations suggest that the dual role of spacer promoter in regulating rRNA transcription can be distinguished by its capacity either to form a loop or to drive IGS rRNA synthesis: in the first case, it stimulates pre-rRNA synthesis; in the second case, it is required for NoRC-mediated rDNA silencing.

4.4 Regulation of rRNA Synthesis by Epigenetic and Chromatin Related-Mechanisms

During evolution, rRNA genes have been tuned to reach appropriate levels of rRNA transcription by using Pol I and by amplifying the number of rRNA genes to hundreds or even thousands of copies per genome. Concomitant with this process, a third mechanism has evolved to keep a large percentage of those rDNA units in a silent state. This led to the proposal that cells might modulate rRNA levels by changing the number of genes to be transcribed. A corollary of this model is that each rRNA gene is a "binary unit" that is either on or off and, if on, is producing rRNA at approximately the same rate as other active genes. However, several lines of evidence indicated that rRNA synthesis is mainly due to the ability of cells to control transcription rate per gene. For example, elongation rates were found to be directly proportional to 45S pre-rRNA synthesis and phosphorylation of UBF was shown to directly regulate elongation by inducing the remodeling of ribosomal gene chromatin (Stefanovsky et al. 2006). In line with this, data indicated that the number of rRNA genes is not fundamental in regulating rRNA transcription. Two yeast strains containing different numbers of rRNA genes (143 and 42 copies) produced the same amount of rRNAs (French et al. 2003). Miller spread analysis showed that in the reduced copy strain, the mean number of Pol I complexes loaded on each gene was twofold higher than in the control strain, suggesting that rRNA synthesis in exponentially growing yeast cells is controlled by the ability of cells to load polymerases and not by the number of open genes. Similarly, maize inbred lines can vary almost tenfold in their rRNA gene content yet have similar morphological characteristics and growth rates (Rivin et al. 1986). The same is true for aneuploid chicken cells that contain different numbers of rRNA copies and display the same levels of rRNA transcription (Muscarella et al. 1985). Taken together, these results indicated that each rRNA gene is not a "binary unit" that is either on or off but that the transcriptional potential of each on gene can be modulated by controlling the rRNA synthesis rate. However, this does not exclude the possibility that cells might be able to silence active genes and to activate silent genes in a dynamic way, according to the metabolic requirement. Initial studies using psoralen crosslinking of rDNA chromatin in mouse cells showed that the proportion of the two ribosomal chromatin structures is similar in interphase and metaphase and is independent of the transcriptional activity of the gene (Conconi et al. 1989). This study led to the proposal that mammalian cells modulate rRNA synthesis by controlling the transcription rates of active (non-nucleosomal) rRNA genes while the inactive (nucleosomal) rRNA gene copies never transcribe. In contrast to vertebrate cells, the proportion of psoralen-accessible (active) genes and psoralen-inaccessible (silent) genes changed in response to variations in environmental conditions (i.e., growing in complex or minimal medium and exponential or stationary phase), suggesting that yeast can regulate rRNA synthesis by varying the number of active gene copies (Dammann et al. 1993). As discussed above, yeast and mammalian active rRNA chromatin might greatly differ in histone occupancy. Thus, in mammalian cells, psoralen analysis might not be sufficient in monitoring changes at rDNA chromatin that affect transcription without a drastic alteration in nucleosome occupancy. Indeed, an increasing body of evidence indicates that in mammals the pool of active ribosomal genes is not static but it can acquire heterochromatic features leading to rDNA silencing. Recent results showed that mammalian rDNA chromatin can be epigenetically modified in response to intracellular energy status (Murayama et al. 2008). Although this study did not analyze nucleosome occupancy at rDNA by psoralen, it clearly indicated that active rRNA genes acquire silent histone modifications under conditions of energy starvation or glucose deprivation. They identified eNoSC, a complex containing nucleomethylin (NML), the NAD^+-dependent deacetylase SIRT1, and H3K9 methyltransferase SUV39H1. Both SIRT1 and SUV39H1 are required for energy-dependent rRNA transcriptional repression, suggesting that a change in the NAD^+/NADH ratio induced by reduction of energy status could activate SIRT1, leading to deacetylation of histone H3 and dimethylation of H3K9 by SUV39H1 and establishment of heterochromatin at the rDNA locus. Importantly, levels of rRNA transcription in the absence of glucose reduced more slowly in cells depleted of NML while the total cellular ATP levels decreased faster than those in control cells. These observations suggested that under low-glucose conditions eNoSC is the key player in repressing transcription of rRNA genes and it plays an important role in restoring energy levels. As NML is a human homolog of yeast Rrp8p, which is involved in cleavage of rRNA in yeast (Bousquet-Antonelli et al. 2000), this opens up the possibility that eNoSC may also connect rRNA processing with intracellular energy status. Recently, the SIRT1-MOF mediated acetylation state of the NoRC complex was also linked to establishment of rDNA silencing according to cellular energy status (Zhou et al. 2009). In the absence of glucose, acetylation of TIP5 at K633, a modification that abolishes pRNA binding, reduced and binding of TIP5 to rDNA increased. However, whether NoRC is required to establish rDNA silencing according to cellular energy status was not determined. Although these studies showed increased levels of silent histone marks at the rDNA locus induced by glucose deprivation, data concerning rDNA CpG methylation were not reported. This is surprising, considering also that NoRC studies have always been accompanied so far by measurements of rDNA methylation. Changes of rDNA chromatin were also reported during several differentiation processes. In adipocyte differentiation of 3T3-L1 cells, downregulation of rRNA transcription is accompanied by an increase in heterochromatic histone modifications, such as hypoacetylation of histone H4 and H3K9me2, shift of nucleosome from the active to the silent positioning, decreased UBF expression, and increased TIP5 levels (Li et al. 2006). Unfortunately, this study did not include results concerning the levels of rDNA methylation during differentiation. Reduced UBF expression is common during the terminal differentiation of many cell types (Alzuherri and White 1999; Datta et al. 1997; Larson et al. 1993; Liu et al. 2007; Poortinga et al. 2004), suggesting that downregulation of UBF is a widespread mechanism for the silencing of active rRNA genes during development. Nucleosome occupancy measurement by psoralen of the murine promyelocytic (MPRO) cell line during terminal differentiation process showed a significant reduction in the number of active genes ($43.7 \pm 2.8\%$ active in day 0 compared with $19.4 \pm 6\%$ active in day 4 of differentiation) and that this occurred in the absence of changes in rDNA promoter methylation (Sanij et al. 2008). Consistent with this, and supporting the notion of the role of UBF in mediating active rDNA chromatin, depletion of UBF led to an increase of psoralen-inaccessible rDNA chromatin accompanied by accumulation of silent histone marks and histone H1 without remarkable changes of CpG methylation levels at the promoter. Importantly, restoration of UBF levels after

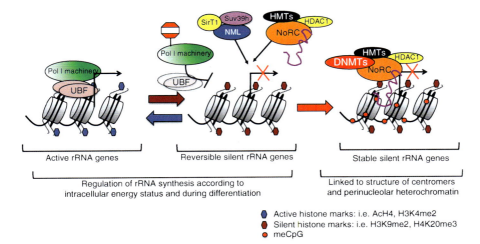

Fig. 4.2 Schema representing the epigenetic features of active, reversible silent, and stable silent rRNA genes in mammalian cells. Active rRNA genes are enriched in histones containing active marks (i.e., AcH4, H3K4me3) and associated with the Pol I transcription machinery and upstream binding factor (UBF). During energy starvation and/or differentiation, rRNA synthesis is downregulated and active rRNA genes acquire histones modified with silent marks (i.e., H3K9me2 and hypoacetylation). Changes in CpG methylation levels were either not reported or not detected. Importantly, during differentiation, the levels of UBF reduced while TIP5 levels increased. This chromatin state is probably reversible and I predict that, by restoration of proper cellular energy levels, these silent copies will reacquire epigenetic features compatible for transcription. Further addition of meCpG moieties produces a stable silent rDNA chromatin structure. This reaction appears to be uniquely mediated by the NoRC complex that associates with DNA methyltransferases 1 and 3B. The stable silent copies are probably not controlling the rRNA synthesis levels but involved in the nucleolus/nuclear architecture of important chromatin domains like centromeres and perinucleolar heterochromatin

knockdown correlates with re-establishment of the number of active genes back to wild-type levels, suggesting that rDNA silencing in response to UBF depletion is reversible. Taken together, all these results indicated that active rRNA genes acquire heterochromatic marks in response to cellular energy status and differentiation and that this process is probably not mediated by CpG methylation. On the basis of these results, it becomes evident that two sub-classes of silent rRNA genes might exist: one representing genes that are transcriptional silent and associated with repressive histone marks but deficient in CpG methylation (reversible silent genes); the other one that contains meCpG sequences and, because CpG methylation represents a relatively stable epigenetic mark, I refer to these genes as "stable" silent copies (Fig. 4.2). Intriguingly, with the exception of loss of rDNA silencing related to pathologies like cancer and knockdown of silencing effectors like TIP5 and DNA methyltransferases (Espada et al. 2007; Guetg et al. 2010), all data published so far described mechanisms of silencing of active genes while the reversed process was never reported. On the basis of all these results, it appears that mammalian active rRNA genes acquire heterochromatic features to downregulate rRNA synthesis. This kind of silencing seems to be reversible, as only histones are modified with silent marks while CpG methylation remains probably unaffected. In contrast, upregulation of rRNA transcription does not seem to operate in increasing the number of active rRNA genes but most probably by modulating their transcription rates.

4.5 Function of Silent rRNA Genes

Formation of specific heterochromatic domains is crucial for genome stability (Grewal and Jia 2007; Peng and Karpen 2008). This is exemplified by the heterochromatin structure of repetitive major satellite (pericentric) and minor satellite (centric) DNA sequences whose maintenance and accurate reproduction throughout multiple cell divisions represents a major challenge to ensure genome stability. In interphase, the centromeric heterochromatin is predominantly located either at the nuclear periphery or around the nucleolus (Fig. 4.3a) (Haaf and Schmid 1991; Pluta et al. 1995). In humans and apes, rRNA genes are located between the short arm and

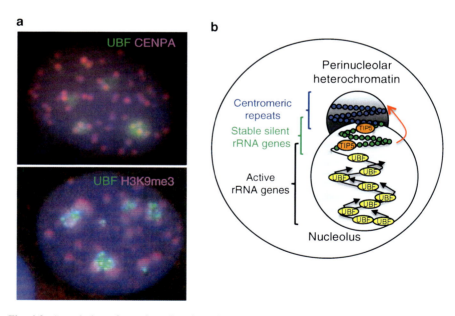

Fig. 4.3 Association of centric and pericentric heterochromatin with the nucleolus. (a) Indirect immunofluorescence showing the core kinetochore protein CENPA (*upper panel*) and centric-pericentric heterochromatin enriched in H3K9me3 (*lower panel*) located close to the nucleolus stained with anti-UBF antibodies. (b) TIP5 mediates heterochromatin formation at nucleolar/perinucleolar associated chromatin. The schema depicts the cellular distribution of active/silent rRNA genes and centromeric heterochromatin within the nucleolus and at the nucleolar periphery

the satellite body of acrocentric chromosomes. In mouse, rRNA genes cluster within the centromeric regions of chromosomes 12, 15, 16, 18, and 19 (Dev et al. 1977; Kurihara et al. 1994). Because of the linear proximity, centromeres of chromosomes bearing rDNA repeats associate with nucleoli. Notably, also some other chromosomes devoid of rRNA genes have their centromeres associated with the nucleolus at a frequency more than that expected for a random distribution (Carvalho et al. 2001). The basis of this association probably relies on the linear proximity along the chromosome and on the repeated nature of DNA sequence, which provides multiple binding sites for specific proteins capable of forming multimeric complexes. Several pieces of evidence indicate that silent rDNA arrays are located in the extranucleolar space, frequently associated with the perinucleolar heterochromatin (Mosgöller 2004). Active rDNA repeats, on the other hand, are located inside the nucleolus within the dense fibrillar components. Consistent with this, CpG-methylated rRNA genes ("stable" silent copies) were shown to assemble adjacent to the perinucleolar heterochromatin composed of centric repeats in mouse neuronal cells, suggesting an intricate relationship between these heterochromatic regions (Akhmanova et al. 2000). Consistent with this, recent data showed that depletion of TIP5 in NIH3T3 cells not only reduced the levels of silent histone marks and CpG methylation at the rDNA locus but also decreased the levels of two typical silent histone modifications (H3K9me3 and H4K20me3) at centric and pericentric heterochromatin (Guetg et al. 2010). Moreover, cells depleted of TIP5 lacked the characteristic perinucleolar heterochromatin, implying an intimate connection that links TIP5 with rDNA silencing and formation of centromeric heterochromatin. These results suggest that the role of silent rRNA genes and TIP5 go beyond regulation of rRNA synthesis and that they can play an important role at the level of nuclear/nucleolus chromatin architecture. Indeed, the presence of silent rDNA copies was also detected in a yeast strain containing about 42 rDNA copies (Merz et al. 2008). Previous electron microscopy analysis showed that inactive genes were rarely seen in the 42 copies strain, with the interpretation that all genes are active (French et al. 2003). However, although the rRNA genes of this strain are highly transcribing to compensate for the absence of about 100 copies, a fraction of 10–20% of rRNA genes persists to remain inaccessible to psoralen (nucleosomal silent rDNA fraction) (Merz et al. 2008). Probably, the presence of this silent rDNA fraction could have been missed in the electron microscopy analysis (French et al. 2003). Indeed, while in the 143-copies strain, active and silent copies are interspersed, the silent genes in the 42-copies strain, if located at the boundary of the rDNA arrays, might have been indistinguishable from the rest of the neighboring chromatin. The presence of silent rRNA copies in a strain where all the *bona fide* rRNA genes should be dedicated to transcription suggests that the presence of silent copies is indispensable and that probably their role goes beyond regulation of rRNA synthesis levels. In mammalian cells, the spatial and linear closeness between rRNA genes and centric repeats may allow TIP5, bound to silent rRNA genes, to interact with centric repeats and to aid in establishing heterochromatic structures using similar mechanisms as used to silence the rDNA locus (Fig. 4.3b) (Guetg et al. 2010; Santoro et al. 2002). The association of TIP5 with the centromeric protein CENPA suggested that this interaction indeed takes place.

Alternatively, the repressive chromatin of silent rRNA copies may affect the centric and pericentric heterochromatin either by spreading mechanisms or by creating a nucleolar/perinucleolar compartment enriched in chromatin repressor complexes. Notably, a role of the perinucleolar compartment in mediating the incorporation of repressive chromatin factors was recently discussed for the establishment of the inactive X-chromosome that contacts the nucleolus during mid-to-late S-phase to faithfully duplicate its epigenetic character (Zhang et al. 2007). Indeed, TIP5 and rDNA silencing seem to play a role during maintenance of inactive X chromosome structure (R. Santoro, unpublished data).

Assembly of DNA repeats into silent chromatin is generally thought to serve as a mechanism ensuring repeat stability by limiting access to the recombination machinery. A large body of evidence indicates that maintenance of silent rDNA chromatin plays an important role for the stability of rRNA repeats. In the yeast *S. cerevisiae*, recruitment of the nucleolar protein complexes RENT (regulator of nucleolar silencing and telophase exit) and Cohibin to rDNA suppresses unequal recombination at the rDNA repeats (Mekhail et al. 2008). This suppression is seemingly linked to the ability of these complexes to induce rDNA silencing. Similarly, segments of rRNA genes and satellite repeat arrays become dispersed in *Drosophila* mutants that are defective in the histone methyltransferase Su(var)3-9, in HP1 also known as Su(var)2-5, or in several genes involved in the RNA interference (RNAi) pathway (Peng and Karpen 2007). Because rRNA gene expression is the driving force for the assembly of nucleoli, one consequence of dispersing rRNA genes is the appearance of multiple nucleoli, instead of the single nucleolus typical of wild-type cells, an indication of accumulation of extrachromosomal DNAs and a typical result of rDNA recombination events. Consistent with this, knockdown of TIP5 in NIH3T3 cells not only impairs formation of heterochromatin at rRNA genes and satellite repeats but also induces specific loss of silent rRNA repeats and of major and minor satellites replicating in late S phase (Guetg et al. 2010). Importantly, by tracking rRNA genes with polymorphic variations, it was shown that TIP5-mediated heterochromatin formation specifically protects CpG methylated (stable) silent rRNA genes from illicit recombination events whereas active genes are not affected. As formation and maintenance of heterochromatic structures is crucial for genome stability, it was proposed that TIP5-mediated heterochromatin has an important role in protecting the genome from inappropriate chromosomal rearrangements and that the structure of "stable" silent rRNA genes plays a role in the nucleolus/nuclear chromatin architecture.

In conclusion, establishment of heterochromatin at rRNA repeats not only plays a role in regulating rRNA transcription but may provide structural organization of the nucleolus and nuclear architecture (Fig. 4.3). In mammalian cells, the differences between these two functions may depend on the specific epigenetic signature of the rRNA copies, particularly the CpG methylation content (Fig. 4.2). When cells require downregulation of rRNA synthesis (i.e., by energy starvation), rRNA genes acquire silent histone modification but not CpG methylation. This silent state can be reversed and I predict that, by restoration of proper cellular energy levels, these silent copies will reacquire epigenetic features compatible for transcription.

In contrast, stable silent rRNA genes containing CpG methylated sequences cannot be easily reversed. Probably, these stable silent rRNA copies localize near the centromeric repeats and function together with TIP5 in the formation perinucleolar heterochromatin. If the heterochromatic structure of these stable silent repeats is lost, these sequences will undergo illicit recombination events, a typical result of genomic instability.

4.6 rDNA Silencing and Cancer

The relationship between the nucleolus and cancer has been the subject of study for many years. Up-regulation of ribosome production might contribute to neoplastic transformation, by affecting the balance of protein translation, thus altering the synthesis of proteins that play an important role in the genesis of cancer (Montanaro et al. 2008). The association of human carcinomas with nucleolar hypertrophy with bad prognoses is worthy of note. Abnormalities in the nucleolar morphology of cancer cells attracted the attention of tumor pathologists as early as the late nineteenth century. From that moment on, a series of studies have been performed to clarify whether these nucleolar changes were a consequence of the cancerous state or, instead, they might represent a cause of neoplastic transformation. As cell proliferation appears to be closely coordinated with nucleolar function, nucleolar structural-functional changes in tumors were considered as a mere consequence of both the proliferative activity of cancer cells and alterations of the mechanisms controlling cancer cell proliferation. In recent years some data have been produced that also suggest an active role of ribosome biogenesis in tumorigenesis. For example, human nontumor lesions characterized by an up-regulation of nucleolar function were found to be associated with an increased risk of neoplastic transformation, and evidence shows that people with inherited diseases characterized by the production of abnormal ribosomes have a very high incidence of cancer.

In this section, I describe recent findings about the role of rDNA CpG methylation in cancer. Hypomethylation of the rRNA genes has been observed in several tumors such as lung cancer, Wilms tumor, and hepatocellular carcinomasa (Ghoshal et al. 2004; Powell et al. 2002; Qu et al. 1999; Shiraishi et al. 1999). Moreover, rDNA CpG methylation levels were found to be higher in ovarian cancer patients with long progression survival as compared with that in patients with short survival, an indication that rDNA silencing levels may influence cell growth properties essential for active tumor proliferation and tumor aggressiveness (Chan et al. 2005; Powell et al. 2002). Decreased CpG methylation of rRNA genes was found in many African–American women, who suffer disproportionately worse outcomes from endometrial cancer (EC) (even after controlling for socioeconomic factors and tumor stage/grade) (Powell et al. 2002). These women possess notably lower rDNA methylation than non-African–American women. Consequently, it was proposed that rDNA methylation changes contribute in numerous ways to endometrial cancer and profiles of such alterations will likely be valuable for prognosis and therapeutic decision making. To note, the more aggressive type II EC tumors possess significantly reduced levels of DNA methylation, as compared with the less aggressive type I ECs, possibly contributing to the type II EC characteristic of genomic instability (Zhou et al. 2007). NIH3T3 cells depleted of TIP5 not only display impairment of rDNA silencing but also undergo genomic instability. Knockdown of TIP5 promotes higher rRNA synthesis and formation of enlarged nucleoli, a typical result of elevated nucleolar activities (Guetg et al. 2010). Consistent with this, depletion of TIP5 and consequent impairment of rDNA silencing promoted ribosome synthesis and enhanced the productivity of recombinant proteins in NIH3T3, CHO, and HEK293 cells (Santoro et al. 2009). Importantly, these cells not only proliferate at higher rates but also grow beyond confluence and display a transformed phenotype. Surprisingly, upregulation of rRNA transcription in TIP5-depleted cells does not depend on the de-repression of silent genes. Whereas the amount of CpG methylated silent genes decreases in these cells, the number of active genes is not affected. It seems, therefore, that TIP5 and/or presence of heterochromatic silent repeats indirectly affects the transcription rate of active genes, probably by enriching the nucleolar compartment of the chromatin repressor complexes. However, the possibility that upregulation of rDNA transcription is a consequence of genome instability that caused the acquisition of aberrant mechanisms of rDNA transcriptional regulation cannot be excluded, thus representing an advantage for the elevated protein synthesis necessary for high proliferative rates.

Taken together, all these studies suggest that controlling the CpG methylation state of rRNA genes may contribute to the aggressiveness of tumor. How and when these epigenetic changes occur will be an issue for future studies.

4.7 Conclusions

Over the past decade, emerging evidence indicated that epigenetic factors control and regulate nuclear processes. Disruption of the balance of the epigenetic networks can cause several major pathologies, that is, cancer, syndromes involving chromosomal instabilities, and mental retardation. Thus, dissecting the cause-and-effect relationship between the epigenetic marks, and determining how all the chromatin modifier complexes can co-ordinate with each other, has great potential for the development of therapies based on the use of inhibitors for enzymes controlling epigenetic modifications. The rRNA genes represent an ideal model to study how epigenetics and chromatin structure can modulate gene expression and how the chromatin and epigenetic information is inherited during cell division. Unraveling the mechanistic insights of how changes in chromatin and epigenetic modification affect rDNA transcription during the cell cycle or during external stimuli as well as how RNA polymerase I and the transcription factors can overcome the chromatin barrier at initiation and during elongation will shed new light on the basic mechanisms of rDNA transcriptional regulation. In addition, it will reveal new strategies to apply in those pathologies where rRNA synthesis is altered.

4 The Epigenetics of the Nucleolus... 77

Acknowledgments I thank Claudio Guetg for critical reading of this work. The work of R.S is funded by SNF and Mäxi-Stiftung.

References

Akhmanova A, Verkerk T, Langeveld A, Grosveld F, Galjart N (2000) Characterisation of transcriptionally active and inactive chromatin domains in neurons. J Cell Sci 113(pt 24):4463–4474

Alzuherri HM, White RJ (1999) Regulation of RNA polymerase I transcription in response to F9 embryonal carcinoma stem cell differentiation. J Biol Chem 274(7):4328–4334

Angelov D, Bondarenko VA, Almagro S, Menoni H, Mongelard F, Hans F, Mietton F, Studitsky VM, Hamiche A, Dimitrov S, Bouvet P (2006) Nucleolin is a histone chaperone with FACT-like activity and assists remodeling of nucleosomes. EMBO J 25(8):1669–1679

Bazett-Jones DP, Leblanc B, Herfort M, Moss T (1994) Short-range DNA looping by the *Xenopus* HMG-box transcription factor, xUBF. Science 264(5162):1134–1137

Berger C, Horlebein A, Gogel E, Grummt F (1997) Temporal order of replication of mouse ribosomal RNA genes during the cell cycle. Chromosoma 106(8):479–484

Birch JL, Tan BC, Panov KI, Panova TB, Andersen JS, Owen-Hughes TA, Russell J, Lee SC, Zomerdijk JC (2009) FACT facilitates chromatin transcription by RNA polymerases I and III. EMBO J 28(7):854–865

Bousquet-Antonelli C, Vanrobays E, Gelugne JP, Caizergues-Ferrer M, Henry Y (2000) Rrp8p is a yeast nucleolar protein functionally linked to Gar1p and involved in pre-rRNA cleavage at site A2. RNA 6(6):826–843

Carvalho C, Pereira HM, Ferreira J, Pina C, Mendonca D, Rosa AC, Carmo-Fonseca M (2001) Chromosomal G-dark bands determine the spatial organization of centromeric heterochromatin in the nucleus. Mol Biol Cell 12(11):3563–3572

Caudy AA, Pikaard CS (2002) *Xenopus* ribosomal RNA gene intergenic spacer elements conferring transcriptional enhancement and nucleolar dominance-like competition in oocytes. J Biol Chem 277(35):31577–31584

Cavellan E, Asp P, Percipalle P, Farrants AK (2006) The WSTF-SNF2h chromatin remodeling complex interacts with several nuclear proteins in transcription. J Biol Chem 281(24):16264–16271

Chan MW, Wei SH, Wen P, Wang Z, Matei DE, Liu JC, Liyanarachchi S, Brown R, Nephew KP, Yan PS, Huang TH (2005) Hypermethylation of 18S and 28S ribosomal DNAs predicts progression-free survival in patients with ovarian cancer. Clin Cancer Res 11(20):7376–7383

Chen D, Belmont AS, Huang S (2004) Upstream binding factor association induces large-scale chromatin decondensation. Proc Natl Acad Sci USA 101(42):15106–15111

Clos J, Normann A, Ohrlein A, Grummt I (1986) The core promoter of mouse rDNA consists of two functionally distinct domains. Nucleic Acids Res 14(19):7581–7595

Conconi A, Widmer RM, Koller T, Sogo JM (1989) Two different chromatin structures coexist in ribosomal RNA genes throughout the cell cycle. Cell 57(5):753–761

Dammann R, Lucchini R, Koller T, Sogo JM (1993) Chromatin structures and transcription of rDNA in yeast *Saccharomyces cerevisiae*. Nucleic Acids Res 21(10):2331–2338

Dammann R, Lucchini R, Koller T, Sogo JM (1995) Transcription in the yeast rRNA gene locus: distribution of the active gene copies and chromatin structure of their flanking regulatory sequences. Mol Cell Biol 15(10):5294–5303

Datta PK, Budhiraja S, Reichel RR, Jacob ST (1997) Regulation of ribosomal RNA gene transcription during retinoic acid-induced differentiation of mouse teratocarcinoma cells. Exp Cell Res 231(1):198–205

De Winter RF, Moss T (1986) Spacer promoters are essential for efficient enhancement of *X. laevis* ribosomal transcription. Cell 44(2):313–318

De Winter RF, Moss T (1987) A complex array of sequences enhances ribosomal transcription in *Xenopus laevis*. J Mol Biol 196(4):813–827

Dev VG, Tantravahi R, Miller DA, Miller OJ (1977) Nucleolus organizers in *Mus musculus* subspecies and in the RAG mouse cell line. Genetics 86(2 pt 1):389–398

Espada J, Ballestar E, Santoro R, Fraga MF, Villar-Garea A, Nemeth A, Lopez-Serra L, Ropero S, Aranda A, Orozco H, Moreno V, Juarranz A, Stockert JC, Langst G, Grummt I, Bickmore W, Esteller M (2007) Epigenetic disruption of ribosomal RNA genes and nucleolar architecture in DNA methyltransferase 1 (Dnmt1) deficient cells. Nucleic Acids Res 35(7):2191–2198

Fan JY, Gordon F, Luger K, Hansen JC, Tremethick DJ (2002) The essential histone variant H2A.Z regulates the equilibrium between different chromatin conformational states. Nat Struct Biol 9(3):172–176

Fasy TM, Inoue A, Johnson EM, Allfrey VG (1979) Phosphorlyation of H1 and H5 histones by cyclic AMP-dependent protein kinase reduces DNA binding. Biochim Biophys Acta 564(2):322–334

Felle M, Exler JH, Merkl R, Dachauer K, Brehm A, Grummt I, Langst G (2010) DNA sequence encoded repression of rRNA gene transcription in chromatin. Nucleic Acids Res 38(16): 5304–5314

Flickinger R (2001) Replication timing and cell differentiation. Differentiation 69(1):18–26

Fraser P (2006) Transcriptional control thrown for a loop. Curr Opin Genet Dev 16(5):490–495

French SL, Osheim YN, Cioci F, Nomura M, Beyer AL (2003) In exponentially growing *Saccharomyces cerevisiae* cells, rRNA synthesis is determined by the summed RNA polymerase I loading rate rather than by the number of active genes. Mol Cell Biol 23(5): 1558–1568

Gadal O, Labarre S, Boschiero C, Thuriaux P (2002) Hmo1, an HMG-box protein, belongs to the yeast ribosomal DNA transcription system. EMBO J 21(20):5498–5507

Gagnon-Kugler T, Langlois F, Stefanovsky V, Lessard F, Moss T (2009) Loss of human ribosomal gene CpG methylation enhances cryptic RNA polymerase II transcription and disrupts ribosomal RNA processing. Mol Cell 35(4):414–425

Ghoshal K, Majumder S, Datta J, Motiwala T, Bai S, Sharma SM, Frankel W, Jacob ST (2004) Role of human ribosomal RNA (rRNA) promoter methylation and of methyl-CpG-binding protein MBD2 in the suppression of rRNA gene expression. J Biol Chem 279(8):6783–6793

Gimelbrant AA, Ensminger AW, Qi P, Zucker J, Chess A (2005) Monoallelic expression and asynchronous replication of p120 catenin in mouse and human cells. J Biol Chem 280(2): 1354–1359

Goldmit M, Bergman Y (2004) Monoallelic gene expression: a repertoire of recurrent themes. Immunol Rev 200:197–214

Goren A, Cedar H (2003) Replicating by the clock. Nat Rev Mol Cell Biol 4(1):25–32

Grewal SI, Jia S (2007) Heterochromatin revisited. Nat Rev Genet 8(1):35–46

Grimaldi G, Di Nocera PP (1988) Multiple repeated units in *Drosophila melanogaster* ribosomal DNA spacer stimulate rRNA precursor transcription. Proc Natl Acad Sci USA 85(15): 5502–5506

Guetg C, Lienemann P, Sirri V, Grummt I, Hernandez-Verdun D, Hottiger MO, Fussenegger M, Santoro R (2010) The NoRC complex mediates the heterochromatin formation and stability of silent rRNA genes and centromeric repeats. EMBO J 29(13):2135–2146

Gunjan A, Sittman DB, Brown DT (2001) Core histone acetylation is regulated by linker histone stoichiometry in vivo. J Biol Chem 276(5):3635–3640

Haaf T, Schmid M (1991) Chromosome topology in mammalian interphase nuclei. Exp Cell Res 192(2):325–332

Harper F, Puvion-Dutilleul F (1979) Non-nucleolar transcription complexes of rat liver as revealed by spreading isolated nuclei. J Cell Sci 40:181–192

Inagaki S, Kakutani T (2010) Control of genic DNA methylation in *Arabidopsis*. J Plant Res 123(3):299–302

Jackson DA, Iborra FJ, Manders EM, Cook PR (1998) Numbers and organization of RNA polymerases, nascent transcripts, and transcription units in HeLa nuclei. Mol Biol Cell 9(6):1523–1536

Keohane AM, Lavender JS, O'Neill LP, Turner BM (1998) Histone acetylation and X inactivation. Dev Genet 22(1):65–73

Kurihara Y, Suh DS, Suzuki H, Moriwaki K (1994) Chromosomal locations of Ag-NORs and clusters of ribosomal DNA in laboratory strains of mice. Mamm Genome 5(4):225–228

Kwon SH, Florens L, Swanson SK, Washburn MP, Abmayr SM, Workman JL (2010) Heterochromatin protein 1 (HP1) connects the FACT histone chaperone complex to the phosphorylated CTD of RNA polymerase II. Genes Dev 24(19):2133–2145

Langst G, Becker PB, Grummt I (1998) TTF-I determines the chromatin architecture of the active rDNA promoter. EMBO J 17(11):3135–3145

Langst G, Blank TA, Becker PB, Grummt I (1997) RNA polymerase I transcription on nucleosomal templates: the transcription termination factor TTF-I induces chromatin remodeling and relieves transcriptional repression. EMBO J 16(4):760–768

Larson DE, Xie W, Glibetic M, O'Mahony D, Sells BH, Rothblum LI (1993) Coordinated decreases in rRNA gene transcription factors and rRNA synthesis during muscle cell differentiation. Proc Natl Acad Sci USA 90(17):7933–7936

Lawrence RJ, Earley K, Pontes O, Silva M, Chen ZJ, Neves N, Viegas W, Pikaard CS (2004) A concerted DNA methylation/histone methylation switch regulates rRNA gene dosage control and nucleolar dominance. Mol Cell 13(4):599–609

Lee DY, Northrop JP, Kuo MH, Stallcup MR (2006) Histone H3 lysine 9 methyltransferase G9a is a transcriptional coactivator for nuclear receptors. J Biol Chem 281(13):8476–8485

Li E, Bestor TH, Jaenisch R (1992) Targeted mutation of the DNA methyltransferase gene results in embryonic lethality. Cell 69(6):915–926

Li J, Langst G, Grummt I (2006) NoRC-dependent nucleosome positioning silences rRNA genes. EMBO J 25(24):5735–5741

Li J, Santoro R, Koberna K, Grummt I (2005) The chromatin remodeling complex NoRC controls replication timing of rRNA genes. EMBO J 24(1):120–127

Liu M, Tu X, Ferrari-Amorotti G, Calabretta B, Baserga R (2007) Downregulation of the upstream binding factor1 by glycogen synthase kinase3beta in myeloid cells induced to differentiate. J Cell Biochem 100(5):1154–1169

Long EO, Dawid IB (1980) Repeated genes in eukaryotes. Annu Rev Biochem 49:727–764

Lucchini R, Sogo JM (1995) Replication of transcriptionally active chromatin. Nature 374(6519):276–280

Mais C, Wright JE, Prieto JL, Raggett SL, McStay B (2005) UBF-binding site arrays form pseudo-NORs and sequester the RNA polymerase I transcription machinery. Genes Dev 19(1):50–64

Mayer C, Neubert M, Grummt I (2008) The structure of NoRC-associated RNA is crucial for targeting the chromatin remodelling complex NoRC to the nucleolus. EMBO Rep 9(8):774–780

Mayer C, Schmitz KM, Li J, Grummt I, Santoro R (2006) Intergenic transcripts regulate the epigenetic state of rRNA genes. Mol Cell 22(3):351–361

Mekhail K, Seebacher J, Gygi SP, Moazed D (2008) Role for perinuclear chromosome tethering in maintenance of genome stability. Nature 456(7222):667–670

Merz K, Hondele M, Goetze H, Gmelch K, Stoeckl U, Griesenbeck J (2008) Actively transcribed rRNA genes in *S. cerevisiae* are organized in a specialized chromatin associated with the high-mobility group protein Hmo1 and are largely devoid of histone molecules. Genes Dev 22(9):1190–1204

Miller OL Jr, Beatty BR (1969) Visualization of nucleolar genes. Science 164(882):955–957

Montanaro L, Trere D, Derenzini M (2008) Nucleolus, ribosomes, and cancer. Am J Pathol 173(2):301–310

Mosgöller W (2004) Nucleolar ultrastructure in vertebrates. Kluwer Academic/Plenum, New York

Moss T (2004) At the crossroads of growth control; making ribosomal RNA. Curr Opin Genet Dev 14(2):210–217

Moss T, Langlois F, Gagnon-Kugler T, Stefanovsky V (2007) A housekeeper with power of attorney: the rRNA genes in ribosome biogenesis. Cell Mol Life Sci 64(1):29–49

Murano K, Okuwaki M, Hisaoka M, Nagata K (2008) Transcription regulation of the rRNA gene by a multifunctional nucleolar protein, B23/nucleophosmin, through its histone chaperone activity. Mol Cell Biol 28(10):3114–3126

Murayama A, Ohmori K, Fujimura A, Minami H, Yasuzawa-Tanaka K, Kuroda T, Oie S, Daitoku H, Okuwaki M, Nagata K, Fukamizu A, Kimura K, Shimizu T, Yanagisawa J (2008) Epigenetic control of rDNA loci in response to intracellular energy status. Cell 133(4):627–639

Muscarella DE, Vogt VM, Bloom SE (1985) The ribosomal RNA gene cluster in aneuploid chickens: evidence for increased gene dosage and regulation of gene expression. J Cell Biol 101(5 pt 1): 1749–1756

Nemeth A, Guibert S, Tiwari VK, Ohlsson R, Langst G (2008) Epigenetic regulation of TTF-I-mediated promoter-terminator interactions of rRNA genes. EMBO J 27(8):1255–1265

O'Sullivan AC, Sullivan GJ, McStay B (2002) UBF binding in vivo is not restricted to regulatory sequences within the vertebrate ribosomal DNA repeat. Mol Cell Biol 22(2):657–668

Okano M, Bell DW, Haber DA, Li E (1999) DNA methyltransferases Dnmt3a and Dnmt3b are essential for de novo methylation and mammalian development. Cell 99(3):247–257

Paalman MH, Henderson SL, Sollner-Webb B (1995) Stimulation of the mouse rRNA gene promoter by a distal spacer promoter. Mol Cell Biol 15(8):4648–4656

Panov KI, Friedrich JK, Russell J, Zomerdijk JC (2006) UBF activates RNA polymerase I transcription by stimulating promoter escape. EMBO J 25(14):3310–3322

Peng JC, Karpen GH (2007) H3K9 methylation and RNA interference regulate nucleolar organization and repeated DNA stability. Nat Cell Biol 9(1):25–35

Peng JC, Karpen GH (2008) Epigenetic regulation of heterochromatic DNA stability. Curr Opin Genet Dev 18(2):204–211

Percipalle P, Fomproix N, Cavellan E, Voit R, Reimer G, Kruger T, Thyberg J, Scheer U, Grummt I, Farrants AK (2006) The chromatin remodelling complex WSTF-SNF2h interacts with nuclear myosin 1 and has a role in RNA polymerase I transcription. EMBO Rep 7(5):525–530

Phillips JE, Corces VG (2009) CTCF: master weaver of the genome. Cell 137(7):1194–1211

Piacentini L, Fanti L, Berloco M, Perrini B, Pimpinelli S (2003) Heterochromatin protein 1 (HP1) is associated with induced gene expression in *Drosophila* euchromatin. J Cell Biol 161(4): 707–714

Pikaard CS (2000) The epigenetics of nucleolar dominance. Trends Genet 16(11):495–500

Pluta AF, Mackay AM, Ainsztein AM, Goldberg IG, Earnshaw WC (1995) The centromere: hub of chromosomal activities. Science 270(5242):1591–1594

Poortinga G, Hannan KM, Snelling H, Walkley CR, Jenkins A, Sharkey K, Wall M, Brandenburger Y, Palatsides M, Pearson RB, McArthur GA, Hannan RD (2004) MAD1 and c-MYC regulate UBF and rDNA transcription during granulocyte differentiation. EMBO J 23(16):3325–3335

Powell MA, Mutch DG, Rader JS, Herzog TJ, Huang TH, Goodfellow PJ (2002) Ribosomal DNA methylation in patients with endometrial carcinoma: an independent prognostic marker. Cancer 94(11):2941–2952

Prior CP, Cantor CR, Johnson EM, Littau VC, Allfrey VG (1983) Reversible changes in nucleosome structure and histone H3 accessibility in transcriptionally active and inactive states of rDNA chromatin. Cell 34(3):1033–1042

Putnam CD, Pikaard CS (1992) Cooperative binding of the *Xenopus* RNA polymerase I transcription factor xUBF to repetitive ribosomal gene enhancers. Mol Cell Biol 12(11):4970–4980

Puvion-Dutilleul F, Bachellerie JP (1979) Ribosomal transcriptional complexes in subnuclear fractions of Chinese hamster ovary cells after short-term actinomycin D treatment. J Ultrastruct Res 66(2):190–199

Qu GZ, Grundy PE, Narayan A, Ehrlich M (1999) Frequent hypomethylation in Wilms tumors of pericentromeric DNA in chromosomes 1 and 16. Cancer Genet Cytogenet 109(1):34–39

Rickards B, Flint SJ, Cole MD, LeRoy G (2007) Nucleolin is required for RNA polymerase I transcription in vivo. Mol Cell Biol 27(3):937–948

Rivin CJ, Cullis CA, Walbot V (1986) Evaluating quantitative variation in the genome of *Zea mays*. Genetics 113(4):1009–1019

Roussel P, Andre C, Comai L, Hernandez-Verdun D (1996) The rDNA transcription machinery is assembled during mitosis in active NORs and absent in inactive NORs. J Cell Biol 133(2): 235–246

Sanij E, Poortinga G, Sharkey K, Hung S, Holloway TP, Quin J, Robb E, Wong LH, Thomas WG, Stefanovsky V, Moss T, Rothblum L, Hannan KM, McArthur GA, Pearson RB, Hannan RD (2008) UBF levels determine the number of active ribosomal RNA genes in mammals. J Cell Biol 183(7):1259–1274

Santoro R (2005) The silence of the ribosomal RNA genes. Cell Mol Life Sci 62(18):2067–2079

Santoro R, De Lucia F (2005) Many players, one goal: how chromatin states are inherited during cell division. Biochem Cell Biol 83(3):332–343

Santoro R, Grummt I (2001) Molecular mechanisms mediating methylation-dependent silencing of ribosomal gene transcription. Mol Cell 8(3):719–725

Santoro R, Grummt I (2005) Epigenetic mechanism of rRNA gene silencing: temporal order of NoRC-mediated histone modification, chromatin remodeling, and DNA methylation. Mol Cell Biol 25(7):2539–2546

Santoro R, Li J, Grummt I (2002) The nucleolar remodeling complex NoRC mediates heterochromatin formation and silencing of ribosomal gene transcription. Nat Genet 32(3):393–396

Santoro R, Lienemann P, Fussenegger M (2009) Epigenetic engineering of ribosomal RNA genes enhances protein production. PLoS One 4(8):e6653

Santoro R, Schmitz KM, Sandoval J, Grummt I (2010) Intergenic transcripts originating from a subclass of ribosomal DNA repeats silence ribosomal RNA genes in trans. EMBO Rep 11(1):52–58

Schlesinger S, Selig S, Bergman Y, Cedar H (2009) Allelic inactivation of rDNA loci. Genes Dev 23(20):2437–2447

Schmid M, Durussel T, Laemmli UK (2004) ChIC and ChEC; genomic mapping of chromatin proteins. Mol Cell 16(1):147–157

Schmitz KM, Mayer C, Postepska A, Grummt I (2010) Interaction of noncoding RNA with the rDNA promoter mediates recruitment of DNMT3b and silencing of rRNA genes. Genes Dev 24(20):2264–2269

Schwartz BE, Ahmad K (2005) Transcriptional activation triggers deposition and removal of the histone variant H3.3. Genes Dev 19(7):804–814

Shiraishi M, Sekiguchi A, Chuu YH, Sekiya T (1999) Tight interaction between densely methylated DNA fragments and the methyl-CpG binding domain of the rat MeCP2 protein attached to a solid support. Biol Chem 380(9):1127–1131

Sogo JM, Ness PJ, Widmer RM, Parish RW, Koller T (1984) Psoralen-crosslinking of DNA as a probe for the structure of active nucleolar chromatin. J Mol Biol 178(4):897–919

Spada F, Haemmer A, Kuch D, Rothbauer U, Schermelleh L, Kremmer E, Carell T, Langst G, Leonhardt H (2007) DNMT1 but not its interaction with the replication machinery is required for maintenance of DNA methylation in human cells. J Cell Biol 176(5):565–571

Stancheva I, Lucchini R, Koller T, Sogo JM (1997) Chromatin structure and methylation of rat rRNA genes studied by formaldehyde fixation and psoralen cross-linking. Nucleic Acids Res 25(9):1727–1735

Stefanovsky V, Langlois F, Gagnon-Kugler T, Rothblum LI, Moss T (2006) Growth factor signaling regulates elongation of RNA polymerase I transcription in mammals via UBF phosphorylation and r-chromatin remodeling. Mol Cell 21(5):629–639

Strohner R, Nemeth A, Jansa P, Hofmann-Rohrer U, Santoro R, Langst G, Grummt I (2001) NoRC – a novel member of mammalian ISWI-containing chromatin remodeling machines. EMBO J 20(17): 4892–4900

Tachibana M, Sugimoto K, Nozaki M, Ueda J, Ohta T, Ohki M, Fukuda M, Takeda N, Niida H, Kato H, Shinkai Y (2002) G9a histone methyltransferase plays a dominant role in euchromatic histone H3 lysine 9 methylation and is essential for early embryogenesis. Genes Dev 16(14): 1779–1791

Tower J, Henderson SL, Dougherty KM, Wejksnora PJ, Sollner-Webb B (1989) An RNA polymerase I promoter located in the CHO and mouse ribosomal DNA spacers: functional analysis and factor and sequence requirements. Mol Cell Biol 9(4):1513–1525

Vakoc CR, Mandat SA, Olenchock BA, Blobel GA (2005) Histone H3 lysine 9 methylation and HP1gamma are associated with transcription elongation through mammalian chromatin. Mol Cell 19(3):381–391

van de Nobelen S, Rosa-Garrido M, Leers J, Heath H, Soochit W, Joosen L, Jonkers I, Demmers J, van der Reijden M, Torrano V, Grosveld F, Delgado MD, Renkawitz R, Galjart N, Sleutels F (2010) CTCF regulates the local epigenetic state of ribosomal DNA repeats. Epigenetics Chromatin 3(1):19

Wright JE, Mais C, Prieto JL, McStay B (2006) A role for upstream binding factor in organizing ribosomal gene chromatin. Biochem Soc Symp (73):77–84

Wu SC, Zhang Y (2010) Active DNA demethylation: many roads lead to Rome. Nat Rev Mol Cell Biol 11(9):607–620

Yuan X, Feng W, Imhof A, Grummt I, Zhou Y (2007) Activation of RNA polymerase I transcription by cockayne syndrome group B protein and histone methyltransferase G9a. Mol Cell 27(4):585–595

Zhang J, Xu F, Hashimshony T, Keshet I, Cedar H (2002) Establishment of transcriptional competence in early and late S phase. Nature 420(6912):198–202

Zhang LF, Huynh KD, Lee JT (2007) Perinucleolar targeting of the inactive X during S phase: evidence for a role in the maintenance of silencing. Cell 129(4):693–706

Zhou XC, Dowdy SC, Podratz KC, Jiang SW (2007) Epigenetic considerations for endometrial cancer prevention, diagnosis and treatment. Gynecol Oncol 107(1):143–153

Zhou Y, Santoro R, Grummt I (2002) The chromatin remodeling complex NoRC targets HDAC1 to the ribosomal gene promoter and represses RNA polymerase I transcription. EMBO J 21(17):4632–4640

Zhou Y, Schmitz KM, Mayer C, Yuan X, Akhtar A, Grummt I (2009) Reversible acetylation of the chromatin remodelling complex NoRC is required for non-coding RNA-dependent silencing. Nat Cell Biol 11(8):1010–1016

Chapter 5
UBF an Essential Player in Maintenance of Active NORs and Nucleolar Formation

Alice Grob, Christine Colleran, and Brian McStay

5.1 Introduction

As long ago as the 1930s, it was realized that the chromosomal loci that give rise to nucleoli have a specialized structure. This chromosomal feature was termed the secondary constriction, the primary constriction being the centromere. Heitz (1931) noted that the number of secondary constrictions observed was the same as the number of nucleoli that reformed at telophase. Then, in 1934 McClintock analyzed nucleolar formation in a strain of *Zea mays* in which the single secondary constriction was divided between two chromosomes as a result of a reciprocal translocation (McClintock 1934). In meiotic cells of this strain, two nucleoli were shown to reform instead of the usual one. Furthermore, nucleoli reformed at each of the chromosomal breakpoints. Thus, secondary constrictions were firmly linked with nucleolar formation and consequently more commonly referred to as nucleolar organizer regions, NORs. Some 30 years later, it was realized that NORs are composed of arrays of ribosomal genes (rDNA) that encode 18S, 28S, and 5.8S ribosomal RNAs (rRNAs) (Brown and Gurdon 1964; Ritossa and Spiegelman 1965; Birnstiel et al. 1966; Ritossa et al. 1966) and the nucleolus was established as the site of ribosome biogenesis (Edstrom et al. 1961; Perry and Errera 1961). More recently, it was realized that in organisms containing multiple NORs in their chromosomes, not all appeared as secondary constrictions or formed nucleoli (Fig. 5.1) (reviewed in McStay and Grummt 2008). A key player in organizing rDNA/NORs in vertebrates is upstream binding factor (UBF). Knockout experiments in the mouse revealed that it is an essential gene (Tom Moss 2010 , personal communication). This chapter focuses on how UBF defines active NORs and facilitates post-metaphase nucleolar reformation in mammals. We consider also the role of UBF in determining the proportion of rDNA repeats that are active in differing cellular contexts.

B. McStay (✉)
Centre for Chromosome Biology, School of Natural Sciences,
National University of Ireland Galway, Galway, Ireland
e-mail: brian.mcstay@nuigalway.ie

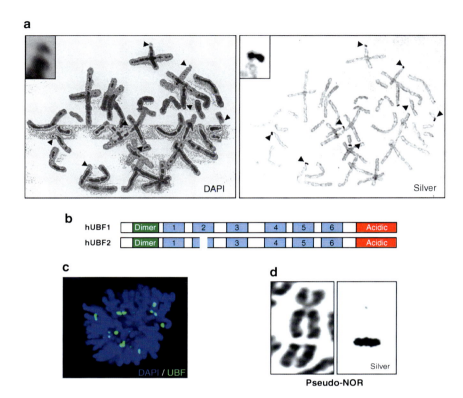

Fig. 5.1 Upstream binding factor (UBF) binding induces secondary constriction. (a) DAPI staining of metaphase chromosomes from human peripheral blood lymphocytes (*left*) reveals the presence of secondary constrictions corresponding to active nucleolar organizer regions (NORs) (*arrowheads*) that can be visualized with silverstaining (*right*). Note that in this spread 8/10 NORs are active. (b) Structural organization of hUBF splice variants: hUBF1 and hUBF2. Both variants contain an amino-terminal dimerization domain ("Dimer" box), six high mobility group (HMG) boxes ("1"–"6" boxes), and a carboxy-terminal acidic tail ("Acid" box). The second HMG box of hUBF2 is alternatively spliced resulting in a deletion of 37 amino acids. (c) Immunostaining of a human metaphase cell with anti-hUBF antibodies reveals mitotic NORs. (d) Insertions of ectopic arrays of UBF binding sites on non-NOR bearing human chromosomes results in artificial structures named pseudo-NORs that mimics endogenous NORs by inducing formation of a novel secondary constriction visualized by DAPI and silver staining (*left* and *right* respectively)

5.2 Chromosomal Organization of rDNA

Mammalian rDNA repeats are typically large, ~43 kb for human and ~45 kb in mice. Sequences encoding the precursor rRNA (pre-rRNA) (13–14 kb) are separated by intergenic spacers (IGS) of ~30 kb (Gonzalez and Sylvester 1995; Grozdanov et al. 2003; Sylvester et al. 2004). The mouse IGS, like that of *Xenopus*

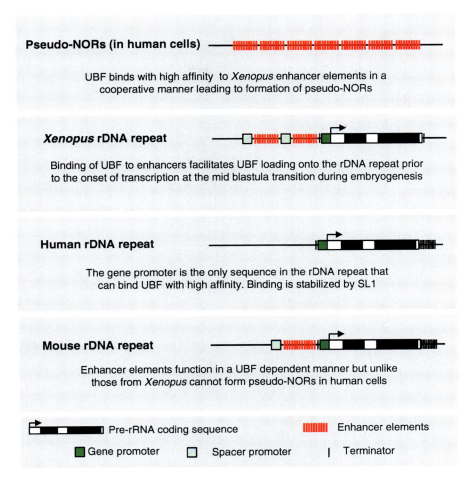

Fig. 5.2 Distribution of regulatory elements in vertebrate rDNA repeats. *Cartoons* show the distribution of promoter, spacer promoter, enhancer, and terminator elements in addition to the pre-rRNA coding sequences in pseudo-NORs, *Xenopus*, human, and mouse rDNA repeats

rDNA, contains a spacer promoter (Moss and Birnstiel 1979; Kuhn and Grummt 1987) and repetitive enhancer elements (Labhart and Reeder 1984; Pikaard et al. 1990) (Fig. 5.2). Curiously, neither spacer promoters nor enhancer elements have yet been identified in the human IGS. The remainder of the mammalian IGS is devoid of regulatory elements and composed of simple sequence repeats and transposable elements (Sylvester et al. 2004). The mammalian gene promoter is bipartite in structure, comprising a core element that spans the transcriptional start site and an upstream control element (UCE) positioned ~100 nucleotides further upstream (Haltiner et al. 1986; Learned et al. 1986). Multiple transcriptional terminators that serve as binding sites for transcription termination factor 1 (TTF1) are found downstream of the pre-rRNA coding sequence (Grummt et al. 1985). A single terminator element T_0 is also situated in a conserved position immediately upstream of the gene promoter (Grummt et al. 1986; Henderson and Sollner Webb 1986) (Fig. 5.2).

In the mouse, NORs are on chromosomes 12, 15, 16, 17, 18, and 19; apart from this, little is known about their organization (Dev et al. 1977). The human genome contains approximately 300 copies of the rDNA repeat (Schmickel 1973) distributed among five NORs located on the short arms of each of the five acrocentric chromosomes 13, 14, 15, 21, and 22 (Henderson et al. 1972). The repetitive nature of both rDNA and adjacent sequences on acrocentric short arms has precluded sequencing of mouse and human NORs. However, estimates of the sizes of NORs can be determined using pulse-field gel electrophoresis of genomic DNA digested with enzymes that do not cut human rDNA, such as EcoRV. Initial experiments with established cell lines revealed a major rDNA band of 3 Mb as well as several minor bands of 1 and 2 Mb (Sakai et al. 1995), implying that most human NORs are composed of ~70 copies of rDNA repeats. A more recent study (Stults et al. 2008) applied the same technique to peripheral blood lymphocytes from a panel of healthy human volunteers and found a striking variability in NOR size between and within individuals, ranging from 50 kb to more than 6 Mb. Furthermore, analysis of multigenerational human families revealed a high degree of meiotic rearrangement. NORs are also recombinational hotspots in human cancers (Stults et al. 2009). Further evidence of the plasticity of human rDNA repeat arrays has come from single-DNA-molecule analysis by molecular combing, which has revealed that human NORs comprise a mosaic of normal and rearranged rDNA repeats (Caburet et al. 2005). As many as one-third of rDNA repeats are rearranged or noncanonical and form what appear to be palindromic structures. The degree to which noncanonical rDNA repeats vary in proportion between individuals or between normal and disease states is at present unclear.

5.3 Active vs. Inactive NORs

During metaphase, active mammalian NORs, like their plant counterparts, can be visualized by the presence of secondary constrictions (Fig. 5.1a). However, a more reliable method for the detection and enumeration of active NORs came from the observation that they can be selectively visualized by staining metaphase chromosomes with silver nitrate (Goodpasture and Bloom 1975). Many studies have investigated the proportion of NORs that silver stains and by inference are active in human lymphocytes. Tabulated results from seven independent studies revealed that on average 8 out of 10 NORs are active in human lymphocytes (Heliot et al. 2000) (Fig. 5.1a). In the cancer cell line HeLa, 7 of the 10 NORs are active (Roussel et al. 1993).

Silverstaining of active NORs is due to argyophilic proteins that remain bound during metaphase. What are these proteins? Obvious candidates include components of the RNA polymerase I (Pol I) transcription machinery. In mammals, two DNA binding transcription factors UBF and a TBP centered complex, SL1 in humans or TIF-IB in mouse, form a stable preinitiation complex over Core and UCE elements of the promoter (see Grummt 2003; Russell and Zomerdijk 2005; Moss et al. 2007 for reviews). This pre-initiation complex recruits an active sub-fraction of Pol I. Pol I recruitment is facilitated by Rrn3/TIF-IA, a factor that contacts both SL1/TIF-IB and Pol I in a highly regulated manner. UBF, SL1, Pol I, and TTF1 have been demonstrated to remain associated with mitotic NORs (Scheer and Rose 1984; Roussel et al. 1993, 1996; Jordan et al. 1996; Sirri et al. 1999). More recent evidence has suggested that at most only a small fraction of Pol I, or a sub-set of its constituent polypeptides, remain at mitotic NORs (Leung et al. 2004). The other hallmark of active NORs is their appearance as secondary constrictions. Chromatin corresponding to active NORs is as much as tenfold less condensed than the chromatin on both their distal and proximal sides (Heliot et al. 1997). This under-condensation results in reduced dye binding when chromosomes are stained, giving rise to an apparent gap in the chromosome. Often, an axis or stalk of condensed DNA, possibly AT-rich, is found within the secondary constriction (Saitoh and Laemmli 1994). The identity of DNA sequences in this stalk is uncertain, but one proposal is that it comprises condensed/silent rDNA repeats whereas undercondensed/active repeats form lateral loops extending from this central axis (Suja et al. 1997). In contrast to active NORs, their silent counterparts do not form an evident secondary constriction, do not stain with silver, and do not have associated Pol I transcription machinery. Furthermore, during interphase silent NORs can be found disassociated from nucleoli and nucleolar proteins (McStay and Grummt 2008).

Biochemical analysis of rDNA in immortalized or cancer cell lines from a number of species reveals that approximately 50% of rDNA repeats are active at any given time (Conconi et al. 1989) (and reviewed in McStay and Grummt 2008). This fact in itself does not mean that mammals have an excess of rDNA over requirements as one could imagine that certain cell types in the context of adult animals or developing embryos might require a higher proportion or all of their rDNA repeats to be active. Robertsonian translocations are the most common form of translocation in humans, about 1/1,000 newborns (Therman et al. 1989; Shaffer and Lupski 2000). These translocations arise from the fusion of two acrocentric chromosomes and result in the loss of 2 out of the 10 NORs. As no phenotype is observed in people that carry this translocation, we can conclude that NORs are in excess of requirement. As the number of inactive NORs present in most cell types is not sufficient to account for the observed 50% of rDNA repeats being silent, it is likely that active NORs are themselves a mosaic of active and silent rDNA repeats. Furthermore, it is probable that the noncanonical rDNA repeats described above fall into this silent class. The role of the promoter proximal terminator, T_0, in establishing the activity status of individual rDNA repeats has been reviewed elsewhere (Grummt and Pikaard 2003; McStay and Grummt 2008) and is further discussed in Chap. 4.

5.4 UBF, Domain Structure, and DNA Binding

The Pol I transcription factor UBF is characterized by an amino-terminal (N-terminal) dimerization domain, multiple HMG (high mobility group) boxes, and a carboxy-terminal (C-terminal) acidic tail (Jantzen et al. 1990, 1992; McStay et al. 1991) (Fig. 5.1b). Mammalian UBF is visualized in western blots as a doublet of bands of comparable intensity corresponding to UBF1 and UBF2. As a consequence of alternative splicing, UBF2 lacks 37 amino acids from the second HMG box (O'Mahony and Rothblum 1991).

In solution, UBF forms dimers. Dimerization of UBF occurs through an N-terminal domain spread over 80 residues (McStay et al. 1991). As will be discussed below, this domain is highly conserved through evolution, suggesting that it may have additional roles in transcription and/or UBF binding to active NORs. UBF dimers bind rDNA via their HMG boxes, so called because of their similarity in sequence and structure to the DNA binding domains present in the nonhistone chromosomal architectural HMGB proteins. Two subclasses of HMG box proteins have been defined according to their DNA binding preferences; that is, sequence-specific transcription factors, such as mammalian testis-determining factor SRY (Werner et al. 1995) and factors such as HMGB proteins that bind DNA with little or no specificity, recognizing DNA structural features instead (Travers 2003). UBF appears to fall into this latter class. Among the multiple HMG boxes present in UBF, the first HMG box is sufficient for UBF binding to rDNA while other HMG boxes enhance this interaction (Jantzen et al. 1990, 1992; McStay et al. 1991; Reeder et al. 1995). In vitro experiments have failed to identify a consensus binding sequence other than a preference for binding GC-rich sequences (Copenhaver et al. 1994). This lack of sequence specificity is at odds with UBF's specific association with rDNA throughout the cell cycle (Roussel et al. 1993).

Like the HMG boxes present in HMGB proteins, UBF's HMG boxes adopt the characteristic twisted L-shape of three α-helices (Xu et al. 2002), which can introduce tight bends into DNA. In vitro, a dimer of *Xenopus* UBF (xUBF) can organize up to 180 bp of nucleosome free rDNA promoter into a 360° loop (Putnam et al. 1994; Bazett-Jones et al. 1994) establishing a structure termed the "enhancesome" (Stefanovsky et al. 2001a) that resembles the core nucleosome. In vivo, it is likely that UBF binds to a nucleosomal template (see below). In common with other HMG boxes proteins, UBF has also an affinity for structured DNA such as cruciforms (Reeder et al. 1995) and cisplatin-DNA adducts (Treiber et al. 1994). In addition, UBF can simultaneously bind two separate DNA molecules (Hu et al. 1994).

The C-terminal domain of UBF is composed of 57 acidic residues (glutamic and aspartic acids) and 23 serines residues. Phosphorylation of serine residues increases the overall negative charge of this essential UBF domain. This acidic domain is key for UBF's role in transcriptional activation (Jantzen et al. 1992; Voit et al. 1992) and is required for SL1 complex recruitment to the promoter via direct protein–protein interaction (Tuan et al. 1999). Additionally, this negatively-charged domain of UBF is likely to be a major contributor to the silverstaining of active NORs.

5.5 Extensive UBF Binding Defines Active NORs

UBF is a highly abundant transcription factor with up to 10^6 UBF molecules in human primary fibroblasts (Sullivan et al. 2001) and is highly concentrated together with rDNA repeats in the fibrillar centers (FC) of nucleoli during interphase. A detailed description of the ultrastructural organization of nucleoli, including the composition of the FC, is provided in Chap. 1. The apparent vast molar excess of UBF within the FC is incompatible with its binding being restricted to the rDNA promoter. It was therefore essential to determine the in vivo distribution of UBF on the rDNA repeat. Chromatin immunoprecipitation (ChIP) is a commonly used technique to address such a question. In standard ChIP, a soluble chromatin fraction is generated by sonication of nuclei prepared from cells treated with 1–2% formaldehyde. However, nucleoli are remarkably dense and resistant to sonication even without prior formaldehyde crosslinking (Muramatsu et al. 1963). Thus, standard ChIP protocols could potentially release an unrepresentative fraction of rDNA chromatin. A specific protocol for releasing nucleolar chromatin has been developed (O'Sullivan et al. 2002). In this altered protocol, nucleoli are isolated from cells treated with a lower percentage of formaldehyde (0.1–0.2%). Nucleolar chromatin is dispersed by addition of detergent and a soluble chromatin fraction is generated by sonication. This nucleolar ChIP revealed that UBF binds not only to regulatory sequences but across the entire rDNA repeat, including pre-rRNA coding sequences and the IGS (O'Sullivan et al. 2002). Extensive UBF binding across the repeat was observed in *Xenopus*, human and mouse cells. Lack of UBF association with satellite sequences also present on human acrocentric short arms highlights its restricted and specific binding to rDNA within nucleoli (O'Sullivan et al. 2002). This in vivo distribution of UBF, and its binding throughout the cell cycle, including mitosis (Fig. 5.1c), is persuasive evidence in support of a structural or architectural role for UBF.

The most compelling evidence in support of an architectural role for UBF comes from the generation of pseudo-NORs. Megabase arrays of a DNA sequence with high affinity for UBF were integrated into nonacrocentric chromosomes (Fig. 5.2). The sequence of choice, XEn (*Xenopus* Enhancer) elements, was derived from the IGS of the *Xenopus* rDNA repeat and consists of blocks of ten 60 or 81 bp repeats each containing 42 bp with significant homology to the gene promoter. In their natural context these elements function as transcriptional enhancers absolutely dependent on UBF binding (Pikaard et al. 1989; McStay et al. 1997). A feature of rDNA and the Pol I transcription machinery is their rapid evolutionary divergence as exemplified by the inability of the human Pol I transcription machinery to functionally interact with the mouse rDNA promoter and vice versa. Despite this evolutionary divergence, there appears to be greater constraints on UBF evolution. This point is elaborated on in the final section of this chapter. However, hUBF can bind to XEn elements and support enhancer function in vitro (McStay et al. 1997). Construction of XEn sequences arrays on non-NOR bearing human chromosomes clearly supports UBF's involvement in specifying NORs morphology (Mais et al.

2005). These XEn arrays ranged from 0.1 to 2 Mb, comparable with the size range of endogenous NORs. XEn arrays efficiently recruit hUBF to sites even outside the nucleolus and throughout cell cycle (Mais et al. 2005). Recruitment of UBF to these arrays results in formation of a chromatin structure apparently identical to true NORs with all their primary characteristics, that is, formation of a novel silver stainable secondary constriction lacking DAPI staining or Q-banding during mitosis (Mais et al. 2005) (Fig. 5.1d). In contrast to endogenous NORs, XEn arrays remain transcriptionally silent throughout the cell cycle as they lack functional promoters. Consequently, these XEn arrays were named pseudo-NORs. In support of a direct role of UBF binding in promoting undercondensation of active NORs, siRNA depletion of UBF leads to both loss of silver staining and secondary constrictions at pseudo-NORs (Prieto and McStay 2007, 2008). Pseudo-NORs leave us in no doubt that UBF plays a fundamental role in defining the morphology of active NORs.

5.6   UBF Loading onto the rDNA Repeat

The ability to generate pseudo-NORs that are distinct from nucleoli demonstrates that the in vivo localization of UBF can be driven solely by its affinity for rDNA and occur in the absence of promoters and transcription (Mais et al. 2005). UBF binds cooperatively and with high affinity to XEn elements in vitro (Putnam and Pikaard 1992). These characteristics of XEn elements may reflect the biology of early *Xenopus* development in which NORs are loaded with UBF during early stages of embryogenesis prior to the onset of transcription at the mid-blastula transition (Newport and Kirschner 1982a, b). Interestingly, UBF binding, and pseudo-NORs formation, is not observed on large arrays of the mouse enhancer repeat integrated into human chromosomes (McStay, unpublished observation), arguing that they have lower affinity for UBF in vivo than those from *Xenopus*. Furthermore no enhancer elements are found in an analogous position in the human rDNA repeat (Sylvester et al. 2004). In humans, the rDNA promoter may be the only DNA sequence element capable of directly recruiting UBF to the human rDNA repeat albeit with the help of SL1 (Fig. 5.2).

UBF binding in vivo is not restricted to regulatory elements such as enhancers and promoters but extends across the pre-rRNA coding region and into the IGS in all the organisms analyzed thus far (O'Sullivan et al. 2002). It is difficult to imagine that low UBF binding affinity sequences from across the mammalian rDNA repeat can effectively compete with the rest of the genome for UBF binding, raising the question of how this spreading is achieved. In a number of respects, this problem is reminiscent of the mysteries surrounding centromere formation. Centromeres are defined by the presence of nucleosomes containing the histone H3 variant CENP-A. However, specificity in CENP-A loading is not due to the underlying sequence of centromeres, as neo-centromeres can form at novel chromosomal sites that do not contain α satellite sequences. Nevertheless, once formed, neo-centromeres are stable structures. Likewise, extensive UBF binding endows on NORs the ability to have their activity status faithfully transmitted through cell division.

As alluded to above, the mechanism of UBF spreading from enhancers or promoters onto DNA sequences with intrinsic low affinity is not currently understood. One key question is whether or not UBF can bind to a nucleosomal template. In vitro evidence demonstrates that UBF can bind to DNA packaged as a nucleosome (Kermekchiev et al. 1997). Furthermore, DNA templates assembled into chromatin can be transcribed in vitro in a UBF-dependent manner (Langst et al. 1997). More recently, a SILAC based proteomic analysis has identified UBF as a nucleosome interactor (Bartke et al. 2010). The fact that rDNA chromatin exists in two states, active and silent, complicates the analysis of its histone composition in vivo. In mammalian cells, indirect methods such as DNA methylation status are used to infer the activity status of ChIP-ed rDNA sequences. Through such analyses, it is now established that a variety of histone modifications, particularly over regulatory elements and the transcribed region, distinguish active from inactive rDNA repeats (see Grummt and Pikaard 2003; McStay and Grummt 2008 for reviews). Thus, both extensive UBF binding and the presence of specific histone modifications correlate with active repeats. Micrococcal nuclease digestion of nuclei from pseudo-NOR containing cells reveals that XEn DNA is packaged as nucleosomes. As pseudo-NORs are uniformly and constitutively loaded with UBF, we can finally conclude that UBF binds a nucleosomal template in vivo.

H1 linker histones and the prototypical HMG box protein, HMGB1, bind to linker DNA in chromatin, in the vicinity of the nucleosome dyad. Binding appears to be mutually exclusive and the two proteins have opposing effects, with H1 stabilizing and HMGB1 destabilizing chromatin, respectively (Stros 2010). UBF and H1 binding also appear to be mutually exclusive. UBF can compete with H1 for binding to nucleosomes in vitro (Kermekchiev et al. 1997) and depletion of UBF in cells correlates with increased H1 binding to rDNA (see below). HMGB1 contains two HMG boxes and an acidic carboxy terminus that may loop back and modulate the DNA binding activity of HMG boxes. The fact that UBF contains four to six HMG boxes and binds as a dimer suggests that it would have greater potential to bind to DNA over the surface of the nucleosome. Remembering that UBF can bend free DNA molecules into a 360° loop (Bazett-Jones et al. 1994; Putnam et al. 1994), one can speculate that DNA on the surface of a nucleosome can act as a pre-bent template.

To understand how UBF recognizes rDNA chromatin with such remarkable specificity, a major goal will be a better understanding of how it binds to a nucleosomal template. Additionally, a description of how histone modifications and/or histone variants contribute to this specificity will be critical. The next question will be how these marks are established in the first place. Transcription could provide a mechanism for facilitating UBF loading along the coding region of the rDNA repeat. Direct interactions between Pol I and UBF and a role for UBF in transcriptional elongation may be relevant in this regard. It is also possible that UBF loading on high affinity sites such as enhancers or promoters can recruit factors that could facilitate UBF spreading to adjacent sequences in the absence of transcription. The spreading of heterochromatin has established a precedent for such a mechanism. This spreading involves a "self-sustaining" loop. Methylated H3-K9 histones bind to HP1, which in turn recruits more H3-K9 histone methyltransferase.

Table 5.1 List of direct UBF interactors with their respective functions

UBF interactor	Function	Experimental evidence	References
PAF53	Pol I subunit	In vitro interaction and coimmunoprecipitation	Hanada et al. (1996); Meraner et al. (2006)
PAF49	Pol I subunit	In vitro interaction	Panov et al. (2006)
TAF$_I$48	SL1 subunit	In vitro interaction	Beckmann et al. (1995)
TBP	SL1 subunit	In vitro interaction	Kwon and Green (1994)
Treacle	Interacts with box C/D snoRNPs	Yeast 2-hybrid and coimmunoprecipitation	Valdez et al. (2004)
SIRT7	Potential ADP-ribosyl transferase and deacetylase	In vitro interaction	Grob et al. (2009)
CTCF	Chromatin organization	In vitro interaction and copurification	van de Nobelen et al. (2010)

Interestingly UBF has already been shown to directly interact with a number of factors capable of modulating local chromatin structure, including Sirtuin 7 and CTCF (see Table 5.1).

5.7 UBF Loaded Chromatin, a Platform for Coordinating Ribosome Biogenesis

The vast majority of Pol I within cells is not engaged in rDNA transcription. Live-cell imaging experiments have provided evidence that only 7–10% of Pol I molecules are actively engaged in transcription (Dundr et al. 2002). In each cell, one can calculate that around 30,000 Pol I molecules are engaged in transcription at any given time, with ~300 active rDNA repeats (50%) each loaded with ~100 Pol I molecules (Miller and Bakken 1972; Puvion-Dutilleul 1983; Scheer and Benavente 1990). Quantitative western blots reveal that HeLa cells contain greater than 10^6 molecules of the RPA43 Pol I subunit (Wright and McStay, unpublished observation). However, the excess of unengaged Pol I still colocalizes with UBF and rDNA in the FC of nucleoli (Raska et al. 2006). Likewise, quantitative western blots also reveal that HeLa cells contain at least two orders of magnitude excess of SL1 over the number of active promoters (Wright and McStay, unpublished observation). Excess SL1 is also localized within the FC. The presence of such large pools of the Pol I transcription machinery within nucleoli is incompatible with a recruitment model solely based on PIC formation at promoters. How then is the Pol I transcription machinery recruited to the FC? Nucleolar ChIP experiments reveal that high levels of both Pol I and SL1 associate with the IGS in human nucleoli (Mais et al. 2005). This raises the probability that Pol I transcription machinery is recruited to the IGS via interaction with UBF. Strong support for this hypothesis comes from pseudo-NORs that appear during interphase as novel nuclear bodies (Mais et al.

2005), reminiscent in many ways of FCs (Raska et al. 2006; Prieto and McStay 2008). During interphase, pseudo-NORs not only recruit UBF but also sequester all the Pol I subunits analyzed so far (RPA43, RPA195, RPA135 and PAF53) together with SL1 subunits TAF_I110 and TBP (Mais et al. 2005). Pol I molecules associated with XEn arrays are likely to be transcriptionally competent as Rrn3/TIF-IA is also highly enriched at the pseudo-NORs. Thus, it appears that every component of Pol I transcription machinery is recruited to UBF-loaded chromatin such as pseudo-NORs. A list of direct interactions between UBF and components of the Pol I transcription machinery that may facilitate this localization is presented in Table 5.1. Importantly, siRNA experiments reveal that recruitment of Pol I transcription machinery to pseudo-NORs is strictly UBF dependent (Prieto and McStay 2007).

Terminal knobs observed by electron microscopy in Miller spreads have provided support for an intimate connection between transcription and early processing of pre-rRNA (Granneman and Baserga 2004). These structures also named SSU (Small Sub-Unit) processomes contain a large collection of factors implicated in 18S rRNA maturation, among which are the U3 snoRNP and the UTPs (U Three Proteins) (Dragon et al. 2002; Grandi et al. 2002; Krogan et al. 2004). Intriguingly, SSU processome components hUTP4, hUTP5, hUTP10, hUTP15, and hUTP17 are highly enriched at pseudo-NORs (Prieto and McStay 2007). These so-called tUTPs (transcription UTPs) are required for both rDNA transcription and SSU processome formation. Other processome components, including U3 snoRNA and Fibrillarin do not associate with pseudo-NORs. This observation together with IGS association of these tUTPs is consistent with their proposed role in coupling rDNA transcription and pre-rRNA maturation machineries (Gallagher et al. 2004). Pseudouridylation and $2'-O$ methylation of pre-rRNA carried out respectively by box H/ACA and box C/D snoRNPs are now also thought to occur cotranscriptionally (Warner and Kim 2010). Nucleolar phosphoproteins Nopp140 and TCOF1/Treacle interacting respectively with box H/ACA snoRNPs (Meier and Blobel 1994) and box C/D snoRNPs (Hayano et al. 2003) are targeted to pseudo-NORs, further suggesting that recruitment to rDNA chromatin provides a means to coordinate transcription and maturation of pre-rRNA. TCOF1/Treacle can directly associate with UBF (Table 5.1) while Nopp140 association is likely mediated by Pol I (Chen et al. 1999). Notably, siRNA experiments also reveal that tUTPs, TCFO1/Treacle, and Nopp140 recruitments to UBF loaded chromatin such as pseudo-NORs are strictly UBF dependent (Prieto and McStay 2007). The pseudo-NOR model has provided evidence that UBF loading on the IGS of endogenous rDNA repeats creates a platform for sequestrating high levels of factors participating in PIC formation and factors coupling pre-rRNA transcription and early processing.

5.8 UBF and Nucleolar Reformation

As mammalian cells enter mitosis, rDNA transcription is repressed and nucleoli disassemble (Sirri et al. 2008). At the exit from mitosis, inactivation of CDK1 following Cyclin B degradation results solely in resumption of Pol I transcription but not in pre-rRNA processing reactivation (Sirri et al. 2000). In order to avoid accumulation of unprocessed pre-rRNA, it is essential for these two processes to be reactivated in a highly coordinated manner. Transcription elongation on rDNA is linked to efficient rRNA processing and ribosome assembly (Schneider et al. 2007). One way of coordinating reactivation of these processes is for NORs to sequester components of both the transcription and the processing machineries prior to resumption of rDNA transcription. This is the case for UBF, SL1, and TCOF1/Treacle that remain associated with rDNA at NORs throughout mitosis (Roussel et al. 1993, 1996; Jordan et al. 1996; Valdez et al. 2004). Furthermore, transcriptionally silent pseudo-NORs have allowed identification of other factors implicated in coordination of ribosome biogenesis that are recruited to NORs independently of ongoing Pol I transcription, that is, tUTPs and Nopp140. The protein composition of pseudo-NORs closely resembles that of FCs, which are considered to be the interphase "counterparts" of mitotic NORs (Goessens 1984). Resumption of rDNA transcription after cell division is likely to be facilitated by rDNA undercondensation and retention of key factors at NORs during mitosis. Pseudo-NORs have highlighted the central role of UBF in maintaining this chromatin state. Thus, extensive binding of UBF over rDNA appears to impart a transcriptional memory so that NOR activity status can be faithfully transmitted through cell division. This important role for UBF makes it a prime target for regulation of ribosomal gene expression.

5.9 UBF and Regulation of Ribosome Biogenesis

The rate of rDNA transcription fluctuates throughout the cell cycle progression and in response to growth factors, stress, and differentiation, through regulation of key components of the Pol I machinery. rDNA transcription is modulated by long- and short-term regulation. NoRC and TTF1 are major targets in long-term regulation of rDNA transcription, which is achieved by regulating the number of active rDNA repeats. Rrn3/TIF-IA is an important target of short-term regulation, which controls the rate of transcription from each active repeat. Here, we focus on UBF, which is subject to short-term regulation by posttranslational modifications and long-term regulation by alterations in absolute levels.

During the cell cycle, UBF is modulated through a series of CDK-dependent phosphorylation (Voit et al. 1999; Voit and Grummt 2001). Acetylation is an additional modification that modulates UBF ability to regulate rDNA transcription during the cell cycle (Meraner et al. 2006). Availability of nutrients and growth factors has a direct effect on the rate of rDNA transcription. Epidermal growth factor (EGF) activation of MAPK/ERK pathway causes ERK1/2 dependent phosphorylation of UBF in the first two HMG boxes, resulting in up-regulation of rDNA transcription by enhancing transcriptional elongation of Pol I (Stefanovsky et al. 2001b). Phosphorylation of UBF HMG boxes 1 and 2 by ERK alters their affinity for linear DNA, thereby facilitating elongation of Pol I machinery (Stefanovsky et al. 2006a, b). The mTOR pathway stimulates transcription through S6-kinase dependent phosphorylation of UBF at the C-terminal tail, promoting UBF-SL1 interactions

(Hannan et al. 2003). Additionally insulin growth factor 1 (IGF1) stimulation causes insulin receptor substrate-1 (IRS-1) to bind to phosphoinositide 3-kinase (PI3K), which directly phosphorylates UBF, thus increasing rDNA transcription (Drakas et al. 2004). In addition to posttranslational modification, UBF levels alter through the cell cycle. Coincident with rDNA replication, UBF levels increase (Junera et al. 1997). This observation provides further support for UBF's structural role and its ability to impart a transcriptional memory.

Absolute levels of UBF are also regulated in response to long term changes in cellular demands as observed in differentiation or cancer. During hypertrophic growth of neonatal and adult cardiomyocytes, the levels of UBF increase, correlating with up-regulation of rDNA transcription (Brandenburger et al. 2003). In contrast, during differentiation the rate of rDNA transcription is down regulated, which is associated with a significant reduction in UBF expression. Reduction in UBF levels is a common characteristic observed in a number of differentiation models, including differentiation of L6 myoblasts to myotubes (Larson et al. 1993), F9 embryonal carcinoma cells to primitive endoderm cells (Datta et al. 1997; Alzuherri and White 1999), 3T3-L1 preadipocyte cells to adipocyctes (Li et al. 2006), and during murine granulocytic differentiation (Poortinga et al. 2004, 2011; Liu et al. 2007; Sanij et al. 2008). In the granuloctye differentiation model, reduction in UBF levels correlates with an increase in the number of silent rDNA repeats, suggesting that UBF levels modulate the proportion of active and inactivate rDNA repeats (Poortinga et al. 2004; Sanij et al. 2008). RNAi-mediated depletion of UBF in NIH3T3 cells confirm that reduced UBF levels are a cause of the reduction in the number of active repeats rather than a consequence (Poortinga et al. 2004; Sanij et al. 2008). Changes in the proportion of active and inactive rDNA repeats during differentiation contradict the notion that the number of silent rDNA repeats is fixed (Conconi et al. 1989; Stefanovsky and Moss 2006). In both the granulocyte differentiation model and the siRNA experiments, reduction in the number of active repeats is not accompanied by increase in DNA methylation of the gene promoter, consistent with NoRC independent silencing. ChIP analyses show increased linker Histone H1 association with previously active repeats, presumably resulting in condensation of the rDNA chromatin (Sanij et al. 2008). Analysis of a human cell line that contains an inducible UBF shRNA provides further support for UBF counteracting H1 mediated chromatin condensation. 3D-immuno FISH reveals that UBF depletion is accompanied by rDNA condensation with NORs initially moving to the periphery and eventually dissociating completely from the nucleoli (Colleran and McStay, unpublished observation).

5.10 UBF is Present Across Animal Phyla

Until recently it was thought that UBF was restricted to vertebrates. This view was fueled by the absence of a UBF-like protein in the genomes of model organisms including *Drosophila*, *Caenorhabditis elegans*, yeast, and *Arabidopsis*. While searches for UBF reveal the presence of many HMG box-containing proteins, no

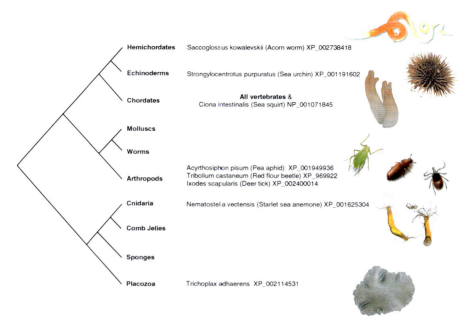

Fig. 5.3 Phylogenetic tree of UBF containing species. UBF is found throughout animal phyla. The names of species containing open reading frames (orfs) with significant homology to UBF (including a conserved amino-terminal dimerization domain, multiple HMG boxes, and an acidic tail) are shown in the appropriate position alongside a phylogenetic tree. The accession number associated with each UBF homolog is also shown

open reading frames (orfs) are identified that fulfill a more strict definition of UBF; that is, the presence of a conserved N-terminal dimerization domain, multiple HMG box motifs, and a C-terminal acidic domain. As the appearance of secondary constrictions at NORs is a widespread, if not a universal, characteristic of eukaryotic chromosomes, either UBF is more prevalent than previously thought or some other protein performs its role. It now seems likely that both are true.

As DNA sequence information becomes available for an increasingly wide variety of eukaryotic organisms it becomes clear that UBF is present in nonvertebrate animals (Fig. 5.3). UBF is found in other chordates such as *Ciona intestinalis*. *Ciona* and human UBFs share 40 and 54% sequence identities in their dimerization and first HMG boxes, respectively. Furthermore, as in humans, *Ciona* UBF contains multiple (at least five) HMG box motifs. More surprising, however, is the finding that UBF is present in a wide variety of insects; for example, dear ticks, pea aphids, and red flower beetles (Fig. 5.3). *Trichoplax adhaerans* is considered to represent a primitive metazoan and is arguably one of the simplest free-living animals, consisting of a layer of multinucleate fiber cells sandwiched between two epithelial layers (Srivastava et al. 2008). Its genome encodes an orf with considerable similarity to UBF. Comparison with *Ciona* UBF reveals 27% identity in their presumed dimerization domains.

5 UBF in Maintenance of Active NORs and Nucleolar Formation

This *Trichoplax* orf homolog also encodes multiple HMG boxes with the box adjacent to the dimerization domain having almost 30% identity with the equivalent domain in *Ciona* UBF. However, the third defining characteristic of UBF, an acidic C-terminal tail is missing. Gene loss appears to have been more extensive in model invertebrates such as *Drosophila* and *C. elegans* than previously assumed (Kortschak et al. 2003). Consequently, some genes formerly thought to be vertebrate inventions were present in more primitive metazoan ancestors. UBF appears to fall into this class.

Fungi and plants do not appear to have UBF. Nevertheless plants at least have prominent secondary constrictions at their NORs. Work in the yeast *Saccharomyces cerevisiae* has revealed that a more distantly related HMG box protein, HMO1, may perform a similar role to UBF in metazoans. HMO1 binds in vivo extensively across the rDNA repeat and its depletion impacts on growth (Gadal et al. 2002; Hall et al. 2006; Merz et al. 2008). Human UBF can partially rescue this growth defect (Olivier Gadal 2010, personal communication). Furthermore, yeast hmo1p targets to NORs throughout the cell cycle when it is introduced into human cells (Colleran and McStay, unpublished observation). These observations lead us to suggest that a conserved HMG protein is responsible for formation of secondary constrictions in the plant species where they were originally described some 80 years ago.

Acknowledgments We would like to thank Tom Moss and Olivier Gadal for communicating results prior to publication. Work in the McStay laboratory is funded by PI grant number 07/IN.1/B924 from Science Foundation Ireland. Alice Grob is funded by a postdoctoral fellowship from IRCSET.

References

Alzuherri HM, White RJ (1999) Regulation of RNA polymerase I transcription in response to F9 embryonal carcinoma stem cell differentiation. J Biol Chem 274:4328–4334

Beckmann H, Chen JL, O'Brien T, Tjian R (1995) Coactivator and promoter-selective properties of RNA polymerase I TAFs. Science 270:1506–1509

Bartke T, Vermeulen M, Xhemalce B, Robson SC, Mann M, Kouzarides T (2010) Nucleosome-interacting proteins regulated by DNA and histone methylation. Cell 143:470–484

Bazett-Jones DP, Leblanc B, Herfort M, Moss T (1994) Short-range DNA looping by the *Xenopus* HMG-box transcription factor, xUBF. Science 264:1134–1137

Birnstiel ML, Wallace H, Sirlin JL, Fischberg M (1966) Localization of the ribosomal DNA complements in the nucleolar organizer region of *Xenopus laevis*. Natl Cancer Inst Monogr 23:431–447

Brandenburger Y, Arthur JF, Woodcock EA, Du XJ, Gao XM, Autelitano DJ, Rothblum LI, Hannan RD (2003) Cardiac hypertrophy in vivo is associated with increased expression of the ribosomal gene transcription factor UBF. FEBS Lett 548:79–84

Brown DD, Gurdon JB (1964) Absence of ribosomal RNA synthesis in the anucleolate mutant of *Xenopus laevis*. Proc Natl Acad Sci USA 51:139–146

Caburet S, Conti C, Schurra C, Lebofsky R, Edelstein SJ, Bensimon A (2005) Human ribosomal RNA gene arrays display a broad range of palindromic structures. Genome Res 15:1079–1085

Chen HK, Pai CY, Huang JY, Yeh NH (1999) Human Nopp 140, which interacts with RNA polymerase I: implications for rRNA gene transcription and nucleolar structural organization. Mol Cell Biol 19:8536–8546

Conconi A, Widmer RM, Koller T, Sogo JM (1989) Two different chromatin structures coexist in ribosomal RNA genes throughout the cell cycle. Cell 57:753–761

Copenhaver GP, Putnam CD, Denton ML, Pikaard CS (1994) The RNA polymerase I transcription factor UBF is a sequence-tolerant HMG-box protein that can recognize structured nucleic acids. Nucleic Acids Res 22:2651–2657

Datta PK, Budhiraja S, Reichel RR, Jacob ST (1997) Regulation of ribosomal RNA gene transcription during retinoic acid-induced differentiation of mouse teratocarcinoma cells. Exp Cell Res 231:198–205

Dev VG, Tantravahi R, Miller DA, Miller OJ (1977) Nucleolus organizers in *Mus musculus* subspecies and in the RAG mouse cell line. Genetics 86:389–398

Dragon F, Gallagher JE, Compagnone-Post PA, Mitchell BM, Porwancher KA, Wehner KA, Wormsley S, Settlage RE, Shabanowitz J, Osheim Y et al (2002) A large nucleolar U3 ribonucleoprotein required for 18S ribosomal RNA biogenesis. Nature 417:967–970

Drakas R, Tu X, Baserga R (2004) Control of cell size through phosphorylation of upstream binding factor 1 by nuclear phosphatidylinositol 3-kinase. Proc Natl Acad Sci USA 101:9272–9276

Dundr M, Hoffmann-Rohrer U, Hu Q, Grummt I, Rothblum LI, Phair RD, Misteli T (2002) A kinetic framework for a mammalian RNA polymerase in vivo. Science 298:1623–1626

Edstrom JE, Grampp W, Schor N (1961) The intracellular distribution and heterogeneity of ribonucleic acid in starfish oocytes. J Biophys Biochem Cytol 11:549–557

Gadal O, Labarre S, Boschiero C, Thuriaux P (2002) Hmo1, an HMG-box protein, belongs to the yeast ribosomal DNA transcription system. EMBO J 21:5498–5507

Gallagher JE, Dunbar DA, Granneman S, Mitchell BM, Osheim Y, Beyer AL, Baserga SJ (2004) RNA polymerase I transcription and pre-rRNA processing are linked by specific SSU processome components. Genes Dev 18:2506–2517

Goessens G (1984) Nucleolar structure. Int Rev Cytol 87:107–158

Gonzalez IL, Sylvester JE (1995) Complete sequence of the 43-kb human ribosomal DNA repeat: analysis of the intergenic spacer. Genomics 27:320–328

Goodpasture C, Bloom SE (1975) Visualization of nucleolar organizer regions in mammalian chromosomes using silver staining. Chromosoma 53:37–50

Grandi P, Rybin V, Bassler J, Petfalski E, Strauss D, Marzioch M, Schafer T, Kuster B, Tschochner H, Tollervey D et al (2002) 90S Pre-ribosomes include the 35S pre-rRNA, the U3 snoRNP, and 40S subunit processing factors but predominantly lack 60S synthesis factors. Mol Cell 10:105–115

Granneman S, Baserga SJ (2004) Ribosome biogenesis: of knobs and RNA processing. Exp Cell Res 296:43–50

Grob A, Roussel P, Wright JE, McStay B, Hernandez-Verdun D, Sirri V (2009) Involvement of SIRT7 in resumption of rDNA transcription at the exit from mitosis. J Cell Sci 122:489–498

Grozdanov P, Georgiev O, Karagyozov L (2003) Complete sequence of the 45-kb mouse ribosomal DNA repeat: analysis of the intergenic spacer. Genomics 82:637–643

Grummt I (2003) Life on a planet of its own: regulation of RNA polymerase I transcription in the nucleolus. Genes Dev 17:1691–1702

Grummt I, Pikaard CS (2003) Epigenetic silencing of RNA polymerase I transcription. Nat Rev Mol Cell Biol 4:641–649

Grummt I, Maier U, Ohrlein A, Hassouna N, Bachellerie JP (1985) Transcription of mouse rDNA terminates downstream of the 3′ end of 28S RNA and involves interaction of factors with repeated sequences in the 3′ spacer. Cell 43:801–810

Grummt I, Kuhn A, Bartsch I, Rosenbauer H (1986) A transcription terminator located upstream of the mouse rDNA initiation site affects rRNA synthesis. Cell 47:901–911

Hall DB, Wade JT, Struhl K (2006) An HMG protein, Hmo1, associates with promoters of many ribosomal protein genes and throughout the rRNA gene locus in *Saccharomyces cerevisiae*. Mol Cell Biol 26:3672–3679

Haltiner MM, Smale ST, Tjian R (1986) Two distinct promoter elements in the human rRNA gene identified by linker scanning mutagenesis. Mol Cell Biol 6:227–235

5 UBF in Maintenance of Active NORs and Nucleolar Formation 99

Hanada K, Song CZ, Yamamoto K, Yano K, Maeda Y, Yamaguchi K, Muramatsu M (1996) RNA polymerase I associated factor 53 binds to the nucleolar transcription factor UBF and functions in specific rDNA transcription. Embo J 15:2217–2226

Hannan KM, Brandenburger Y, Jenkins A, Sharkey K, Cavanaugh A, Rothblum L, Moss T, Poortinga G, McArthur GA, Pearson RB et al (2003) mTOR-dependent regulation of ribosomal gene transcription requires S6K1 and is mediated by phosphorylation of the carboxy-terminal activation domain of the nucleolar transcription factor UBF. Mol Cell Biol 23: 8862–8877

Hayano T, Yanagida M, Yamauchi Y, Shinkawa T, Isobe T, Takahashi N (2003) Proteomic analysis of human Nop56p-associated pre-ribosomal ribonucleoprotein complexes. Possible link between Nop56p and the nucleolar protein treacle responsible for Treacher Collins syndrome. J Biol Chem 278:34309–34319

Heitz E (1931) Die ursache der gesetzmassigen zahl, lage, form und grosse pflanzlicher nukleolen. Planta 12:775–844

Heliot L, Kaplan H, Lucas L, Klein C, Beorchia A, Doco-Fenzy M, Menager M, Thiry M, O'Donohue MF, Ploton D (1997) Electron tomography of metaphase nucleolar organizer regions: evidence for a twisted-loop organization. Mol Biol Cell 8:2199–2216

Heliot L, Mongelard F, Klein C, O'Donohue MF, Chassery JM, Robert-Nicoud M, Usson Y (2000) Nonrandom distribution of metaphase AgNOR staining patterns on human acrocentric chromosomes. J Histochem Cytochem 48:13–20

Henderson S, Sollner Webb B (1986) A transcriptional terminator is a novel element of the promoter of the mouse ribosomal RNA gene. Cell 47:891–900

Henderson AS, Warburton D, Atwood KC (1972) Location of ribosomal DNA in the human chromosome complement. Proc Natl Acad Sci USA 69:3394–3398

Hu CH, McStay B, Jeong SW, Reeder RH (1994) xUBF, an RNA polymerase I transcription factor, binds crossover DNA with low sequence specificity. Mol Cell Biol 14:2871–2882

Jantzen HM, Admcn A, Bell SP, Tjian R (1990) Nucleolar transcription factor hUBF contains a DNA-binding motif with homology to HMG proteins. Nature 344:830–836

Jantzen HM, Chow AM, King DS, Tjian R (1992) Multiple domains of the RNA polymerase I activator hUBF interact with the TATA-binding protein complex hSL1 to mediate transcription. Genes Dev 6:1950–1963

Jordan P, Mannervik M, Tora L, Carmo-Fonseca M (1996) In vivo evidence that TATA-binding protein/SL1 colocalizes with UBF and RNA polymerase I when rRNA synthesis is either active or inactive. J Cell Biol 133:225–234

Junera HR, Masson C, Geraud G, Suja J, Hernandez-Verdun D (1997) Involvement of in situ conformation of ribosomal genes and selective distribution of upstream binding factor in rRNA transcription. Mol Biol Cell 8:145–156

Kermekchiev M, Workman JL, Pikaard CS (1997) Nucleosome binding by the polymerase I transactivator upstream binding factor displaces linker histone H1. Mol Cell Biol 17:5833–5842

Kortschak RD, Samuel G, Saint R, Miller DJ (2003) EST analysis of the cnidarian *Acropora millepora* reveals extensive gene loss and rapid sequence divergence in the model invertebrates. Curr Biol 13:2190–2195

Krogan NJ, Peng WT, Cagney G, Robinson MD, Haw R, Zhong G, Guo X, Zhang X, Canadien V, Richards DP et al (2004) High-definition macromolecular composition of yeast RNA-processing complexes. Mol Cell 13:225–239

Kuhn A, Grummt I (1987) A novel promoter in the mouse rDNA spacer is active in vivo and in vitro. EMBO J 6:3487–3492

Kwon H, Green MR (1994) The RNA polymerase I transcription factor, upstream binding factor, interacts directly with the TATA box-binding protein. J Biol Chem 269:30140–30146

Labhart P, Reeder RH (1984) Enhancer-like properties of the 60/81 bp elements in the ribosomal gene spacer of *Xenopus laevis*. Cell 37:285–289

Langst G, Blank TA, Becker PB, Grummt I (1997) RNA polymerase I transcription on nucleosomal templates: the transcription termination factor TTF-I induces chromatin remodeling and relieves transcriptional repression. EMBO J 16:760–768

Larson DE, Xie W, Glibetic M, O'Mahony D, Sells BH, Rothblum LI (1993) Coordinated decreases in rRNA gene transcription factors and rRNA synthesis during muscle cell differentiation. Proc Natl Acad Sci USA 90:7933–7936

Learned RM, Learned TK, Haltiner MM, Tjian RT (1986) Human rRNA transcription is modulated by the coordinate binding of two factors to an upstream control element. Cell 45:847–857

Leung AK, Gerlich D, Miller G, Lyon C, Lam YW, Lleres D, Daigle N, Zomerdijk J, Ellenberg J, Lamond AI (2004) Quantitative kinetic analysis of nucleolar breakdown and reassembly during mitosis in live human cells. J Cell Biol 166:787–800

Li J, Langst G, Grummt I (2006) NoRC-dependent nucleosome positioning silences rRNA genes. EMBO J 25:5735–5741

Liu M, Tu X, Ferrari-Amorotti G, Calabretta B, Baserga R (2007) Downregulation of the upstream binding factor1 by glycogen synthase kinase3beta in myeloid cells induced to differentiate. J Cell Biochem 100:1154–1169

Mais C, Wright JE, Prieto JL, Raggett SL, McStay B (2005) UBF-binding site arrays form pseudo-NORs and sequester the RNA polymerase I transcription machinery. Genes Dev 19:50–64

McClintock B (1934) The relationship of a particular chromosomal element to the development of the nucleoli in *Zea mays*. Zeit Zellforsch Mik Anat 21:294–328

McStay B, Grummt I (2008) The epigenetics of rRNA genes: from molecular to chromosome biology. Annu Rev Cell Dev Biol 24:131–157

McStay B, Frazier MW, Reeder RH (1991) xUBF contains a novel dimerization domain essential for RNA polymerase I transcription. Genes Dev 5:1957–1968

McStay B, Sullivan GJ, Cairns C (1997) The *Xenopus* RNA polymerase I transcription factor, UBF, has a role in transcriptional enhancement distinct from that at the promoter. EMBO J 16:396–405

Meier UT, Blobel G (1994) NAP57, a mammalian nucleolar protein with a putative homolog in yeast and bacteria. J Cell Biol 127:1505–1514

Meraner J, Lechner M, Loidl A, Goralik-Schramel M, Voit R, Grummt I, Loidl P (2006) Acetylation of UBF changes during the cell cycle and regulates the interaction of UBF with RNA polymerase I. Nucleic Acids Res 34:1798–1806

Merz K, Hondele M, Goetze H, Gmelch K, Stoeckl U, Griesenbeck J (2008) Actively transcribed rRNA genes in *S. cerevisiae* are organized in a specialized chromatin associated with the high-mobility group protein Hmo1 and are largely devoid of histone molecules. Genes Dev 22:1190–1204

Miller OL Jr, Bakken AH (1972) Morphological studies of transcription. Acta Endocrinol Suppl (Copenh) 168:155–177

Moss T, Birnstiel ML (1979) The putative promoter of a *Xenopus laevis* ribosomal gene is reduplicated. Nucleic Acids Res 6:3733–3743

Moss T, Langlois F, Gagnon-Kugler T, Stefanovsky V (2007) A housekeeper with power of attorney: the rRNA genes in ribosome biogenesis. Cell Mol Life Sci 64:29–49

Muramatsu M, Smetana K, Busch H (1963) Quantitative aspects of isolation of nucleoli of the Walker carcinosarcoma and liver of the rat. Cancer Res 23:510–522

Newport J, Kirschner M (1982a) A major developmental transition in early *Xenopus* embryos: I. characterization and timing of cellular changes at the midblastula stage. Cell 30:675–686

Newport J, Kirschner M (1982b) A major developmental transition in early *Xenopus* embryos: II. Control of the onset of transcription. Cell 30:687–696

O'Mahony DJ, Rothblum LI (1991) Identification of two forms of the RNA polymerase I transcription factor UBF. Proc Natl Acad Sci USA 88:3180–3184

O'Sullivan AC, Sullivan GJ, McStay B (2002) UBF binding in vivo is not restricted to regulatory sequences within the vertebrate ribosomal DNA repeat. Mol Cell Biol 22:657–668

Panov KI, Panova TB, Gadal O, Nishiyama K, Saito T, Russell J, Zomerdijk JC (2006) RNA polymerase I-specific subunit CAST/hPAF49 has a role in the activation of transcription by upstream binding factor. Mol Cell Biol 26:5436–5448

5 UBF in Maintenance of Active NORs and Nucleolar Formation

Perry RP, Errera M (1961) The role of the nucleolus in ribonucleic acid-and protein synthesis. I. Incorporation of cytidine into normal and nucleolar inactivated HeLa cells. Biochim Biophys Acta 49:47–57

Pikaard CS, McStay B, Schultz MC, Bell SP, Reeder RH (1989) The *Xenopus* ribosomal gene enhancers bind an essential polymerase I transcription factor, xUBF. Genes Dev 3:1779–1788

Pikaard CS, Pape LK, Henderson SL, Ryan K, Paalman MH, Lopata MA, Reeder RH, Sollner WB (1990) Enhancers for RNA polymerase I in mouse ribosomal DNA. Mol Cell Biol 10:4816–4825

Poortinga G, Hannan KM, Snelling H, Walkley CR, Jenkins A, Sharkey K, Wall M, Brandenburger Y, Palatsides M, Pearson RB et al (2004) MAD1 and c-MYC regulate UBF and rDNA transcription during granulocyte differentiation. EMBO J 23:3325–3335

Poortinga G, Wall M, Sanij E, Siwicki K, Ellul J, Brown D, Holloway TP, Hannan RD, McArthur GA (2011) c-MYC coordinately regulates ribosomal gene chromatin remodeling and Pol I availability during granulocyte differentiation. Nucleic Acids Res 39:3267–3281

Prieto JL, McStay B (2007) Recruitment of factors linking transcription and processing of pre-rRNA to NOR chromatin is UBF-dependent and occurs independent of transcription in human cells. Genes Dev 21:2041–2054

Prieto JL, McStay B (2008) Pseudo-NORs: a novel model for studying nucleoli. Biochim Biophys Acta 1783:2116–2123

Putnam CD, Pikaard CS (1992) Cooperative binding of the *Xenopus* RNA polymerase I transcription factor xUBF to repetitive ribosomal gene enhancers. Mol Cell Biol 12:4970–4980

Putnam CD, Copenhaver GP, Denton ML, Pikaard CS (1994) The RNA polymerase I transactivator upstream binding factor requires its dimerization domain and high-mobility-group (HMG) box 1 to bend, wrap, and positively supercoil enhancer DNA. Mol Cell Biol 14:6476–6488

Puvion-Dutilleul F (1983) Morphology of transcription at cellular and molecular levels. Int Rev Cytol 84:57–101

Raska I, Shaw PJ, Cmarko D (2006) Structure and function of the nucleolus in the spotlight. Curr Opin Cell Biol 18:325–334

Reeder RH, Pikaard CS, McStay B (1995) UBF, an architectural element for RNA polymerase I promoters. In: Eckstein F, Lilley DMJ (eds) Nucleic acids and molecular biology. Springer, Berlin, pp 251–263

Ritossa FM, Spiegelman S (1965) Localization of DNA complementary to ribosomal RNA in the nucleolus organizer region of *Drosophila melanogaster*. Proc Natl Acad Sci USA 53:737–745

Ritossa FM, Atwood KC, Lindsley DL, Spiegelman S (1966) On the chromosomal distribution of DNA complementary to ribosomal and soluble RNA. Natl Cancer Inst Monogr 23:449–472

Roussel P, Andre C, Masson C, Geraud G, Hernandez VD (1993) Localization of the RNA polymerase I transcription factor hUBF during the cell cycle. J Cell Sci 104:327–337

Roussel P, Andre C, Comai L, Hernandez-Verdun D (1996) The rDNA transcription machinery is assembled during mitosis in active NORs and absent in inactive NORs. J Cell Biol 133:235–246

Russell J, Zomerdijk JC (2005) RNA-polymerase-I-directed rDNA transcription, life and works. Trends Biochem Sci 30:87–96

Saitoh Y, Laemmli UK (1994) Metaphase chromosome structure: bands arise from a differential folding path of the highly AT-rich scaffold. Cell 76:609–622

Sakai K, Ohta T, Minoshima S, Kudoh J, Wang Y, de Jong PJ, Shimizu N (1995) Human ribosomal RNA gene cluster: identification of the proximal end containing a novel tandem repeat sequence. Genomics 26:521–526

Sanij E, Poortinga G, Sharkey K, Hung S, Holloway TP, Quin J, Robb E, Wong LH, Thomas WG, Stefanovsky V et al (2008) UBF levels determine the number of active ribosomal RNA genes in mammals. J Cell Biol 183:1259–1274

Scheer U, Benavente R (1990) Functional and dynamic aspects of the mammalian nucleolus. Bioessays 12:14–21

Scheer U, Rose KM (1984) Localization of RNA polymerase I in interphase cells and mitotic chromosomes by light and electron microscopic immunocytochemistry. Proc Natl Acad Sci USA 81:1431–1435

Schmickel RD (1973) Quantitation of human ribosomal DNA: hybridization of human DNA with ribosomal RNA for quantitation and fractionation. Pediatr Res 7:5–12

Schneider DA, Michel A, Sikes ML, Vu L, Dodd JA, Salgia S, Osheim YN, Beyer AL, Nomura M (2007) Transcription elongation by RNA polymerase I is linked to efficient rRNA processing and ribosome assembly. Mol Cell 26:217–229

Shaffer LG, Lupski JR (2000) Molecular mechanisms for constitutional chromosomal rearrangements in humans. Annu Rev Genet 34:297–329

Sirri V, Roussel P, Hernandez-Verdun D (1999) The mitotically phosphorylated form of the transcription termination factor TTF-1 is associated with the repressed rDNA transcription machinery. J Cell Sci 112:3259–3268

Sirri V, Roussel P, Hernandez-Verdun D (2000) In vivo release of mitotic silencing of ribosomal gene transcription does not give rise to precursor ribosomal RNA processing. J Cell Biol 148:259–270

Sirri V, Urcuqui-Inchima S, Roussel P, Hernandez-Verdun D (2008) Nucleolus: the fascinating nuclear body. Histochem Cell Biol 129:13–31

Srivastava M, Begovic E, Chapman J, Putnam NH, Hellsten U, Kawashima T, Kuo A, Mitros T, Salamov A, Carpenter ML et al (2008) The *Trichoplax* genome and the nature of placozoans. Nature 454:955–960

Stefanovsky V, Moss T (2006) Regulation of rRNA synthesis in human and mouse cells is not determined by changes in active gene count. Cell Cycle 5:735–739

Stefanovsky VY, Pelletier G, Bazett-Jones DP, Crane-Robinson C, Moss T (2001a) DNA looping in the RNA polymerase I enhancesome is the result of non-cooperative in-phase bending by two UBF molecules. Nucleic Acids Res 29:3241–3247

Stefanovsky VY, Pelletier G, Hannan R, Gagnon-Kugler T, Rothblum LI, Moss T (2001b) An immediate response of ribosomal transcription to growth factor stimulation in mammals is mediated by ERK phosphorylation of UBF. Mol Cell 8:1063–1073

Stefanovsky V, Langlois F, Gagnon-Kugler T, Rothblum LI, Moss T (2006a) Growth factor signaling regulates elongation of RNA polymerase I transcription in mammals via UBF phosphorylation and r-chromatin remodeling. Mol Cell 21:629–639

Stefanovsky VY, Langlois F, Bazett-Jones D, Pelletier G, Moss T (2006b) ERK modulates DNA bending and enhancesome structure by phosphorylating HMG1-boxes 1 and 2 of the RNA polymerase I transcription factor UBF. Biochemistry 45:3626–3634

Stros M (2010) HMGB proteins: interactions with DNA and chromatin. Biochim Biophys Acta 1799:101–113

Stults DM, Killen MW, Pierce HH, Pierce AJ (2008) Genomic architecture and inheritance of human ribosomal RNA gene clusters. Genome Res 18:13–18

Stults DM, Killen MW, Williamson EP, Hourigan JS, Vargas HD, Arnold SM, Moscow JA, Pierce AJ (2009) Human rRNA gene clusters are recombinational hotspots in cancer. Cancer Res 69: 9096–9104

Suja JA, Gebrane-Younes J, Geraud G, Hernandez-Verdun D (1997) Relative distribution of rDNA and proteins of the RNA polymerase I transcription machinery at chromosomal NORs. Chromosoma 105:459–469

Sullivan GJ, Bridger JM, Cuthbert AP, Newbold RF, Bickmore WA, McStay B (2001) Human acrocentric chromosomes with transcriptionally silent nucleolar organizer regions associate with nucleoli. EMBO J 20:2867–2874

Sylvester JE, Gonzales IL, Mougey EB (2004) Structure and organisation of vertebrate ribosomal DNA. In: Olson MO (ed) The nucleolus. Kluwer Academic/Plenum, New York, pp 58–72

Therman E, Susman B, Denniston C (1989) The nonrandom participation of human acrocentric chromosomes in Robertsonian translocations. Ann Hum Genet 53:49–65

Travers AA (2003) Priming the nucleosome: a role for HMGB proteins? EMBO Rep 4:131–136

5 UBF in Maintenance of Active NORs and Nucleolar Formation

Treiber DK, Zhai X, Jantzen HM, Essigmann JM (1994) Cisplatin-DNA adducts are molecular decoys for the ribosomal RNA transcription factor hUBF (human upstream binding factor). Proc Natl Acad Sci USA 91:5672–5676

Tuan JC, Zhai W, Comai L (1999) Recruitment of TATA-binding protein-TAFI complex SL1 to the human ribosomal DNA promoter is mediated by the carboxy-terminal activation domain of upstream binding factor (UBF) and is regulated by UBF phosphorylation. Mol Cell Biol 19:2872–2879

Valdez BC, Henning D, So RB, Dixon J, Dixon MJ (2004) The Treacher Collins syndrome (TCOF1) gene product is involved in ribosomal DNA gene transcription by interacting with upstream binding factor. Proc Natl Acad Sci USA 101:10709–10714

van de Nobelen S, Rosa-Garrido M, Leers J, Heath H, Soochit W, Joosen L, Jonkers I, Demmers J, van der Reijden M, Torrano V, Grosveld F, Delgado MD, Renkawitz R, Galjart N, Sleutels F (2010) CTCF regulates the local epigenetic state of ribosomal DNA repeats. Epigenetics Chromatin 3:19

Voit R, Grummt I (2001) Phosphorylation of UBF at serine 388 is required for interaction with RNA polymerase I and activation of rDNA transcription. Proc Natl Acad Sci USA 98:13631–13636

Voit R, Schnapp A, Kuhn A, Rosenbauer H, Hirschmann P, Stunnenberg HG, Grummt I (1992) The nucleolar transcription factor mUBF is phosphorylated by casein kinase II in the C-terminal hyperacidic tail which is essential for transactivation. EMBO J 11:2211–2218

Voit R, Hoffmann M, Grummt I (1999) Phosphorylation by G1-specific cdk-cyclin complexes activates the nucleolar transcription factor UBF. EMBO J 18:1891–1899

Warner JR, Kim HS (2010) The fast track is cotranscriptional. Mol Cell 37:745–746

Werner MH, Huth JR, Gronenborn AM, Clore GM (1995) Molecular basis of human 46X, Y sex reversal revealed from the three-dimensional solution structure of the human SRY-DNA complex. Cell 81:705–714

Xu Y, Yang W, Wu J, Shi Y (2002) Solution structure of the first HMG box domain in human upstream binding factor. Biochemistry 41:5415–5420

Part II
Role of the Nucleolus
in Ribosome Biogenesis

Chapter 6
The RNA Polymerase I Transcription Machinery

Renate Voit and Ingrid Grummt

6.1 Introduction

The synthesis of rRNA, the first event in ribosome biogenesis, essentially determines the cell's capacity to grow and proliferate. The genes that encode rRNA (rDNA) are efficiently transcribed by RNA polymerase I (Pol I) and rRNA synthesis is intricately regulated to be responsive to both general metabolism and specific environmental challenges. In fact, almost all signaling pathways that affect cell growth and proliferation directly regulate rRNA synthesis, their downstream effectors converging at the Pol I transcription machinery. These topics have been reviewed in the past, and readers are referred to some articles for further reading (Russell and Zomerdijk 2005; Moss et al. 2007; Grummt 2010).

Vertebrate cells contain several hundred copies of tandemly repeated rRNA genes per haploid genome, ranging from fewer than 100 to more than 10,000. Mammalian rDNA clusters are alternating modules of an intergenic spacer of approximately 30 kb and a pre-rRNA coding region of approximately 14 kb. In higher vertebrates, each rRNA gene encodes a precursor transcript (47 S pre-rRNA) that is either co- or posttranscriptionally processed and modified by snoRNPs (small nucleolar ribonucleoproteins) to generate one molecule each of 18 S, 5.8 S, and 28 S rRNA, the backbone of the ribosome. Each unit also contains important *cis*-acting sequence elements that regulate pre-rRNA synthesis, such as the rDNA promoter, enhancers, spacer promoters, and several transcription terminators (Fig. 6.1). RNA polymerase I is unique in that in most eukaryotes its sole function is the transcription of genes encoding rRNAs. Like Pol II and Pol III, transcription by Pol I requires auxiliary factors that mediate promoter recognition, promote transcription elongation, and facilitate transcription termination.

I. Grummt (✉)
Division of Molecular Biology of the Cell II, German Cancer Research Center,
DKFZ-ZMBH Alliance 69120, Heidelberg, Germany
e-mail: i.grummt@dkfz.de

M.O.J. Olson (ed.), *The Nucleolus*, Protein Reviews 15,
DOI 10.1007/978-1-4614-0514-6_6, © Springer Science+Business Media, LLC 2011

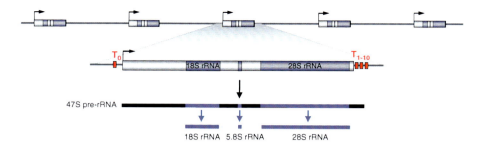

Fig. 6.1 Structural organization of mammalian rDNA. The diagram at the *top* depicts the "head-to tail" arrangement of tandem rDNA repeats showing the pre-rRNA coding regions that are separated by long intergenic spacer sequences. The *arrows* mark the Pol I transcription initiation site. A single rDNA transcription unit comprising transcribed intragenic spacer sequences (*gray*) and regions encoding 18 S rRNA, 5.8 S rRNA, and 28 rRNA (*blue*) is shown below. In subsequent cleaving reactions, the primary transcript (47 S pre-rRNA) is processed via distinct intermediates into mature ribosomal RNAs. The *red boxes* represent terminator elements that are located downstream of the transcription unit (T_1-T_{10}) and upstream of the rDNA promoter (T_0 at position −170)

6.2 Components of the Pol I Transcription Machinery

6.2.1 Structure and Function of RNA Polymerase I

Pol I is the most complex protein of the preinitiation complex, comprising a 10-subunit catalytic core and four associated subunits. The structure of yeast Pol I has been resolved showing specific structural features that are conserved among class I, II, and III DNA-dependent RNA polymerases (Kuhn et al. 2007; Cremer et al. 2008). Shared and homologous core subunits are involved in basic functions of RNA polymerases, including transcription start site selection, promoter melting, nucleotide binding, initiation, elongation, and termination (Table 6.1). Sequence alignment and structural analysis of the two largest subunits of nuclear RNA polymerases revealed homology to the β´ and β subunit of bacterial RNA polymerases, suggesting that the two large subunits are functionally equivalent. The structure of mammalian Pol I is similar to that of yeast, consisting of the 10 subunit catalytic core, and a peripheral heterodimeric subcomplex comprising A14/43, and a Pol I-specific A49/34.5 dimer. A14/43 interacts with the basal transcription initiation factor TIF-IA/Rrn3, thus mediating transcription initiation complex formation. The A49/34.5 dimer contacts DNA, mediates Pol I-intrinsic RNA cleavage, and stimulates processivity of Pol I transcription (Geiger et al. 2010). Thus, Pol I-specific subunits create a surface that facilitates interactions with basal transcription factors and regulatory proteins.

Mammalian Pol I exists in two distinct forms, Pol Iα and Pol Iβ, both of which are catalytically active; however, only Pol Iβ can assemble into productive transcription initiation complexes (Miller et al. 2001). At least two Pol I-specific subunits, PAF53 and PAF49, serve important functions in transcription regulation. PAF53 interacts directly with UBF (Hanada et al. 1996) and this interaction is modulated by the phosphorylation and acetylation state of UBF (Voit and Grummt 2001; Meraner et al. 2006). hPAF49, the human ortholog of the yeast subunit A34.4 and originally

6 The RNA Polymerase I Transcription Machinery

Table 6.1 RNA polymerase I: subunit composition and conservation

		Yeast	Human	
	Pol I subunits	Unique/shared subunits in Pols	Homolog in Pol II	Pol I subunits
Core subunits	RPA190 (A190)	I	Rpb1	hRPA190
	RPA135 (A135)	I	Rpb2	hRPA135
	RPA40 (AC40)	I, III	Rpb3	hRPA40 (AC40/hRPA5)
	Rbp5 (ABC27)	I, II, III	Rbp5	hRbp5
	Rbp6 (ABC23)	I, II, III	Rpb6	hRpb6
	RPA19 (AC19)	I, III	Rpb11	hRPA19 (AC19)
	Rbp8 (ABC14.5)	I, II, III	Rpb8	hRbp8
	RPA12 (A12.2)	I	Rpb9	hRPA12.2
	Rpb10 (ABC10a)	I, II, III	Rbp10	hRbp10
	Rpb12 (ABC10ß)	I, II, III	Rpb12	hRpb12
Rpb4/7-like	RPA14 (A14)	I	Rpb4	hRpb4
	RPA43 (A43)	I	Rpb7	hRPA43
TFIIF-like	RPA49 (A49)	I	(TFIIF/Rap74)	hRPA49 (hPAF53)
	RPA34 (A34.5)	I	(TFIIF/Rap30)	CAST (ASE1/hPAF49/ hRPA34.5)

termed ASE-1 (antisense to ERCC1) and CAST (CD3-associated signal transducer), interacts specifically with UBF and with the TAF_I48 subunit of SL1/TIF-IB, (Panov et al. 2006b; Yamamoto et al. 2004). Though hPAF49 associates with both Pol Iα and Pol Iβ, the initiation-competent Pol Iβ contains hPAF49 phosphorylated at Y82, suggesting a stimulatory role of tyrosine kinases on Pol I transcription.

Apart from these specific subunits, numerous proteins have been identified that are associated with Pol I, including the growth-dependent transcription initiation factor TIF-IA/Rrn3, protein kinase CK2, the chromatin modifiers PCAF, G9a, WSTF and SNF2h, nuclear actin, and myosin (NM1), as well as proteins involved in DNA repair and replication, such as topoisomerases I and IIα, Ku70/80, PCNA, TFIIH, and CSB, which were shown to be associated with Pol Iβ. These findings are compatible with a mechanism by which Pol I is recruited to the rDNA promoter as a multiprotein complex that acts as a scaffold to coordinate rRNA synthesis and maturation as well as chromatin modification and DNA repair.

6.2.2 Basal Factors Required for Transcription Initiation

Transcription initiation is a stepwise process that begins with the recruitment and assembly of Pol I and other transcription factors into a specific multi-protein pre-initiation

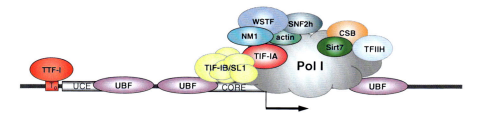

Fig. 6.2 Numerous proteins assemble at the rDNA promoter to form a productive transcription initiation complex. The ellipsoids show the factors that are associated with the rDNA promoter and Pol I, respectively, and are required for efficient pre-rRNA synthesis. The structural organization of the murine rDNA promoter comprising the core promoter (CORE), the upstream control element (UCE), and the upstream terminator T_0 is indicated (see text for details)

complex (PIC) at the rDNA promoter (Fig. 6.2). With only a few exceptions, rDNA promoters share a common modular organization, consisting of a transcription start site proximal core promoter and an upstream control element (UCE). The stereo-specific alignment and orientation of both sequence elements is crucial for efficient transcription initiation. Analysis of structural parameters of ribosomal gene promoters from human to lower plants revealed that conservation of specific structural features, rather than nucleotide sequence, is fundamental for promoter function (Marilley and Pasero 1996). Apparently, a structural code, in addition to primary sequence, directs specific DNA–protein interactions at the rDNA promoter and may serve an important function in transcriptional control.

In mammals, the preinitiation complex is assembled at the rDNA promoter by the synergistic action of two DNA binding Pol I-specific factors, the upstream binding factor UBF (Jantzen et al. 1990) and the promoter selectivity factor, termed SL1 in humans and TIF-IB in mice (Learned et al. 1985; Clos et al. 1986). UBF is an abundant nucleolar protein that contains several high mobility group (HMG) domains involved in DNA binding. UBF activates rDNA transcription by several means, for example, by stabilizing binding of TIF-IB/SL1 and Pol I at the rDNA promoter, and by displacing nonspecific DNA binding proteins, such as histone H1 (Kuhn and Grummt 1992; Kuhn et al. 1993). Additional roles have been ascribed to UBF, including promoter escape and transcription elongation (Panov et al. 2006a; Stefanovsky et al. 2006a). Recent data suggest that UBF is also involved in determining the number of active rRNA genes (Sanij et al. 2008). A comprehensive review on the role of UBF in nucleolus formation and maintenance of active NORs is provided in Part 1, Chap. 5.

Promoter specificity is brought about by SL1/TIF-IB, a multiprotein complex that binds to the core promoter and nucleates transcription complex assembly. SL1/TIF-IB comprises the TATA-box binding protein (TBP) and five TBP-associated factors (TAF$_I$s), including TAF$_I$110/95, TAF$_I$68, TAF$_I$48, TAF$_I$41, and TAF$_I$12 (Comai et al. 1992; Zomerdijk et al. 1994; Heix et al. 1997; Denissov et al. 2007; Gorski et al. 2007). The TAF$_I$ subunits perform important tasks in transcription

6 The RNA Polymerase I Transcription Machinery

complex assembly, mediating specific interactions between the rDNA promoter and Pol I, thereby recruiting Pol I – together with the essential transcription initiation factor TIF-IA and a collection of Pol I-associated factors – to rDNA.

On assembly of a productive transcription initiation complex, the promoter is opened and the first internucleotide bond is formed. Interestingly, promoter clearance and escape of Pol I from the promoter is the rate-limiting step in rDNA transcription. UBF has been shown to exert its stimulatory effect on RNA synthesis after PIC formation, promoter opening, and formation of the first phosphodiester bond, demonstrating that this basal transcription factor activates Pol I transcription during transition between initiation and elongation, that is, at promoter escape and clearance by Pol I (Panov et al. 2006a). This mechanism enables UBF to activate transcription after PIC assembly both from previously inactive promoters and from SL1-engaged promoters at each round of transcription.

6.3 The RNA Polymerase I-Dependent Transcription Cycle

6.3.1 *Dynamics of Transcription Complex Assembly*

The classical view of transcription initiation complex formation is that of an ordered stepwise assembly of multiple proteins on the promoter via specific protein–protein interactions or, alternatively, binding of a preassembled ready-to-use multiprotein complex, often termed "factory." This view has been challenged by a study that analyzed the kinetics of assembly and elongation of mammalian Pol I by the FRAP (*f*luorescent *r*ecovery *a*fter *p*hotobleaching) technique. This method makes use of green fluorescent protein (GFP)-tagged proteins that permits their observation in living cells. The data revealed that the Pol I transcription machinery is a highly dynamic complex that assembles in a stochastic manner from freely diffusible subunits. Each of the components is steadily and rapidly exchanged between the nucleoplasm and the nucleolus, indicating that Pol I subunits enter the nucleolus as distinct subunits rather than as a preassembled complex. Thus, the assembly of Pol I into a functional initiation complex appears to proceed in a sequential manner via metastable intermediates, which increasingly stabilize as more subunits are added (Dundr et al. 2002). A "hit-and-run" mechanism was proposed, in which transcriptional factors quickly exchange between individual rDNA promoters. Calculations of the FRAP data indicated that transcription initiation occurs on average every ~1.4 s, Pol I subunits reside in the pool for ~9 to ~37 s, and the residence time of elongating Pol I is 2–3 min. Although one can question whether imaging and mathematical models can provide such an unambiguous picture of assembly, the estimated numbers are indeed very similar to those obtained by French et al. (2003) who calculated the elongation rate of yeast Pol I directly from the rate of rRNA synthesis and the number of Pol I molecules per rRNA gene.

6.3.2   Elongation of Pol I Transcription

In contrast to transcription initiation and termination, the process of Pol I transcription elongation is poorly understood. After formation of the first few internucleotide bonds, Pol I must contend with nucleoprotein complexes in chromatin that may impede elongation of the nascent transcripts. Recent studies have shown that proteins with chromatin-remodeling activities, such as nucleolin, nucleophosmin (B23), and FACT (facilitates chromatin transcription), promote Pol I transcription elongation (Rickards et al. 2007; Murano et al. 2008; Birch et al. 2009). Consistent with the finding that transcription elongation by Pol I on nucleosomal templates requires structural rearrangements that are mediated by the histone chaperone FACT, subunits of FACT are associated with mammalian Pol I and the transcribed part of rDNA. Ablation of FACT by RNAi significantly reduces pre-rRNA levels without affecting the synthesis of the first 40 nucleotides of pre-rRNA, underscoring the role of FACT in transcription elongation (Birch et al. 2009).

6.3.3   Termination of Pol I Transcription

Termination is a multistep process involving Pol I pausing, release of both pre-rRNA, and Pol I and 3'-end processing of the primary transcript. Studies on Pol I transcription termination in mouse, rat, humans, frog, and yeast have revealed that the mechanism of Pol I transcription termination has been conserved during evolution. All characterized Pol I terminators are orientation sensitive; that is, reversal of the terminator elements relative to the direction of transcription prevents termination. Terminators are recognized by a sequence-specific DNA-binding protein that contacts the elongating RNA polymerase and mediates the termination reaction. In mouse, 10 terminator elements, termed "Sal boxes" because they contain a recognition site for the endonuclease $SalI$, are clustered over several hundred base pairs downstream of the pre-rRNA coding region and are flanked by long pyrimidine stretches, not uncommon for a eukaryotic transcription terminator (Grummt et al. 1985). A similar terminator element, defined as T_0, is located immediately upstream of the rDNA promoter (Grummt et al. 1986a). Mutational analysis and footprinting experiments on human and mouse terminators have shown that a nucleolar factor, designated TTF-I (for transcription termination factor) binds to the "Sal box" elements and stops the elongating RNA polymerase I (Grummt et al. 1986b). Alterations in the "Sal box" that reduce binding of TTF-I also impair transcription termination. The cDNAs for murine and human TTF-I have been cloned and deletion analysis has revealed functionally distinct domains of the protein (Evers et al. 1995; Evers and Grummt 1995). Interestingly, the DNA binding activity of recombinant TTF-I is masked in the intact protein. Removal of the N-terminal part of TTF-I greatly augments DNA binding, indicating that the N-terminus of TTF-I inhibits DNA binding via intermolecular protein–protein interactions. Consistent with this idea, the N-terminal 184 amino acids of TTF-I can form stable oligomers in solution and repress DNA binding when fused to a heterologous DNA binding domain (Smid et al. 1992; Sander et al. 1996).

TTF-I is a multifunctional protein that mediates transcription termination and replication fork arrest (Gerber et al. 1997; Grummt et al. 1986b). Termination probably involves pausing of the elongation complex coupled with release of both the transcript and Pol I from the template. Though TTF-I bound to the Sal box element is sufficient for the arrest of elongating Pol I, complete termination, that is, release of Pol I and nascent RNA, requires the presence of 5'-flanking sequences and a TTF-I-associated factor, termed PTRF (Polymerase and Transcript Release Factor), that binds to transcripts containing the 3' end of pre-rRNA and is capable of dissociating ternary transcription complexes (Jansa et al. 1998).

On the basis of the properties of the termination factor, the features of the DNA element, and the requirement for accessory factors, the following model for Pol I transcription termination emerges. Specific binding of TTF-I to its target sequence leads to bending the DNA. The approaching Pol I recognizes this DNA structure, contacts TTF-I and pauses upstream of the "Sal box" terminator. By analogy to *E. coli* RNA polymerase, which changes its conformation at Rho-independent terminators, Pol I may undergo conformational changes either prior to or after it has been paused by TTF-I. This conformational change, in turn, could supply the energy required for dissociation of the elongation complex by PTRF.

6.3.4 The Upstream Terminator T_0 is an Essential Promoter Element

Murine and human rDNA promoters are flanked at their 5' ends by a sequence motif, termed T_0, which is almost identical to the terminator elements T_{1-10} downstream of the rRNA coding sequence. The finding that a binding site for a Pol I transcription terminator protein is located adjacent to the gene promoter suggested that TTF-I may also exert some essential function in transcription initiation. Indeed, binding of TTF-I (or the frog homolog Rib2) to the promoter-proximal terminator stimulated Pol I transcription in vivo (Henderson and Sollner-Webb 1986; McStay and Reeder 1990). Subsequent in vitro studies showed that TTF-I binding to the upstream terminator triggered structural alterations of the chromatin on preassembled nucleosomal templates, and these changes in chromatin structure correlated with activation of Pol I transcription in vitro (Längst et al. 1997, 1998). These results suggested that TTF-I recruits chromatin remodeling activities to rDNA that modify the promoter-bound nucleosome, thereby facilitating or preventing the access of transcription factors and Pol I.

The occurrence of the same binding site for the transcription factor TTF-I upstream and downstream of rRNA genes raises the possibility that TTF-I can interact with both sequences simultaneously, thus bringing the terminator in the vicinity of the gene promoter by looping out the pre-rRNA coding sequence. The "ribomotor model" proposed by Planta and colleagues (Kulkens et al. 1992) implies that interaction between the upstream and downstream terminators of the same or adjacent transcription units can be juxtaposed thereby allowing the Pol I molecules, having terminated at the downstream terminator, to be transferred directly from the 3' end of the gene to the promoter of the adjacent rDNA unit without entering the free pool. This model is supported by the observation that TTF-I oligomerizes in vitro and is capable to link two DNA fragments in *trans* (Sander and Grummt 1997). Moreover, a chromosome conformation capture (3 C) method has been applied to provide evidence that the 5'- and 3'-terminal parts of active rRNA genes are in close spatial proximity (Németh et al. 2008). Apparently, looping out of the transcribed region is crucial in establishing an open chromatin domain and activating transcription. Pol I is known to exist in large macromolecular machines, termed "factories," that interact with DNA within the structural contexts imposed by both chromatin and higher-order nuclear organization. Whether the interaction between the upstream and downstream terminators is mediated exclusively by TTF-I, or whether it involves additional proteins that may anchor the rDNA to the nucle(ol)ar matrix in a highly ordered, linear manner is not known. What emerges is an increasingly complex view of how the multifarious functions of the nucleus are embedded in a dynamic and complex nuclear architecture.

6.4 Nuclear Actin and Myosin Promote Pol I Transcription

Several studies have demonstrated that the traditionally "cytoplasmic" actin has important functions within the nucleus, and is involved in diverse processes, such as chromatin remodeling, transcription, RNA processing, and nuclear export (for reviews, see Bettinger et al. 2004; Grummt 2006). Given that actin usually works in conjunction with myosin motor proteins, it is not surprising that nuclei contain also a specific isoform of myosin I, termed NM1, a monomeric, single-headed myosin that possesses a unique 16-amino acid N-terminal extension required for nuclear localization. The finding that nuclei contain both actin and myosin (Nowak et al. 1997; Pestic-Dragovich et al. 2000), along with the observation that both actin and NM1 co-localize at sites of active transcription and are associated with RNA polymerases (Fomproix and Percipalle 2004; Kysela et al. 2005), suggests a functional link between nuclear actin, NM1, and transcriptional activity (Fig. 6.3a). Indeed, depletion or inhibition of actin or NM1 decreased transcription *in vivo* and *in vitro*, indicating that both proteins serve important functions in the transcription process (Fig. 6.3b). The association of actin and myosin with rDNA and the Pol I transcription apparatus requires the motor function of NM1. Mutants that are deficient in either ATPase activity or actin binding do not interact with Pol I and their association with rDNA is severely impaired (Ye et al. 2008). Significantly, the association of actin and NM1 with Pol I was abolished in the presence of ATP and stabilized by ADP, an observation that implicates that actin and myosin function by means of the same mechanism in both the nucleus and cytoplasm, supporting the view that nuclear actomyosin complexes act as molecular motors that facilitates transcription.

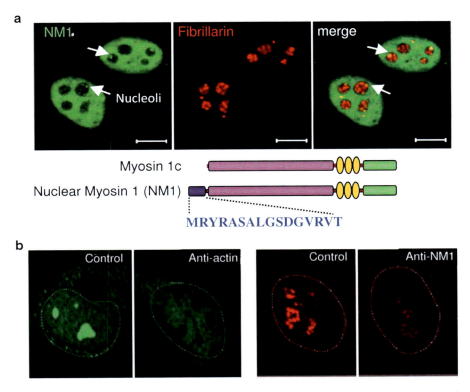

Fig. 6.3 (a) NM1 localizes in the nucleoplasm and in nucleoli. HeLa cells were immunostained with antibodies against NM1 and fibrillarin, and NM1 and fibrillarin were visualized by confocal microscopy. The figure is from the study by Fromproix and Percipalle (2004) and reproduced with permission of Elsevier Science. The scheme below shows the structure of NM1. The N-terminal part responsible for the nuclear localization of NM1 is colored *purple*, the head is colored *pink*, the neck including the IQ motifs is colored *yellow*, and the tail is colored *green*. (b) Actin and NM1 are required for rDNA transcription. Confocal images of transcription sites in HeLa nuclei microinjected with dextran (control), anti-actin or anti-NM1 antibodies as indicated. Br-UTP was incorporated into permeabilized cells for 10 min, and nascent BrU-labeled RNA was visualized with an anti-BrdU antibody conjugated to Alexa Fluor 488 (*green*) or with an anti-BrdU antibody and a Cy3-conjugated secondary antibody (*red*)

These results, together with previous findings demonstrating that different anti-NM1 antibodies do or do not recognize NM1 in the transcribed region (Philimonenko et al. 2004), indicate that NM1 in the initiation complex has a different conformation than NM1 functioning in transcription elongation.

Although both actin and NM1 are required for Pol I transcription, they appear to serve distinct functions. Analysis of spatial distribution at the ultrastructural level revealed the presence of NM1 mainly at nucleolar transcription foci in a transcription-dependent manner, colocalizing with nascent rRNA (Fomproix and Percipalle 2004).

Actin is present at transcriptionally active and inactive regions of the nucleolus, and is unaffected by actinomycin D-induced inhibition of Pol I transcription, suggesting an additional role of actin in maintaining the nucleolar structure (Philimonenko et al. 2010). The association of actin and NM1 with the transcription apparatus might trigger a conformational change of Pol I, and this structural change could be important for efficient transcription elongation. This model is supported by previous studies demonstrating that TIF-IA dissociates from Pol I at early steps of elongation (Bierhoff et al. 2008), the reversible formation and disruption of the Pol I/TIF-IA complex representing a molecular target for regulation of pre-rRNA synthesis. A nucleolar actin-NM1 complex may facilitate this switch from initiation to elongation, possibly in concert with a supramolecular structure that leads to the correct positioning of rRNA genes at distinct functional zones within nucleoli. There is substantial evidence that NM1 facilitates Pol I transcription at the chromatin level. The chromatin remodeling complex WSTF-SNF2h, which is associated with the rDNA promoter and the entire transcribed region, has been shown to interact with NM1. RNAi of WSTF (Williams Syndrome Transcription Factor) decreased pre-rRNA levels indicating that the NM1-WSTF-SNF2h complex promotes Pol I transcription at chromatin (Percipalle et al. 2006).

Recent data suggests that NM1 and actin function together as a molecular motor that drives RNA polymerase movement (Ye et al. 2008). According to this model, NM1 binds the DNA backbone through its positively charged tail domain, while the head interacts with actin bound to RNA polymerase. Anchoring NM1 to DNA, and actin to RNA polymerase, is supposed to generate an ATP-dependent force that powers the sliding of RNA polymerase relative to DNA, with the implication that nuclear myosin I and polymerized actin function like a classical ATP–dependent actomyosin-like motor to power transcription. Related issues include how the actomyosin-like complex mechanistically impacts on Pol I, and whether this complex might also play a role in creating the force needed to kick the finished transcript away from the DNA template at the transcription termination site.

6.5 Regulation of Pol I Transcription

The synthesis of rRNA, the rate-limiting step in ribosome synthesis, is an energy-consuming process that is carefully tuned to match external conditions and accommodate the cell's requirements for protein synthesis, while preventing overinvestment of biosynthetic resources in energetically costly ribosomes. The current notion is that short-term regulation in response to growth factor signaling, nutrients, or stress occurs by altering the transcription rate at euchromatic, active genes, whereas the establishment of a cell-specific ratio of active versus silent rDNA copies during development and differentiation is mediated by mechanisms involving more stable chromatin modifications (see Chap. 4 of this volume). These overlapping mechanisms of transcriptional and epigenetic control have complicated the identification of the major pathways that impart proliferation- and metabolism-dependent control

6 The RNA Polymerase I Transcription Machinery

of rDNA transcription. Nevertheless, work over the last few years has greatly contributed to understanding the molecular mechanisms that adapt Pol I transcription to different growth conditions and environmental cues. There is evidence that almost all basal Pol I transcription factors, for example, TIF-IA, SL1/TIF-IB, and UBF, are modulated by different signaling pathways. We summarize the major principles of transcriptional regulation during the cell cycle, in response to nutrient availability, growth factor and stress signaling, as well as oncogenes and tumor suppressors.

6.5.1 TIF-IA Links Pol I Transcription to Cell Proliferation

Transcription of rDNA is efficiently regulated to be responsive to both general metabolism and specific environmental challenges (Moss 2004). Conditions that impair cellular metabolism, such as nutrient starvation, oxidative stress, inhibition of protein synthesis or cell confluence, will downregulate rDNA transcription, whereas growth factors and agents that stimulate growth and proliferation will upregulate Pol I transcription. The key factor that transfers extracellular signals to the Pol I apparatus is TIF-IA, the mammalian homolog of yeast Rrn3p (Moorefield et al. 2000; Bodem et al. 2000). TIF-IA associates with Pol Iβ by interaction with RPA43, a unique subunit of Pol I. TIF-IA also interacts with specific subunits of SL1/TIF-IB, that is, $TAF_I95/110$ and TAF_I68, thereby bridging Pol I with promoter-bound SL1/TIF-IB and facilitating the assembly of productive transcription initiation complexes (Miller et al. 2001; Yuan et al. 2002). TIF-IA is phosphorylated at several serine/threonine residues, and specific phosphorylation in response to certain metabolic and environmental cues affects the interaction with Pol I and/ or TIF-IB/SL1, thereby regulating the assembly of productive transcription initiation complexes.

Recent studies have established that CK2 is present at the rDNA promoter and is physically associated with the initiation-competent Pol Iβ complex, suggesting that CK2 promotes early steps in Pol I transcription (Lin et al. 2006; Panova et al. 2006). After transcription initiation, CK2 phosphorylates TIF-IA at two serine residues, Ser170 and Ser172, and phosphorylation at Ser170/172 determines whether or not TIF-IA is capable to interact with Pol I, initiate transcription initiation, escape from the promoter, and proceed transcription elongation (Bierhoff et al. 2008). The interaction of TIF-IA with Pol I and the assembly into productive transcription initiation complexes require TIF-IA that is unphosphorylated at Ser170/172. Phosphorylation at Ser170/172 weakens the interaction between TIF-IA and Pol I, leading to dissociation of TIF-IA from Pol I and release from the elongation complex. After release, TIF-IA is dephosphorylated by the phosphatase FCP1, and dephosphorylated TIF-IA is capable of re-associating with Pol I (Fig. 6.4). Thus, phosphorylation and dephosphorylation of TIF-IA at Ser170/172 occurs during each round of transcription, restricting the association of TIF-IA with Pol I to transcription initiation and early steps of elongation and promoting multiple rounds of transcription.

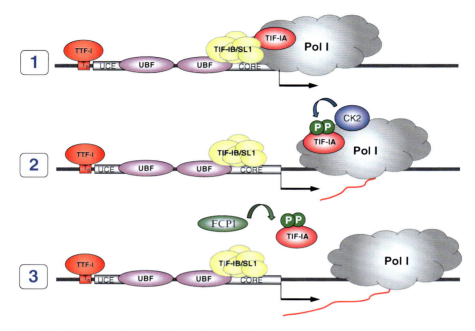

Fig. 6.4 Phosphorylation by CK2 facilitates rDNA transcription by promoting dissociation of TIF-IA from elongating Pol I. After transcription initiation, Pol I-associated CK2 phosphorylates TIF-IA at serines 170 and 172 (Ser170/172). This phosphorylation triggers switching Pol I from the initiation into the elongation phase by promoting dissociation of TIF-IA from Pol I. Dephosphorylation of Ser170/172 by the phosphatase FCP1 mediates re-association of TIF-IA with Pol I, allowing a new round of transcription

In addition to the CK2-mediated "housekeeping" phosphorylation of TIF-IA at Ser170/172, the activity of TIF-IA is regulated by a complex pattern of activating and inactivating phosphorylations that ultimately fine-tune the transcriptional output in response to diverse signaling events, which will be described below (Fig. 6.5).

6.5.2 Growth Factor-Dependent Regulation of rDNA Transcription

On mitogen stimulation, the epidermal growth factor receptor (EGFR) triggers a signaling cascade involving the GTPase Ras, and the mitogen-activated protein kinases (MAPKs) Raf, MEK, ERK, and RSK. Consistent with their positive effects on cell growth and proliferation, MAPKs were found to activate rRNA synthesis by targeting components of the nucleolar transcription apparatus. Transcription activation on mitogenic stimulation correlates with ERK-dependent phosphorylation of UBF at two threonine residues (Thr117 and Thr201), both of them being essential for Pol I transcription elongation (Stefanovsky et al. 2001, 2006a). Moreover, phosphorylation by ERK influences the interaction of UBF

6 The RNA Polymerase I Transcription Machinery

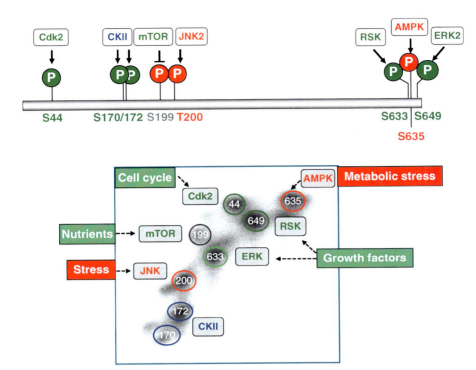

Fig. 6.5 Multiple signaling pathways up- and down-regulate the activity of the transcription initiation factor TIF-IA. The scheme depicts phosphorylation sites that activate (*green*) or inhibit (*red*) TIF-IA activity. mTOR activates TIF-IA indirectly by promoting hypophosphorylation of S199 (*grey*). A two-dimensional tryptic phosphopeptide map of metabolically labeled TIF-IA is shown below. The encircled numbers indicate the positions of the phosphorylated serine or threonine residues contained in the respective tryptic peptides

with DNA, suggesting that dynamic phosphorylation and dephosphorylation events promote the passage of Pol I through an altered UBF-DNA complex, presumably immediately downstream of the transcription start site (Stefanovsky et al. 2006b). In addition, ERK and RSK phosphorylate TIF-IA at two serine residues (Ser633 and Ser649). Replacement of Ser649 by aspartic acid activates TIF-IA and accelerates cell proliferation, whereas the respective alanine mutation leads to retardation of cell growth, underscoring the importance of ERK/RSK-mediated phosphorylation of TIF-IA in regulating rRNA synthesis and nucleolar activity (Zhao et al. 2003). Thus, the MAPK signaling cascade targets two basal Pol I transcription factors, TIF-IA and UBF, leading to upregulation of rDNA transcription, a process that is necessary for enhanced cell proliferation.

As cells have to double in size before dividing, cell growth correlates with rRNA synthesis. The type 1 insulin-like growth factor receptor (IGF-IR) and its docking protein, insulin receptor substrate-1 (IRS-1), control cell size in mammals and flies. Activation of the type I insulin-like IGF-IR by IGF-I stimulates rDNA transcription (Wu et al. 2005), increased transcription correlating with UBF1 activation

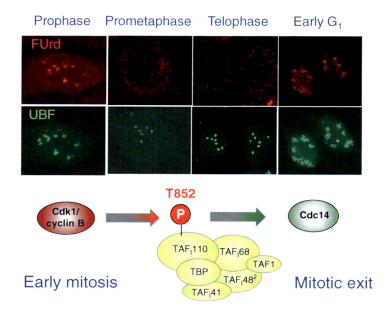

Fig. 6.6 Pol I transcription is inactivated during mitosis. Transcripts were pulse-labeled with fluorouridine (FUrd) and visualized by staining with anti-BrdU antibody (*red, upper panel*). Nucleoli and mitotic NORs were visualized by immunostaining of UBF (*green, lower panel*). The cartoon below illustrates that TAF$_I$110 is phosphorylated during prometaphase at threonine 852 (T852) by Cdk1/cyclin B, and this phosphorylation inactivates TIF-IB/SL1. At the exit from mitosis, T852 is dephosphorylated by hCdc14B and nucleolar transcription is recovered

by phosphorylation of the acidic, serine-rich C terminus. In addition, IRS-1 signaling stabilizes UBF1, demonstrating that IGF signaling increases both the amount and the activity of UBF1 (James and Zomerdijk 2004).

6.5.3 *Transcriptional Regulation During the Cell Cycle*

The synthesis of rRNA oscillates during cell cycle progression. Transcription is silenced during mitosis, gradually increases during G$_1$-phase, and reaches maximal levels at S- and G$_2$-phase (Weisenberger and Scheer 1995; Kuhn et al. 1998; Klein and Grummt 1999). Mitotic silencing and reactivation of rDNA transcription on mitotic exit are controlled at multiple levels, mostly by posttranslational modification of basal transcription factors. At the entry into mitosis, SL1/TIF-IB is inactivated by phosphorylation of TAF$_I$110 at a single threonine residue (Thr852) by Cdk1/cyclin B (Fig. 6.6). This phosphorylation impairs the interaction between SL1 and UBF and prevents the assembly of pre-initiation complexes at the rDNA promoter

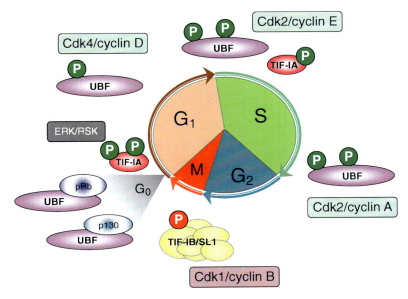

Fig. 6.7 Pol I transcription oscillates during cell cycle progression. During G_1-phase, UBF is activated by phosphorylation of Ser484 by Cdk4/cyclin D and TIF-IA by phosphorylation of Ser649 and Ser 633 by ERK and RSK. During S-phase, a further increase in UBF activity is achieved by phosphorylation at serine 388 by Cdk2/cyclin E&A, and this phosphorylation increases the interaction of UBF with Pol I. At the entry into mitosis, phosphorylation of TAF_I110, the large subunit of the promoter selectivity factor TIF-IB/SL1, at threonine 852 by Cdk1/cyclin B inactivates TIF-IB/SL1, leading to shut-off of Pol I transcription during mitosis. In quiescent G_0-phase cells, UBF is inactivated by association with pRb and p130. Re-entry into G_1 requires Cdk6/4-dependent dissociation of pRb/p130, phosphorylation of UBF at S484, and MAPK/RSK-mediated phosphorylation of TIF-IA at serine 649

(Heix et al. 1998; Kuhn et al. 1998). In addition, mitotic hyperphosphorylation of UBF increases the residence time of UBF on mitotic NORs (Chen et al. 2005; Olsen et al. 2010). Like UBF, SL1/TIF-IB and TTF-I remain associated with mitotic chromosomes, whereas Pol I is transiently released during metaphase (Leung et al. 2004; Chen et al. 2005). A recent quantitative phosphoproteomic study in HeLa cells revealed mitotic phosphorylation of a large number of nucleolar proteins that are involved in ribosome biogenesis, including not only components of the Pol I transcription apparatus, but also proteins involved in pre-rRNA processing, as well as coactivators and corepressors (Olsen et al. 2010). This indicates that phosphorylation ensures efficient shutdown of rRNA synthesis during mitosis.

As cells progress through G_1- and S-phase, rDNA transcription is restored by dephosphorylation of SL1/TIF-IB and hyperphosphorylation of UBF by Cdk4/cyclin D1 and Cdk2/cyclin E or A (Fig. 6.7). Phosphorylation of UBF at serine residues S484 and S388 stimulates transcription pSer388 promoting interaction of UBF with the Pol I subunit PAF53 (Voit et al. 1999; Voit and Grummt 2001). The association of UBF with PAF53/Pol I is further augmented by PCAF-dependent acetylation of

UBF during S-phase (Meraner et al. 2006). A FRAP-based survey of the association of the Pol I transcription apparatus with the rDNA promoter in G_1- and S-phase revealed a clear correlation between the residence times of individual factors and transcriptional output (Gorski et al. 2008). Upregulation of rRNA synthesis in S-phase is accompanied by increased promoter binding and prolonged promoter residence of Pol I and TIF-IA, whereas the dynamics of UBF was not affected (Gorski et al. 2008). These results indicate that increased capturing of components of the transcription apparatus contribute to transcription complex formation and upregulation of rDNA transcription during cell cycle progression.

6.5.4 Transcriptional Regulation by Reversible Acetylation of Transcription Factors

Acetylation of lysine residues has proven to be a key mechanism that alters the structure and functional properties of proteins. Lysine acetylation preferentially targets macromolecular complexes, such as chromatin modifiers, cell cycle regulators, and proteins involved in nuclear transport. Therefore, it is not surprising that reversible acetylation modulates Pol I transcription in direct and indirect ways. Indeed, all important cellular HATs and HAT-complexes have been implicated in modifying basal components of the Pol I transcription machinery. PCAF, p300, and CBP have been shown to target UBF (Hirschler-Laszkiewicz et al. 2001; Pelletier et al. 2000; Meraner et al. 2006). Acetylation of UBF peaks at G_1/S, when UBF activity is high, suggesting that PCAF contributes to cell cycle-dependent fluctuations of rRNA synthesis (Meraner et al. 2006). In addition, UBF directly interacts with the acetyltransferase CBP leading to acetylation of UBF both in vitro and in vivo (Pelletier et al. 2000). This study has suggested an acetylation–deacetylation "flip-flop" mechanism that involves upregulation of UBF by CBP, which in turn prevents recruitment of pRb and HDAC and therefore counteracts repression of Pol I transcription.

Acetylation also regulates the activity of SL1/TIF-IB. Acetylation of TAF_I68 by the histone acetyltransferase PCAF stimulates the interaction of TAF_I68 with the rDNA promoter, thereby increasing SL1/TIF-IB activity and transcription initiation (Muth et al. 2001). PCAF-dependent acetylation of TAF_I68 is counteracted by SIRT1, the founding member of a family of conserved NAD^+-dependent histone deacetylases, termed Sirtuins. SIRT1 is conserved from bacteria to humans and regulates a wide range of biological processes, such as gene silencing, aging, differentiation, and metabolism (Blander and Guarente 2004). SIRT1-dependent deacetylation of TAF_I68 leads to transcriptional repression, underscoring the functional relevance of reversible acetylation in regulating Pol I transcription. In contrast, another member of the Sirtuin family, SIRT7, activates Pol I transcription and plays a major role in cell survival. SIRT7 is associated with active rDNA repeats, interacts with both Pol I and UBF, and augments rDNA occupancy of Pol I (Ford et al. 2006). In addition, SIRT7 is required for resumption of Pol I transcription after exit from

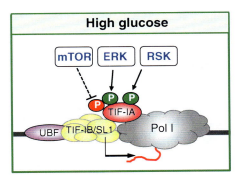 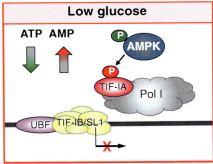

Fig. 6.8 Glucose deprivation downregulates pre-rRNA synthesis. In nutrient-rich medium, TIF-IA is phosphorylated by RSK and ERK at Ser649 and Ser633, and hypophosphorylated at Ser199 in a mTOR-dependent manner. Respectively, and these phosphorylations are required for Pol I transcription. Upon glucose deprivation, elevation of the cellular AMP/ATP ratio activates the AMP-dependent protein kinase AMPK. AMPK-dependent phosphorylation of TIF-IA at Ser635 prevents the interaction of TIF-IA with TIF-IB/SL1, thus impairing the recruitment of the TIF-IA/Pol I complex to preinitiation complex

mitosis (Grob et al. 2009). RNAi-induced depletion of SIRT7 leads to decreased pre-rRNA synthesis and apoptosis. Conversely, SIRT7 is overexpressed in tumor cells, such as breast and prostate cancer (Ashraf et al. 2006). Thus, Sirtuins play important but divergent roles in rDNA transcription regulation, with SIRT1 repressing and SIRT7 stimulating Pol I transcription.

6.5.5 Pol I Transcription is Linked to the Cellular Energy Supply

One of the most important environmental variables is the availability of nutrients, so it makes sense that rDNA transcription is tightly linked to the metabolic state of a cell. It has been known for a long time that a given nutritional state gives rise to an equilibrium in which the synthesis of ATP and GTP is balanced by their use in protein synthesis. Accordingly, rDNA promoter activity is regulated by the intracellular levels of ATP, consistent with the growth-dependent control and homeostatic regulation of rRNA synthesis (Grummt and Grummt 1976). Superimposed on this regulation is deacetylation of TAF_I68, a subunit of the Pol I promoter selectivity factor SL1/TIF-IB, by the NAD^+-dependent deacetylase SIRT1. Deacetylation of TAF_I68 impairs binding of SL1/TIF-IB to the rDNA promoter and leads to transcriptional repression (Muth et al. 2001).

The key enzyme that translates changes in energy levels into adaptive cellular responses is the AMP-activated protein kinase (AMPK). If energy levels are low and the intracellular AMP/ATP ratio is high, AMPK is activated, switching on energy-producing pathways and switching off energy-consuming pathways to restore cellular ATP levels. Therefore, under conditions of nutrient shortage, transcription of rRNA genes is downregulated (Fig. 6.8). In vitro and in vivo phosphorylation experiments combined with in vitro transcription assays revealed that activation of AMPK triggers phosphorylation of TIF-IA at a single serine residue, Ser635, which leads to inactivation of TIF-IA and downregulation of rRNA synthesis (Hoppe et al. 2009). AMPK-mediated phosphorylation of TIF-IA at Ser635 does not compromise binding of TIF-IA to Pol I but abrogates the interaction between promoter-bound SL1/TIF-IB and TIF-IA, which in turn impairs the assembly of productive transcription initiation complexes. This result adds another level of regulation of Pol I transcription, in which TIF-IA not only senses external signals but also translates changes in intracellular energy supply into AMPK-dependent phosphorylation of TIF-IA, which ultimately prevents Pol I transcription initiation (Grummt and Voit 2010).

6.5.6 TOR Signaling Adapts rRNA Synthesis to Nutrient Availability

Another pathway that regulates Pol I transcription in response to nutrient availability is the TOR (Target of Rapamycin) kinase pathway. TOR proteins are members of the phosphatidylinositol 3-kinase (PI3K) superfamily, and have been implicated in the nutrient regulation of cell growth and proliferation in yeast and mammalian cells. Proteins in the mTOR family all have a C-terminal kinase domain that phosphorylates serine and threonine residues. mTOR signaling controls diverse readouts, all of which are related to cell growth, including transcription, translation, PKC signaling, protein degradation, membrane traffic, or actin organization (for review, see Wullschleger et al. 2006). The number and diversity of growth-related readouts controlled by mTOR indicate that this functionally conserved kinase may not be simply part of a single, linear growth-controlling pathway, but can be regarded as a central player that integrates cell physiology and environment thus ensuring balanced growth. The critical role of TOR in linking environmental queues to ribosome biogenesis provides an efficient means by which cells alter their overall protein biosynthetic capacity. Nearly all functions of TOR are specifically inhibited by the natural product rapamycin, an immunosuppressive macrolide that inhibits the PI3K-like kinases TOR1 and TOR2 in yeast and mTOR in mammals, usually in complexes with the prolyl isomerase FKBP12.

Studies in yeast, Drosophila, and mammalian cells indicate that regulation of rRNA synthesis is a conserved TOR function, the control of ribosome biosynthesis by the TOR pathway being surprisingly complex (Claypool et al. 2004; Hannan et al. 2003; James and Zomerdijk 2004; Lin et al. 2006; Mayer et al. 2004). Early studies have established that rRNA synthesis in mammalian cells is regulated by the availability of nutrients, especially amino acids (Grummt et al. 1976). This finding, together with the observation that rDNA transcription is rapamycin-sensitive, indicated that rRNA synthesis is controlled by mTOR. Inactivation of mTOR either by nutrient deprivation or treatment of cells with the mTOR inhibitor rapamycin leads to reduced pre-rRNA synthesis and decreased ribosome production. Both in yeast and mammals, TOR controls Pol I transcription via the transcription factor

Rrn3p/TIF-IA (Claypool et al. 2004; Mayer et al. 2004). Inhibition of mTOR signaling inactivates TIF-IA by two means. It activates the phosphatase PP2A that dephosphorylates Ser44 and enhances phosphorylation at Ser199, and these changes in TIF-IA phosphorylation impair transcription complex formation. Phosphorylation of S44 and S199 affects TIF-IA activity in opposite ways. While S44 phosphorylation is required for TIF-IA activity, phosphorylation at S199 inactivates TIF-IA. This indicates that mTOR-responsive kinase(s) and phosphatase(s) modulate the activity of TIF-IA in different ways and implies that antagonizing phosphorylations play a key role in mTOR-dependent regulation of Pol I transcription. Interestingly, mTOR signaling not only controls the activity but also the intracellular localization of TIF-IA. Once inactivated by rapamycin treatment, a significant part of TIF-IA translocates from the nucleus into the cytoplasm. Presumably, relocating just TIF-IA rather than the entire Pol I machinery is advantageous under conditions where transcription repression has to be both immediate and reversible. mTOR-sensitive sequestration of TIF-IA in the cytoplasm is reminiscent of studies in yeast that have shown that the TOR signaling pathway broadly controls nutrient metabolism by sequestering several transcription factors in the cytoplasm (Beck and Hall 1999). Together, these results demonstrate that inhibition of mTOR signaling downregulates Pol I transcription by three interrelated mechanisms that involve hypophosphorylation of S44, hyperphosphorylation of S199, and shuttling of TIF-IA from the nucle(ol)us into the cytoplasm. The functional interplay of these mechanisms may provide a mechanistic explanation of the possible role of TOR in regulating rRNA synthesis in response to environmental queues. Although the mechanisms for starvation-induced inactivation of mTOR are not completely understood, it is known that increased AMP/ATP ratio, for example, on nutrient deprivation, inhibits mTOR activity via activation of the LKB1-AMPK pathway (Hardie 2007). Overall, a complex signaling network that integrates mTOR, PI3K (phosphatidylinositol 3-kinase), MAPK (mitogen-activated protein kinase), and AMPK pathways regulate ribosome subunit production in response to changes in nutrient levels.

6.5.7 Pol I Transcription Responds to Genotoxic Stress

The nucleolus, long regarded as a mere ribosome-producing factory, plays a key role in monitoring and responding to cellular stress. Cells rapidly and efficiently shut down rRNA synthesis after exposure to extra- or intracellular stress. This inhibition of Pol I transcription process requires the relay of intracellular signals through JNK2 (c-jun N-terminal protein kinase 2), a ubiquitously expressed member of the JNK family that is activated by multiple cellular stresses. Stress-induced activation of JNK2 triggers phosphorylation of TIF-IA at a single threonine residue at position 200 (Mayer et al. 2005). Phosphorylation at Thr200 has two effects. First, it impairs the ability of TIF-IA to interact with Pol I and with SL/TIF-IB, thus preventing the formation of the transcription initiation complex at the rDNA promoter.

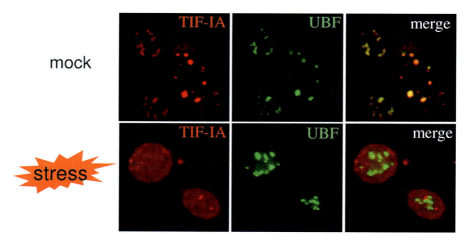

Fig. 6.9 Ribotoxic stress leads to accumulation of TIF-IA in the nucleoplasm. Immunostaining of TIF-IA and UBF in MEFs that were untreated (mock) or treated with 10 μM anisomycin for 60 min (stress). A merged image is shown on the *right*

Second, phosphorylation at Thr200 causes TIF-IA to move from the nucleolus to the nucleoplasm where it is sequestered from Pol I (Fig. 6.9). Mutation of threonine 200 prevents inactivation of TIF-IA by JNK2-mediated phosphorylation and leads to stress-resistance of Pol I transcription. These findings highlight the important role of JNK2 in protecting rRNA synthesis against the harmful consequences of cellular stress, reinforcing the idea that nucleoli orchestrate the chain of events the cell needs to properly respond to stress signals.

Impairment of nucleolar function in response to stress is accompanied by perturbation of nucleolar structure, cell cycle arrest, and stabilization of p53, widely dubbed as "the guardian of the genome." This functional intimate link between Pol I activity, nucleolar integrity, and p53-mediated damage control has also been observed after genetic inactivation of TIF-IA, placing the Pol I transcription machinery in the center of control pathways that are influenced by p53. Abrogation of Pol I transcription either by treatment with actinomycin D or by disrupting the *TIF-IA* gene by Cre-dependent homologous recombination led to disintegration of the nucleolus and p53-dependent apoptosis, reinforcing the central role of p53 in surveying cellular health (Yuan et al. 2005). In TIF-IA-deficient cells, p53 levels are strongly enhanced, most likely due to inhibition of MDM2/HDM2. Under normal conditions, MDM2/HDM2 controls the abundance of p53 by ubiquitinylation, marking p53 for proteasome-dependent proteolysis. Under conditions of nucleolar stress, for example after genetic inactivation of TIF-IA, the p53-MDM2 complex is disrupted and p53-dependent pathways are activated. One possible mechanism for the regulation of p53 in response to nucleolar stress is that proteins that interact with MDM2, including ARF or ribosomal proteins, such as L5, L11, or L23, are released and stabilize p53 by inhibiting the E3 ligase activity of MDM2 (for review, see Boulon et al. 2010).

6 The RNA Polymerase I Transcription Machinery

The finding that inhibition of Pol I transcription induces the apoptotic program raises the exciting possibility that cell-specific inactivation of TIF-IA in proliferating cells may be a powerful approach to trigger cell- or tissue-specific cell suicide. Indeed, targeted disruption of the TIF-IA gene in the developing nervous system has been shown to lead to chronic neurodegeneration in mice. Mutant mice are born alive but die shortly after birth, lacking the entire brain because of selective activation of the apoptotic machinery in neural and glial progenitors (Parlato et al. 2008). Moreover, Cre-loxP-mediated excision of the TIF-IA gene in dopaminergic neurons leads to mice displaying a remarkable spectrum of Parkinsonian symptoms with relentless chronic progression, neurodegeneration correlating with increased levels of p53 and apoptosis in dopaminergic neurons (Rieker et al. 2011). Thus, targeted inactivation of TIF-IA promises to represent a novel and successful strategy not only to establish animal models for specific diseases but also to specifically perturb nucleolar function and to induce apoptosis in defined cells and tissues. Together, the striking correlation between perturbation of nucleolar function, elevated levels of p53, and induction of cell suicide suggests that, depending on the gravity of the nucleolar stress, cells face the decision whether to arrest cell cycle progression and initiate repair mechanisms, or to commit to the p53-dependent apoptotic pathway.

6.6 Oncogenes and Tumor Suppressors Control Pol I Transcription

Increased rRNA synthesis is a hallmark of neoplastic transformation. Consistently, proto-oncogenes and tumor suppressors have been shown to directly target basal Pol I transcription factors and regulate rRNA synthesis. For example, the proto-oncogene c-Myc augments transcription by all three classes of RNA polymerases (Gomez-Roman et al. 2006). c-Myc upregulates rRNA synthesis by increasing the level of UBF, whereas the c-myc antagonist Mad1 downregulates the UBF promoter (Poortinga et al. 2004). Thus, c-Myc and Mad1 coordinate ribosome biogenesis and cell growth under conditions of sustained growth inhibition, for example, cell differentiation. In addition, c-myc interacts with the rDNA promoter via several E-box binding motifs, facilitating recruitment of the TBP-TAF$_I$-complex SL1 to the rDNA promoter and increasing histone acetylation (Arabi et al. 2005; Grandori et al. 2005).

Consistent with their growth inhibiting function, tumor suppressors repress Pol I transcription. The tumor suppressor ARF, an upstream regulator of p53, is located in nucleoli and counteracts hyperproliferative signals induced by oncogenic stimuli. Induction of ARF activity raises p53 levels by binding to HDM2 and inhibiting p53 degradation, thereby repressing Pol I transcription. Likewise, binding of ARF to nucleophosmin (B23) is essential for stabilizing and maintaining basal levels of ARF in nucleoli. Oncogenic signals increase cellular ARF levels promoting ubiquitylation and degradation of nucleophosmin, which in turn blocks a specific step in the maturation of rRNA (Itahana et al. 2003). ARF also regulates rRNA synthesis by interacting with the transcription termination factor TTF-1. Binding of ARF to

TTF-1 masks the nucleolar localization domain of TTF-1, excluding TTF-I from the nucleolus. Moreover, knockdown of TTF-1 inhibits pre-rRNA processing, indicating that ARF mediates pre-rRNA processing through its interaction with TTF-1 (Lessard et al. 2010).

Biochemical and genetic data have established that Pol I transcription is also regulated by pRb, a member of the retinoblastoma protein family, comprising the pocket proteins pRb, p107, and p130. pRb accumulates in the nucleolus on cell confluence or during differentiation and represses rDNA transcription. Transcriptional repression is brought about by interaction of pRb with UBF, preventing UBF from recruiting SL1 and from binding to the coactivator CBP. CBP acetylates UBF, thereby stimulating UBF activity (Cavanaugh et al. 1995; Voit et al. 1997; Pelletier et al. 2000). Notably, p130 but not p107, serves a similar role as pRb in serum starved cells and during cell differentiation, consistent with overlapping and specific functions of individual members of the retinoblastoma protein family (Ciarmatori et al. 2001). pRb-and p130-dependent repression of Pol I transcription is abolished by point mutations in the pocket domain, underscoring the importance of the integrity of the pocket domain in transcriptional repression. Thus, pRb and p130 help to ensure that the output of Pol I is throttled under inappropriate growth conditions.

Finally, PTEN, another tumor suppressor that counteracts PI3/Akt signaling and whose function is frequently abrogated in cancer, downregulates Pol I transcription by targeting SL1. PTEN-induced disruption of the SL1 complex leads to release of TBP, TAF_I110, and TAF_I48 from the rDNA promoter, without affecting promoter occupancy of UBF and TAF_I68 (Zhang et al. 2005). Similarly, downregulation of Pol I transcription during differentiation of mouse F9 cells occurs because of selective disruption of SL1/TIF-IB (Alzuherri and White 1999).

6.7 Perspectives

Recent years have seen several important advances in our understanding of the Pol I transcription machinery and the mechanisms that have evolved to guarantee the efficiency and regulation of transcription. Many of the components required for rDNA transcription and ribosome biogenesis have been characterized and the advent of proteomics will undoubtedly identify more proteins that control the maintenance of a balanced ribosome supply. However, although our understanding of the signaling cascades that transmit information on the cellular growth state to the Pol I transcription apparatus has advanced considerably, many questions remain to be answered. For example, the molecular mechanisms that regulate elongation of Pol I transcription are poorly understood. Likewise, the link between Pol I transcription and processing of pre-rRNA, the topology of transcriptionally active and silent rRNA genes, the functional relevance of specific posttranslational modifications of components of the Pol I transcription apparatus, as well as the cross-talk of Pol I with histone modifying enzymes that facilitate elongation on chromatin templates represent challenging and rewarding subjects of research.

Changes in ribosome biogenesis correlate with ribosome-related diseases, such as Diamond-Blackfan anemia, demonstrating that deregulation of rRNA synthesis can have an enormous impact on the ability of cells to sustain life. Therefore, the elucidation of the molecular pathways that transmit information on the growth state of a cell population to the Pol I transcription apparatus not only is of great scientific interest but also holds in store the potential discovery of novel therapeutic strategies that restrain cell proliferation by selectively targeting proteins involved in ribosome biogenesis. Understanding the intimate link between deregulated rRNA synthesis and tumorigenesis will be instrumental for the development of specific and selective inhibitors of rRNA synthesis, aiming to combat cancer through targeted downregulation of Pol I transcription. Although the area of targeting anticancer drugs to the Pol I transcription machinery is still in its infancy, it promises to be a provocative and emerging field of research.

Acknowledgments Our work has been supported by grants from the Deutsche Forschungsgemeinschaft, the EU Network ‚The Epigenome', the BMBF ('EpiSys'), an ERC Advanced Grant, and the Fonds der Chemischen Industrie.

References

Alzuherri HM, White RJ (1999) Regulation of RNA polymerase I transcription in response to F9 embryonal carcinoma stem cell differentiation. J Biol Chem 274:4328–4334

Arabi A, Wu S, Ridderstrale K, Bierhoff H, Shiue C et al (2005) c-Myc associates with ribosomal DNA and activates RNA polymerase I transcription. Nat Cell Biol 7:303–310

Ashraf N, Zino S, Macintyre A, Kingsmore D, Payne AP, George WD, Shiels PG (2006) Altered sirtuin expression is associated with node-positive breast cancer. Br J Cancer 95:1056–1061

Beck T, Hall MN (1999) The TOR signaling pathway controls nuclear localization of nutrient-regulated transcription factors. Nature 402:689–692

Bettinger BT, Gilbert DM, Amberg DC (2004) Actin up in the nucleus. Nat Rev Mol Cell Biol 5:410–415

Bierhoff H, Dundr M, Michels A, Grummt I (2008) Phosphorylation by CK2 facilitates rDNA transcription by promoting dissociation of TIF-IA from elongating RNA polymerase. Mol Cell Biol 28:4988–4998

Birch JL, Tan BC, Panov KI, Panova TB, Andersen JS, Owen-Hughes TA, Russell J, Lee SC, Zomerdijk JC (2009) FACT facilitates chromatin transcription by RNA polymerases I and III. EMBO J 28:854–865

Blander G, Guarente L (2004) The Sir2 family of protein deacetylases. Ann Rev Biochem 73:417–435

Bodem J, Dobreva G, Hoffmann-Rohrer U, Iben S, Zentgraf H, Delius H, Vingron M, Grummt I (2000) TIF-IA, the factor mediating growth-dependent control of ribosomal RNA synthesis, is the mammalian homolog of yeast Rrn3p. EMBO Rep 1:171–175

Boulon S, Westman BJ, Hutten S, Boisvert FM, Lamond AI (2010) The nucleolus under stress. Mol Cell 40:216–227

Cavanaugh AH, Hempel WM, Taylor LJ, Rogalsky V, Todorov G, Rothblum LI (1995) Activity of RNA polymerase I transcription factor UBF blocked by Rb gene product. Nature 374:177–180

Chen D, Dundr M, Wang C, Leung A, Lamond A, Misteli T, Huang S (2005) Condensed mitotic chromatin is accessible to transcription factors and chromatin structural proteins. J Cell Biol 168:41–54

Ciarmatori S, Scott PH, Sutcliffe JE, McLees A, Alzuherri HM, Dannenberg JH, te Riele H, Grummt I, Voit R, White RJ (2001) Overlapping functions of the pRb family in the regulation of rRNA synthesis. Mol Cell Biol 21:5806–5814

Claypool JA et al (2004) Tor pathway regulates Rrn3p-dependent recruitment of yeast RNA polymerase I to the promoter but does not participate in alteration of the number of active genes. Mol Biol Cell 15:946–956

Clos J, Buttgereit D, Grummt I (1986) A purified transcription factor (TIF-IB) binds to essential sequences of the mouse rDNA promoter. Proc Natl Acad Sci 83:604–608

Comai L, Tanese N, Tjian R (1992) The TATA-binding protein and associated factors are integral components of the RNA polymerase I transcription factor, SL1. Cell 68:965–976

Cremer P, Armache KJ, Baumli S, Benkert S, Brueckner F, Buchen C, Damsma GE, Dengl S, Geiger SR, Jasiak AJ, Jawhari A, Jennebach S, Kamenski T, Kettenberger H, Kuhn CD, Lehmann E, Leike K, Sydow JF, Vannini A (2008) Structure of eukaryotic RNA polymerases. Ann Rev Biophys 37:337–352

Denissov S, van Driel M, Voit R, Hekkelman M, Hulsen T, Hernandez N, Grummt I, Wehrens R, Stunnenberg H (2007) Identification of novel functional TBP-binding sites and general factor repertoires. EMBO J 26:944–954

Dundr M, Hoffmann-Rohrer U, Hu Q, Grummt I, Rothblum LI, Phair RD, Misteli T (2002) A kinetic framework for a mammalian RNA polymerase in vivo. Science 298:1623–1626

Evers R, Grummt I (1995) Molecular coevolution of mammalian ribosomal gene terminator sequences and the transcription termination factor TTF-I. Proc Natl Acad Sci USA 92: 5827–3581

Evers R, Smid A, Rudloff U, Lottspeich F, Grummt I (1995) Different domains of the murine RNA polymerase I-specific termination factor mTTF-I serve distinct functions in transcription termination. EMBO J 14:1248–1256

Fomproix N, Percipalle P (2004) An actin-myosin complex on actively transcribing genes. Exp Cell Res 294:140–148

Ford E, Voit R, Liszt G, Magin C, Grummt I, Guarente L (2006) Mammalian Sir2 homolog SIRT7 is an activator of RNA polymerase I transcription. Genes Dev 20:1075–1080

French SL, Osheim YN, Cioci F, Nomura M, Beyer AL (2003) In exponentially growing Saccharomyces cerevisiae cells, rRNA synthesis is determined by the summed RNA polymerase I loading rate rather than by the number of active genes. Mol Cell Biol 23: 1558–1568

Geiger SR, Lorenzen K, Schreieck A, Hanecker P, Kostrewa D, Heck AJR, Cramer P (2010) RNA polymerase I contains a TFIIF-related DNA-binding subcomplex. Mol Cell 39:583–594

Gerber JK, Gögel E, Berger C, Wallisch M, Müller F, Grummt I, Grummt F (1997) Termination of mammalian rDNA replication: polar arrest of replication fork movement by transcription termination factor TTF-I. Cell 90:559–567

Gomez-Roman N, Felton-Edkins ZA, Kenneth NS, Goodfellow SJ, Athineos D, Zhang J, Ramsbottom BA, Innes F, Kantidakis T, Kerr ER, Brodie G, Grandori C, White RJ (2006) Activation by c-myc of transcription by RNA polymerases I, II and III. Biochem Soc Symp 73: 141–154

Gorski JJ, Pathak S, Panov K, Kasciukovic T, Panova T, Russell J, Zomerdijk JCBM (2007) A novel TBP-associated factor of SL1 functions in RNA polymerase I transcription. EMBO J 26:1560–1568

Gorski SA, Snyder SK, John S, Grummt I, Misteli T (2008) Modulation of RNA polymerase assembly dynamics in transcription regulation. Mol Cell 30:486–497

Grandori C, Gomez-Roman N, Felton-Edkins ZA, Ngouenet C, Galloway DA et al (2005) c-Myc binds to human ribosomal DNA and stimulates transcription of rRNA genes by RNA polymerase I. Nat Cell Biol 7:311–318

Grob A, Roussel P, Wright JE, McStay B, Hernandez-Verdun D, Sirri V (2009) Involvement of SIRT7 in resumption of rDNA transcription at the exit from mitosis. J Cell Sci 122:489–498

Grummt I (2006) Actin and myosin as transcription factors. Curr Opin Biol Genet Dev 16: 191–196

Grummt I (2010) Wisely chosen paths - regulation of rRNA synthesis. FEBS J 277:4626–2639

6 The RNA Polymerase I Transcription Machinery

Grummt I, Grummt F (1976) Control of nucleolar RNA synthesis by the intracellular pool sizes of ATP and GTP. Cell 7:447–453

Grummt I, Voit R (2010) Linking rDNA transcription to the cellular energy supply. Cell Cycle 9:18–19

Grummt I, Smith VA, Grummt F (1976) Amino acid starvation affects the initiation frequency of nucleolar RNA polymerase. Cell 7:439–445

Grummt I, Maier U, Öhrlein A, Hassouna N, Bachellerie JP (1985) Transcription of mouse rDNA terminates downstream of the 3' end of 28 S RNA and involves interaction of factors with repeated sequences in the 3' spacer. Cell 43:801–810

Grummt I, Kuhn A, Bartsch I, Rosenbauer H (1986a) A transcription terminator located upstream of the mouse rDNA initiation site affects rRNA synthesis. Cell 47:901–911

Grummt I, Rosenbauer H, Niedermeyer I, Maier U, Öhrlein A (1986b) A repeated 18 bp sequence motif in the mouse rDNA spacer mediates binding of a nuclear factor and transcription termination. Cell 45:837–846

Hanada K, Song CZ, Yamamoto K, Yano K, Maeda Y, Yamaguchi K, Muramatsu M (1996) RNA polymerase I associated factor 53 binds to the nucleolar transcription factor UBF and functions in specific rDNA transcription. EMBO J 15:2217–2226

Hannan KM, Brandenburger Y, Jenkins A, Sharkey K, Cavanaugh A et al (2003) mTOR-dependent regulation of ribosomal gene transcription requires S6K1 and is mediated by phosphorylation of the carboxy-terminal activation domain of the nucleolar transcription factor UBF. Mol Cell Biol 23:8862–8877

Hardie DG (2007) AMP-activated/SNF1 protein kinases: conserved guardians of cellular energy. Nat Rev Mol Cell Biol 8:774–785

Heix J, Zomerdijk JCBM, Ravanpay A, Tjian R, Grummt I (1997) Cloning of murine RNA polymerase I- specific TAFs: Conserved interactions between the four sub- units of the species-specific transcription factor TIF-IB/SL1. Proc Natl Acad Sci 94:1733–1738

Heix J, Vente A, Voit R, Budde A, Michaelidis TM, Grummt I (1998) Mitotic silencing of human rRNA synthe- sis: Inactivation of the promoter selectivity factor SL1 by cdc2/cyclin B-mediated phosphorylation. EMBO J 17:7373–7381

Henderson S, Sollner-Webb B (1986) A transcriptional terminator is a novel element of the promoter of the mouse ribosomal RNA gene. Cell 47:891–900

Hirschler-Laszkiewicz I, Cavanaugh A, Hu Q, Catania J, Avantaggiati ML, Rothblum LI (2001) The role of acetylation in rDNA transcription. Nucleic Acids Res 29:4114–4124

Hoppe S, Bierhoff H, Cado I, Weber A, Tiebe M, Grummt I, Voit R (2009) AMP-activated protein kinase adapts rRNA synthesis to cellular energy supply. Proc Natl Acad Sci USA 106: 17781–17786

Itahana K, Bhat KP, Jin A, Itahana Y, Hawke D, Kobayasi R, Zhang Y (2003) Tunor suppressor ARF degrades B23, a nucleolar protein involved in ribosome biogenesis and cell proliferation. Mol Cell 12:1151–1161

James MJ, Zomerdijk JC (2004) Phosphatidylinositol 3-kinase and mTOR signaling pathways regulate RNA polymerase I transcription in response to IGF-1 and nutrients. J Biol Chem 279:8911–8918

Jansa P, Mason SW, Hoffmann-Rohrer U, Grummt I (1998) Cloning and functional characterization of PTRF, a novel protein which induces dissociation of paused ternary transcription complexes. EMBO J 17:2855–2864

Jantzen HM, Admon A, Bell SP, Tjian R (1990) Nucleolar transcription factor hUBF contains a DNA-binding motif with homology to HMG protein. Nature 344:830–836

Klein J, Grummt I (1999) Cell cycle-dependent regulation of RNA polymerase I transcription: The nucleolar transcription factor UBF is inactive in mitosis and early G1. Proc Natl Acad Sci 96:6095–6101

Kuhn A, Grummt I (1992) Dual role of the nucleolar transcription factor UBF: trans-activator and antirepressor. Proc Natl Acad Sci 89:7340–7344

Kuhn A, Stefanovsky V, Grummt I (1993) The nucleolar transcription activator UBF relieves Ku antigen-mediated repression of mouse ribosomal gene transcription. Nucleic Acids Res 21: 2057–2063

Kuhn A, Vente A, Dorée M, Grummt I (1998) Mitotic phosphorylation of the TBP-containing factor SL1 represses ribosomal gene transcription. J Mol Biol 284:1–5

Kuhn CD, Geiger SR, Baumli S, Gartmann M, Gerber J, Jennebach S, Mielke T, Tschochner H, Beckmann R, Cramer P (2007) Functional architecture of RNA polymerase I. Cell 131:1260–1272

Kulkens T, van der Sande CA, Dekker AF, van Heerikhuizen H, Planta RJ (1992) A system to study transcription by yeast RNA polymerase I within the chromosomal context: functional analysis of the ribosomal DNA enhancer and the RBP1/REB1 binding sites. EMBO J 11:4665–4674

Kysela K, Philimonenko AA, Philimonenko VV, Janacek J, Kahle M, Hozak P (2005) Nuclear distribution of actin and myosin I depends on transcriptional activity of the cell. Histochem Cell Biol 124:347–358

Längst G, Blank TA, Becker PB, Grummt I (1997) RNA polymerase I transcription on nucleosomal templates: the transcription termination factor TTF-I induces chromatin remodeling and relieves transcriptional repression. EMBO J 16:760–768

Längst G, Becker PB, Grummt I (1998) TTF-I determines the chromatin architecture of the active rDNA promoter. EMBO J 17:3135–3145

Learned RM, Cordes S, Tjian R (1985) Purification and characterization of a transcription factor that confers promoter specificity to human RNA poymerase I. Mol Cell Biol 5:1358–1369

Lessard F, Morin F, Ivanchuk S, Langlois F, Stefanovsky V, Rutka J, Moss T (2010) The ARF tumor suppressor controls ribosome biogenesis by regulating the RNA polymerase I transcription factor TTF-I. Mol Cell 38:539–550

Leung AK, Gerlich D, Miller G, Lyon C, Lam YW, Lleres D, Daigle N, Zomerdijk J, Ellenberg J, Lamond AI (2004) Quantitative linetic analysis of nucleolar breakdown and reassembly during mitosis in live human cells. J Cell Biol 166:787–800

Lin CY, Navarro S, Reddy S, Comai L (2006) CK2-mediated stimulation of Pol I transcription by stabilization of UBF-SL1 interaction. Nucleic Acids Res 34:4752–4766

Marilley M, Pasero P (1996) Common DNA structural features exhibited by eukaryotic ribosomal gene promoters. Nucleic Acids Res 24:2204–2211

Mayer C, Zhao J, Yuan X, Grummt I (2004) mTOR-dependent activation of the transcription factor TIF-IA links rRNA synthesis to nutrient availability. Genes Dev 18:423–434

Mayer C, Bierhoff H, Grummt I (2005) The nucleolus as a stress sensor: JNK2 inactivates the transcription factor TIF-IA and down-regulates rRNA synthesis. Genes Dev 19:933–941

McStay B, Reeder RH (1990) An RNA polymerase I termination site can stimulate the adjacent ribosomal gene promoter by two distinct mechanisms in Xenopus laevis. Genes Dev 4:1240–1251

Meraner J, Lechner M, Loidl A, Goralik-Schramel M, Voit R, Grummt I, Loidl P (2006) Acetylation of UBF changes during the cell cycle and regulates the interaction of UBF with RNA polymerase I. Nucleic Acids Res 34:1798–1806

Miller G, Panov KI, Friedrich JK, Trinkle-Mulcahy L, Lamond AI, Zomerdijk JC (2001) hRRN3 is essential in the SL1-mediated recruitment of RNA Polymerase I to rRNA gene promoters. EMBO J 20:1373–1382

Moorefield B, Greene EA, Reeder RH (2000) RNA polymerase I transcription factor Rrn3 is functionally conserved between yeast and human. Proc Natl Acad Sci USA 97:4724–4729

Moss T (2004) At the crossroads of growth control: making ribosomal RNA. Curr Opin Genet Dev 14:210–217

Moss T, Langlois F, Gagnon-Kugler T, Stefanovsky V (2007) A housekeeper with power of attorney: the rRNA genes in ribosome biogenesis. Cell Mol Life Sci 64:29–49

Murano K, Okuwaki M, Hisaoka M, Nagata K (2008) Transcription regulation of the rRNA gene by a multifunctional nucleolar protein, B23/nucleophosmin, through its histone chaperone activity. Mol Cell Biol 28:3114–3126

Muth V, Nadaud S, Grummt I, Voit R (2001) Acetylation of TAFI68, a subunit of TIF-IB/SL1, activates RNA polymerase I transcription. EMBO J 20:1353–1362

Németh A, Guibert S, Tiwari VK, Ohlsson R, Längst G (2008) Epigenetic regulation of TTF-I-mediated promoter-terminator interactions of rRNA genes. EMBO J 27:1255–1265

Nowak G, Pestic-Dragovich L, Hozák P, Philimonenko A, Simerly C, Schatten G, de Lanerolle P (1997) Evidence for the presence of myosin I in the nucleus. J Biol Chem 272:17176–17181

Olsen JV, Vermeulen M, Santamaria A, Kumar C, Miller ML, Jensen LJ, Gnad F, Cox J, Jensen TS, Nigg EA, Brunak S, Mann M (2010) Quantitative phosphoproteomics reveals widespread full phosphorylation site occupancy during mitosis. Sci Signal 3:ra3

Panov KI, Friedrich JK, Russell J, Zomerdijk JC (2006a) UBF activates RNA polymerase I transcription by stimulating promoter escape. EMBO J 25:3310–3322

Panov KI, Panova TB, Gadal O, Nishiyama K, Saito T, Russell J, Zomerdijk JC (2006b) RNA polymerase I-specific subunit CAST/hPAF49 has a role in the activation of transcription by upstream binding factor. Mol Cell Biol 26:5436–5448

Panova TB, Panov KI, Russell J, Zomerdijk JC (2006) Casein kinase 2 associates with initiation-competent RNA polymerase I and has multiple roles in ribosomal DNA transcription. Mol Cell Biol 26:5957–5968

Parlato R, Kreiner G, Erdmann G, Rieker C, Stotz S et al (2008) Activation of an endogenous suicide response after perturbation of rRNA synthesis leads to neurodegeneration in mice. J Neurosci 28:12759–12764

Pelletier G, Stefanovsky VY, Faubladier M, Hirschler-Laszkiewicz I, Savard J, Rothblum LI, Côté J, Moss T (2000) Competitive recruitment of CBP and Rb-HDAC regulates UBF acetylation and ribosomal transcription. Mol Cell 6:1059–1066

Percipalle P, Fomproix N, Cavellan E, Voit R, Reimer G, Kruger T, Thyberg J, Scheer U, Grummt I, Farrants AK (2006) The chromatin remodelling complex WSTF–SNF2h interacts with nuclear myosin 1 and has a role in RNA polymerase I transcription. EMBO Rep 7:525–530

Pestic-Dragovich L, Stojiljkovic L, Philimonenko AA, Nowak G, Ke Y, Settlage RE, Shabanowitz J, Hunt DF, Hozák P, de Lanerolle P (2000) A myosin I isoform in the nucleus. Science 290:337–341

Philimonenko VV, Zhao J, Iben S, Dingova H, Kysela K, Kahle M, Zentgraf H, Hofmann WA, de Lanerolle P, Hozak P, Grummt I (2004) Nuclear actin and myosin I are required for RNA polymerase I transcription. Nat Cell Biol 6:1165–1172

Philimonenko VV, Janacek J, Harata M, Hozak P (2010) Transcription-dependent rearrangements of actin and nuclear myosin I in the nucleolus. Histochem Cell Biol 133:607–626

Poortinga G, Hannan KM, Snelling H, Walkley CR, Jenkins A, Sharkey K, Wall M, Brandenburger Y, Palatsides M, Pearson RB, McArthur GA, Hannan RD (2004) MAD1 and c-MYC regulate UBF and rDNA transcription during granulocyte differentiation. EMBO J 23:3325–3335

Rickards B, Flint SJ, Cole MD, Leroy G (2007) Nucleolin is required for RNA polymerase I transcription in vivo. Mol Cell Biol 27:937–948

Rieker C, Engblom D, Kreiner G, Schober A, Stotz S, Neumann M, Yuan X, Grummt I, Schütz G, Parlato R (2011) Nucleolar disruption in dopaminergic neurons leads to oxidative damage and Parkinsonism through repression of mTOR signaling. J Neurosci 31:453–460

Russell J, Zomerdijk JC (2005) RNA-polymerase-I-directed rDNA transcription, life and works. Trends Biochem Sci 30:87–96

Sander EE, Grummt I (1997) Oligomerization of the transcription termination factor TTF-I: implications for the structural organization of ribosomal transcription units. Nucleic Acids Res 25:1142–1147

Sander EE, Mason SW, Munz C, Grummt I (1996) The amino-terminal domain of the transcription termination factor TTF-I causes protein oligomerization and inhibition of DNA binding. Nucleic Acids Res 24:3677–3684

Sanij E, Poortinga G, Sharkey K, Hung S, Holloway TP, Quin J, Robb E, Wong LH, Thomas WG, Stefanovsky V, Moss T, Rothblum L, Hannan KM, McArthur GA, Pearson RB, Hannan RD (2008) UBF levels determine the number of active ribosomal RNA genes in mammals. J Cell Biol 183:1259–1274

Smid A, Finsterer M, Grummt I (1992) Limited proteolysis unmasks specific DNA-binding of the murine RNA polymerase I-specific transcription termination factor TTFI. J Mol Biol 227:635–647

Stefanovsky VY, Pelletier G, Hannan R, Gagnon-Kugler T, Rothblum LI, Moss T (2001) An immediate response of ribosomal transcription to growth factor stimulation in mammals is mediated by ERK phosphorylation of UBF. Mol Cell 8:1063–1073

Stefanovsky V, Langlois F, Gagnon-Kugler T. Rothblum LI, Moss T (2006a) Growth factor signaling regulates elongation of RNA polymerase I transcription in mammals via UBF phosphorylation and r-chromatin remodeling. Mol Cell 21:629–639

Stefanovsky VY, Langlois F, Bazett-Jones D, Pelletier G, Moss T (2006b) ERK modulates DNA bending and enhancesome structure by phosphorylating HMG1-boxes 1 and 2 of the RNA polymerase I transcription factor UBF. Biochemistry 45:3626–3634

Voit R, Grummt I (2001) Phosphorylation of UBF at serine 388 is required for interaction with RNA polymerase I and activation of rDNA transcription. Proc Natl Acad Sci USA 98: 13631–13636

Voit R, Schäfer K, Grummt I (1997) Mechanism of repression of RNA polymerase I transcription by the retinoblastoma protein. Mol Cell Biol 17:4230–4237

Voit R, Hoffmann M, Grummt I (1999) Phosphorylation by G1-specific cdk-cyclin complexes activates the nucleolar transcription factor UBF. EMBO J 18:1891–1899

Weisenberger D, Scheer U (1995) A possible mechanism for the inhibition of ribosomal RNA gene transcription during mitosis. J Cell Biol 129:561–575

Wu A, Tu X, Prisco M, Baserga R (2005) Regulation of upstream binding factor 1 activity by insulin-like growth factor I receptor signaling. J Biol Chem 280:2863–2872

Wullschleger S, Loewith R, Hall MN (2006) TOR signaling in growth and metabolism. Cell 124:471–484

Yamamoto K, Yamamoto M, Hanada K, Nogi Y, Matsuyama T, Muramatsu M (2004) Multiple protein-protein interactions by RNA polymerase I-associated factor PAF49 and role of PAF49 in rRNA transcription. Mol Cell Biol 24:6338–6340

Ye J, Zhao J, Hoffmann-Rohrer U, Grummt I (2008) Nuclear myosin I acts in concert with polymeric actin to drive RNA polymerase I transcription. Genes Dev 22:322–330

Yuan X, Zhao J, Zentgraf W, Hoffmann-Rohrer U, Grummt I (2002) Multiple interactions between RNA polymerase I, TIF-IA and TAF(I) subunits regulate preinitiation complex assembly at the ribosomal gene promoter. EMBO Rep 3:1082–1087

Yuan X, Zhou Y, Casanova E, Chai M, Kiss E, Gröne HJ, Schütz G, Grummt I (2005) Genetic inactivation of the transcription factor TIF-IA leads to nucleolar disruption, cell cycle arrest and p53-mediated apoptosis. Mol Cell 19:77–89

Zhang C, Comai L, Johnson DL (2005) PTEN represses RNA polymerase I transcription by disrupting the SL1 complex. Mol Cell Biol 25:6899–6911

Zhao J, Yuan X, Frödin M, Grummt I (2003) ERK-dependent phosphorylation of the transcription initiation factor TIF-IA is required for RNA polymerase I transcription and cell growth. Mol Cell 11:405–413

Zomerdijk JC, Beckmann H, Comai L, Tjian R (1994) Assembly of transcriptionally active RNA polymerase I initiation factor SL1 from recombinant subunits. Science 266:2015–2018

Chapter 7
Small Ribonucleoproteins in Ribosome Biogenesis

Franziska Bleichert and Susan Baserga

7.1 Introduction

The nucleolus harbors one of the most abundant set of non-coding RNAs, called small nucleolar RNAs (snoRNAs). They can be allocated into different classes on the basis of conserved sequence motifs in the RNAs that were defined more than a decade ago, the two major ones being box H/ACA and box C/D snoRNAs. The snoRNAs associate with different sets of proteins to form small nucleolar ribonucleoprotein (snoRNP) complexes. As such, they play important roles in the processing and maturation of the ribosomal RNA (rRNA) during ribosome biogenesis (Fig. 7.1).

7.2 Nucleotide Modification in rRNAs Catalyzed by snoRNPs

The nucleotides present in the mature rRNAs that make up ribosomes contain numerous chemical modifications. The most abundant modifications are pseudouridylations, that is, the isomerization of uridine into pseudouridine, and $2'-O$-ribose methylations, with up to ~100 modified sites in the *Saccharomyces cerevisiae* ribosome and up to ~200 modified sites in vertebrate ribosomes (Decatur and Fournier 2002; Maden and Hughes 1997; Samarsky and Fournier 1999). Base methylations also occur but are, with ~10 modifications per ribosome, much less frequent. While base methylations are catalyzed by stand-alone protein enzymes, the catalysis of the isomerization of uridine into pseudouridine is performed by box H/ACA snoRNPs and the methylation of ribose is performed by box C/D snoRNPs in a

---

S. Baserga (✉)
Departments of Molecular Biophysics and Biochemistry, Genetics,
and Therapeutic Radiology, Yale University, New Haven, CT, USA
e-mail: susan.baserga@yale.edu

M.O.J. Olson (ed.), *The Nucleolus*, Protein Reviews 15,
DOI 10.1007/978-1-4614-0514-6_7, © Springer Science+Business Media, LLC 2011

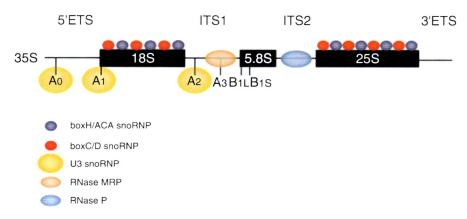

Fig. 7.1 Small nucleolar RNPs (snoRNPs) involved in ribosome biogenesis. The primary yeast 35S pre-rRNA transcript encoding the 18S, 25S, and 5.8S rRNAs is shown and snoRNPs involved in different aspects of rRNA maturation are indicated. The processing sites that require small RNPs are also specified, whereas the remaining processing sites are omitted for clarity. B_{1L} and B_{1S} sites mark alternative 5' ends of the 5.8S rRNA, with the latter occurring if RNase MRP cleaves at site A_3. The U3 snoRNP is required for cleavages at sites A_0, A_1, and A_2, and most likely a single copy of the U3 snoRNP is sufficient for all three processing steps

snoRNA-guided manner in eukaryotes (Cavaille et al. 1996; Ganot et al. 1997; Kiss-Laszlo et al. 1996; Ni et al. 1997; Tycowski et al. 1996). This mode of modification is evolutionary conserved in archaea, where related RNPs (sRNP for small ribonucleoprotein because of the lack of nucleoli in archaea) perform these tasks (Omer et al. 2000). In eubacteria, on the other hand, all modifications are catalyzed by stand-alone protein enzymes instead of RNPs. The reasons for this dichotomy are not completely understood, but the dramatic increase in the number of pseudoruridines and 2'-O-methylated ribose residues in the ribosome during evolution (only few nucleotide modifications occur in eubacterial rRNA as compared to several hundred in rRNA of higher eukaryotes) may have required this alternative mechanism to ensure the site-specificity of these modifications.

Efforts to determine the location of these modifications within the sequence of the rRNA have revealed that they are enriched in rRNA regions with highly evolutionarily conserved secondary structures (Maden 1990), implying that the modifications are critical for ribosome function. The subsequent mapping of the modification sites onto the structure of the prokaryotic ribosome as it became available further supported the importance of the nucleotide modifications; they are located in regions functionally important to protein synthesis, such as the peptidyl transferase center, the A, P, and E sites, the decoding center, the interaction interfaces between the small and large ribosomal subunits, and the nascent polypeptide chain exit tunnel (Decatur and Fournier 2002).

So why are there so many nucleotide modifications and how do they contribute to ribosome function? Seminal studies in *S. cerevisiae* established that a universal loss of either pseudouridylations or 2'-O-ribose methylations in rRNAs does not

7 Small Ribonucleoproteins in Ribosome Biogenesis

support cell growth (Lafontaine et al. 1998; Tollervey et al. 1993). However, loss of individual modifications, and thereby loss of individual snoRNPs responsible for their catalysis, was found to have no substantial effect on cell growth (King et al. 2003; Liang et al. 2009b). Still, synergistic effects between different modifications are common so that loss of two or more nucleotide modifications results in a slow growth phenotype, reduced protein synthesis, reduced ribosome stability, and structural alterations in the ribosome (King et al. 2003; Liang et al. 2007b). This suggests that some modifications perform redundant functions or are necessary only under conditions that may challenge ribosome performance. Interestingly, artificially introducing modifications into rRNAs at sites usually not modified can result in the same detrimental phenotype as the loss of modifications (Liu et al. 2008).

While it is now obvious that pseudouridylations and 2'-O-ribose methylations are essential for optimal ribosome function, the detailed molecular mechanism by which these are achieved and the impact of these modifications on the molecular structure of the ribosome are less well understood. Generally, it has been proposed that modified nucleotides stabilize secondary and tertiary RNA structures (Ishitani et al. 2008). Pseudouridine has an additional hydrogen bond to donate, when compared to uridine, and is thought to enhance base stacking interactions. 2'-O-ribose methylations also increase base stacking interactions, increase the hydrophobicity of the RNA, block the formation of hydrogen bonds, and render the RNA less susceptible to hydrolysis. In accordance with the greater RNA stability introduced by these modifications, the number of pseudouridines and 2'-O-ribose methylations increases with the growth temperature in archaeal hyperthermophilic organisms (Dennis et al. 2001).

7.3 A Subset of snoRNPs is Directly Involved in Pre-rRNA Cleavage Events

While the primary function of the vast majority of snoRNPs is the catalysis of nucleotide modifications, a small subset of eukaryotic, but not archaeal, snoRNPs stands out of this group, differing in several key aspects from conventional snoRNPs: (1) the snoRNPs are essential for pre-rRNA cleavage events, and hence for cell growth, (2) if they also catalyze nucleotide modifications, this modification activity is not obligatory for their function in pre-rRNA cleavage events, and (3) these snoRNPs contain additional snoRNA sequence elements that are essential for pre-rRNA cleavage and that base-pair with the pre-rRNA. This subset of snoRNPs includes the U3, U14, snR30/E1/U17, snR10, U8, and U22 snoRNPs.

The U3 snoRNP is central to the biogenesis of the small ribosomal subunit and its function is conserved from yeast to mammals. As a box C/D snoRNP, it does not catalyze nucleotide modifications but instead acts as a chaperone during ribosome biogenesis to facilitate pre-rRNA folding and subsequent pre-rRNA cleavage (Steitz and Tycowski 1995). To achieve this, the 5' portion of the U3 snoRNA engages in Watson–Crick base-pairing interactions with four different regions in the pre-rRNA, two in the 5' external transcribed spacer and two in the 18S coding region

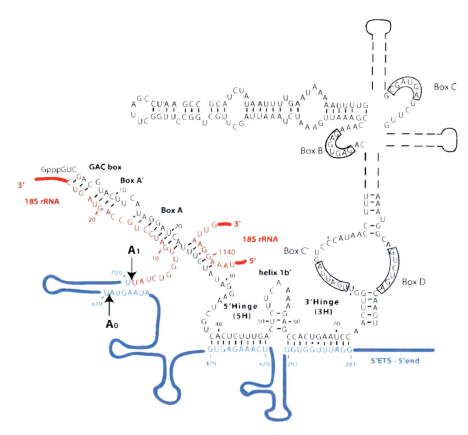

Fig. 7.2 The U3 small nucleolar RNAs (snoRNA) base-pairing interactions with the pre-ribosomal RNA (pre-rRNA) in yeast. The 5′ in most regions of the U3 snoRNA (*black*) base-pairs with two distinct regions in the 18S rRNA (*red*), one at the 5′ end of the 18S and the other in the central region of the 18S. In addition, the 5′ and 3′ hinge regions of the U3 snoRNA base-pair with two distinct regions in the 5′ ETS (*blue*) of the pre-rRNA (figure from Dutca et al. 2011)

(Beltrame and Tollervey 1992, 1995; Borovjagin and Gerbi 2000; Dutca et al. 2011; Sharma and Tollervey 1999; Tyc and Steitz 1992) (Fig. 7.2). These base-pairing interactions likely ensure accurate formation of the 5′ end pseudoknot in the 18S rRNA, a highly conserved structural feature in the small ribosomal subunit (Hughes 1996). In addition, the U3:pre-rRNA base-pairing interactions are essential for the pre-rRNA cleavage events that liberate the rRNA destined for the small ribosomal subunit. However, to date there is no direct evidence that the U3 snoRNA actively participates in the catalysis of the pre-rRNA cleavage events as a ribozyme. Instead, the U3 snoRNP likely provides an anchor to recruit other processing factors to the pre-rRNA in the form of the SSU processome, including putative endonucleases such as Utp24 that could catalyze the necessary pre-rRNA cleavages (Bleichert et al. 2006; Dragon et al. 2002; Grandi et al. 2002).

In contrast to the U3 snoRNA, U14, a box C/D snoRNA, and snR10, a box H/ACA snoRNA, perform dual functions during ribosome biogenesis. Both snoRNAs are, similar to U3, required for pre-rRNA cleavage events, specifically for processing of the small subunit RNA (Li et al. 1990; Tollervey 1987). It is likely that base-pairing events between the snoRNAs and the pre-rRNA play a critical role for this function. In addition, both snoRNAs also guide nonessential nucleotide modifications in the context of their respective snoRNPs: U14 guides $2'$-O-ribose methylation of the 18S rRNA, whereas snR10 guides pseudouridylation in the 25S rRNA (Samarsky and Fournier 1999). Yeast snR30, which is known as E1/U17 in vertebrates, is the fourth snoRNA essential for pre-18S rRNA processing (Morrissey and Tollervey 1993). Instead of guiding pseudouridylation, this box H/ACA snoRNA contains conserved $3'$ sequence elements that directly base-pair with the 18S region (Atzorn et al. 2004; Fayet-Lebaron et al. 2009).

While the above snoRNAs are conserved throughout eukaryotes, two box C/D snoRNAs required for pre-rRNA cleavage events in higher eukaryotes do not exist in yeast. These include the U8 and U22 snoRNAs. Interestingly, whereas U22 is required for pre-18S processing (Tycowski et al. 1994), U8 is the only snoRNA known to date that is essential for processing of the large ribosomal subunit RNAs, 28S and 5.8S (Peculis and Steitz 1993). This requires base-pairing of the $5'$ end of U8 with the $5'$ end of 28S. Intriguingly, the latter usually interacts with the $3'$ end of the 5.8S in the mature ribosome, suggesting that U8 prevents premature formation of this interaction between these rRNAs during ribosome biogenesis (Peculis 1997). This raises the question as to how yeast solves this same problem in the absence of the U8 snoRNP. It has been proposed that parts of the internal transcribed spacer region 2 (ITS2) of the pre-rRNA perform this function in *cis*, making U8 dispensable in yeast (Cote et al. 2002).

In summary, only a small subset of snoRNAs (4 out of ~100 in yeast and 6 out of ~200 in mammals) are essential and are required for pre-rRNA cleavage events. The most likely mode of action of these "processing" snoRNAs is to chaperone certain RNA conformations that are a prerequisite for RNA cleavage events, ultimately liberating the individual rRNAs from the precursor, rather than to directly catalyze these cleavages. However, future research will be necessary to decipher the exact molecular mechanism by which these "processing" snoRNAs perform their functions.

7.4 The Architecture of snoRNPs

A common hallmark of both box H/ACA and box C/D snoRNPs is that they are composed of a snoRNA with conserved secondary structure elements and sets of proteins that are specific to each class of snoRNP. Hence, different snoRNAs within the same class are bound by the same set of core proteins that recognize the class-specific structural features of the snoRNA. One of these common snoRNP proteins acts as the catalytic subunit and is directed to the site of modification by base-pairing

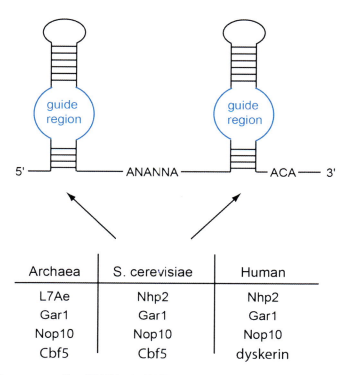

Fig. 7.3 Components of box H/ACA s(no)RNPs. The secondary structure of a two-hairpin box H/ACA s(no)RNA is shown and conserved sequence elements are indicated. The guide regions that base-pair with the substrate RNA are highlighted in *blue*. Four core proteins are conserved and are thought to bind to each hairpin

of the snoRNA with the substrate RNA and by interaction with the remaining common core proteins. Studies of eukaryotic snoRNPs initially provided some insight into the architecture of snoRNPs. However, our understanding of the structure of snoRNPs leaped forward with the establishment of *in vitro* reconstitution techniques of catalytically active sRNPs from archaea, which enabled structural studies of these assemblies.

7.4.1 Conserved Secondary Structures of Small Nucleolar RNAs

Box H/ACA and box C/D snoRNAs adopt distinctive class-specific secondary structures that are maintained in most, if not all, snoRNAs within the respective class. In eukaryotes, box H/ACA snoRNAs are characterized by two RNA hairpins, connected by a single stranded hinge region with box H (ANANNA in sequence) (Balakin et al. 1996) (Fig. 7.3). The terminal hairpin is followed by a single stranded tail containing the conserved box ACA. Each hairpin contains an internal bulge, the

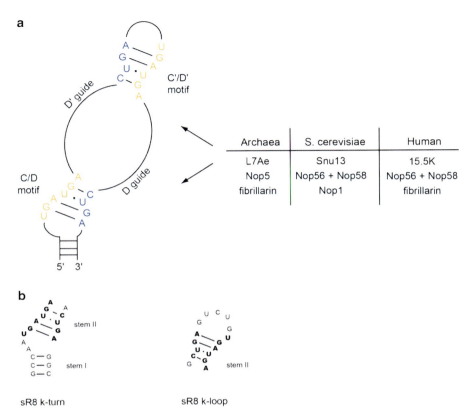

Fig. 7.4 Components of box C/D s(no)RNPs. (**a**) Secondary structure and conserved sequence elements of box C/D s(no)RNAs are shown. The guide regions that base-pair to substrate RNAs reside in the spacer regions bridging the C/D and C'/D' motifs. Three to four conserved core proteins bind to each conserved C/D and C'/D' motif. (**b**) Secondary structures of a k-turn (box C/D) and a k-loop (C'/D') motif

pseudouridylation pocket, which serves as a bipartite guide sequence, base-pairing with substrate RNA to direct isomerization of an internal unpaired substrate uridine (Ganot et al. 1997; Ni et al. 1997). This pseudouridylation occurs ~15 nucleotides upstream of box ACA (Ni et al. 1997). In addition to dual-hairpin box H/ACA sRNAs, single- or triple-hairpin box H/ACA sRNAs also occur in archaeal organisms.

Typical features of box C/D snoRNAs include the conserved sequence elements, boxes C (consensus RUGAUGA) and D (consensus CUGA), respectively, which fold into an asymmetric stem-internal loop-stem structure, termed the kink-turn or k-turn motif, introducing a sharp bent into the RNA (Kiss-Laszlo et al. 1996, 1998; Moore et al. 2004; Tycowski et al. 1996; Watkins et al. 2000) (Fig. 7.4). Besides terminal boxes C and D, variations of them can be found internally as boxes C' and D' (Kiss-Laszlo et al. 1998). However, most C'/D' motifs fold into k-loops rather than k-turns because of the lack of a stem I (Nolivos et al. 2005). Both spacer regions between boxes C and D′ as well as boxes C′ and D encompass 10–21 guide nucleotides that are complementary to substrate RNAs, specifying the rRNA nucleotide base-paired to the fifth nucleotide upstream of box D or box D′ as the nucleotide targeted for methylation (D plus 5 rule) (Kiss-Laszlo et al. 1996). A fourth hallmark of box C/D snoRNAs is a terminal stem formed by base-pairing of nucleotides 5′ of box C and 3′ of box D, which is important for snoRNA stability *in vivo* (Huang et al. 1992). These general features of box C/D snoRNAs are shared by both eukaryotic and archaeal box C/D s(no)RNAs (Gaspin et al. 2000; Omer et al. 2000). However, in archaeal box C/D sRNAs boxes C′ and D′ are more similar if not identical to boxes C and D in sequence and D and D′ spacer lengths are constrained to an average length of ~12 nucleotides (Omer et al. 2000; Tran et al. 2005).

7.4.2 Box H/ACA s(no)RNPs

Box H/ACA snoRNAs associate with four common box H/ACA core proteins, Nhp2 (L7Ae in archaea), Cbf5 (also named Nap57 and dyskerin), Nop10, and Gar1 (Kiss et al. 2010) (Fig. 7.3). Cbf5 is the catalytic subunit of box H/ACA s(no)RNPs and is homologous to *Escherichia coli* TruB pseudouridine synthase. To understand the architecture of box H/ACA snoRNPs, the key questions that need to be answered are how the pseudouridine synthase Cbf5/Nap57/dyskerin interacts with the snoRNA and what role the remaining three core proteins play in this process. Studies of eukaryotic box H/ACA snoRNPs revealed that Cbf5/Nap57/dyskerin does not directly interact with the H/ACA snoRNA in the absence of the remaining core proteins, suggesting that the latter proteins are responsible for recruitment of the catalytic subunit to the snoRNA (Wang and Meier 2004). Consistent with that, several protein–protein interactions have been described, including interactions of each Nop10 and Gar1 with Cbf5/Napf57/dyskerin, the formation of a stable heterotrimeric complex of Cbf5, Nop10, and Gar1 in yeast, as well as of Nap57/dyskerin, Nop10, and Nhp2 in mammals (Henras et al. 2004; Wang and Meier 2004). These could reflect snoRNP assembly intermediates as suggested by the observation that specific box H/ACA snoRNA binding in mammals requires at least the core heterotrimeric Nap57/dyskerin-Nop10-Nhp2 complex (Wang and Meier 2004). Moreover, a heterotetrameric complex of all four core proteins in the absence of the snoRNA has been observed (Henras et al. 2004; Wang and Meier 2004), suggesting an extensive network of protein–protein interactions within the RNP. Unfortunately, to date eukaryotic box H/ACA snoRNPs have been refractory to *in vitro* reconstitution of catalytically active RNPs from purified recombinant components, which has impeded further in depth structural studies of these complexes.

In contrast to the situation in eukaryotes, archaeal box H/ACA sRNPs can be successfully reconstituted *in vitro* (Baker et al. 2005). This greatly facilitated the determination of high-resolution crystal structures of archaeal box H/ACA sRNPs in the *apo*-form and in the substrate-bound form, which have provided great insight

7 Small Ribonucleoproteins in Ribosome Biogenesis

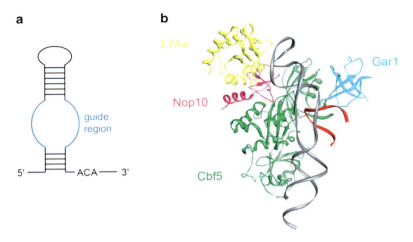

Fig. 7.5 Architecture of archaeal box H/ACA sRNPs. (**a**) Single-hairpin box H/ACA sRNA. (**b**) Crystal structure of *Pyrococcus furiosus* box H/ACA sRNP containing all common box H/ACA proteins, L7Ae, Nop10, Gar1, and Cbf5 and a single hairpin box H/ACA sRNA. The box H/ACA sRNA is shown in *gray* and is bound to a short substrate RNA (PDB code 3HAY, Duan et al. 2009)

into fundamental questions of box H/ACA s(no)RNP architecture and function (Duan et al. 2009; Li and Ye 2006; Liang et al. 2007a, 2009a) (Fig. 7.5). The crystal structure of a single-hairpin box H/ACA sRNA associated with all H/ACA proteins, L7Ae (the archaeal Nhp2 homolog), Nop10, Gar1, and Cbf5 showed that L7Ae, Nop10, and the catalytic domain of Cbf5 bind to the upper stem of the hairpin RNA, whereas the PUA domain of Cbf5 binds the lower stem and also directly interacts with the conserved ACA sequence element. This positions the guide sequences and the pseudouridylation pocket near the active site cleft of the catalytic subunit Cbf5 (Li and Ye 2006). Gar1 does not directly contact the RNA but instead is recruited into the complex via interaction with the thumb region of Cbf5. The protein–protein and RNA–protein contacts made in the crystal structure rationalize previous biochemical observations, such as the direct interaction of archaeal Cbf5 with the sRNA, the requirement of the conserved box ACA element for Cbf5 binding, and the specific RNA binding activity of L7Ae to the k-turn motif in the upper stem of the hairpin (Baker et al. 2005). Eukaryotic box H/ACA snoRNAs do not contain k-turn motifs, implying that Nhp2 is recruited to the RNP via protein–protein interaction (Rozhdestvensky et al. 2003). Accordingly, mammalian Nhp2 forms a stable trimeric complex with Nop10 and Nap57/dyskerin that has specific RNA binding activity, whereas archaeal L7Ae does not and is recruited to the RNA independently (Baker et al. 2005; Wang and Meier 2004).

Comparison of apo- and substrate-bound archaeal box H/ACA sRNP structures provides insights into the mechanism of substrate RNA binding and pseudouridylation. Upon substrate binding, the guide sequences of the H/ACA RNA become more structured. In addition, Cbf5 makes numerous contacts with both the guide and substrate RNAs in a mainly sequence-independent manner (Duan et al. 2009; Liang et al. 2009a). Interestingly, in a crystal structure of a substrate-bound box H/ACA sRNP lacking L7Ae, the target uridine was positioned >10 Å away from the active site of Cbf5, indicating that L7Ae plays a critical role in positioning the target nucleotide by anchoring the upper stem of the RNA hairpin (Liang et al. 2007a). In contrast to the other proteins, Gar1 does not contact the H/ACA RNA or the substrate directly (Duan et al. 2009; Li and Ye 2006). Gar1 is also not required for correct positioning of the substrate uridine in the active site of Cbf5. However, Gar1 is thought to regulate substrate loading, product release, and enzyme turnover, most likely by promoting conformational changes in the thumb region of Cbf5 (Duan et al. 2009). Nevertheless, the mechanism of substrate release and sRNP turnover is not explained by the current structures and questions as to how exactly the product is released and whether this involves dissociation of protein components remain (Hamma and Ferre-D'Amare 2010).

Conservation of the secondary structure of the s(no)RNA, of the proteins, and of some RNP interactions between archaea and eukaryotes suggests that the overall architecture of the archaeal single-hairpin box H/ACA sRNP is conserved and that it is also reflected in the assembly of the eukaryotic core proteins onto one of the hairpins in eukaryotic H/ACA snoRNAs (Li and Ye 2006). Therefore, eukaryotic dual-hairpin box H/ACA snoRNPs likely contain two copies (one per hairpin) of each of the core proteins, and box H may substitute for box ACA in binding Cbf5/Nap57/dyskerin in the 5′ hairpin. This symmetrical assembly and the stoichiometry of the core protein components would be consistent with electron microscopic studies of purified yeast box H/ACA snoRNPs (Watkins et al. 1998). However, one puzzling finding to date that cannot be explained by the existing data is that two intact hairpins are required for pseudouridylation directed by either hairpin in eukaryotic box H/ACA snoRNPs (Bortolin et al. 1999), suggesting inter-hairpin communication, possibly mediated by protein–protein interactions between core proteins associated with the two different hairpins. Unfortunately, the exact molecular and structural basis for the hairpin interdependency remains unclear and will require structural studies of the eukaryotic RNPs.

7.4.3 Box C/D snoRNPs

Eukaryotic box C/D snoRNPs contain four common core proteins, 15.5K/Snu13, Nop56, Nop58, and fibrillain/Nop1 (Reichow et al. 2007). Fibrillarin/Nop1 is the catalytic subunit within the RNP that catalyzes the methyltransferase reaction using *S*-adenosylmethionine as a cofactor (Wang et al. 2000). Nop56 and Nop58 are highly homologous but not redundant proteins that evolved from a common ancestor by gene duplication, as archaeal organisms have a single homolog (Gautier et al. 1997; Omer et al. 2000). Interestingly, 15.5K/Snu13 is related to eukaryotic Nhp2 and both proteins have L7Ae as a common homolog in archaea

(Kuhn et al. 2002; Rozhdestvensky et al. 2003). However, while L7Ae is a component of both archaeal box H/ACA sRNPs and box C/D sRNPs, 15.5K/Snu13 cannot replace Nhp2 in eukaryotic box H/ACA snoRNPs and vice versa (Watkins et al. 1998, 2000).

Early experiments reconstituting partial eukaryotic box C/D snoRNPs in cell extracts or in *Xenopus* oocytes established that box C/D snoRNPs assemble in a hierarchical manner. Binding of Snu13/15.5K to the box C/D k-turn motif is a prerequisite for the binding of the remaining core proteins (Watkins et al. 2000, 2002). Interestingly, Snu13/15.5K does not bind the related C'/D' motif, at least not *in vitro* (Szewczak et al. 2002, 2005). This asymmetric architecture is further substantiated by *in vivo* crosslinking experiments revealing direct contact sites of Nop58 with box C, of Nop56 with box C', and of fibrillarin with box D, C', and D' (Cahill et al. 2002). However, in cell extracts, all proteins can assemble on a snoRNA containing the C/D motif only, but not on a snoRNA that only contains the C'/D' motif (Watkins et al. 2002), emphasizing the importance of the terminal C/D motif for snoRNP assembly and function. This is also consistent with *in vivo* studies that determined that the terminal C/D motif, as well as the directly contacting proteins, are strictly required for snoRNA stability *in vivo* (Lafontaine and Tollervey 1999; Watkins et al. 2000).

Contrary to eukaryotic box C/D snoRNPs, archaeal box C/D sRNPs have been reconstituted *in vitro* and are catalytically active for 2'-O-ribose methylation of substrate RNAs (Omer et al. 2000), which greatly stimulated efforts to obtain their structures. Initially, crystal structures of all protein components and of partial sRNPs were determined, and recently both an EM reconstruction and a crystal structure of fully assembled and enzymatically active box C/D sRNPs have been revealed (Aittaleb et al. 2003; Bleichert et al. 2009; Lin et al. 2011; Oruganti et al. 2007; Xue et al. 2010; Ye et al. 2009). Surprisingly, both full-complex structures strikingly differ in their architecture (Figs. 7.6 and 7.7).

The EM structure of catalytically active, fully assembled *Methanocaldococcus jannaschii* sR8 sRNP revealed a dimeric box C/D sRNP. Interpretation of the EM structure in light of previous crystal structures of individual protein components and partial RNPs suggested a structural model of this dimeric archaeal box C/D sRNP. Accordingly, L7Ae binds to both the box C/D k-turn and the box C'/D' k-loop, and each creates a combined sRNA-L7Ae binding platform for the C-terminus of Nop5. Nop5 homodimerizes via a coiled-coil domain and also interacts with fibrillarin via its N-terminus to form a heterotetramer (Aittaleb et al. 2003; Ye et al. 2009). Strikingly, the EM density can accommodate four molecules of each protein and likely contains two sRNAs (Bleichert et al. 2009). Since neither L7Ae nor the Nop5-fibrillarin heterotetramer is known to dimerize further, it was suggested that the two sRNAs must orchestrate di-sRNP assembly by bridging two Nop5-fibrillarin heterotetramers to form the dimeric RNP. This idea is also consistent with the observed additional EM density that may reflect the presence of the sRNA guide sequences. These findings contrast with a longstanding monomeric sRNP model that argues that archaeal box C/D sRNPs contain two sets of

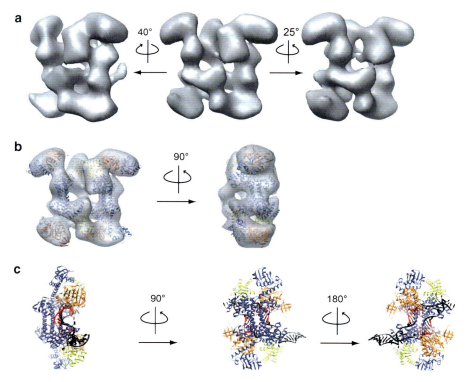

Fig. 7.6 Structures of archaeal box C/D sRNPs. (**a**) EM reconstruction of a dimeric box C/D sRNP from *Methanocaldococcus jannaschii* with fitted crystal structures of *P. furiosus* Nop5-fibrillarin (PDB 2nnw) and *M. jannaschii* L7Ae (PDB 1xbi) in (**b**). Figures in (**a**) and (**b**) are reprinted from Bleichert et al. (2009). (**c**) Crystal structure of a monomeric box C/D sRNP with bound substrate from *Sulfolobus solfataricus* (PDB 3pla, Lin et al. 2011). L7Ae – *yellow*, Nop5 – *blue*, fibrillarin – *orange*, sRNA – *black*, substrate RNA – *red*

core proteins and one sRNA, with the sRNA binding to one Nop5-fibrillarin heterotetramer (Aittaleb et al. 2003).

Besides the EM structure of a *M. jannaschii* box C/D sRNP, the crystal structure of a catalytically active, substrate-bound *Sulfolobus solfataricus* box C/D sRNP has been reported recently (Lin et al. 2011). The crystallized RNP contains all core proteins and a box C/D sRNA mimic. The structure represents a traditional mono-sRNP where the sRNA engages with a single Nop5-fibrillarin heterotetramer. The ribose of the substrate nucleotide is positioned within the active site of fibrillarin, which is achieved by conformational changes of the N-terminal Nop5 domain that recruits fibrillarin. These changes are not unexpected as previous crystal structures also found the N-terminal domain of Nop5 differentially positioned with respect to the Nop5 coiled-coil domains (Aittaleb et al. 2003; Lin et al. 2011; Oruganti et al. 2007; Ye et al. 2009).

7 Small Ribonucleoproteins in Ribosome Biogenesis

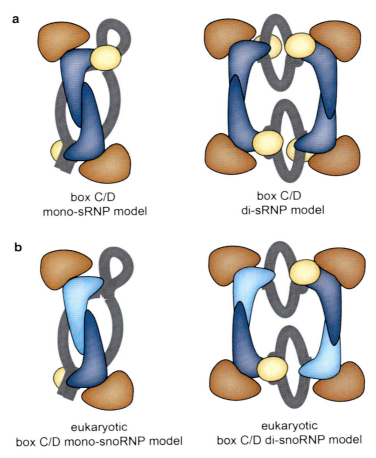

Fig. 7.7 Architectural models of (**a**) archaeal box C/D sRNPs and (**b**) eukaryotic box C/D snoRNPs. Colors for archaeal proteins are as in Fig. 7.6. Colors for eukaryotic proteins are as follows: 15.5K/Snu13 – *yellow*, Nop56 – *light blue*, Nop58 – *dark blue*, and fibrillarin/Nop1 – *orange*. The RNAs are depicted in *gray*

It is not yet understood why two quite different organizations of *in vitro* reconstituted archaeal box C/D sRNPs exist. Recent sizing experiment of *M. jannaschii* and *Pyrococcus furiosus* box C/D sRNPs reconstituted with different naturally occurring sRNAs suggests a primarily dimeric apo-sRNP (Bleichert and Baserga 2010a; Bleichert et al. 2009; Ghalei et al. 2010). In contrast, reconstitution of an enzymatically active *S. solfataricus* box C/D sRNP with substrate RNA resulted in both RNP monomers and probably RNP dimers, of which only the RNP monomer crystallized (Lin et al. 2011). It is interesting that the crystallized RNP contains an artificial RNA composed of two different ribo-oligonucleotides, resulting in a symmetric box C/D sRNA containing two classical k-turns rather than a k-turn and a k-loop as found in naturally occurring sRNAs (Nolivos et al. 2005). Whether this RNA conformation influences box C/D sRNP architecture remains to be tested in future studies. Analysis of archaeal box C/D sRNPs both *in vitro* and *in vivo* will also be important to resolve the controversy between mono- and di-sRNP.

So what does the knowledge of the structure of archaeal box C/D sRNPs teach us about the architecture and function of box C/D snoRNPs in eukaryotes? Integration of available biochemical data on box C/D snoRNP protein–protein and protein–snoRNA interactions into archaeal models of box C/D sRNP architecture leads to two possible models of eukaryotic box C/D snoRNP organization, a mono- and a di-snoRNP, respectively (Bleichert and Baserga 2010b). In either model, only the C/D motif would be associated with Snu13/15.5K. The archaeal Nop5s contacting the C/D motif and C′/D′ motif, respectively, would be replaced by Nop58 and Nop56, resulting in an asymmetric organization of the core proteins on a single snoRNA as has been observed by *in vivo* RNA–protein crosslinking and by *in vitro* binding studies (Cahill et al. 2002; Szewczak et al. 2002, 2005). However, the asymmetry of eukaryotic box C/D snoRNPs has recently been challenged by Qu et al. (2011), who observed the association of Snu13 with a snoRNA lacking a C/D motif *in vivo* in yeast, concluding that Snu13 binds the C′/D′ motif *in vivo*. The basis for the discrepancies between this *in vivo* study and previous *in vitro* studies is not clear but could be reconciled by a di-snoRNP, where the artificial snoRNA lacking the C/D motif is incorporated into a di-snoRNP that contains a typical endogenous snoRNA with an intact C/D motif that directly binds Snu13/15.5K. Whether this is indeed the case is not yet known, as it is not yet known whether eukaryotic box C/D snoRNPs are indeed di-snoRNPs. The sizing information on eukaryotic box C/D snoRNPs is not precise enough to differentiate between the models.

7.5 The Role of RNA Helicases in snoRNP Function

Intrinsic to the mechanism of snoRNP function, whether in catalyzing RNA modifications or in participating in pre-rRNA processing, is the formation of Watson–Crick base-pairing interactions with the target RNA (Beltrame and Tollervey 1992; Cavaille et al. 1996; Fayet-Lebaron et al. 2009; Ganot et al. 1997; Kiss-Laszlo et al. 1996; Ni et al. 1997; Tycowski et al. 1996). Typically, duplexes of 10–21 bp are formed co-transcriptionally between the snoRNA guide and the nascent pre-rRNA (Osheim et al. 2004). However, after successfully performing their task, snoRNPs would need to be released from the pre-rRNA as they are not part of the mature ribosome and their presence would most likely interfere with correct ribosome assembly. Both active and passive mechanisms of snoRNP release can be envisioned. First, RNA helicases have been implicated to directly unwind snoRNA–pre-rRNA duplexes. Second, conformational changes in the pre-rRNA during ribosome biogenesis could weaken or disrupt base-pairing interactions of the snoRNA with the pre-rRNA. Third, binding of ribosomal proteins or nonhelicase pre-rRNA processing factors could displace snoRNAs from pre-rRNAs. Fourth, snoRNP components could be directly involved in enzyme turnover such that conformational changes in the snoRNP would be incompatible with base-pairing between snoRNA and substrate RNA, resulting in substrate RNA release.

Models of active snoRNA–preRNA duplex unwinding are supported by observations that some snoRNAs remain associated with pre-ribosomes upon depletion of certain RNA helicases. Examples include Dbp4, Has1, and Rok1 in yeast and Ddx51 in mammals (Bohnsack et al. 2008; Kos and Tollervey 2005; Liang and Fournier 2006; Srivastava et al. 2010). However, in no case has direct unwinding of a snoRNA–pre-rRNA duplex been shown. It cannot be excluded that snoRNA sequestration into preribosomes is indirectly related to RNA helicase depletion and results as a consequence of impairing critical steps in ribosome biogenesis unrelated to snoRNA release. Furthermore, it is unlikely that the 20 nucleolar RNA helicases in yeast can actively release all ~100 snoRNAs that are predicted to base-pair to the pre-rRNA at various steps during ribosome biogenesis (Bleichert and Baserga 2007; Cordin et al. 2006), indicating the likelihood of existing alternative mechanisms. Thus, both direct and indirect mechanisms most likely complement each other *in vivo* to release snoRNAs from their substrates and to ensure proper ribosome maturation.

7.6 RNase MRP snoRNP

The third class of snoRNPs is RNase MRP. The RNAse MRP snoRNP is the only member of this class, and unlike box H/ACA and box C/D snoRNPs, does not catalyze chemical modifications of RNA nucleotides. Instead, the RNase MRP snoRNP acts as a site-specific endonuclease. As such, it is involved in several cellular processes and acts on quite different target RNAs: (1) In mitochondria, RNAse MRP has been suggested to cleave the RNA primer involved in mitochondrial DNA replication (Chang and Clayton 1987). (2) In the nucleolus, RNase MRP functions in pre-rRNA processing. In yeast, it cleaves the pre-rRNA in the ITS1 at site A_3, initiating maturation of the short form of the 5.8S rRNA (Fig. 7.1) (Chu et al. 1994; Lygerou et al. 1996; Schmitt and Clayton 1993), and (3) in processing of certain mRNAs, including CLB2 and viperin (Gill et al. 2004; Mattijssen et al. 2011). Studies with CLB2 mRNA indicate that RNase MRP regulates mRNA turnover, as endonucleotytic cleavage of this mRNA in the 5′ UTR provides an entry site for mRNA degrading exonucleases. Most likely, the function of RNase MRP in mRNA cleavage is responsible for it being essential for cell viability, as RNAse MRP concentration in mitochondria is low and loss of mitochondrial DNA replication and loss of A_3 cleavage of the pre-rRNA are not essential for cell growth (Kiss and Filipowicz 1992; Pham et al. 2006; Schmitt and Clayton 1993).

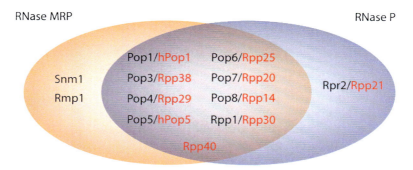

Fig. 7.8 Protein composition of RNase MRP (*orange*) and RNase P (*purple*). Protein distribution reflects subunits common to both RNPs in the center and protein subunits specific to either RNP at the periphery. Yeast proteins are in *black* and corresponding human homologs are in *red*

RNase MRP is exclusively found in eukaryotes. In contrast, its close relative, RNase P, is found in all domains of life, albeit the fact that the RNP differs greatly in complexity, ranging from one RNA and one protein in bacteria to one RNA and ten proteins in eukaryotes. A well-established function for RNase P is the endonucleolytic maturation of the 5' end of tRNAs. In addition, there is some evidence that RNase P also participates in ribosome biogenesis in the nucleolus (Jacobson et al. 1997): (1) RNase P plays a role in maturation of intron encoded box C/D snoRNAs (Coughlin et al. 2008) and (2) RNase P participates in pre-rRNA processing in ITS2 and in 3' end formation of the 5.8S rRNA (Chamberlain et al. 1996; Lee et al. 1996) (Fig. 7.1).

Despite their distinct functions and substrates, RNases MRP and P are highly evolutionary related. Both RNPs contain an RNA component of similar secondary structure (Forster and Altman 1990) and share the majority of protein components (Fig. 7.8). Interestingly, many MRP and P RNAs contain a k-turn or k-loop motif (Rosenblad et al. 2006), which most likely acts as a protein binding site as it does in box C/D and box H/ACA snoRNPs. Unfortunately, to date no structures exist of archaeal or eukaryotic RNase MRP or RNase P holoenzymes. In contrast, the crystal structure of bacterial RNase P holoenzyme has been determined (Reiter et al. 2010). However, because of the much greater complexity of archaeal and eukaryotic counterparts, their architecture and the specific role of all protein components remain elusive. Nonetheless, biochemical studies probing the RNA secondary structure, protein–protein interactions, and protein–RNA interactions provide some insights into the architecture of these RNPs (Aspinall et al. 2007; Houser-Scott et al. 2002; Pluk et al. 1999; Welting et al. 2004) (Fig. 7.9). Moreover, some proteins appear to be differentially or only transiently associated with RNase MRP because they are not found associated with MRP RNA in 60–80S preribosomes (Welting et al. 2006). The functional consequences of these differential associations are currently unknown.

7 Small Ribonucleoproteins in Ribosome Biogenesis

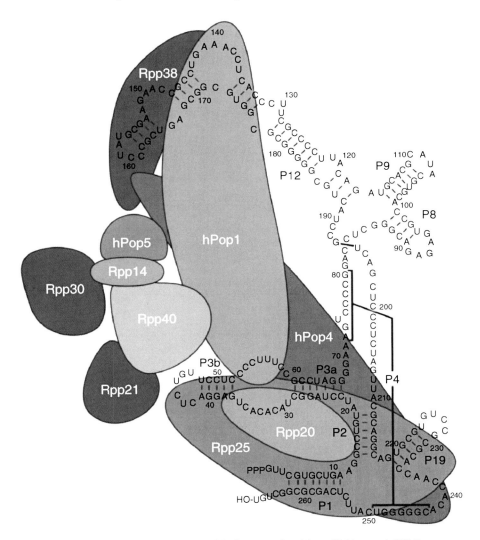

Fig. 7.9 Architecture of human RNase MRP (figure reprinted from Welting et al. 2004)

7.7 Conclusions

Research over the last several decades has provided exciting and important insights into both the structure and function of small RNPs residing in the nucleolus. Future research will undoubtedly further enhance our mechanistic understanding of how these RNPs perform their tasks and how this is linked to fundamental mechanisms of cell growth and survival. This has widespread implications for the pathogenesis and treatment of diseases that result from deregulation of ribosome biogenesis.

References

Aittaleb M, Rashid R, Chen Q, Palmer JR, Daniels CJ, Li H (2003) Structure and function of archaeal box C/D sRNP core proteins. Nat Struct Biol 10:256–263

Aspinall TV, Gordon JM, Bennett HJ, Karahalios P, Bukowski JP, Walker SC, Engelke DR, Avis JM (2007) Interactions between subunits of *Saccharomyces cerevisiae* RNase MRP support a conserved eukaryotic RNase P/MRP architecture. Nucleic Acids Res 35:6439–6450

Atzorn V, Fragapane P, Kiss T (2004) U17/snR30 is a ubiquitous snoRNA with two conserved sequence motifs essential for 18S rRNA production. Mol Cell Biol 24:1769–1778

Baker DL, Youssef OA, Chastkofsky MI, Dy DA, Terns RM, Terns MP (2005) RNA-guided RNA modification: functional organization of the archaeal H/ACA RNP. Genes Dev 19:1238–1248

Balakin AG, Smith L, Fournier MJ (1996) The RNA world of the nucleolus: two major families of small RNAs defined by different box elements with related functions. Cell 86:823–834

Beltrame M, Tollervey D (1992) Identification and functional analysis of two U3 binding sites on yeast pre-ribosomal RNA. EMBO J 11:1531–1542

Beltrame M, Tollervey D (1995) Base pairing between U3 and the pre-ribosomal RNA is required for 18S rRNA synthesis. EMBO J 14:4350–4356

Bleichert F, Baserga SJ (2007) The long unwinding road of RNA helicases. Mol Cell 27:339–352

Bleichert F, Baserga SJ (2010a) Dissecting the role of conserved box C/D sRNA sequences in di-sRNP assembly and function. Nucleic Acids Res 38:8295–8305

Bleichert F, Baserga SJ (2010b) Ribonucleoprotein multimers and their functions. Crit Rev Biochem Mol Biol 45:331–350

Bleichert F, Granneman S, Osheim YN, Beyer AL, Baserga SJ (2006) The PINc domain protein Utp24, a putative nuclease, is required for the early cleavage steps in 18S rRNA maturation. Proc Natl Acad Sci USA 103:9464–9469

Bleichert F, Gagnon KT, Brown BA II, Maxwell ES, Leschziner AE, Unger VM, Baserga SJ (2009) A dimeric structure for archaeal box C/D small ribonucleoproteins. Science 325:1384–1387

Bohnsack MT, Kos M, Tollervey D (2008) Quantitative analysis of snoRNA association with pre-ribosomes and release of snR30 by Rok1 helicase. EMBO Rep 9:1230–1236

Borovjagin AV, Gerbi SA (2000) The spacing between functional Cis-elements of U3 snoRNA is critical for rRNA processing. J Mol Biol 300:57–74

Bortolin ML, Ganot P, Kiss T (1999) Elements essential for accumulation and function of small nucleolar RNAs directing site-specific pseudouridylation of ribosomal RNAs. EMBO J 18:457–469

Cahill NM, Friend K, Speckmann W, Li ZH, Terns RM, Terns MP, Steitz JA (2002) Site-specific cross-linking analyses reveal an asymmetric protein distribution for a box C/D snoRNP. EMBO J 21:3816–3828

Cavaille J, Nicoloso M, Bachellerie JP (1996) Targeted ribose methylation of RNA in vivo directed by tailored antisense RNA guides. Nature 383:732–735

Chamberlain JR, Pagan R, Kindelberger DW, Engelke DR (1996) An RNase P RNA subunit mutation affects ribosomal RNA processing. Nucleic Acids Res 24:3158–3166

Chang DD, Clayton DA (1987) A novel endoribonuclease cleaves at a priming site of mouse mitochondrial DNA replication. EMBO J 6:409–417

Chu S, Archer RH, Zengel JM, Lindahl L (1994) The RNA of RNase MRP is required for normal processing of ribosomal RNA. Proc Natl Acad Sci USA 91:659–663

Cordin O, Banroques J, Tanner NK, Linder P (2006) The DEAD-box protein family of RNA helicases. Gene 367:17–37

Cote CA, Greer CL, Peculis BA (2002) Dynamic conformational model for the role of ITS2 in pre-rRNA processing in yeast. RNA 8:786–797

Coughlin DJ, Pleiss JA, Walker SC, Whitworth GB, Engelke DR (2008) Genome-wide search for yeast RNase P substrates reveals role in maturation of intron-encoded box C/D small nucleolar RNAs. Proc Natl Acad Sci USA 105:12218–12223

7 Small Ribonucleoproteins in Ribosome Biogenesis

Decatur WA, Fournier MJ (2002) rRNA modifications and ribosome function. Trends Biochem Sci 27:344–351

Dennis PP, Omer A, Lowe T (2001) A guided tour: small RNA function in Archaea. Mol Microbiol 40:509–519

Dragon F, Gallagher JE, Compagnone-Post PA, Mitchell BM, Porwancher KA, Wehner KA, Wormsley S, Settlage RE, Shabanowitz J, Osheim Y et al (2002) A large nucleolar U3 ribonucleoprotein required for 18S ribosomal RNA biogenesis. Nature 417:967–970

Duan J, Li L, Lu J, Wang W, Ye K (2009) Structural mechanism of substrate RNA recruitment in H/ACA RNA-guided pseudouridine synthase. Mol Cell 34:427–439

Dutca LM, Gallagher JE, Baserga SJ (2011) The initial U3 snoRNA: pre-rRNA base pairing interaction required for pre-18S rRNA folding revealed by in vivo chemical probing. Nucleic Acids Res 39:5164–5180

Fayet-Lebaron E, Atzorn V, Henry Y, Kiss T (2009) 18S rRNA processing requires base pairings of snR30 H/ACA snoRNA to eukaryote-specific 18S sequences. EMBO J 28:1260–1270

Forster AC, Altman S (1990) Similar cage-shaped structures for the RNA components of all ribonuclease P and ribonuclease MRP enzymes. Cell 62:407–409

Ganot P, Bortolin ML, Kiss T (1997) Site-specific pseudouridine formation in preribosomal RNA is guided by small nucleolar RNAs. Cell 89:799–809

Gaspin C, Cavaille J, Erauso G, Bachellerie JP (2000) Archaeal homologs of eukaryotic methylation guide small nucleolar RNAs: lessons from the *Pyrococcus* genomes. J Mol Biol 297: 895–906

Gautier T, Berges T, Tollervey D, Hurt E (1997) Nucleolar KKE/D repeat proteins Nop56p and Nop58p interact with Nop1p and are required for ribosome biogenesis. Mol Cell Biol 17:7088–7098

Ghalei H, Hsiao HH, Urlaub H, Wahl MC, Watkins NJ (2010) A novel Nop5-sRNA interaction that is required for efficient archaeal box C/D sRNP formation. RNA 16:2341–2348

Gill T, Cai T, Aulds J, Wierzbicki S, Schmitt ME (2004) RNase MRP cleaves the CLB2 mRNA to promote cell cycle progression: novel method of mRNA degradation. Mol Cell Biol 24:945–953

Grandi P, Rybin V, Bassler J, Petfalski E, Strauss D, Marzioch M, Schafer T, Kuster B, Tschochner H, Tollervey D et al (2002) 90S Pre-ribosomes include the 35S pre-rRNA, the U3 snoRNP, and 40S subunit processing factors but predominantly lack 60S synthesis factors. Mol Cell 10:105–115

Hamma T, Ferre-D'Amare AR (2010) The box H/ACA ribonucleoprotein complex: interplay of RNA and protein structures in post-transcriptional RNA modification. J Biol Chem 285: 805–809

Henras AK, Capeyrou R, Henry Y, Caizergues-Ferrer M (2004) Cbf5p, the putative pseudouridine synthase of H/ACA-type snoRNPs, can form a complex with Gar1p and Nop10p in absence of Nhp2p and box H/ACA snoRNAs. RNA 10:1704–1712

Houser-Scott F, Xiao S, Millikin CE, Zengel JM, Lindahl L, Engelke DR (2002) Interactions among the protein and RNA subunits of *Saccharomyces cerevisiae* nuclear RNase P. Proc Natl Acad Sci USA 99:2684–2689

Huang GM, Jarmolowski A, Struck JC, Fournier MJ (1992) Accumulation of U14 small nuclear RNA in *Saccharomyces cerevisiae* requires box C, box D, and a 5′, 3′ terminal stem. Mol Cell Biol 12:4456–4463

Hughes JM (1996) Functional base-pairing interaction between highly conserved elements of U3 small nucleolar RNA and the small ribosomal subunit RNA. J Mol Biol 259:645–654

Ishitani R, Yokoyama S, Nureki O (2008) Structure, dynamics, and function of RNA modification enzymes. Curr Opin Struct Biol 18:330–339

Jacobson MR, Cao LG, Taneja K, Singer RH, Wang YL, Pederson T (1997) Nuclear domains of the RNA subunit of RNase P. J Cell Sci 110(pt 7):829–837

King TH, Liu B, McCully RR, Fournier MJ (2003) Ribosome structure and activity are altered in cells lacking snoRNPs that form pseudouridines in the peptidyl transferase center. Mol Cell 11:425–435

Kiss T, Filipowicz W (1992) Evidence against a mitochondrial location of the 7-2/MRP RNA in mammalian cells. Cell 70:11–16

Kiss T, Fayet-Lebaron E, Jady BE (2010) Box H/ACA small ribonucleoproteins. Mol Cell 37:597–606

Kiss-Laszlo Z, Henry Y, Bachellerie JP, Caizergues-Ferrer M, Kiss T (1996) Site-specific ribose methylation of preribosomal RNA: a novel function for small nucleolar RNAs. Cell 85:1077–1088

Kiss-Laszlo Z, Henry Y, Kiss T (1998) Sequence and structural elements of methylation guide snoRNAs essential for site-specific ribose methylation of pre-rRNA. EMBO J 17:797–807

Kos M, Tollervey D (2005) The putative RNA helicase Dbp4p is required for release of the U14 snoRNA from preribosomes in *Saccharomyces cerevisiae*. Mol Cell 20:53–64

Kuhn JF, Tran EJ, Maxwell ES (2002) Archaeal ribosomal protein L7 is a functional homolog of the eukaryotic 15.5kD/Snu13p snoRNP core protein. Nucleic Acids Res 30:931–941

Lafontaine DL, Tollervey D (1999) Nop58p is a common component of the box C+D snoRNPs that is required for snoRNA stability. RNA 5:455–467

Lafontaine DL, Bousquet-Antonelli C, Henry Y, Caizergues-Ferrer M, Tollervey D (1998) The box H+ACA snoRNAs carry Cbf5p, the putative rRNA pseudouridine synthase. Genes Dev 12:527–537

Lee B, Matera AG, Ward DC, Craft J (1996) Association of RNase mitochondrial RNA processing enzyme with ribonuclease P in higher ordered structures in the nucleolus: a possible coordinate role in ribosome biogenesis. Proc Natl Acad Sci USA 93:11471–11476

Li L, Ye K (2006) Crystal structure of an H/ACA box ribonucleoprotein particle. Nature 443:302–307

Li HD, Zagorski J, Fournier MJ (1990) Depletion of U14 small nuclear RNA (snR128) disrupts production of 18S rRNA in *Saccharomyces cerevisiae*. Mol Cell Biol 10:1145–1152

Liang XH, Fournier MJ (2006) The helicase Has1p is required for snoRNA release from pre-rRNA. Mol Cell Biol 26:7437–7450

Liang B, Xue S, Terns RM, Terns MP, Li H (2007a) Substrate RNA positioning in the archaeal H/ACA ribonucleoprotein complex. Nat Struct Mol Biol 14:1189–1195

Liang XH, Liu Q, Fournier MJ (2007b) rRNA modifications in an intersubunit bridge of the ribosome strongly affect both ribosome biogenesis and activity. Mol Cell 28:965–977

Liang B, Zhou J, Kahen E, Terns RM, Terns MP, Li H (2009a) Structure of a functional ribonucleoprotein pseudouridine synthase bound to a substrate RNA. Nat Struct Mol Biol 16:740–746

Liang XH, Liu Q, Fournier MJ (2009b) Loss of rRNA modifications in the decoding center of the ribosome impairs translation and strongly delays pre-rRNA processing. RNA 15:1716–1728

Lin J, Lai S, Jia R, Xu A, Zhang L, Lu J, Ye K (2011) Structural basis for site-specific ribose methylation by box C/D RNA protein complexes. Nature 469:559–563

Liu B, Liang XH, Piekna-Przybylska D, Liu Q, Fournier MJ (2008) Mis-targeted methylation in rRNA can severely impair ribosome synthesis and activity. RNA Biol 5:249–254

Lygerou Z, Allmang C, Tollervey D, Seraphin B (1996) Accurate processing of a eukaryotic precursor ribosomal RNA by ribonuclease MRP in vitro. Science 272:268–270

Maden BE (1990) The numerous modified nucleotides in eukaryotic ribosomal RNA. Prog Nucleic Acid Res Mol Biol 39:241–303

Maden BE, Hughes JM (1997) Eukaryotic ribosomal RNA: the recent excitement in the nucleotide modification problem. Chromosoma 105:391–400

Mattijssen S, Hinson ER, Onnekink C, Hermanns P, Zabel B, Cresswell P, Pruijn GJ (2011) Viperin mRNA is a novel target for the human RNase MRP/RNase P endoribonuclease. Cell Mol Life Sci 68:2469–2480

Moore T, Zhang Y, Fenley MO, Li H (2004) Molecular basis of box C/D RNA-protein interactions; cocrystal structure of archaeal L7Ae and a box C/D RNA. Structure 12:807–818

Morrissey JP, Tollervey D (1993) Yeast snR30 is a small nucleolar RNA required for 18S rRNA synthesis. Mol Cell Biol 13:2469–2477

Ni J, Tien AL, Fournier MJ (1997) Small nucleolar RNAs direct site-specific synthesis of pseudouridine in ribosomal RNA. Cell 89:565–573

Nolivos S, Carpousis AJ, Clouet-d'Orval B (2005) The K-loop, a general feature of the *Pyrococcus* C/D guide RNAs, is an RNA structural motif related to the K-turn. Nucleic Acids Res 33:6507–6514

Omer AD, Lowe TM, Russell AG, Ebhardt H, Eddy SR, Dennis PP (2000) Homologs of small nucleolar RNAs in Archaea. Science 288:517–522

Oruganti S, Zhang Y, Li H, Robinson H, Terns MP, Terns RM, Yang W (2007) Alternative conformations of the archaeal Nop56/58-fibrillarin complex imply flexibility in box C/D RNPs. J Mol Biol 371:1141–1150

Osheim YN, French SL, Keck KM, Champion EA, Spasov K, Dragon F, Baserga SJ, Beyer AL (2004) Pre-18S ribosomal RNA is structurally compacted into the SSU processome prior to being cleaved from nascent transcripts in *Saccharomyces cerevisiae*. Mol Cell 16:943–954

Peculis BA (1997) The sequence of the 5′ end of the U8 small nucleolar RNA is critical for 5.8S and 28S rRNA maturation. Mol Cell Biol 17:3702–3713

Peculis BA, Steitz JA (1993) Disruption of U8 nucleolar snRNA inhibits 5.8S and 28S rRNA processing in the *Xenopus* oocyte. Cell 73:1233–1245

Pham XH, Farge G, Shi Y, Gaspari M, Gustafsson CM, Falkenberg M (2006) Conserved sequence box II directs transcription termination and primer formation in mitochondria. J Biol Chem 281:24647–24652

Pluk H, van Eenennaam H, Rutjes SA, Pruijn GJ, van Venrooij WJ (1999) RNA-protein interactions in the human RNase MRP ribonucleoprotein complex. RNA 5:512–524

Qu G, van Nues RW, Watkins NJ, Maxwell ES (2011) The spatial-functional coupling of box C/D and C′/D′ RNPs is an evolutionarily conserved feature of the eukaryotic box C/D snoRNP nucleotide modification complex. Mol Cell Biol 31:365–374

Reichow SL, Hamma T, Ferre-D'Amare AR, Varani G (2007) The structure and function of small nucleolar ribonucleoproteins. Nucleic Acids Res 35:1452–1464

Reiter NJ, Osterman A, Torres-Larios A, Swinger KK, Pan T, Mondragon A (2010) Structure of a bacterial ribonuclease P holoenzyme in complex with tRNA. Nature 468:784–789

Rosenblad MA, Lopez MD, Piccinelli P, Samuelsson T (2006) Inventory and analysis of the protein subunits of the ribonucleases P and MRP provides further evidence of homology between the yeast and human enzymes. Nucleic Acids Res 34:5145–5156

Rozhdestvensky TS, Tang TH, Tchirkova IV, Brosius J, Bachellerie JP, Huttenhofer A (2003) Binding of L7Ae protein to the K-turn of archaeal snoRNAs: a shared RNA binding motif for C/D and H/ACA box snoRNAs in Archaea. Nucleic Acids Res 31:869–877

Samarsky DA, Fournier MJ (1999) A comprehensive database for the small nucleolar RNAs from *Saccharomyces cerevisiae*. Nucleic Acids Res 27:161–164

Schmitt ME, Clayton DA (1993) Nuclear RNase MRP is required for correct processing of pre-5.8S rRNA in *Saccharomyces cerevisiae*. Mol Cell Biol 13:7935–7941

Sharma K, Tollervey D (1999) Base pairing between U3 small nucleolar RNA and the 5′ end of 18S rRNA is required for pre-rRNA processing. Mol Cell Biol 19:6012–6019

Srivastava L, Lapik YR, Wang M, Pestov DG (2010) Mammalian DEAD box protein Ddx51 acts in 3′ end maturation of 28S rRNA by promoting the release of U8 snoRNA. Mol Cell Biol 30:2947–2956

Steitz JA, Tycowski KT (1995) Small RNA chaperones for ribosome biogenesis. Science 270:1626–1627

Szewczak LB, DeGregorio SJ, Strobel SA, Steitz JA (2002) Exclusive interaction of the 15.5 kD protein with the terminal box C/D motif of a methylation guide snoRNP. Chem Biol 9:1095–1107

Szewczak LB, Gabrielsen JS, Degregorio SJ, Strobel SA, Steitz JA (2005) Molecular basis for RNA kink-turn recognition by the h15.5K small RNP protein. RNA 11:1407–1419

Tollervey D (1987) A yeast small nuclear RNA is required for normal processing of pre-ribosomal RNA. EMBO J 6:4169–4175

Tollervey D, Lehtonen H, Jansen R, Kern H, Hurt EC (1993) Temperature-sensitive mutations demonstrate roles for yeast fibrillarin in pre-rRNA processing, pre-rRNA methylation, and ribosome assembly. Cell 72:443–457

Tran E, Zhang X, Lackey L, Maxwell ES (2005) Conserved spacing between the box C/D and C'/D' RNPs of the archaeal box C/D sRNP complex is required for efficient 2'-O-methylation of target RNAs. RNA 11:285–293

Tyc K, Steitz JA (1992) A new interaction between the mouse 5' external transcribed spacer of pre-rRNA and U3 snRNA detected by psoralen crosslinking. Nucleic Acids Res 20:5375–5382

Tycowski KT, Shu MD, Steitz JA (1994) Requirement for intron-encoded U22 small nucleolar RNA in 18S ribosomal RNA maturation. Science 266:1558–1561

Tycowski KT, Smith CM, Shu MD, Steitz JA (1996) A small nucleolar RNA requirement for site-specific ribose methylation of rRNA in *Xenopus*. Proc Natl Acad Sci USA 93:14480–14485

Wang C, Meier UT (2004) Architecture and assembly of mammalian H/ACA small nucleolar and telomerase ribonucleoproteins. EMBO J 23:1857–1867

Wang H, Boisvert D, Kim KK, Kim R, Kim SH (2000) Crystal structure of a fibrillarin homologue from *Methanococcus jannaschii*, a hyperthermophile, at 1.6 A resolution. EMBO J 19:317–323

Watkins NJ, Gottschalk A, Neubauer G, Kastner B, Fabrizio P, Mann M, Luhrmann R (1998) Cbf5p, a potential pseudouridine synthase, and Nhp2p, a putative RNA-binding protein, are present together with Gar1p in all H BOX/ACA-motif snoRNPs and constitute a common bipartite structure. RNA 4:1549–1568

Watkins NJ, Segault V, Charpentier B, Nottrott S, Fabrizio P, Bachi A, Wilm M, Rosbash M, Branlant C, Luhrmann R (2000) A common core RNP structure shared between the small nucleoar box C/D RNPs and the spliceosomal U4 snRNP. Cell 103:457–466

Watkins NJ, Dickmanns A, Luhrmann R (2002) Conserved stem II of the box C/D motif is essential for nucleolar localization and is required, along with the 15.5K protein, for the hierarchical assembly of the box C/D snoRNP. Mol Cell Biol 22:8342–8352

Welting TJ, van Venrooij WJ, Pruijn GJ (2004) Mutual interactions between subunits of the human RNase MRP ribonucleoprotein complex. Nucleic Acids Res 32:2138–2146

Welting TJ, Kikkert BJ, van Venrooij WJ, Pruijn GJ (2006) Differential association of protein subunits with the human RNase MRP and RNase P complexes. RNA 12:1373–1382

Xue S, Wang R, Yang F, Terns RM, Terns MP, Zhang X, Maxwell ES, Li H (2010) Structural basis for substrate placement by an archaeal box C/D ribonucleoprotein particle. Mol Cell 39:939–949

Ye K, Jia R, Lin J, Ju M, Peng J, Xu A, Zhang L (2009) Structural organization of box C/D RNA-guided RNA methyltransferase. Proc Natl Acad Sci USA 106:13808–13813

Chapter 8
Crosstalk Between Ribosome Synthesis and Cell Cycle Progression and Its Potential Implications in Human Diseases

Marie Gérus, Michèle Caizergues-Ferrer, Yves Henry, and Anthony Henras

8.1 Introduction

Ribosome biogenesis necessitates the coordinated expression of several hundreds of genes encoding ribosome components and numerous trans-acting factors and small nucleolar RNAs involved in the assembly, maturation, and nuclear export of the preribosomal particles. In eukaryotic cells, this process mobilizes the majority of the cellular activities involved in gene expression, namely transcription, splicing, nucleo-cytoplasmic transport, and translation. Ribosome biogenesis is therefore one of the most energy-consuming processes in the cells and must be intimately adjusted to cell growth and proliferation. Early analyses in yeast have shown that activation of ribosome synthesis following a nutritional shift induces a rapid increase in cell size and division (Johnston et al. 1979). These observations suggested that regulatory mechanisms ensure communications between early steps of ribosome biogenesis and some aspects of cell cycle progression in yeast cells. Since then, these communications have been the focus of numerous studies both in yeast and mammalian cells and two main conserved features emerge. The first is that several factors involved in ribosome synthesis also function directly in specific stages of cell cycle progression both in yeast and mammalian cells. The biological significance of the dual functionality of these factors remains unclear. They may function independently in ribosome synthesis and cell cycle progression or alternatively, may allow a concerted regulation of these two essential processes. The second conserved feature is that a ribosome synthesis surveillance mechanism seems to operate during

---

Y. Henry • A. Henras (✉)
Laboratoire de Biologie Moléculaire Eucaryote, CNRS,
118 route de Narbonne, 31062, Toulouse Cedex 9, France
e-mail: henry@biotoul.fr; henras@biotoul.fr

the G1 phase of the cell cycle and to communicate with the G1–S transition machinery, allowing cells to commit or not to cell division depending on the vigor of the ribosome synthesis process. Interestingly, mutations in genes encoding ribosomal proteins or factors involved in ribosome biogenesis have been associated with several human diseases collectively referred to as ribosomopathies, characterized by some recurrent symptoms including hematopoietic defects, developmental abnormalities, and predisposition to cancer. Although the origin of these symptoms remains unclear, one current hypothesis is that they may result from impaired proliferation and apoptosis as a consequence of ribosome synthesis defects.

8.2 An Evolutionarily Conserved Mechanism Monitors Ribosome Biogenesis During the G1 Phase of the Cell Cycle and Directly Influences the G1–S Transition

A series of studies presented below have suggested that, both in yeast and mammalian cells, defects in ribosome synthesis induce the accumulation of cells in the G1 phase of the cell cycle before they result in a shortage of functional ribosomes and in a reduction of the cellular translational capacity. These observations raised the hypothesis that a ribosome synthesis surveillance mechanism operates during the G1 phase and conditions commitment to cell division through a communication with the G1–S transition machinery.

8.2.1 Communications Between Ribosome Synthesis and the G1–S Transition (Start) in Yeast Cells

In yeast cells, both cell cycle duration and cell size homeostasis depend on the G1 phase and the G1–S transition called "Start." During the G1 phase, cells grow until they reach a critical size, which is one of the parameters that determine commitment to cell division and passage through Start (Johnston et al. 1977). Active translation is required to support cell growth and indeed, treatment of yeast cells with low doses of cycloheximide, a translation inhibitor, induces a rapid accumulation of unbudded cells in the G1 phase of the cell cycle and therefore delays passage through Start (Popolo et al. 1982). However, translation per se may not be the only parameter monitored during the G1 phase. Growing evidence indeed suggests the existence of a direct connection between early steps of ribosome biogenesis and the G1–S transition of the cell cycle in yeast. Depletion of several factors required for ribosome biogenesis in yeast such as Sda1p (Zimmerman and Kellogg 2001), SSU processome components (Bernstein and Baserga 2004), Pwp2p (Bernstein et al. 2007), Rpl3p (Rosado et al. 2007), or Mak11 (Saveanu et al. 2007) has been shown to result in the accumulation of cells in the G1 phase of the cell cycle. However, as mentioned previously, long term depletion of ribosome biogenesis factors ultimately affects ribosome content and translation, which eventually inhibits growth and results in the accumulation of cells in G1. Some of these reported cell cycle defects could therefore have indirectly resulted from an inhibition of translation. However, in the *pwp2* or *rpl3* conditional mutant strains, the analysis of cell cycle progression shortly after transfer to restrictive conditions revealed that the increase in the proportion of unbudded G1 cells significantly precedes the depletion of mature ribosomes (Rosado et al. 2007) and the appearance of global translation defects (Bernstein et al. 2007). These data suggest that defects in ribosome biogenesis induce a rapid cell cycle arrest in G1 before they result in a shortage of functional ribosomes and a reduction in global translation. Yeast cells may therefore have developed a mechanism detecting defects in ribosome biogenesis and inhibiting passage through the G1–S transition to delay deleterious reductions in the translation rate. The molecular mechanisms underlying this surveillance remain to be elucidated. The negative regulator of Start, Whi5p (Costanzo et al. 2004; de Bruin et al. 2004), which displays functional similarities with the metazoan retinoblastoma (Rb) protein, is required for cell cycle arrest in response to impaired ribosome synthesis suggesting that Whi5p may be implicated in this surveillance (Bernstein et al. 2007).

Communication between ribosome synthesis and the G1–S transition of the cell cycle in yeast, important for coupling cell growth with cell division, has also been reported by the Tyers laboratory. This group carried out a systematic analysis of cell size distribution of haploid strains bearing individual deletions of all nonessential genes, and of diploid strains lacking one allele of all essential genes (in search for phenotypes resulting from haploinsufficiency) (Jorgensen et al. 2002). Abolishing or reducing the expression of many ribosomal proteins or many factors involved in ribosome biogenesis induces a so-called "whi" phenotype characterized by a reduced cell size. As cell size homeostasis is set at Start, these observations suggest that Start occurs before the critical cell size has been reached in these mutant strains and therefore that cell division is partially uncoupled from growth. In all these strains, however, ribosome biogenesis is impaired to different extents, and the reduction in cell size could simply result from an inhibition of cell growth due to a decreased ribosome content and translation rate. Surprisingly, however, several strains bearing deletions of genes encoding ribosome biogenesis factors display a severe reduction in cell size that cannot be solely accounted for by their moderately reduced growth. In these mutant strains, passage through Start probably occurs aberrantly early with respect to cell size, suggesting that these specific factors function in the coupling between cell growth and division. Two of the most striking phenotypes result from the deletion of the genes encoding Sfp1p or Sch9p required for maximal transcription of numerous genes encoding ribosome assembly factors and ribosomal proteins (Jorgensen et al. 2002, 2004). A more detailed analysis focused on Sfp1p and Sch9p revealed that they display characteristics of *bona fide* negative regulators of Start as all the Start related events, that is, the onset of SBF/MBF transcription, DNA replication and bud emergence, occur earlier in the absence of these factors (Jorgensen et al. 2004).

8.2.2 Unproductive Ribosome Synthesis in Mammalian Cells Induces a p53-Dependent Cell Cycle Arrest in the G1 Phase

A crosstalk between ribosome biogenesis and the G1–S transition of the cell cycle seems to operate also in multicellular eukaryotes. Inhibition of ribosomal DNA (rDNA) transcription in different mammalian cell lines, through the inactivation of the RNA polymerase I (RNA Pol I) transcription initiation factors TIF-IA or UBF (Rubbi and Milner 2003; Yuan et al. 2005) or through treatment with low doses of actinomycin D (Bhat et al. 2004; Dai and Lu 2004; Dai et al. 2004; Gilkes et al. 2006; Jin et al. 2004; Zhang et al. 2003), results in the activation of p53, a transcription factor promoting cell cycle arrest in G1 and/or apoptosis. In mouse cells, the inactivation of nucleolar factors involved with the maturation of the pre-60S preribosomal particles such as Bop1 (Strezoska et al. 2000), Pes1 (Lapik et al. 2004; Lerch-Gaggl et al. 2002), WDR12 (Holzel et al. 2005), and WDR55 (Iwanami et al. 2008), induces cell cycle arrest in the G1 phase (Grimm et al. 2006; Holzel et al. 2005; Lapik et al. 2004; Pestov et al. 2001; Strezoska et al. 2002). No general defect in bulk protein synthesis can be detected when this cell cycle arrest occurs, suggesting that it does not result from a reduced translational capacity. Instead, as in yeast, a mechanism directly inhibiting passage through the G1–S transition in response to deficient ribosome biogenesis more likely operates in mouse cells. The cell cycle arrest resulting from defects in the function of Bop1, Pes1, or WDR12 in mouse cells is abrogated in the absence of p53, strongly suggesting that p53 activation is involved in this ribosome synthesis surveillance mechanism (Holzel et al. 2005; Lapik et al. 2004; Pestov et al. 2001). Such observations have also been made more recently in human cell lines in culture. Indeed, depletion of factors required for the synthesis of the 18S rRNA such as ribosomal protein RPS9 (Lindstrom and Nister 2010; Lindstrom and Zhang 2008) or WDR3 (McMahon et al. 2010), 1A6/DRIM (Peng et al. 2010; Wang et al. 2007) and hUTP18 (Holzel et al. 2010), the homologues of yeast Utp12p, Utp20p, and Utp18p, respectively induces a p53-dependent cell cycle arrest in the G1 phase. Similarly, depletion of Las1L (Castle et al. 2010), ribosomal proteins RPL29, RPL30, and RPL37 (Llanos and Serrano 2010; Sun et al. 2010), nucleostemin (Dai et al. 2008; Ma and Pederson 2007), or PAK1IP1 (Yu et al. 2011), all required for the synthesis of the 28S rRNA and the production of the large ribosomal subunit, also results in the activation of p53 and a cell cycle arrest in the G1 phase. Altogether, these observations in mammalian cells have led to the notion that perturbations in ribosome synthesis induce so-called "nucleolar stress" or "ribosomal stress" which is under a surveillance mechanism implicating p53. More generally, Rubbi and Milner suggested that the vast majority of the stressors known to activate p53 actually perturb nucleolar integrity and ribosome synthesis, making the nucleolus a major cellular stress sensor and altered ribosome synthesis a major cause of p53 activation in the cell. In support of this concept, localized UV irradiations of cell nuclei excluding the nucleolar region induce DNA damage in the nucleoplasm but do not result in p53 accumulation (Rubbi and Milner 2003). Although mostly inferred from in vitro studies using mouse or human cell lines in culture, this ribosome synthesis surveillance mechanism is supported by more physiological evidence in mice and fish. Conditional deletion of one allele of the gene encoding the small ribosomal subunit protein RPS6 in mouse embryos leads to impaired production of the 40S ribosomal subunit (Panic et al. 2006). This defect is associated with a failure to develop beyond the gastrulation stage. Strikingly, inactivation of p53 partially overcomes this arrest and allows further developmental stages to occur. This result suggests that ribosome synthesis defects in mouse embryonic cells prevent cell proliferation through a p53-dependent mechanism. In *RPS6*-heterozygous embryos lacking p53, abundant proteins such as actin or ribosomal protein RPL11 accumulate at levels indistinguishable from control embryos suggesting that the developmental arrest during gastrulation does not result from a bulk decrease in the translational capacity. Another series of experiments showed that conditional deletion of both alleles of the gene encoding ribosomal protein RPS6 in the liver of adult mice prevents cell proliferation and liver mass regeneration after partial hepatectomy (Volarevic et al. 2000). Again, these defects do not seem to result from a decreased translational capacity as RPS6-deficient liver cells retain the ability to increase in size normally in response to nutrients following a fasting period, suggesting that translation is not limiting in these cells (Volarevic et al. 2000). One allele of the *RPS6* gene has also been specifically inactivated in early T lymphocytes in the thymus of mice (Sulic et al. 2005). This inactivation results in a reduction in the number of mature T cells in the spleen and lymph nodes, a phenotype which is alleviated in p53-deficient mice. *RPS6* haploinsufficiency does not prevent cell growth when these purified T cells are in vitro-stimulated with antigens but induces a p53-dependent cell cycle arrest and therefore inhibits proliferation. Altogether, these results suggest that defects in the synthesis of the 40S ribosomal subunit in mouse embryos or in different tissues of adult mice inhibit cell proliferation through a p53-dependent mechanism. In zebrafish, reduced production of ribosomal proteins of the 40S and 60S subunits has been shown to recapitulate some of the phenotypes observed in patients suffering from Diamond–Blackfan anemia (DBA) (see Sect. 8.5.1) (Danilova et al. 2008; Uechi et al. 2008). Interestingly, these ribosomal protein deficiencies and the ribosome synthesis defects they induce lead to the activation of the p53 network and down-regulation of this network partially alleviates the mutant phenotypes (Danilova et al. 2008). In another fish model, medaka, point mutations within the gene encoding nucleolar protein WDR55, shown to induce the accumulation of aberrant preribosomal RNA (pre-rRNA) intermediates, result in p53 activation and early developmental defects (Iwanami et al. 2008). Taken together, these results suggest that as in mice, ribosome synthesis in zebrafish and medaka is under the control of a p53-dependent surveillance mechanism that regulates cell cycle progression.

8.2.3 Model of p53 Activation in Response to Ribosomal Stress in Mammalian Cells

How defects in ribosome synthesis are detected and communicate with the p53 pathway has been the focus of numerous studies in the past few years. In proliferating cells, p53 accumulation is maintained at reduced levels through a proteasome-mediated degradation mechanism implicating the E3 ubiquitin ligase MDM2 (MDM2 in mouse or HDM2 in humans but systematically referred to as MDM2 in the rest of this chapter for simplicity). Under a variety of stress conditions, MDM2 is inhibited, which results in p53 accumulation and transcriptional activation of genes involved in cell cycle arrest, the stress response, and/or apoptosis. Interestingly, several ribosomal proteins have been shown to interact with MDM2 and to inhibit MDM2-mediated p53 degradation when overexpressed in different mammalian cell lines in culture, which results in a rapid cell cycle arrest in G1. These include the 60S ribosomal subunit proteins RPL5 (Dai and Lu 2004; Marechal et al. 1994), RPL11 (Lohrum et al. 2003; Zhang et al. 2003), RPL23 (Dai et al. 2004; Jin et al. 2004), and RPL26 (Zhang et al. 2010) and the 40S ribosomal subunit proteins RPS3 (Yadavilli et al. 2009) and RPS7 (Chen et al. 2007; Zhu et al. 2009). In addition, several nucleolar factors involved in early stages of ribosome synthesis such as nucleophosmin (NPM) (Kurki et al. 2004), nucleolin (Saxena et al. 2006), nucleostemin (Dai et al. 2008), and PAK1IP1 (Yu et al. 2011) have also been shown to inhibit MDM2 and induce a p53-dependent cell cycle arrest in G1. The current model, depicted in Fig. 8.1, postulates that under conditions of ribosomal stress, some ribosomal proteins and factors involved in ribosome synthesis become less mobilized within nascent preribosomal particles in the nucleoli and accumulate as free proteins in the nucleoplasm where they inhibit the activity of MDM2 and thereby activate p53. In agreement with this model, an increase in the nucleoplasmic accumulation of some ribosomal proteins, which correlates with an increase in their association with MDM2, has been observed in response to growth inhibitory signals such as contact inhibition, serum starvation, or treatment with low doses of actinomycin D, all of which result in a reduction in ribosome synthesis (e.g., see Bhat et al. 2004). The cell cycle arrest induced by ribosomal stress is unlikely to result from a reduced translational capacity as it is not observed in p53-deficient cells, which continue to proliferate presumably until the pool of functional ribosomes becomes limiting. RNAi-mediated depletion of RPL5 (Dai and Lu 2004), RPL11 (Bhat et al. 2004), RPL23 (Dai et al. 2004), RPL26 (Zhang et al. 2010), or RPS7 (Zhu et al. 2009) is sufficient to significantly attenuate p53 activation and cell cycle arrest following ribosomal stress, suggesting that this specific subset of ribosomal proteins may have a preferential role in the inhibition of MDM2. Overexpression of both RPL11 and RPL5 is required for strong MDM2 inhibition, suggesting that these ribosomal proteins, and possibly also others, may cooperate in this inhibition (Horn and Vousden 2008). In support of this model, MDM2-dependent p53 polyubiquitination is not abrogated in response to RPL11 overexpression in cultured cells expressing altered versions of MDM2 that do not interact with RPL5 and RPL11 (Lindstrom et al. 2007). In a more physiological context, inhibition of rRNA synthesis in the skin of mice expressing one

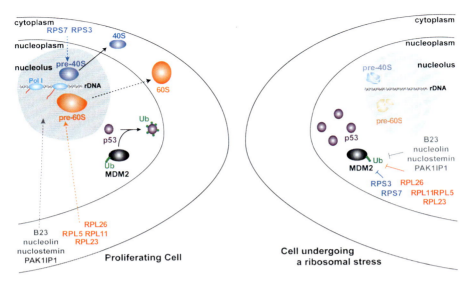

Fig. 8.1 A ribosome synthesis surveillance mechanism communicates with the p53 machinery in mammalian cells. In proliferating cells characterized by active ribosome synthesis (*left*), ribosomal proteins and ribosome assembly factors translated in the cytoplasm are massively recruited to the nucleoli where they get incorporated into nascent pre-ribosomal particles. Following early pre-rRNA cleavages, pre-40S particles are rapidly exported from the nucleolus to the cytoplasm, whereas pre-60S particles undergo maturation events in the nucleoplasm before they transit through the nuclear pore complexes. In the nucleoplasm, the E3-ubiquitin ligase MDM2 ensures p53 degradation and cell cycle progression. Defects in rDNA transcription, pre-rRNA processing, or pre-ribosomal particle assembly induce nucleolar or ribosomal stress which can be correlated with a disorganization of the nucleoli (*right*). Under these conditions, some ribosomal proteins and ribosome assembly factors become less mobilized in the nucleoli and accumulate in the nucleoplasm where they interact with MDM2 and inhibit its activity, resulting in p53 activation, cell cycle arrest in the G1 phase, and/or apoptosis

such altered version of MDM2 (MDM2$^{C305F/C305F}$), through topical treatment with actinomycin D, results in a reduced stabilization of p53 in comparison to control mice expressing wild-type MDM2 (Macias et al. 2010). The response to nucleolar stress is impaired in these mutant mice, whereas in contrast, the p53-dependent response to DNA damage induced by whole body γ-irradiations is not affected, suggesting that the mutation introduced in *MDM2* specifically alters the p53-dependent nucleolar stress-signaling pathway.

Ribosomal protein RPL26 and nucleolin seem to provide another level of complexity in this pathway. These proteins have indeed been proposed to influence the translation of the p53-encoding mRNA in several human cell lines exposed to irradiation, through a direct interaction with the 5′ untranslated region (5′ UTR) of this transcript (Takagi et al. 2005). In addition, under unstressed conditions, part of the pool of the RPL26 protein is subjected to degradation via a MDM2-dependent polyubiquitination mechanism, presumably to prevent low levels of free RPL26 protein from activating p53 mRNA translation (Takagi et al. 2005).

8.3 The Tumor Suppressor ARF Inhibits both Ribosome Synthesis and Cell Cycle Progression in Response to Oncogenic Signals in Mammalian Cells

In mammalian cells, the tumor suppressor ARF (p19^{ARF} in mouse; p14^{ARF} in humans) is another interesting example of a factor connecting ribosome biogenesis, the p53 pathway, and the G1–S transition of the cell cycle. Induction of p19^{ARF} expression following oncogenic signals in mouse cells induces a rapid cell cycle arrest in G1. ARF inhibits MDM2 and prevents proteasome-dependent p53 destruction (Zhang et al. 1998). ARF also inhibits cell cycle progression in the absence of p53 (Zhang et al. 1998) and several reports have suggested that this property of the protein relies on its ability to inhibit ribosome synthesis at multiple levels. In human cells, p14^{ARF} has been shown to bind rDNA promoters (Ayrault et al. 2004), to interact with topoisomerase I (Ayrault et al. 2003; Karayan et al. 2001), and to inhibit several components of the RNA Pol I transcription initiation machinery such as the transcription factors UBF and E2F1 (Ayrault et al. 2006a, b). More recently, ARF was shown to interact with TTF1, an RNA Pol I transcription termination factor and to prevent its nucleolar localization, which also contributes to the inhibition of RNA Pol I transcription (Lessard et al. 2010). In addition, p19^{ARF} induction in mouse cells also rapidly affects the posttranscriptional steps of ribosome synthesis, by inhibiting the processing of the 47S pre-rRNA precursor and the production of the mature 18S and 28S rRNAs. Immunoprecipitation experiments showed that both in human and mouse cells, ARF interacts with NPM, a nucleolar endoribonuclease that functions in the processing of the pre-rRNA (Bertwistle et al. 2004; Itahana et al. 2003; Savkur and Olson 1998) and this interaction is correlated with a rapid degradation of NPM, which does not depend on MDM2 and p53 (Itahana et al. 2003). The pre-rRNA processing defects arising from p19^{ARF} expression may therefore result at least in part from NPM inactivation. Taken together, these data provide a compelling example of a concerted inhibition of both the G1–S transition of the cell cycle and ribosome synthesis in response to oncogenic signals.

8.4 A Subset of Ribosome Synthesis Factors Function Directly in Specific Stages of Cell Cycle Progression in Yeast and Mammals

The conclusions presented in Sect. 8.2, indicating that ribosome synthesis is monitored during the G1 phase of the cell cycle and regulates passage through the G1–S transition, imply that inactivation of any factor required for ribosome synthesis should result in an accumulation of cells in the G1 phase of the cell cycle. Unexpectedly however, depletion of a subset of ribosome assembly factors in yeast and mammalian cells has been shown to affect progression through other stages of the cell cycle, mainly S phase and mitosis, in addition to delaying the G1–S transition.

8 Crosstalk Between Ribosome Synthesis and Cell Cycle Progression...

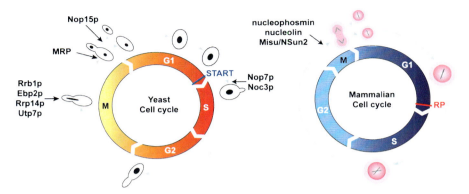

Fig. 8.2 Ribosome synthesis factors participate directly in cell cycle progression in yeast and mammalian cells. In yeast cells (*left*), the Nop7p and Noc3p proteins have been shown to interact with the ORC complex and to function directly in the initiation of DNA replication at the beginning of S phase. Inactivation Rrb1p, Ebp2p, and Rrp14p affects elongation or positioning of the mitotic spindle during mitosis and impairs chromosome segregation. Utp7p is a component of the kinetochores also required for chromosome segregation. The MRP endonuclease takes part in the rapid degradation of the Clb2p-encoding mRNA at the end of mitosis and contributes to mitosis exit. Nop15p is required for formation of the actin ring at the bud neck allowing daughter cell separation at cytokinesis. In mammalian cells (*right*), nucleophosmin (NPM) and possibly also nucleolin participate in the regulation of centrosome duplication, and altering their function results in supernumerary centrosomes and the establishment of aberrant multipolar spindles during mitosis. The methyltransferase Misu/NSun2 accumulates in the nucleoli in interphase and takes part in the organization of the microtubule spindle required for chromosome segregation during mitosis. *RP*, Restriction point

These results, reviewed in the next section and summarized in Fig. 8.2, suggest that several factors involved in ribosome biogenesis are also directly required for proper progression through S phase or mitosis and therefore display a dual functionality.

8.4.1 Several Ribosome Assembly Factors are Required for Proper Progression Through S Phase or Mitosis in Yeast Cells

The yeast Nop7p and Noc3p proteins have been originally characterized as components of pre-60S preribosomal particles (Bassler et al. 2001; Harnpicharnchai et al. 2001) required for the maturation of the large ribosomal subunits (Adams et al. 2002; Milkereit et al. 2001; Oeffinger et al. 2002). Surprisingly, immunoprecipitation experiments have shown that Nop7p and Noc3p are physically associated with the ORC complex and the MCM proteins involved in the initiation of DNA replication (Du and Stillman 2002; Zhang et al. 2002). In addition, Noc3p is specifically associated with autonomous replication sequences (ARSs), where DNA replication is initiated, as assessed by chromatin immunoprecipitation (ChIP) assays (Zhang et al. 2002). These observations suggested that Nop7p and Noc3p are required for the initiation of DNA replication and indeed, depletion of either protein affects S phase progression, suggesting a failure to replicate DNA. This phenotype is manifested very early after the inactivation of Nop7p and Noc3p (a conditional degron system (Sanchez-Diaz et al. 2004) is used in both studies) and is therefore unlikely to result from an indirect defect in the overall rate of protein synthesis. It is tempting to speculate from these data that Nop7p and Noc3p, in addition to their function in ribosome biogenesis, also directly participate in the initiation of DNA replication.

MRP endoribonuclease is another interesting example of a factor required for cell cycle progression in addition to its well characterized function in the processing of the pre-rRNA at site A_3. Temperature sensitive mutations in the *NME1* or *SNM1* genes, encoding respectively the RNA component and a specific core protein of the ribonucleoprotein (RNP) particle, induce a cell cycle delay at the end of mitosis (Cai et al. 2002). At the restrictive temperature, a significant proportion of mutant cells accumulate as large budded cells with separated, dumbbell-shaped nuclei and extended mitotic spindles, a characteristic phenotype of impaired exit from mitosis at the end of telophase. The following observations strongly suggest that this cell cycle delay does not result indirectly from impaired ribosome biogenesis. One particular mutation in *SNM1* (*snm1-172*) affects the processing of the 5.8S rRNA and induces a severe temperature-sensitive phenotype (Cai et al. 2002). This phenotype is suppressed by over expression of *CDC5*, a gene encoding a polo-like kinase required for proper mitotic progression, repeatedly identified as a suppressor of mutations affecting exit from mitosis. The *snm1-172* strain over expression *CDC5* displays an aberrant pattern of accumulation of the 5.8S rRNA species, strictly identical to the one observed in the original *snm1-172* strain. These results indicate that the mitotic delay resulting from impaired MRP function is not a consequence of the defects in 5.8S rRNA processing. Instead, MRP more likely takes part directly in the regulation of mitosis, and several reports provided convincing evidence that MRP functions in the posttranscriptional regulation of the expression of *CLB2*, a gene encoding the major B-type mitotic cyclin in yeast (Cai et al. 2002; Gill et al. 2004, 2006). MRP is indeed required for the rapid degradation of the Clb2p-encoding mRNA at the end of mitosis, through the introduction of endonucleolytic cleavages within the 5′ UTR of this transcript. This rapid destruction of Clb2p at the end of mitosis is required for proper exit from mitosis and entry into a new G1 phase. Altogether, these data show that MRP endonuclease carries out two separate, independent functions in ribosome biogenesis and cell cycle control.

In addition to MRP, several other factors have also been reported to be required for proper progression through mitosis besides their well established function in the maturation of the large ribosomal subunit. Conditional mutations in the genes encoding Ebp2p (Ionescu et al. 2004), Rrb1p (Killian et al. 2004), and Rrp14p (Oeffinger et al. 2007; Yamada et al. 2007) induce a significant accumulation of large budded cells in which DNA has been replicated but fails to segregate between mother and daughter cells. In each case, impaired chromosome segregation likely results from defects in the elongation and/or positioning of the mitotic spindle as

(1) G2-arrested mutant cells display abnormal mitotic spindles and (2) the cell cycle arrest is abrogated in the absence of Mad2p or Bub2p, in the case of mutations in *EBP2* or *RRP14*, respectively. These factors participate in the spindle assembly checkpoint that monitors the attachment of chromosomes to microtubules and delays the metaphase–anaphase transition when this process is defective. Mutations in the gene encoding Utp7p, a component of the SSU processome required for the production of the 40S ribosomal subunit (Dragon et al. 2002), also result in a significant accumulation of large budded cells arrested in mitosis in addition to the presence of unbudded, G1-arrested cells (Jwa et al. 2008). Besides its nucleolar localization, Utp7p is also present at kinetochores, as assessed by ChIP experiments, and is important for the organization of these structures and proper chromosome segregation during mitosis (Jwa et al. 2008). Depletion of another ribosome assembly factor, Nop15p, also delays mitosis in addition to the G1–S transition impairment (Oeffinger and Tollervey 2003). The absence of Nop15p has been shown to prevent formation of the contractile actin ring at the bud neck required for the separation of mother and daughter cells during cytokinesis (Oeffinger and Tollervey 2003). The mitosis defects observed in the absence of these factors are probably not indirect consequences of impaired translation. Indeed, the cell cycle phenotypes resulting from mutations in *RRB1* and *NOP15* begin rapidly after transfer of the mutant cells to restrictive conditions and therefore, very probably precede depletion of functional ribosomes (Killian et al. 2004; Oeffinger and Tollervey 2003). In the case of Ebp2p, in addition, the ribosome biogenesis function of the protein can be uncoupled from its function in mitosis by different mutations in the encoding gene (Ionescu et al. 2004), suggesting that Ebp2p carries out independent functions in these two processes. Maybe even more compelling is the observation that although mutations in both *EBP2* and *RRP14* affect chromosome segregation, the precise stage of mitosis at which cell cycle arrest occurs in each case seems to be slightly different. The arrest resulting from impaired Ebp2p function is abrogated in the absence of Mad2p but not Bub2p (Ionescu et al. 2004) and conversely, the absence of Bub2p but not Mad2p overrides the arrest induced by mutations in *RRP14* (Oeffinger et al. 2007).

8.4.2 Several Ribosome Biogenesis Factors are Required for Progression Through Mitosis in Mammalian Cells

In mammalian cells, several proteins involved in early steps of ribosome synthesis have also been attributed additional functions in cell cycle progression. Nucleophosmin (NPM, also coined B23) is an abundant nucleolar protein associated with ribonuclease activity in vitro (Herrera et al. 1995) and proposed to function in the processing of the pre-rRNAs within ITS2 (Savkur and Olson 1998). Interestingly, in addition to its nucleolar localization, NPM was detected at unduplicated centrosomes in mouse cells and has been proposed to prevent inappropriate duplication of these structures and thus to coordinate centrosome duplication with cell cycle progression (Okuda et al. 2000). Consistent with this function, mouse embryonic fibroblasts lacking NPM contain supernumerary centrosomes, which results in frequent aberrant mitotic figures characterized by multipolar spindles (Grisendi et al. 2005). Taken together, these data strongly suggest that NPM fulfills an extra-nucleolar function in the regulation of centrosome duplication, and defects in this function compromise mitosis. Similarly, down regulation of another abundant nucleolar protein, nucleolin/C23, required for early processing of the pre-rRNAs (Ginisty et al. 1998; Roger et al. 2003), results in supernumerary centrosomes, multipolar spindles, and defects in progression through mitosis in human cells (Ugrinova et al. 2007). Although nucleolin has not been detected at the centrosome, these observations suggest that like NPM, nucleolin could also regulate some aspects of centrosome duplication. The RNA methyltransferase Misu/NSun2, homologous to yeast Trm4p and Nop2p proteins, accumulates in the nucleoli in interphase and relocalizes to the microtubule spindle during mitosis in human cells (Hussain et al. 2009). Down regulation of Misu/NSun2 affects the organization of the microtubule spindle and impairs chromosome segregation in mitosis, suggesting that this protein imparts mitotic spindle stability (Hussain et al. 2009).

8.5 Defective Ribosome Biogenesis and Human Diseases

It has become apparent in the past decade that several human diseases are caused by alterations of factors required for ribosome biogenesis, including Diamond–Blackfan anemia (DBA), 5q- syndrome, Treacher Collins syndrome (TCS), and Shwachman–Diamond syndrome (SDS). These diseases are characterized by marked pleiotropy, affecting several different tissues, and a great variability in the severity and spectrum of symptoms depending on the patients. The characteristics of these diseases can include, in diverse combinations, hematopoietic defects, developmental anomalies, and cancer predisposition. These diseases due to alterations of genes encoding constituents of mature ribosomes and/or factors involved in ribosome biogenesis have been termed "ribosomopathies" (Narla and Ebert 2010) (Table 8.1).

8.5.1 Diseases Linked to Ribosomal Protein Deficiency: Diamond–Blackfan Anemia and Human 5q- Syndrome

DBA is usually diagnosed early in infancy and affects approximately seven newborns per million live births (Aguissa-Toure et al. 2009; Da Costa et al. 2001; Flygare and Karlsson 2007; Gazda and Sieff 2006; Vlachos et al. 2008). The classical form of the disease is defined by the occurrence of anemia associated with macrocytosis (enlargement of red blood cells), reticulocytopenia (reticulocyte decrease), and a paucity of red cell precursors in the bone marrow, while the differentiation of other hematopoietic lineages is little affected (Diamond et al. 1976; Lipton and Ellis 2009;

8 Crosstalk Between Ribosome Synthesis and Cell Cycle Progression... 169

Table 8.1 Human diseases resulting from mutations in genes encoding ribosome components or factors involved in ribosome synthesis

Disease	Genes mutated	Impaired molecular functions	Clinical features
Diamond–Blackfan anemia (DBA)	*RPS7, RPS10, RPS17, RPS19, RPS24, RPS26, RPL5, RPL11, RPL35a*	Maturation of the 40S and 60S ribosomal subunits	Macrocytic anemia, reticulocytopenia, bone marrow failure, short stature, physical malformations, cancer predisposition
5q- syndrome	*RPS14*	Maturation of the 40S ribosomal subunit	Severe macrocytic anemia, propensity to develop acute myeloid leukemia, morphological abnormalities of megakaryocytes
Treacher Collins syndrome (TCS)	*TCOF1, POLR1C, POLR1D*	rDNA transcription and rRNA methylation	Craniofacial abnormalities
Shwachman–Diamond syndrome (SDS)	*SBDS*	Maturation of the 60S ribosomal subunit	Neutropenia, exocrine pancreatic insufficiency, skeletal abnormalities, aplastic anemia, leukemia

Vlachos et al. 2008). Patients also often suffer from short stature and several malformations of the head, thumb, upper limbs, heart and/or the genitourinary system (Lipton and Ellis 2009; Vlachos et al. 2008). They also have an increased predisposition to certain cancers (Lipton and Ellis 2009; Vlachos et al. 2008). The hematopoietic phenotype of DBA is closely related to that of the 5q- syndrome, caused by an interstitial deletion of the long arm of chromosome 5. This syndrome is indeed characterized by a severe macrocytic anemia and a propensity to develop acute myeloid leukemia (Boultwood et al. 2010; Van den Berghe et al. 1974). Unlike DBA, however, morphological abnormalities of megakaryocytes are also a characteristic feature of the 5q- syndrome.

A major breakthrough in the understanding of DBA came with the demonstration that in 25% of reported cases, it is caused by heterozygous mutations of the gene encoding the RPS19 protein of the small ribosomal subunit (Draptchinskaia et al. 1999). Since then, heterozygous mutations were detected in different subsets of DBA patients in the genes encoding either small (RPS7, RPS10, RPS17, RPS24, RPS26) or large (RPL5, RPL11, RPL35a) ribosomal subunit proteins (Cmejla et al. 2007, 2009; Doherty et al. 2010; Farrar et al. 2008; Gazda et al. 2006, 2008). The DBA-associated mutations of the *RPS19* gene, the best studied example, have been shown or are expected to impair RPS19 production, accumulation, or association with ribosomal subunits and hence to lead to RPS19 haploinsufficiency (Aguissa-Toure et al. 2009; Angelini et al. 2007; Campagnoli et al. 2008; Gazda et al. 2004).

Altogether, these studies firmly establish DBA as a disease caused by ribosomal protein deficiency. This conclusion is reinforced by the phenotypes of artificial RPS19 depletion in cell culture or whole animals. RNA interference strategies that reduce the accumulation of the *RPS19* transcript impair the proliferation and differentiation of human erythroid progenitor cells in culture (Ebert et al. 2005; Flygare et al. 2005). Reduced RPS19 expression in zebrafish embryos achieved by injection of antisense morpholino oligonucleotides leads to a shortening of the body, craniofacial defects, and defective proliferation and differentiation of erythroid progenitors, as seen in DBA patients (Danilova et al. 2008; Uechi et al. 2008). Finally, mice carrying a heterozygous missense mutation in *RPS19* display a reduced birth weight as well as depressed reticulocyte count, mild reduction in red blood cell count, and increased apoptosis in bone marrow progenitor cells, reminiscent of a mild DBA phenotype (McGowan et al. 2008). Strikingly, similar recent evidence suggests that haploinsufficiency of the *RPS14* gene encoding the RPS14 ribosomal protein (present on the portion of chromosome 5 deleted in 5q- syndrome patients) contributes to the phenotype of the 5q- syndrome. Accumulation of *RPS14* mRNA is reduced by 40% in patient cells (Boultwood et al. 2007). In addition, lentivirally expressed shRNAs targeting *RPS14* inhibit the differentiation of human hematopoietic progenitor cells into erythroid cells, while overexpression of *RPS14* rescues the erythroid differentiation of hematopoietic progenitor cells obtained from 5q- syndrome patients (Ebert et al. 2008).

Ribosomal proteins, as constituents of mature cytoplasmic ribosomes are, needless to say, required for translation. DBA is probably not caused by a very specific translational defect, however, as the ribosomal proteins linked to DBA are not restricted to one type of subunit and do not obviously cluster on a given subunit (Ben-Shem et al. 2010). Ribosomal proteins also play diverse and important roles in pre-rRNA processing, ribosomal subunit assembly, and transport to the cytoplasm and are therefore essential for nucleolar integrity and the production of mature cytoplasmic ribosomal subunits (Choesmel et al. 2007, 2008; Doherty et al. 2010; Farrar et al. 2008; Flygare et al. 2007; Gazda et al. 2008; Idol et al. 2007; O'Donohue et al. 2010; Robledo et al. 2008). Cell lines derived from DBA patients carrying mutations in the genes encoding RPS7, RPS10, RPS19, RPS24, RPS26, RPL5, RPL11, or RPL35a display similar, albeit sometimes milder, pre-rRNA processing defects to those observed following depletion of these proteins by RNA interference in non-DBA cells (Choesmel et al. 2007, 2008; Doherty et al. 2010; Farrar et al. 2008; Flygare et al. 2007; Gazda et al. 2008). Likewise, bone marrow cells from 5q- syndrome patients display similar pre-rRNA processing defects to those obtained following artificial RPS14 depletion (Ebert et al. 2008). Hence, defects in ribosome biogenesis are likely to play a key role in the establishment of the DBA and 5q- phenotypes. DBA is unlikely to be triggered by very specific ribosome biogenesis defects as, depending on the patients, the synthesis of either the small or large ribosomal subunit is affected and the biogenesis defect may occur early or late in the pathway (O'Donohue et al. 2010; Robledo et al. 2008). The disease phenotype is more probably the result of an overall shortage of functional 80S ribosomes, of a cellular stress due to partial ribosome biogenesis failure and/or nucleolar disorganization. As detailed in Sect. 8.2, it is currently envisaged that one or several of these phenomena lead to activation of the p53 pathway, and hence to cell cycle arrest and apoptosis. Recent data suggest that indeed p53 activation plays a crucial role in both DBA and the 5q- syndrome. Morpholino oligonucleotide-induced depletion of RPS19 in wild-type zebrafish embryos leads to a DBA-like phenotype correlated with increased *P53* transcription, while RPS19 depletion has little effect on the phenotype and survival of p53$^{-/-}$ zebrafish embryos (Danilova et al. 2008). Similar results are obtained in mouse; that is, a heterozygous mutation of *RPS19* leads to increased p53 accumulation and the DBA-like hematological phenotype of this mutation is corrected by inactivation of the *P53* gene (McGowan et al. 2008). Finally, p53 levels are elevated in bone marrow cells of a mouse model of the human 5q- syndrome, featuring a monoallelic deletion of the *Cd74-Nid67* chromosomal interval encompassing *RPS14*, and the hematopoietic defects induced by this deletion are reversed by *P53* gene disruption (Barlow et al. 2010; Pellagatti et al. 2010).

8.5.2 Disease Linked to Defective Pre-rRNA Accumulation: Treacher Collins syndrome

TCS is a syndrome of craniofacial development (cleft palate, hypoplasia of the facial bones, downward slanting of the palpebral fissures, deformity of the external ear) affecting one newborn in every 50,000 live births (Dixon et al. 2007; Sakai and Trainor 2009; Trainor et al. 2009). In the great majority of cases, it is due to heterozygous loss of function mutations of the *TCOF1* gene, encoding the phosphoprotein treacle that accumulates within the dense fibrillar component of the nucleolus (Dixon et al. 2007; Group 1996; Isaac et al. 2000; Sakai and Trainor 2009). Treacle contains a central domain related to a domain found in the nucleolar phosphoprotein NOPP140 (Dixon et al. 1997; Wise et al. 1997). Like NOPP140, treacle seems to be phosphorylated by casein kinase 2, with which it interacts (Isaac et al. 2000), and is associated with the box C/D snoRNP components NOP56 and fibrillarin (Gonzales et al. 2005; Hayano et al. 2003). Unlike NOPP140, treacle does not associate with the box H/ACA snoRNP components NAP57/dyskerin and GAR1 (Isaac et al. 2000). Consistent with this selective interaction with box C/D snoRNP components, treacle is required for 2'O ribose methylation of C463 of mouse 18S rRNA but is not involved in pseudouridylation of 18S rRNA U1642 (Gonzales et al. 2005). As other positions were not tested, it is not yet known whether treacle has a global role in pre-rRNA methylation, although this is likely as the positions analyzed were apparently chosen at random. Treacle probably also has a role in pre-rRNA synthesis as (1) it interacts with RNA Pol I (Lin and Yeh 2009) and with the RNA Pol I transcription factor UBF (Valdez et al. 2004), (2) it is found in close proximity to the rDNA promoter by ChIP analysis (Gonzales et al. 2005; Lin and Yeh 2009), and (3) its depletion leads to a reduction in 47S pre-rRNA synthesis (Valdez et al. 2004). Strikingly, it has been very recently demonstrated that a subset of TCS patients that display wild-type *TCOF1* alleles carry mutations in either the *POLR1C* or *POLR1D* genes encoding two subunits of both RNA Pol I and RNA Pol III (Dauwerse et al. 2011). This reinforces the notion that TCS is caused by defective rDNA transcription.

Tcof1[+/-] mouse embryos have been generated that recapitulate the facial anomalies found in severe cases of TCS (Dixon et al. 2006). These abnormalities arise as a consequence of apoptosis and reduced proliferation of neuroepithelial cells leading to reduced production of cranial neural crest cells. These defects are associated with reduced accumulation of mature ribosomal RNAs in the neuroepithelium and craniofacial mesenchyme (Dixon et al. 2006; Jones et al. 2008). As in the cases of DBA and the 5q- syndrome, p53 activation seems to play a key role in the development of TCS. *Tcof1*[+/-] mouse embryos display increased accumulation of p53 specifically in neuroepithelial cells associated with a strong increase in cell apoptosis. Loss of the *P53* genes in a *Tcof1*[+/-] background abrogates this increase in neuroepithelial apoptosis and suppresses the craniofacial abnormalities typical of TCS (Jones et al. 2008). Interestingly, these authors claim that ribosome biogenesis is not improved by the *P53* gene deletions, although this claim, based only on in situ detection of 28S rRNAs, needs to be substantiated with more thorough analyses. Were this proposal indeed true, it would indicate that TCS results from p53 activation and not from defects in ribosome biogenesis and/or a general decrease in mature ribosomal subunit levels as such.

8.5.3 Disease Linked to Defective 60S Ribosomal Subunit Biogenesis: Shwachman–Diamond syndrome

This syndrome is characterized by neutropenia, exocrine pancreatic insufficiency, and skeletal abnormalities (metaphyseal dysostosis). Patients often develop aplastic anemia and leukemia (Burroughs et al. 2009; Dror 2005; Shimamura 2006). SDS is an autosomal recessive disease caused in 90% of cases by mutations within the *SBDS* (Shwachman–Bodian–Diamond syndrome) gene (Boocock et al. 2003). The genomic locus containing *SBDS* has been locally duplicated and the duplicon contains a pseudogene, *SBDSP*, sharing 97% sequence identity with *SBDS*. Most disease-linked mutations result from recombination events between *SBDS* and *SBDSP*. SBDS is a ubiquitously expressed protein essential for early mammalian embryonic development (Zhang et al. 2006) and orthologs are found in archaea and yeast, suggesting that SBDS fulfills a housekeeping function. SBDS and its yeast ortholog, Sdo1p, are detected both in the nucleus (including the nucleolus) and the cytoplasm (Austin et al. 2005; Huh et al. 2003). SBDS and Sdo1p cofractionate on gradients with (pre-)60S particles (Ganapathi et al. 2007; Menne et al. 2007) and interact with protein factors involved in ribosome biogenesis (Ball et al. 2009; Ganapathi et al. 2007; Hesling et al. 2007; Krogan et al. 2006; Luz et al. 2009; Savchenko et al. 2005). Furthermore, SBDS has been shown to specifically interact with 28S rRNA (Ganapathi et al. 2007), while Sdo1p has been demonstrated to associate with both mature rRNAs and pre-rRNAs (Luz et al. 2009). These data suggest that SBDS

8 Crosstalk Between Ribosome Synthesis and Cell Cycle Progression... 173

functions in 60S ribosomal subunit formation and/or activation. Most interestingly, it has been reported that the severe growth defect of yeast cells lacking Sdo1p can be suppressed by mutations within the gene encoding the pre-60S ribosomal particle factor Tif6p (Menne et al. 2007). Tif6p interacts with pre-60 particles in the nucleus, is released from these particles in the cytoplasm, and is re-imported in the nucleus (Henras et al. 2008). In the absence of Sdo1p, Tif6p accumulates in the cytoplasm, whereas suppressor Tif6p proteins accumulate normally in the nucleus. This aberrant Tif6p localization in absence of Sdo1p may result from the retention of Tif6p in cytoplasmic 60S particles. Menne et al. propose that Sdo1p and Efl1p cooperate to release Tif6p from cytoplasmic 60S particles to allow their association with the 43S preinitiation complex (Menne et al. 2007). In cells from SDS patients, eIF6, the Tif6p ortholog, is redistributed to the cytoplasm (Menne et al. 2007), consistent with the retention model just outlined. Thus, one cause of SDS may be partial inhibition of translation initiation. This proposal is consistent with the decrease in global translation observed in cells depleted of SBDS (Ball et al. 2009). In addition, bone marrow cells from SDS patients exhibit increased apoptosis (Dror and Freedman 2001) as well as increased p53 levels (Elghetany and Alter 2002). The apoptotic phenotype is most probably the result of SBDS deficiency, as depletion of SBDS by RNA interference leads to growth inhibition associated with accelerated spontaneous apoptosis (Nihrane et al. 2009; Rujkijyanont et al. 2008). Moreover, SBDS depletion in murine hematopoietic progenitors leads to several hematopoietic defects, including impaired granulopoiesis, which is a typical feature of SDS (Rawls et al. 2007). Thus, as in the case of the previously described syndromes, some aspects of the phenotype of SDS may in part be caused by increased apoptosis of the progenitors of some specific hematopoietic lineages.

8.6 Perspectives

From the above review of the literature, it appears that ribosome biogenesis and cell cycle progression are interconnected processes in yeast and mammalian cells. Although the mechanisms underlying these connections are probably extremely complex and some aspects remain to be elucidated, two conserved trends emerge from the reported data. One of these is that several factors involved in ribosome synthesis are also directly required for specific steps of cell cycle progression and therefore display a dual functionality. The biological relevance of this observation is not clear. The biochemical activity of some of these factors may have been mobilized independently by two cellular processes during evolution and these factors may not mediate any sort of connection between these processes. Alternatively, sharing the function of a specific factor between two different cellular processes may be a means to coordinate them through the concerted regulation of its activity. In the cases investigated, however, it proved possible to uncouple the two functions of the proteins by mutagenesis, suggesting that these specific factors function independently in ribosome synthesis and cell cycle progression (Cai et al. 2002; Ionescu et al. 2004; our unpublished observations). Characterization of the precise biochemical activities of these factors and mutational analyzes should help understand the relevance of their function to both ribosome biogenesis and cell cycle progression. In mammalian cells, ribosome synthesis is inhibited in mitosis from late prophase to telophase (Prescott and Bender 1962; Roussel et al. 1996) and it was shown more recently that in yeast cells as well, rDNA transcription ceases during anaphase (Clemente-Blanco et al. 2009). Some components or modules of the preribosomal particles could therefore be released during mitosis and fulfill specific functions in spindle assembly, chromosome segregation, or cytokinesis.

The second trend inferred from the reported data is that both in yeast and mammals, defects in ribosome biogenesis directly inhibit passage through the G1–S transition, suggesting that ribosome synthesis is under a surveillance mechanism communicating with this cell cycle transition. Such a mechanism may allow cell cycle progression when ribosome synthesis is active and on the contrary ensure cell cycle arrest under conditions that threaten the cellular translational capacity. During the G1 phase in yeast, cells sense environmental and intra cellular signals to determine if conditions are suitable for cell division. In addition to the rate of translation per se, the activity of the ribosome synthesis machinery is very likely another cellular parameter that conditions commitment to cell division. A mechanism that monitors ribosome biogenesis per se, and not solely variations in the translation rate, may allow cells to anticipate deleterious translation defects and provide more reactivity to adapt to rapid modifications of the environment. The precise aspect of the ribosome biogenesis pathway that is monitored remains to be determined as well as the nature of the sensors involved. In mammalian cells, the current model proposes that ribosomal stress induces the accumulation of a specific subset of ribosomal proteins and ribosome assembly factors in the nucleoplasm where they inhibit MDM2-dependent p53 degradation and thereby result in cell cycle arrest in G1. This cell cycle arrest appears reversible in cultured cells (Grimm et al. 2006; Holzel et al. 2005; Lapik et al. 2004; Pestov et al. 1998, 2001), suggesting that arrested cells have the ability to resume proliferation when optimal ribosome production is restored. Interestingly, amino acid substitutions in MDM2 that disrupt the interaction with ribosomal proteins RPL5 and RPL11 have been detected in human cancers (Schlott et al. 1997). Transgenic mice expressing one such altered version of MDM2 (MDM2[C305F/C305F]) and expressing in addition c-MYC constitutively in the B cell lineage (Eμ-*Myc* mice) develop lymphomas more rapidly than control Eμ-*Myc* mice expressing wild-type MDM2 (Macias et al. 2010 and references therein). These observations suggest that the ribosome biogenesis surveillance mechanism prevents proliferation also in a cellular context where ribosome synthesis is activated through MYC expression and inactivation of this process may be required for malignant transformation.

Mutations in genes encoding ribosome components or factors involved in ribosome synthesis have been associated with several human diseases termed "ribosomopathies." It appears that one common aspect of all the described diseases is increased cell cycle arrest and apoptosis of specific progenitors caused by p53 activation. This activation is itself the result of impaired ribosome biogenesis. Such apoptotic process is responsible for the defective accumulation of specific cell types, generating the disease phenotype. While it is probable that reduced protein synthesis capacity also plays a role in impairing cell proliferation, its importance has been difficult to assess. Moreover, a host of questions remain regarding the mode of p53 activation and its consequences. As the ribosome biogenesis defects seem to affect more severely specific tissues and hematopoietic lineages, one may wonder whether p53 activation is restricted to specific progenitors, and if so why, and/or whether specific cell types are more sensitive to p53 activation than others. Another key and as yet unanswered question is why, if the genetic defects underlying the various diseases all lead to impaired ribosome biogenesis leading to p53 activation, do they preferentially affect different cellular lineages depending on the disease considered. Is it the case that with p53 activation, the cell types in which it occurs and its consequences vary depending on the precise nature of the ribosome biogenesis defect? Some of these issues have started to be addressed, in particular in the case of ribosomal protein deficiencies. Regarding lineage specificity and mode of p53 activation, Dutt and collaborators have found that decreased *RPS14* or *RPS19* expression leads to strong p53 accumulation in erythroid lineage cells in culture, while myeloid or megakaryocyte lineages do not display such p53 accumulation (Dutt et al. 2011). Moreover, they suggest that p53 activation in RPS14-depleted cells results from increased binding of RPL11 to MDM2, leading to p53 stabilization. This may not be the sole mode of p53 activation in the case of DBA, given that this syndrome can be caused by RPL11 haploinsufficiency (see Sect. 8.5.1). Indeed it is striking to note that, in a seemingly contradictory way, three of the genes (*RPL5*, *RPL11*, and *RPS7*) mutated in DBA encode ribosomal proteins known to bind to MDM2 and to participate in p53 activation (Zhang and Lu 2009). Clearly, the mode of p53 activation in DBA needs to be clarified. Another seemingly contradictory feature of DBA, SDS, and 5q- syndrome is the increased cancer risk associated with these diseases in spite of the observed characteristic p53 induction that should function as a tumor suppression mechanism. One hypothesis to explain this cancer predisposition is that over time, in the cellular lineages faced with increased apoptosis, rare clones displaying apoptosis-contravening mutations, for example in *P53*, emerge; these have a growth advantage and are prone to malignant transformation.

Acknowledgments Marie Gérus is the recipient of a graduate fellowship from the *Fondation pour la Recherche Médicale*. Our research is supported by the *CNRS*, the *Ligue Nationale Contre le Cancer (équipe labellisée)*, the *Agence Nationale de la Recherche*, the International Human Frontier Science Program Organization (to AH) and the *Université Paul Sabatier*.

References

Adams CC, Jakovljevic J, Roman J, Harnpicharnchai P, Woolford JL Jr (2002) *Saccharomyces cerevisiae* nucleolar protein Nop7p is necessary for biogenesis of 60S ribosomal subunits. RNA 8:150–165

Aguissa-Toure AH, Da Costa L, Leblanc T, Tchernia G, Fribourg S, Gleizes PE (2009) [Diamond-Blackfan anemia reveals the dark side of ribosome biogenesis]. Med Sci (Paris) 25:69–76

Angelini M, Cannata S, Mercaldo V, Gibello L, Santoro C, Dianzani I, Loreni F (2007) Missense mutations associated with Diamond-Blackfan anemia affect the assembly of ribosomal protein S19 into the ribosome. Hum Mol Genet 16:1720–1727

Austin KM, Leary RJ, Shimamura A (2005) The Shwachman-Diamond SBDS protein localizes to the nucleolus. Blood 106:1253–1258

Ayrault O, Karayan L, Riou JF, Larsen CJ, Seite P (2003) Delineation of the domains required for physical and functional interaction of p14ARF with human topoisomerase I. Oncogene 22:1945–1954

Ayrault O, Andrique L, Larsen CJ, Seite P (2004) Human Arf tumor suppressor specifically interacts with chromatin containing the promoter of rRNA genes. Oncogene 23:8097–8104

Ayrault O, Andrique L, Fauvin D, Eymin B, Gazzeri S, Seite P (2006a) Human tumor suppressor p14ARF negatively regulates rRNA transcription and inhibits UBF1 transcription factor phosphorylation. Oncogene 25:7577–7586

Ayrault O, Andrique L, Seite P (2006b) Involvement of the transcriptional factor E2F1 in the regulation of the rRNA promoter. Exp Cell Res 312:1185–1193

Ball HL, Zhang B, Riches JJ, Gandhi R, Li J, Rommens JM, Myers JS (2009) Shwachman-Bodian Diamond syndrome is a multi-functional protein implicated in cellular stress responses. Hum Mol Genet 18:3684–3695

Barlow JL, Drynan LF, Hewett DR, Holmes LR, Lorenzo-Abalde S, Lane AL, Jolin HE, Pannell R, Middleton AJ, Wong SH, Warren AJ, Wainscoat JS, Boultwood J, McKenzie AN (2010) A p53-dependent mechanism underlies macrocytic anemia in a mouse model of human 5q- syndrome. Nat Med 16:59–66

Bassler J, Grandi P, Gadal O, Lessmann T, Petfalski E, Tollervey D, Lechner J, Hurt E (2001) Identification of a 60S preribosomal particle that is closely linked to nuclear export. Mol Cell 8:517–529

Ben-Shem A, Jenner L, Yusupova G, Yusupov M (2010) Crystal structure of the eukaryotic ribosome. Science 330:1203–1209

Bernstein KA, Baserga SJ (2004) The small subunit processome is required for cell cycle progression at G1. Mol Biol Cell 15:5038–5046

Bernstein KA, Bleichert F, Bean JM, Cross FR, Baserga SJ (2007) Ribosome biogenesis is sensed at the Start cell cycle checkpoint. Mol Biol Cell 18:953–964

Bertwistle D, Sugimoto M, Sherr CJ (2004) Physical and functional interactions of the Arf tumor suppressor protein with nucleophosmin/B23. Mol Cell Biol 24:985–996

Bhat KP, Itahana K, Jin A, Zhang Y (2004) Essential role of ribosomal protein L11 in mediating growth inhibition-induced p53 activation. EMBO J 23:2402–2412

Boocock GR, Morrison JA, Popovic M, Richards N, Ellis L, Durie PR, Rommens JM (2003) Mutations in SBDS are associated with Shwachman-Diamond syndrome. Nat Genet 33:97–101

Boultwood J, Pellagatti A, Cattan H, Lawrie CH, Giagounidis A, Malcovati L, Della Porta MG, Jadersten M, Killick S, Fidler C, Cazzola M, Hellstrom-Lindberg E, Wainscoat JS (2007) Gene expression profiling of CD34+ cells in patients with the 5q- syndrome. Br J Haematol 139:578–589

Boultwood J, Pellagatti A, McKenzie AN, Wainscoat JS (2010) Advances in the 5q- syndrome. Blood 116:5803–5811

Burroughs L, Woolfrey A, Shimamura A (2009) Shwachman-Diamond syndrome: a review of the clinical presentation, molecular pathogenesis, diagnosis, and treatment. Hematol Oncol Clin North Am 23:233–248

Cai T, Aulds J, Gill T, Cerio M, Schmitt ME (2002) The *Saccharomyces cerevisiae* RNase mitochondrial RNA processing is critical for cell cycle progression at the end of mitosis. Genetics 161:1029–1042

Campagnoli MF, Ramenghi U, Armiraglio M, Quarello P, Garelli E, Carando A, Avondo F, Pavesi E, Fribourg S, Gleizes PE, Loreni F, Dianzani I (2008) RPS19 mutations in patients with Diamond-Blackfan anemia. Hum Mutat 29:911–920

Castle CD, Cassimere EK, Lee J, Denicourt C (2010) Las1L is a nucleolar protein required for cell proliferation and ribosome biogenesis. Mol Cell Biol 30:4404–4414

Chen D, Zhang Z, Li M, Wang W, Li Y, Rayburn ER, Hill DL, Wang H, Zhang R (2007) Ribosomal protein S7 as a novel modulator of p53-MDM2 interaction: binding to MDM2, stabilization of p53 protein, and activation of p53 function. Oncogene 26:5029–5037

Choesmel V, Bacqueville D, Rouquette J, Noaillac-Depeyre J, Fribourg S, Cretien A, Leblanc T, Tchernia G, Da Costa L, Gleizes PE (2007) Impaired ribosome biogenesis in Diamond-Blackfan anemia. Blood 109:1275–1283

Choesmel V, Fribourg S, Aguissa-Toure AH, Pinaud N, Legrand P, Gazda HT, Gleizes PE (2008) Mutation of ribosomal protein RPS24 in Diamond-Blackfan anemia results in a ribosome biogenesis disorder. Hum Mol Genet 17:1253–1263

Clemente-Blanco A, Mayan-Santos M, Schneider DA, Machin F, Jarmuz A, Tschochner H, Aragon L (2009) Cdc14 inhibits transcription by RNA polymerase I during anaphase. Nature 458:219–222

Cmejla R, Cmejlova J, Handrkova H, Petrak J, Pospisilova D (2007) Ribosomal protein S17 gene (RPS17) is mutated in Diamond-Blackfan anemia. Hum Mutat 28:1178–1182

Cmejla R, Cmejlova J, Handrkova H, Petrak J, Petrtylova K, Mihal V, Stary J, Cerna Z, Jabali Y, Pospisilova D (2009) Identification of mutations in the ribosomal protein L5 (RPL5) and ribosomal protein L11 (RPL11) genes in Czech patients with Diamond-Blackfan anemia. Hum Mutat 30:321–327

Costanzo M, Nishikawa JL, Tang X, Millman JS, Schub O, Breitkreuz K, Dewar D, Rupes I, Andrews B, Tyers M (2004) CDK activity antagonizes Whi5, an inhibitor of G1/S transcription in yeast. Cell 117:899–913

Da Costa L, Willig TN, Fixler J, Mohandas N, Tchernia G (2001) Diamond-Blackfan anemia. Curr Opin Pediatr 13:10–15

Dai MS, Lu H (2004) Inhibition of MDM2-mediated p53 ubiquitination and degradation by ribosomal protein L5. J Biol Chem 279:44475–44482

Dai MS, Zeng SX, Jin Y, Sun XX, David L, Lu H (2004) Ribosomal protein L23 activates p53 by inhibiting MDM2 function in response to ribosomal perturbation but not to translation inhibition. Mol Cell Biol 24:7654–7668

Dai MS, Sun XX, Lu H (2008) Aberrant expression of nucleostemin activates p53 and induces cell cycle arrest via inhibition of MDM2. Mol Cell Biol 28:4365–4376

Danilova N, Sakamoto KM, Lin S (2008) Ribosomal protein S19 deficiency in zebrafish leads to developmental abnormalities and defective erythropoiesis through activation of p53 protein family. Blood 112:5228–5237

Dauwerse JG, Dixon J, Seland S, Ruivenkamp CA, van Haeringen A, Hoefsloot LH, Peters DJ, Boers AC, Daumer-Haas C, Maiwald R, Zweier C, Kerr B, Cobo AM, Toral JF, Hoogeboom AJ, Lohmann DR, Hehr U, Dixon MJ, Breuning MH, Wieczorek D (2011) Mutations in genes encoding subunits of RNA polymerases I and III cause Treacher Collins syndrome. Nat Genet 43:20–22

de Bruin RA, McDonald WH, Kalashnikova TI, Yates J III, Wittenberg C (2004) Cln3 activates G1-specific transcription via phosphorylation of the SBF bound repressor Whi5. Cell 117:887–898

Diamond LK, Wang WC, Alter BP (1976) Congenital hypoplastic anemia. Adv Pediatr 22:349–378

Dixon J, Edwards SJ, Anderson I, Brass A, Scambler PJ, Dixon MJ (1997) Identification of the complete coding sequence and genomic organization of the Treacher Collins syndrome gene. Genome Res 7:223–234

Dixon J, Jones NC, Sandell LL, Jayasinghe SM, Crane J, Rey JP, Dixon MJ, Trainor PA (2006) Tcof1/Treacle is required for neural crest cell formation and proliferation deficiencies that cause craniofacial abnormalities. Proc Natl Acad Sci USA 103:13403–13408

Dixon J, Trainor P, Dixon MJ (2007) Treacher Collins syndrome. Orthod Craniofac Res 10:88–95

Doherty L, Sheen MR, Vlachos A, Choesmel V, O'Donohue MF, Clinton C, Schneider HE, Sieff CA, Newburger PE, Ball SE, Niewiadomska E, Matysiak M, Glader B, Arceci RJ, Farrar JE, Atsidaftos E, Lipton JM, Gleizes PE, Gazda HT (2010) Ribosomal protein genes RPS10 and RPS26 are commonly mutated in Diamond-Blackfan anemia. Am J Hum Genet 86:222–228

Dragon F, Gallagher JE, Compagnone-Post PA, Mitchell BM, Porwancher KA, Wehner KA, Wormsley S, Settlage RE, Shabanowitz J, Osheim Y, Beyer AL, Hunt DF, Baserga SJ (2002) A large nucleolar U3 ribonucleoprotein required for 18S ribosomal RNA biogenesis. Nature 417:967–970

Draptchinskaia N, Gustavsson P, Andersson B, Pettersson M, Willig TN, Dianzani I, Ball S, Tchernia G, Klar J, Matsson H, Tentler D, Mohandas N, Carlsson B, Dahl N (1999) The gene encoding ribosomal protein S19 is mutated in Diamond-Blackfan anaemia. Nat Genet 21:169–175

Dror Y (2005) Shwachman-Diamond syndrome. Pediatr Blood Cancer 45:892–901

Dror Y, Freedman MH (2001) Shwachman-Diamond syndrome marrow cells show abnormally increased apoptosis mediated through the Fas pathway. Blood 97:3011–3016

Du YC, Stillman B (2002) Yph1p, an ORC-interacting protein: potential links between cell proliferation control, DNA replication, and ribosome biogenesis. Cell 109:835–848

Dutt S, Narla A, Lin K, Mullally A, Abayasekara N, Megerdichian C, Wilson FH, Currie T, Khanna-Gupta A, Berliner N, Kutok JL, Ebert BL (2011) Haploinsufficiency for ribosomal protein genes causes selective activation of p53 in human erythroid progenitor cells. Blood 117(9):2567–2576

Ebert BL, Lee MM, Pretz JL, Subramanian A, Mak R, Golub TR, Sieff CA (2005) An RNA interference model of RPS19 deficiency in Diamond-Blackfan anemia recapitulates defective hematopoiesis and rescue by dexamethasone: identification of dexamethasone-responsive genes by microarray. Blood 105:4620–4626

Ebert BL, Pretz J, Bosco J, Chang CY, Tamayo P, Galili N, Raza A, Root DE, Attar E, Ellis SR, Golub TR (2008) Identification of RPS14 as a 5q- syndrome gene by RNA interference screen. Nature 451:335–339

Elghetany MT, Alter BP (2002) p53 Protein overexpression in bone marrow biopsies of patients with Shwachman-Diamond syndrome has a prevalence similar to that of patients with refractory anemia. Arch Pathol Lab Med 126:452–455

Farrar JE, Nater M, Caywood E, McDevitt MA, Kowalski J, Takemoto CM, Talbot CC Jr, Meltzer P, Esposito D, Beggs AH, Schneider HE, Grabowska A, Ball SE, Niewiadomska E, Sieff CA, Vlachos A, Atsidaftos E, Ellis SR, Lipton JM, Gazda HT, Arceci RJ (2008) Abnormalities of the large ribosomal subunit protein, Rpl35a, in Diamond-Blackfan anemia. Blood 112:1582–1592

Flygare J, Karlsson S (2007) Diamond-Blackfan anemia: erythropoiesis lost in translation. Blood 109:3152–3154

Flygare J, Kiefer T, Miyake K, Utsugisawa T, Hamaguchi I, Da Costa L, Richter J, Davey EJ, Matsson H, Dahl N, Wiznerowicz M, Trono D, Karlsson S (2005) Deficiency of ribosomal protein S19 in CD34+ cells generated by siRNA blocks erythroid development and mimics defects seen in Diamond-Blackfan anemia. Blood 105:4627–4634

Flygare J, Aspesi A, Bailey JC, Miyake K, Caffrey JM, Karlsson S, Ellis SR (2007) Human RPS19, the gene mutated in Diamond-Blackfan anemia, encodes a ribosomal protein required for the maturation of 40S ribosomal subunits. Blood 109:980–986

Ganapathi KA, Austin KM, Lee CS, Dias A, Malsch MM, Reed R, Shimamura A (2007) The human Shwachman-Diamond syndrome protein, SBDS, associates with ribosomal RNA. Blood 110:1458–1465

Gazda HT, Sieff CA (2006) Recent insights into the pathogenesis of Diamond-Blackfan anaemia. Br J Haematol 135:149–157

Gazda HT, Zhong R, Long L, Niewiadomska E, Lipton JM, Ploszynska A, Zaucha JM, Vlachos A, Atsidaftos E, Viskochil DH, Niemeyer CM, Meerpohl JJ, Rokicka-Milewska R, Pospisilova D, Wiktor-Jedrzejczak W, Nathan DG, Beggs AH, Sieff CA (2004) RNA and protein evidence for haplo-insufficiency in Diamond-Blackfan anaemia patients with RPS19 mutations. Br J Haematol 127:105–113

Gazda HT, Grabowska A, Merida-Long LB, Latawiec E, Schneider HE, Lipton JM, Vlachos A, Atsidaftos E, Ball SE, Orfali KA, Niewiadomska E, Da Costa L, Tchernia G, Niemeyer C, Meerpohl JJ, Stahl J, Schratt G, Glader B, Backer K, Wong C, Nathan DG, Beggs AH, Sieff CA (2006) Ribosomal protein S24 gene is mutated in Diamond-Blackfan anemia. Am J Hum Genet 79:1110–1118

Gazda HT, Sheen MR, Vlachos A, Choesmel V, O'Donohue MF, Schneider H, Darras N, Hasman C, Sieff CA, Newburger PE, Ball SE, Niewiadomska E, Matysiak M, Zaucha JM, Glader B, Niemeyer C, Meerpohl JJ, Atsidaftos E, Lipton JM, Gleizes PE, Beggs AH (2008) Ribosomal protein L5 and L11 mutations are associated with cleft palate and abnormal thumbs in Diamond-Blackfan anemia patients. Am J Hum Genet 83:769–780

Gilkes DM, Chen L, Chen J (2006) MDMX regulation of p53 response to ribosomal stress. EMBO J 25:5614–5625

Gill T, Cai T, Aulds J, Wierzbicki S, Schmitt ME (2004) RNase MRP cleaves the CLB2 mRNA to promote cell cycle progression: novel method of mRNA degradation. Mol Cell Biol 24:945–953

Gill T, Aulds J, Schmitt ME (2006) A specialized processing body that is temporally and asymmetrically regulated during the cell cycle in *Saccharomyces cerevisiae*. J Cell Biol 173:35–45

Ginisty H, Amalric F, Bouvet P (1998) Nucleolin functions in the first step of ribosomal RNA processing. EMBO J 17:1476–1486

Gonzales B, Henning D, So RB, Dixon J, Dixon MJ, Valdez BC (2005) The Treacher Collins syndrome (TCOF1) gene product is involved in pre-rRNA methylation. Hum Mol Genet 14:2035–2043

Grimm T, Holzel M, Rohrmoser M, Harasim T, Malamoussi A, Gruber-Eber A, Kremmer E, Eick D (2006) Dominant-negative Pes1 mutants inhibit ribosomal RNA processing and cell proliferation via incorporation into the PeBoW-complex. Nucleic Acids Res 34:3030–3043

Grisendi S, Bernardi R, Rossi M, Cheng K, Khandker L, Manova K, Pandolfi PP (2005) Role of nucleophosmin in embryonic development and tumorigenesis. Nature 437:147–153

Group T.T.C.S.C. [The Treacher Collins Syndrome Collaborative Group] (1996) Positional cloning of a gene involved in the pathogenesis of Treacher Collins syndrome. Nat Genet 12: 130–136

Harnpicharnchai P, Jakovljevic J, Horsey E, Miles T, Roman J, Rout M, Meagher D, Imai B, Guo Y, Brame CJ, Shabanowitz J, Hunt DF, Woolford JL Jr (2001) Composition and functional characterization of yeast 66S ribosome assembly intermediates. Mol Cell 8:505–515

Hayano T, Yanagida M, Yamauchi Y, Shinkawa T, Isobe T, Takahashi N (2003) Proteomic analysis of human Nop56p-associated pre-ribosomal ribonucleoprotein complexes. Possible link between Nop56p and the nucleolar protein treacle responsible for Treacher Collins syndrome. J Biol Chem 278:34309–34319

Henras AK, Soudet J, Gerus M, Lebaron S, Caizergues-Ferrer M, Mougin A, Henry Y (2008) The post-transcriptional steps of eukaryotic ribosome biogenesis. Cell Mol Life Sci 65:2334–2359

Herrera JE, Savkur R, Olson MO (1995) The ribonuclease activity of nucleolar protein B23. Nucleic Acids Res 23:3974–3979

Hesling C, Oliveira CC, Castilho BA, Zanchin NI (2007) The Shwachman-Bodian-Diamond syndrome associated protein interacts with HsNip7 and its down-regulation affects gene expression at the transcriptional and translational levels. Exp Cell Res 313:4180–4195

Holzel M, Rohrmoser M, Schlee M, Grimm T, Harasim T, Malamoussi A, Gruber-Eber A, Kremmer E, Hiddemann W, Bornkamm GW, Eick D (2005) Mammalian WDR12 is a novel member of the Pes1-Bop1 complex and is required for ribosome biogenesis and cell proliferation. J Cell Biol 170:367–378

Holzel M, Orban M, Hochstatter J, Rohrmoser M, Harasim T, Malamoussi A, Kremmer E, Langst G, Eick D (2010) Defects in 18S or 28S rRNA processing activate the p53 pathway. J Biol Chem 285:6364–6370

Horn HF, Vousden KH (2008) Cooperation between the ribosomal proteins L5 and L11 in the p53 pathway. Oncogene 27:5774–5784

Huh WK, Falvo JV, Gerke LC, Carroll AS, Howson RW, Weissman JS, O'Shea EK (2003) Global analysis of protein localization in budding yeast. Nature 425:686–691

Hussain S, Benavente SB, Nascimento E, Dragoni I, Kurowski A, Gillich A, Humphreys P, Frye M (2009) The nucleolar RNA methyltransferase Misu (NSun2) is required for mitotic spindle stability. J Cell Biol 186:27–40

Idol RA, Robledo S, Du HY, Crimmins DL, Wilson DB, Ladenson JH, Bessler M, Mason PJ (2007) Cells depleted for RPS19, a protein associated with Diamond Blackfan Anemia, show defects in 18S ribosomal RNA synthesis and small ribosomal subunit production. Blood Cells Mol Dis 39:35–43

Ionescu CN, Origanti S, McAlear MA (2004) The yeast rRNA biosynthesis factor Ebp2p is also required for efficient nuclear division. Yeast 21:1219–1232

Isaac C, Marsh KL, Paznekas WA, Dixon J, Dixon MJ, Jabs EW, Meier UT (2000) Characterization of the nucleolar gene product, treacle, in Treacher Collins syndrome. Mol Biol Cell 11:3061–3071

Itahana K, Bhat KP, Jin A, Itahana Y, Hawke D, Kobayashi R, Zhang Y (2003) Tumor suppressor ARF degrades B23, a nucleolar protein involved in ribosome biogenesis and cell proliferation. Mol Cell 12:1151–1164

Iwanami N, Higuchi T, Sasano Y, Fujiwara T, Hoa VQ, Okada M, Talukder SR, Kunimatsu S, Li J, Saito F, Bhattacharya C, Matin A, Sasaki T, Shimizu N, Mitani H, Himmelbauer H, Momoi A, Kondoh H, Furutani-Seiki M, Takahama Y (2008) WDR55 is a nucleolar modulator of ribosomal RNA synthesis, cell cycle progression, and teleost organ development. PLoS Genet 4:e1000171

Jin A, Itahana K, O'Keefe K, Zhang Y (2004) Inhibition of HDM2 and activation of p53 by ribosomal protein L23. Mol Cell Biol 24:7669–7680

Johnston GC, Pringle JR, Hartwell LH (1977) Coordination of growth with cell division in the yeast *Saccharomyces cerevisiae*. Exp Cell Res 105:79–98

Johnston GC, Ehrhardt CW, Lorincz A, Carter BL (1979) Regulation of cell size in the yeast *Saccharomyces cerevisiae*. J Bacteriol 137:1–5

Jones NC, Lynn ML, Gaudenz K, Sakai D, Aoto K, Rey JP, Glynn EF, Ellington L, Du C, Dixon J, Dixon MJ, Trainor PA (2008) Prevention of the neurocristopathy Treacher Collins syndrome through inhibition of p53 function. Nat Med 14:125–133

Jorgensen P, Nishikawa JL, Breitkreutz BJ, Tyers M (2002) Systematic identification of pathways that couple cell growth and division in yeast. Science 297:395–400

Jorgensen P, Rupes I, Sharom JR, Schneper L, Broach JR, Tyers M (2004) A dynamic transcriptional network communicates growth potential to ribosome synthesis and critical cell size. Genes Dev 18:2491–2505

Jwa M, Kim JH, Chan CS (2008) Regulation of Sli15/INCENP, kinetochore, and Cdc14 phosphatase functions by the ribosome biogenesis protein Utp7. J Cell Biol 182:1099–1111

Karayan L, Riou JF, Seite P, Migeon J, Cantereau A, Larsen CJ (2001) Human ARF protein interacts with topoisomerase I and stimulates its activity. Oncogene 20:836–848

Killian A, Le Meur N, Sesboue R, Bourguignon J, Bougeard G, Gautherot J, Bastard C, Frebourg T, Flaman JM (2004) Inactivation of the RRB1-Pescadillo pathway involved in ribosome biogenesis induces chromosomal instability. Oncogene 23:8597–8602

Krogan NJ, Cagney G, Yu H, Zhong G, Guo X, Ignatchenko A, Li J, Pu S, Datta N, Tikuisis AP, Punna T, Peregrin-Alvarez JM, Shales M, Zhang X, Davey M, Robinson MD, Paccanaro A, Bray JE, Sheung A, Beattie B, Richards DP, Canadien V, Lalev A, Mena F, Wong P, Starostine A, Canete MM, Vlasblom J, Wu S, Orsi C, Collins SR, Chandran S, Haw R, Rilstone JJ, Gandi K, Thompson NJ, Musso G, St Onge P, Ghanny S, Lam MH, Butland G, Altaf-Ul AM, Kanaya S, Shilatifard A, O'Shea E, Weissman JS, Ingles CJ, Hughes TR, Parkinson J, Gerstein M, Wodak SJ, Emili A, Greenblatt JF (2006) Global landscape of protein complexes in the yeast *Saccharomyces cerevisiae*. Nature 440:637–643

Kurki S, Peltonen K, Latonen L, Kiviharju TM, Ojala PM, Meek D, Laiho M (2004) Nucleolar protein NPM interacts with HDM2 and protects tumor suppressor protein p53 from HDM2-mediated degradation. Cancer Cell 5:465–475

Lapik YR, Fernandes CJ, Lau LF, Pestov DG (2004) Physical and functional interaction between Pes1 and Bop1 in mammalian ribosome biogenesis. Mol Cell 15:17–29

Lerch-Gaggl A, Haque J, Li J, Ning G, Traktman P, Duncan SA (2002) Pescadillo is essential for nucleolar assembly, ribosome biogenesis, and mammalian cell proliferation. J Biol Chem 277:45347–45355

Lessard F, Morin F, Ivanchuk S, Langlois F, Stefanovsky V, Rutka J, Moss T (2010) The ARF tumor suppressor controls ribosome biogenesis by regulating the RNA polymerase I transcription factor TTF-I. Mol Cell 38:539–550

Lin CI, Yeh NH (2009) Treacle recruits RNA polymerase I complex to the nucleolus that is independent of UBF. Biochem Biophys Res Commun 386:396–401

Lindstrom MS, Nister M (2010) Silencing of ribosomal protein S9 elicits a multitude of cellular responses inhibiting the growth of cancer cells subsequent to p53 activation. PLoS One 5:e9578

Lindstrom MS, Zhang Y (2008) Ribosomal protein S9 is a novel B23/NPM-binding protein required for normal cell proliferation. J Biol Chem 283:15568–15576

Lindstrom MS, Jin A, Deisenroth C, White Wolf G, Zhang Y (2007) Cancer-associated mutations in the MDM2 zinc finger domain disrupt ribosomal protein interaction and attenuate MDM2-induced p53 degradation. Mol Cell Biol 27:1056–1068

Lipton JM, Ellis SR (2009) Diamond-Blackfan anemia: diagnosis, treatment, and molecular pathogenesis. Hematol Oncol Clin North Am 23:261–282

Llanos S, Serrano M (2010) Depletion of ribosomal protein L37 occurs in response to DNA damage and activates p53 through the L11/MDM2 pathway. Cell Cycle 9:4005–4012

Lohrum MA, Ludwig RL, Kubbutat MH, Hanlon M, Vousden KH (2003) Regulation of HDM2 activity by the ribosomal protein L11. Cancer Cell 3:577–587

Luz JS, Georg RC, Gomes CH, Machado-Santelli GM, Oliveira CC (2009) Sdo1p, the yeast orthologue of Shwachman-Bodian-Diamond syndrome protein, binds RNA and interacts with nuclear rRNA-processing factors. Yeast 26:287–298

Ma H, Pederson T (2007) Depletion of the nucleolar protein nucleostemin causes G1 cell cycle arrest via the p53 pathway. Mol Biol Cell 18:2630–2635

Macias E, Jin A, Deisenroth C, Bhat K, Mao H, Lindstrom MS, Zhang Y (2010) An ARF-independent c-MYC-activated tumor suppression pathway mediated by ribosomal protein-Mdm2 Interaction. Cancer Cell 18:231–243

Marechal V, Elenbaas B, Piette J, Nicolas JC, Levine AJ (1994) The ribosomal L5 protein is associated with mdm-2 and mdm-2-p53 complexes. Mol Cell Biol 14:7414–7420

McGowan KA, Li JZ, Park CY, Beaudry V, Tabor HK, Sabnis AJ, Zhang W, Fuchs H, de Angelis MH, Myers RM, Attardi LD, Barsh GS (2008) Ribosomal mutations cause p53-mediated dark skin and pleiotropic effects. Nat Genet 40:963–970

McMahon M, Ayllon V, Panov KI, O'Connor R (2010) Ribosomal 18S RNA processing by the IGF-I-responsive WDR3 protein is integrated with p53 function in cancer cell proliferation. J Biol Chem 285:18309–18318

Menne TF, Goyenechea B, Sanchez-Puig N, Wong CC, Tonkin LM, Ancliff PJ, Brost RL, Costanzo M, Boone C, Warren AJ (2007) The Shwachman-Bodian-Diamond syndrome protein mediates translational activation of ribosomes in yeast. Nat Genet 39:486–495

Milkereit P, Gadal O, Podtelejnikov A, Trumtel S, Gas N, Petfalski E, Tollervey D, Mann M, Hurt E, Tschochner H (2001) Maturation and intranuclear transport of pre-ribosomes requires Noc proteins. Cell 105:499–509

Narla A, Ebert BL (2010) Ribosomopathies: human disorders of ribosome dysfunction. Blood 115:3196–3205

Nihrane A, Sezgin G, Dsilva S, Dellorusso P, Yamamoto K, Ellis SR, Liu JM (2009) Depletion of the Shwachman-Diamond syndrome gene product, SBDS, leads to growth inhibition and increased expression of OPG and VEGF-A. Blood Cells Mol Dis 42:85–91

O'Donohue MF, Choesmel V, Faubladier M, Fichant G, Gleizes PE (2010) Functional dichotomy of ribosomal proteins during the synthesis of mammalian 40S ribosomal subunits. J Cell Biol 190:853–866

Oeffinger M, Tollervey D (2003) Yeast Nop15p is an RNA-binding protein required for pre-rRNA processing and cytokinesis. EMBO J 22:6573–6583

Oeffinger M, Leung A, Lamond A, Tollervey D (2002) Yeast Pescadillo is required for multiple activities during 60S ribosomal subunit synthesis. RNA 8:626–636

Oeffinger M, Fatica A, Rout MP, Tollervey D (2007) Yeast Rrp14p is required for ribosomal subunit synthesis and for correct positioning of the mitotic spindle during mitosis. Nucleic Acids Res 35:1354–1366

Okuda M, Horn HF, Tarapore P, Tokuyama Y, Smulian AG, Chan PK, Knudsen ES, Hofmann IA, Snyder JD, Bove KE, Fukasawa K (2000) Nucleophosmin/B23 is a target of CDK2/cyclin E in centrosome duplication. Cell 103:127–140

Panic L, Tamarut S, Sticker-Jantscheff M, Barkic M, Solter D, Uzelac M, Grabusic K, Volarevic S (2006) Ribosomal protein S6 gene haploinsufficiency is associated with activation of a p53-dependent checkpoint during gastrulation. Mol Cell Biol 26:8880–8891

Pellagatti A, Marafioti T, Paterson JC, Barlow JL, Drynan LF, Giagounidis A, Pileri SA, Cazzola M, McKenzie AN, Wainscoat JS, Boultwood J (2010) Induction of p53 and up-regulation of the p53 pathway in the human 5q- syndrome. Blood 115:2721–2723

Peng Q, Wu J, Zhang Y, Liu Y, Kong R, Hu L, Du X, Ke Y (2010) 1A6/DRIM, a novel t-UTP, activates RNA polymerase I transcription and promotes cell proliferation. PLoS One 5:e14244

Pestov DG, Grzeszkiewicz TM, Lau LF (1998) Isolation of growth suppressors from a cDNA expression library. Oncogene 17:3187–3197

Pestov DG, Strezoska Z, Lau LF (2001) Evidence of p53-dependent cross-talk between ribosome biogenesis and the cell cycle: effects of nucleolar protein Bop1 on G(1)/S transition. Mol Cell Biol 21:4246–4255

Popolo L, Vanoni M, Alberghina L (1982) Control of the yeast cell cycle by protein synthesis. Exp Cell Res 142:69–78

Prescott DM, Bender MA (1962) Synthesis of RNA and protein during mitosis in mammalian tissue culture cells. Exp Cell Res 26:260–268

Rawls AS, Gregory AD, Woloszynek JR, Liu F, Link DC (2007) Lentiviral-mediated RNAi inhibition of Sbds in murine hematopoietic progenitors impairs their hematopoietic potential. Blood 110:2414–2422

Robledo S, Idol RA, Crimmins DL, Ladenson JH, Mason PJ, Bessler M (2008) The role of human ribosomal proteins in the maturation of rRNA and ribosome production. RNA 14:1918–1929

Roger B, Moisand A, Amalric F, Bouvet P (2003) Nucleolin provides a link between RNA polymerase I transcription and pre-ribosome assembly. Chromosoma 111:399–407

Rosado IV, Kressler D, de la Cruz J (2007) Functional analysis of *Saccharomyces cerevisiae* ribosomal protein Rpl3p in ribosome synthesis. Nucleic Acids Res 35:4203–4213

Roussel P, Andre C, Comai L, Hernandez-Verdun D (1996) The rDNA transcription machinery is assembled during mitosis in active NORs and absent in inactive NORs. J Cell Biol 133:235–246

Rubbi CP, Milner J (2003) Disruption of the nucleolus mediates stabilization of p53 in response to DNA damage and other stresses. EMBO J 22:6068–6077

Rujkijyanont P, Watanabe K, Ambekar C, Wang H, Schimmer A, Beyene J, Dror Y (2008) SBDS-deficient cells undergo accelerated apoptosis through the Fas-pathway. Haematologica 93:363–371

Sakai D, Trainor PA (2009) Treacher Collins syndrome: unmasking the role of Tcof1/treacle. Int J Biochem Cell Biol 41:1229–1232

Sanchez-Diaz A, Kanemaki M, Marchesi V, Labib K (2004) Rapid depletion of budding yeast proteins by fusion to a heat-inducible degron. Sci STKE 2004:PL8

Savchenko A, Krogan N, Cort JR, Evdokimova E, Lew JM, Yee AA, Sanchez-Pulido L, Andrade MA, Bochkarev A, Watson JD, Kennedy MA, Greenblatt J, Hughes T, Arrowsmith CH, Rommens JM, Edwards AM (2005) The Shwachman-Bodian-Diamond syndrome protein family is involved in RNA metabolism. J Biol Chem 280:19213–19220

Saveanu C, Rousselle JC, Lenormand P, Namane A, Jacquier A, Fromont-Racine M (2007) The p21-activated protein kinase inhibitor Skb15 and its budding yeast homologue are 60S ribosome assembly factors. Mol Cell Biol 27:2897–2909

Savkur RS, Olson MO (1998) Preferential cleavage in pre-ribosomal RNA by protein B23 endoribonuclease. Nucleic Acids Res 26:4508–4515

Saxena A, Rorie CJ, Dimitrova D, Daniely Y, Borowiec JA (2006) Nucleolin inhibits Hdm2 by multiple pathways leading to p53 stabilization. Oncogene 25:7274–7288

Schlott T, Reimer S, Jahns A, Ohlenbusch A, Ruschenburg I, Nagel H, Droese M (1997) Point mutations and nucleotide insertions in the MDM2 zinc finger structure of human tumours. J Pathol 182:54–61

Shimamura A (2006) Shwachman-Diamond syndrome. Semin Hematol 43:178–188

Strezoska Z, Pestov DG, Lau LF (2000) Bop1 is a mouse WD40 repeat nucleolar protein involved in 28S and 5.8S RRNA processing and 60S ribosome biogenesis. Mol Cell Biol 20:5516–5528

Strezoska Z, Pestov DG, Lau LF (2002) Functional inactivation of the mouse nucleolar protein Bop1 inhibits multiple steps in pre-rRNA processing and blocks cell cycle progression. J Biol Chem 277:29617–29625

Sulic S, Panic L, Barkic M, Mercep M, Uzelac M, Volarevic S (2005) Inactivation of S6 ribosomal protein gene in T lymphocytes activates a p53-dependent checkpoint response. Genes Dev 19:3070–3082

Sun XX, Wang YG, Xirodimas DP, Dai MS (2010) Perturbation of 60S ribosomal biogenesis results in ribosomal protein L5- and L11-dependent p53 activation. J Biol Chem 285:25812–25821

Takagi M, Absalon MJ, McLure KG, Kastan MB (2005) Regulation of p53 translation and induction after DNA damage by ribosomal protein L26 and nucleolin. Cell 123:49–63

Trainor PA, Dixon J, Dixon MJ (2009) Treacher Collins syndrome: etiology, pathogenesis and prevention. Eur J Hum Genet 17:275–283

Uechi T, Nakajima Y, Chakraborty A, Torihara H, Higa S, Kenmochi N (2008) Deficiency of ribosomal protein S19 during early embryogenesis leads to reduction of erythrocytes in a zebrafish model of Diamond-Blackfan anemia. Hum Mol Genet 17:3204–3211

Ugrinova I, Monier K, Ivaldi C, Thiry M, Storck S, Mongelard F, Bouvet P (2007) Inactivation of nucleolin leads to nucleolar disruption, cell cycle arrest and defects in centrosome duplication. BMC Mol Biol 8:66

Valdez BC, Henning D, So RB, Dixon J, Dixon MJ (2004) The Treacher Collins syndrome (TCOF1) gene product is involved in ribosomal DNA gene transcription by interacting with upstream binding factor. Proc Natl Acad Sci USA 101:10709–10714

Van den Berghe H, Cassiman JJ, David G, Fryns JP, Michaux JL, Sokal G (1974) Distinct haematological disorder with deletion of long arm of no. 5 chromosome. Nature 251:437–438

Vlachos A, Ball S, Dahl N, Alter BP, Sheth S, Ramenghi U, Meerpohl J, Karlsson S, Liu JM, Leblanc T, Paley C, Kang EM, Leder EJ, Atsidaftos E, Shimamura A, Bessler M, Glader B, Lipton JM (2008) Diagnosing and treating Diamond Blackfan anaemia: results of an international clinical consensus conference. Br J Haematol 142:859–876

Volarevic S, Stewart MJ, Ledermann B, Zilberman F, Terracciano L, Montini E, Grompe M, Kozma SC, Thomas G (2000) Proliferation, but not growth, blocked by conditional deletion of 40S ribosomal protein S6. Science 288:2045–2047

Wang Y, Liu J, Zhao H, Lu W, Zhao J, Yang L, Li N, Du X, Ke Y (2007) Human 1A6/DRIM, the homolog of yeast Utp20, functions in the 18S rRNA processing. Biochim Biophys Acta 1773:863–868

Wise CA, Chiang LC, Paznekas WA, Sharma M, Musy MM, Ashley JA, Lovett M, Jabs EW (1997) TCOF1 gene encodes a putative nucleolar phosphoprotein that exhibits mutations in Treacher Collins syndrome throughout its coding region. Proc Natl Acad Sci USA 94:3110–3115

Yadavilli S, Mayo LD, Higgins M, Lain S, Hegde V, Deutsch WA (2009) Ribosomal protein S3: a multi-functional protein that interacts with both p53 and MDM2 through its KH domain. DNA Repair (Amst) 8:1215–1224

Yamada H, Horigome C, Okada T, Shirai C, Mizuta K (2007) Yeast Rrp14p is a nucleolar protein involved in both ribosome biogenesis and cell polarity. RNA 13:1977–1987

Yu W, Qiu Z, Gao N, Wang L, Cui H, Qian Y, Jiang L, Luo J, Yi Z, Lu H, Li D, Liu M (2011) PAK1IP1, a ribosomal stress-induced nucleolar protein, regulates cell proliferation via the p53-MDM2 loop. Nucleic Acids Res 39(6):2234–2248

Yuan X, Zhou Y, Casanova E, Chai M, Kiss E, Grone HJ, Schutz G, Grummt I (2005) Genetic inactivation of the transcription factor TIF-IA leads to nucleolar disruption, cell cycle arrest, and p53-mediated apoptosis. Mol Cell 19:77–87

Zhang Y, Lu H (2009) Signaling to p53: ribosomal proteins find their way. Cancer Cell 16:369–377

Zhang Y, Xiong Y, Yarbrough WG (1998) ARF promotes MDM2 degradation and stabilizes p53: ARF-INK4a locus deletion impairs both the Rb and p53 tumor suppression pathways. Cell 92:725–734

Zhang Y, Yu Z, Fu X, Liang C (2002) Noc3p, a bHLH protein, plays an integral role in the initiation of DNA replication in budding yeast. Cell 109:849–860

Zhang Y, Wolf GW, Bhat K, Jin A, Allio T, Burkhart WA, Xiong Y (2003) Ribosomal protein L11 negatively regulates oncoprotein MDM2 and mediates a p53-dependent ribosomal-stress checkpoint pathway. Mol Cell Biol 23:8902–8912

Zhang S, Shi M, Hui CC, Rommens JM (2006) Loss of the mouse ortholog of the Shwachman-Diamond syndrome gene (Sbds) results in early embryonic lethality. Mol Cell Biol 26:6656–6663

Zhang Y, Wang J, Yuan Y, Zhang W, Guan W, Wu Z, Jin C, Chen H, Zhang L, Yang X, He F (2010) Negative regulation of HDM2 to attenuate p53 degradation by ribosomal protein L26. Nucleic Acids Res 38:6544–6554

Zhu Y, Poyurovsky MV, Li Y, Biderman L, Stahl J, Jacq X, Prives C (2009) Ribosomal protein S7 is both a regulator and a substrate of MDM2. Mol Cell 35:316–326

Zimmerman ZA, Kellogg DR (2001) The Sda1 protein is required for passage through start. Mol Biol Cell 12:201–219

Chapter 9
The Multiple Properties and Functions of Nucleolin

Rong Cong, Sadhan Das, and Philippe Bouvet

9.1 Introduction

Nucleolin, one of the most abundant proteins of the nucleolus, was first described in rat liver by Orrick et al. (1973). Homologous nucleolin proteins and their corresponding genes were then identified in other rodents (Bourbon et al. 1988), humans (Srivastava et al. 1989), chicken (Maridor and Nigg 1990), and *Xenopus laevis* (Caizergues-Ferrer et al. 1989; Rankin et al. 1993). All these proteins share the same structural organization. In other eukaryotic species, several nucleolar proteins that exhibit more or less similar structural organization and properties were called "nucleolin-like proteins." They include NucMs1 (Bogre et al. 1996), Pea nucleolin (Tong et al. 1997), Nop64A (de Carcer et al. 1997), FMV3bp (Didier and Klee 1992), Nopp52 (McGrath et al. 1997), gar2 (Gulli et al. 1995a), and NSR1 (Lee et al. 1992). The *Arabidopsis* genome possesses two genes related to nucleolin, encoding AtNUC-L1 and AtNUC-L2 proteins, which are similar to nucleolin-like proteins in other plants. AtNUC-L1 has only two RNA recognition motifs (RRMs), like the yeast NSR1 (Saez-Vasquez et al. 2004). *Arabidopsis* PARL1 is also highly similar to the yeast NSR1 and may therefore have similar functional roles (Petricka and Nelson 2007). A new rat nucleolin like protein (NRP) was recently identified in testicular germ cells (Chathoth et al. 2009). This protein lacks the acidic stretches in its N-terminal domain and is encoded in rat chromosome 15 by a gene that presents a different genomic organization compared to that of human nucleolin.

Although nucleolin was discovered almost 40 years ago, its function in the cell remains poorly understood. Its predominant localization in the nucleolar

---

P. Bouvet (✉)
Université de Lyon, Ecole Normale Supérieure de Lyon, CNRS USR 3010,
Laboratoire Joliot-Curie, 46 Allée d'Italie 69007, Lyon, France
e-mail: pbouvet@ens-lyon.fr

compartment initially led researchers to investigate the role of nucleolin in ribosome biogenesis. During the last 10 years, there have been more reports on the functions of nucleolin that seem rather unrelated to the nucleolus and ribosome biogenesis. However, it now seems also clear, that the nucleolus is involved in functions other than ribosome biogenesis. Multifunctional proteins like nucleolin may be important to link regulation of ribosome biogenesis to other cellular processes.

The ability of nucleolin to be involved in many cellular processes is probably related to its structural organization and its capability to form many different interactions with other proteins. In this review, we first describe the properties of the different structural domains of nucleolin, then analyze their posttranslation modifications, and finally describe the main known functions of nucleolin within the nucleolus and nucleoplasm.

9.2 Properties of Nucleolin Domains

The primary sequences of the different nucleolin and nucleolin-like proteins highlight the organization of the protein into three main structural domains (Fig. 9.1): N-terminal, central, and C-terminal domains (Lapeyre and Amalric 1985).

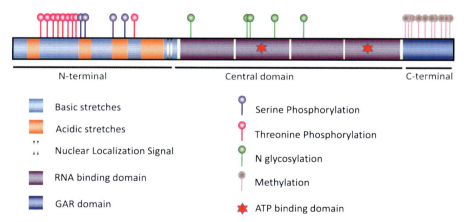

Fig. 9.1 Schematic representation of the Nucleolin structure and its identified posttranslational modifications. The functional domains and target sites of posttranslational modifications known thus far are shown at the upper side of the protein structure. Threonine residues 58, 75, 83, 91, 98, 105, 120, 128, and 219 have been shown to be phosphorylated by the CDK1 kinase (Belenguer et al. 1990). Four serine residues 143, 156, 187, and 209 are phosphorylated *in vivo* by CK II (Mamrack et al. 1979). Possible methylation sites were at position 655, 659, 665, 669, 673, 679, 681, 687, 691, and 694 in the C-terminal GAR domain. Five possible glycosylation sites are at position 317–319, 399–401, 403–405, 477–479, and 491–493 (Lapeyre et al. 1987). Nucleolin seems also to be ADP-ribosylated but the residues that are modified have not been identified (Leitinger and Wesierska-Gadek 1993). Two potential ATP binding sites have been also located in the RNA binding domains (Miranda et al. 1995)

The N-terminal domain (~300 residues) contains highly charged acidic glutamate/aspartate sequence repeats that vary in number depending on the species, separated by basic stretches (Bugler et al. 1987). This region accounts for its relatively low isoelectric pH at 5.5 (Lischwe et al. 1981, 1979; Mamrack et al. 1979). These N-terminal acidic stretches determine the argyrophilic properties of nucleolin (Derenzini 2000; Roussel et al. 1992). The high correlation between intense silver-staining properties of nucleoli and rates of preribosomal RNA biosynthesis (Derenzini et al. 1995) is therefore mainly dependant on the accumulation of nucleolin and of another abundant nucleolar protein, nucleophosmin (Lischwe et al. 1979). The N-terminal domain is the site of multiple protein–protein interactions (Ginisty et al. 1999). Acidic sequences of nucleolin interact with histone H1 and induce chromatin decondensation. This behavior of nucleolin suggests that it can be regarded as an HMG-like protein (Erard et al. 1988). This N-terminal domain is also the site of numerous posttranslational modifications including many cdk1 and CK2 protein kinase phosphorylation sites (Belenguer et al. 1990; Caizergues-Ferrer et al. 1987; Peter et al. 1990) suggesting that this N-terminal domain will be important for the cell-cycle regulation of nucleolin function.

The central region of nucleolin contains four conserved RNA binding domains (RBDs), which are also known as RRMs that allow the specific interaction with nucleic acid sequences. The RNA-binding specificity of mouse nucleolin has been extensively studied and reviewed (Ginisty et al. 1999) and see below Sect. 9.2.1). The number of RBDs within this central domain varies from yeast to humans. Nucleolin-like proteins NSR1, GAR2, NucMs1, pea nucleolin, and FMV3bp possess two RBDs whereas nucleolins from hamster, mouse, rat, humans, chicken, and $X.$ $laevis$ have four RBDs. Although these RBDs are able to give a strong RNA affinity and specificity to nucleolin, recent knockout experiments suggest that these domains harbor redundant functions and that, in $vivo$, specific RNA binding activity may not be required for the vital function of nucleolin (Storck et al. 2009).

The C-terminal domain of nucleolin is rich in glycine, arginine, and phenyla-lanine residues, so it was called the GAR (Glycine- and Arginine-Rich) or RGG (Arg-Gly-Gly) domain. The length of this nucleolin domain is variable among different species, but it is relatively well conserved. Infrared spectroscopic studies reveal that this domain can adopt repeated β-turns (Ghisolfi et al. 1992). Nonspecific interaction of GAR with nucleic acids appears to play a role in strengthening the RBD-specific binding of nucleolin to RNA (Ghisolfi et al. 1992). The GAR domain of pea nucleolin shows ATP-dependent DNA helicase activities (Nasirudin et al. 2005). This C-terminal domain is also a protein–protein interaction domain (Bouvet et al. 1998; Ginisty et al. 1999). Numerous arginine residues of this domain are subject to posttranslational methylation modifications (Lischwe et al. 1982). N^G,N^G-dimethylarginines are the predominant methylated residues, but traces of N^G-monomethylarginine can also be found in nucleolin.

9.2.1 Nucleic Acid Binding Properties of Nucleolin

It was known that nucleolin interacts with nucleic acids before the identification and characterization of the different domains required for these interactions (Herrera and Olson 1986; Olson et al. 1983). Studies of nucleolin interaction with preribosomal RNA have identified two major RNA motifs present in pre-rRNA as the targets of nucleolin (Ghisolfi et al. 1992; Ginisty et al. 2000). The first motif called NRE (nucleolin recognition element) forms a stem-loop structure that contains the consensus sequence UCCCGA in the loop. Specific and high affinity binding of nucleolin with this motif requires the joint action of the first two RBDs (Serin et al. 1997). These stem-loop structures are distributed along the pre-rRNA (Serin et al. 1996). Binding of nucleolin with these motifs during transcription may allow the correct folding of the pre-rRNA required for pre-rRNA processing and assembly of the preribosomal particles (Allain et al. 2000a; Roger et al. 2003). The second motif present in the pre-rRNA was called ECM (evolutionary conserved motif). Nucleolin interaction with this short sequence present just downstream of the first cleavage site (Craig et al. 1991) requires all four RBDs (Ginisty et al. 2001). This interaction of nucleolin with the ECM is required for the assembly of the processing complex for the first cleavage of the pre-rRNA (Ginisty et al. 2001). Different combinations of RBDs can therefore provide different RNA binding specificity to the protein (Ginisty et al. 2001).

Until now, it has not been possible to get a three-dimensional structure of the full protein or even of the four RBDs. But different NMR studies have succeeded in the resolution of the structures of different individual RBDs alone or in complex with nucleic acid sequences of hamster and human nucleolin (Allain et al. 2000b; Arumugam et al. 2010; Finger et al. 2004; Johansson et al. 2004). In particular, RBD1 and 2 were studied in great detail as they bind specifically to the NRE (Ghisolfi-Nieto et al. 1996). The solution structure of hamster nucleolin RBD1 + 2 was determined in its free form and in interaction with a NRE stem-loop of 22 nt selected by SELEX (sNRE). The structure of the sNRE was also determined in the free and bound form (Bouvet et al. 2001). Each RBD adopts the classical $\beta\alpha\beta\beta\alpha\beta$ fold conformation, but each is structurally different from the others (RBD1 has a longer α-helix 1 and shorter β2-β3 loop than RBD2) allowing a specific interaction of each of them with distinct features of the NRE. RBD1 + 2 interacts with the RNA loop via its β-sheets. Johansson et al. (2004) described the solution structure of the 28 kDa complex formed by the two N-terminal RBD1 + 2 and a natural pre-rRNA target, b2NRE. The interaction of RBD1 + 2 with this natural RNA target is less stable than with sNRE (Johansson et al. 2004).

Several reports have also shown that nucleolin is able to interact specifically with the 3′ untranslated region (3′ UTR) of several mRNAs with a major effect on mRNA stability. For instance, RBD1 + 2 binds to a 40-nucleotide region upstream of the bcl-2 AU-rich element (ARE (bcl-2)) and this interaction seems to increase bcl-2 mRNA half-life (Ishimaru et al. 2010). Similarly, nucleolin stabilizes the Bcl-X$_L$ mRNA by binding to the ARE elements in the 3′-UTR *in vitro* and *in vivo*.

This Bcl-X$_L$ mRNA stabilization by nucleolin could be explained by an interaction of the RGG domain of nucleolin with the poly (A) binding protein (PABP), and this interaction could prevent the digestion of the poly(A) tail by a poly(A) RNase (PARN) (Zhang et al. 2008). Interaction of nucleolin with the 3′ UTR of β-globin mRNA and with GADD45α mRNA seems also important for the stability of these mRNAs *in vivo* (Jiang et al. 2006; Zhang et al. 2006). In addition, the binding of NSR1 to the 3′ UTR in the tombusvirus RNA inhibited the *in vitro* replication of the viral RNA in a yeast cell-free assay, probably by interfering with the recruitment of the viral RNA for replication (Jiang et al. 2010b). Interaction of nucleolin with some selenocysteine insertion sequence (SECIS) elements, which contain stem-loop structures in the 3′ UTR of different mRNAs, also seems to be required for the optimal expression of certain selenoproteins (Miniard et al. 2010).

Nucleolin can also bind to the 5′ UTR of p53 mRNA. This interaction with the 5′ UTR affects the efficiency of translation. Overexpression of nucleolin decreases p53 expression whereas down-regulation increases the p53 expression level (Takagi et al. 2005).

Apart from the interaction with RNA, nucleolin is also able to bind different DNA sequences. For instance, nucleolin can bind to denatured single-stranded DNA (ssDNA) (Sapp et al. 1986), matrix-attachment regions (MARs) (Dickinson and Kohwi-Shigematsu 1995), and a number of viral DNA sequences like parvovirus MVMp DNA (Barrijal et al. 1992), human papillomavirus 16 (HPV16) (Sato et al. 2009), and HPV18 (Grinstein et al. 2002). A common feature of many DNA sequences to which nucleolin binds seems to be their richness in guanosine such as guanosine-rich oligonucleotides found in intergenic spacer region of rDNA (Olson et al. 1983), telomeric DNA (Ishikawa et al. 1993; Pollice et al. 2000), and switch regions of immunoglobulin genes (Hanakahi et al. 1997). Guanosine-rich oligodeoxynucle-otides (GROs) that can form G-quartets can be also bound to nucleolin (Bates et al. 1999). One of these nucleolin GRO aptamers, AS1411, was developed for the treatment of cancer, although it is not really clear how the interaction of nucleolin with this aptamer can regulate cell proliferation (Mongelard and Bouvet 2010).

In vitro and *in vivo*, it was found that nucleolin can bind genomic G-rich DNA sequences that have the potency to form G-quartets, like in the gene coding for the human vascular endothelial growth factor (VEGF) gene (Sun et al. 2011) and in the c-MYC promoter (Brooks and Hurley 2010; Gonzalez and Hurley 2010). Binding of nucleolin to this c-MYC promoter region represses transcription of this gene.

9.2.2 Protein–Protein Interactions

One important feature of nucleolin is its ability to interact both with nucleic acids and with a large number of proteins, suggesting that nucleolin may form different complexes that could also explain its numerous functions. Although extensive proteomic studies of nucleolin-interacting proteins have not been yet published, we summarize here some of the better characterized interactions. As nucleolin is predominantly present in the nucleolus and participates in the assembly of preribosomal particles, it was not surprising to find that nucleolin interacts with a subset of ribosomal proteins (Bouvet et al. 1998). Both the RGG and the N-terminal domains are important for these interactions (Bouvet et al. 1998; Sicard et al. 1998). It was also therefore not surprising that in human embryonic kidney (HEK) 293 cells, nucleolin was found associated with a ribonucleoprotein (RNP) complex that contains mainly ribosomal proteins (Yanagida et al. 2001) although it seems that most of these interactions are through an RNA component.

Nucleolin is involved in many protein–protein interactions that play vital role in DNA metabolism. For instance, nucleolin interacts with the N-terminal region of topoisomerase I (Bharti et al. 1996; Edwards et al. 2000), replication protein A (RPA) (Daniely and Borowiec 2000; Wang et al. 2001), p53 (Daniely et al. 2002), YB-1 (Gaudreault et al. 2004), proliferating cell nuclear antigen (PCNA) (Yang et al. 2009), Rad51 (De et al. 2006), the human cytomegalovirus (HCMV) DNA polymerase accessory subunit UL44 (Strang et al. 2010), hepatitis C virus (HCV) NS5B protein, which is an RNA-dependent RNA polymerase important for HCV replication (Hirano et al. 2003), and with influenza A virus non-structural protein 1 (NS1), an important viral regulatory factor that controls cellular processes to facilitate viral replication (Murayama et al. 2007).

The cell-cycle-dependent interaction of nucleolin with different proteins has also been described. Nucleolin and nucleophosmin interact during interphase and cytokinesis but not in prometaphase and metaphase cells (Liu and Yung 1999). The function of this interaction is not known. A nucleolin-retinoblastoma protein (Rb) complex forms in the G1 phase through the growth inhibitory domain of Rb (Grinstein et al. 2006). Interaction of Rb with nucleolin inhibits the DNA binding activity of nucleolin, as shown on the HPV18 enhancer, and therefore mediates the repression of the HPV18 oncogenes. In addition, in epithelial cells, the intracellular distribution of nucleolin is Rb-dependent and loss of Rb results in an altered nucleolin localization in human cancerous tissue (Grinstein et al. 2006).

Recently, several reports have implicated nucleolin as a receptor for several proteins. For instance, nucleolin can act as a receptor for growth factor midkine (MK) (Hovanessian 2006) and pleiotrophin (PTN) (Said et al. 2005), which inhibit HIV infection. It was also proposed that nucleolin is a receptor for endostatin, and that nucleolin mediates the antiangiogenic and antitumor activities of endostatin (Shi et al. 2007). Nucleolin has also been reported to affect ErbB dimerization, which leads to enhanced cell growth (Farin et al. 2009). The C-terminal domain of nucleolin is sufficient for the interaction with ErbB1 and with Ras proteins (H-, N-, and K-Ras) (Farin et al. 2011). Its binding to Ca^{2+} has been proposed to affect the chromatin structure during apoptosis (Gilchrist et al. 2002).

The binding of nucleolin to other proteins can also affect their subcellular localization. For example, by binding to the zinc finger motif of the GZF1 protein, nucleolin ensures the proper subcellular distribution of GZF1 and hence plays a vital role in transcription and cell proliferation (Dambara et al. 2007). Nucleolin can also change the subcellular localization of telomerase by binding to its reverse transcriptase subunit (hTERT) (Khurts et al. 2004).

9.3 Posttranslational Modifications of Nucleolin

The multiple functions of nucleolin can be achieved not only through different protein complexes, but also probably through a complex code of posttranslational modifications (Fig. 9.1). There are no comprehensive studies that describe in detail all possible modifications of nucleolin, but several reports indeed show that nucleolin can be heavily phosphorylated (Olson et al. 1975; Rao et al. 1982), methylated (Lischwe et al. 1982), ADP-ribosylated (Leitinger and Wesierska-Gadek 1993), and glycosylated (Lapeyre et al. 1987). The consequences of these modifications on nucleolin function are still not completely known.

9.3.1 Phosphorylation of Nucleolin

The best described posttranslational modification of nucleolin is its phosphorylation. Several serine and threonine residues of nucleolin are highly phosphorylated by different kinases (Bourbon et al. 1983; Mamrack et al. 1979; Rao et al. 1982). Serine residues predominantly in two highly acidic stretches of the N-terminal domain are phosphorylated by casein kinase II (CK2) during interphase (Caizergues-Ferrer et al. 1987). Cdk1 phosphorylates nucleolin during mitosis on threonine residues within the basic TPXKK repeat (Belenguer et al. 1989; Peter et al. 1990). All potential cdk1 phosphorylation sites are not phosphorylated with the same efficiency *in vivo* as *in vitro* (Belenguer et al. 1989). Nucleolin has also been shown to be a specific substrate of protein kinase C-ζ (PKC-ζ) (Zhou et al. 1997), PI3K (Tediose et al. 2010), and a kinase whose activity depends on ROCK (Rho-associated kinase) (Garcia et al. 2011).

In plants and yeast, these N-terminal phosphorylation sites have been diversely conserved. In plants, the nucleolin-like protein alfalfa NucMsl exhibits consensus CK2 phosphorylation sites, and is highly phosphorylated by this kinase (Bogre et al. 1996). Onion nucleolin-like protein NopA64 was moderately phosphorylated *in vitro* by exogenous CK2 and cdk1, whereas NopA61 was highly phosphorylated by CK2 (de Carcer et al. 1997). In yeast, the *Schizosaccharomyces pombe* GAR2 protein contains several potential CK2 phosphorylation sites and a single one for cdk1 phosphorylation. Indeed GAR2 is phoshorylated *in vitro* by both CK2 and cdk1 and *in vivo* by cdk1 (Gulli et al. 1997).

Although it is known since its discovery that nucleolin is a highly phosphorylated protein, the regulatory roles of these phosphorylations are still not well understood.

Apart from a putative role in the regulation of nucleolin proteolysis and RNA polymerase I transcription (see Sect. 9.4.1), recent data suggest that phosphorylation may regulate the intracellular localization of nucleolin. For example, in *X. laevis*, cytoplasmic localization of nucleolin coincides with massive phosphorylation by cdk1, and nuclear translocation is accompanied by net dephosphorylation (Schwab and Dreyer 1997). Nucleolin phosphorylated by cdk1, could be specifically recognized by the monoclonal antibody TG-3. This TG-3 epitope is absent in interphase and appears abruptly during the prophase and early prometaphase where it is associated with chromosomes through metaphase and then it disappears during separation of chromosomes and exit from mitosis (Dranovsky et al. 2001). It was also found that phosphorylated nucleolin, as revealed by the TG-3 epitope, was associated with the spindle poles from prometaphase to anaphase in HeLa cells (Ma et al. 2007).

It is also likely that phosphorylation modulates the interaction of nucleolin with nucleic acids and with other proteins during cell cycle. For example, after a genotoxic stress, the nucleolin RNA-binding activity is increased by the stress-activated protein kinase p38 (Yang et al. 2002). In malignant cells, phosphorylated nucleolin (probably by PI3K and/or PKC) can also compete for binding in the promoter of genes with REST (restrictive silencer factor), leading to the transcriptional activation of these genes (Tediose et al. 2010). The interplay between REST and phosphorylated nucleolin seems to be a key mechanism for gene activation by PKC, and also for the regulation of cellular proliferation and apoptosis (Tediose et al. 2010).

In ES cells, it was also shown that phosphorylated nucleolin interacts with Tpt1 (translationally controlled tumor protein) during mitosis, and with the transcription factor Oct4 during interphase, suggesting a role for phosphorylated nucleolin in transcription (Johansson et al. 2010). As the Oct4/phosphorylated nucleolin complex formation increases during the early stages of spontaneous differentiation of human ES cell, phosphorylated nucleolin may be involved in the initial differentiation events of mammalian development.

If numerous kinases have been implicated in nucleolin phosphorylation, much less is known about phosphatases that should be also involved in the regulation of nucleolin's phosphorylation status during cell cycle. Protein phosphatase type 1δ (PP1δ) was associated with nucleolin in the nucleolus of MG63 and Saos-2 cells and it was therefore proposed that nucleolin could be a substrate for PP1δ (Morimoto et al. 2002). It has been also reported that nucleolin could interact with the protein tyrosine phosphatase of regenerating liver-3 (PRL-3) (Semba et al. 2010). The PRL-3 phosphatase activity was required for the suppression of the phosphorylated nucleolin in the cytoplasm and the accumulation of nucleolin in the nucleolus (Semba et al. 2010) suggesting that PRL-3 could dephosphorylate nucleolin in the cytoplasm, thereby regulating its nucleolar distribution.

9.3.2 Methylation of Nucleolin

Nucleolin contains relatively large amounts of N^G,N^G-dimethylarginine and traces of N^G-monomethylarginine. It was estimated that about one-third of nucleolin arginine residues are methylated, probably making nucleolin one of the most abundant methylated nuclear proteins (Lischwe et al. 1985, 1982). Dimethylarginine can modulate the interaction of nucleolin with nucleic acids (Raman et al. 2001).

The functional significance of the methylation of nucleolin could be similar to that of the methylation of heterogenous nuclear ribonucleo-protein (Chang et al. 2011; Chen et al. 2008).

The GAR domain of nucleolin is the main substrate for methylation. Asymmetrical dimethylation can be produced by type I protein arginine methyltransferase PRMT1 (Cimato et al. 2002). Teng et al (2007) showed also that nucleolin specifically interacts with PRMT5 in DU145 prostate cancer cells. The nucleolin-PRMT5 complex contains symmetrical ω-N^G,N'^G-dimethylated arginine (sDMA) (Teng et al. 2007). Posttranslational modifications of nucleolin by PRMT5 and localization of PRMT5 associated nucleolin complex can be affected by nucleolin-targeted aptamer AS1411.

PRMT5 complex contains the RioK1 protein, which directly interacts with PRMT5. RioK functions as an adapter protein by recruiting nucleolin to the PRMT5 complex and thereby facilitating its methylation (Guderian et al. 2011).

To address the effect of GAR domain methylation on nucleolar association of nucleolin, 10 arginine residues were substituted by lysine residues within the GAR domain of CHO nucleolin (Pellar and DiMario 2003). Interestingly, this lysine-substituted nucleolin, which was not modified by the yeast methyltransferase Hmt1p/Rmt1, remained in the nucleoli. Hence, methylation of the nucleolin GAR domain is not necessary for proper nucleolar localization (Pellar and DiMario 2003).

9.3.3 ADP-Ribosylation and Glycosylation of Nucleolin

ADP-ribosylation plays an important role for the maintenance of chromatin structure including its organization, DNA replication, or repair. In exponentially growing HeLa cells, it was shown that nucleolin could be modified by ADP-ribosylation (Leitinger and Wesierska-Gadek 1993). However, it is still not clear which residues are modified, and what are the functional consequences for nucleolin.

In CHO cell nucleolin, five potential N-glycosylation sites were found in the central domain, within the sequences Asn-Xaa-Ser and Asn-Xaa-Thr (Lapeyre et al. 1987). Nucleolin is also glycosylated in human U397 cells (Salazar et al. 2000). A small fraction of nucleolin found on the surface of different cell type seems to be the target of N- and O-glycosylation (Carpentier et al. 2005). Two N-glycosylation sites were identified, N317 and N492 that reside in RBD1 and 3, respectively. Inhibition of N-glycosylation by treatment of cells with tunicamycin prevents the expression of nucleolin on the cell surface (Losfeld et al. 2009), showing that this posttranslational modification is absolutely required for this cellular localization. As surface nucleolin serves as a receptor for various extracellular ligands involved in cell proliferation, differentiation, adhesion, mitogenesis, and angiogenesis, it is also therefore possible that glycosylation of nucleolin may be required for these interactions.

9.4 Nucleolar Functions of Nucleolin

Nucleolin is a multifunctional protein which plays important roles not only in ribosome biogenesis but also in RNA polymerase II transcription, DNA metabolism, and cell proliferation. Recently, different experimental systems using siRNA knockdown in different cell types, together with the conditional nucleolin knockout in chicken DT40 cells generated using the MerCreMer/LoxP recombination system, were used to explore the multifunctional properties of nucleolin. In the following chapters, we will introduce these multiple functions with emphasis on the functions related to the nucleolus.

9.4.1 *Role in Polymerase I Transcription*

Since its first description as one of the most abundant nucleolar proteins in the nucleolus, nucleolin has been shown to be associated with chromatin (Olson et al. 1975; Olson and Thompson 1983; Rickards et al. 2007). Indeed, nucleolin interacts with different DNA sequences (Olson et al. 1983; Sapp et al. 1986; Hanakahi et al. 1999), histone H1 (Erard et al. 1988), and histone H3 and H4 tails (Choi et al. 2007; Heo et al. 2007). Taken together, this suggested that nucleolin could play an important role in the regulation of chromatin structure and function, and especially for the transcription of rDNA genes by RNA polymerase I. Indeed, it has been reported that nucleolin could be involved in both the activation and the repression of Pol I transcription in different experimental systems (Derenzini et al. 1995; Egyhazi et al. 1988; Rickards et al. 2007; Roger et al. 2002; Storck et al. 2009; Ugrinova et al. 2007).

For example, in *Chironomus tentans* salivary glands, the pre-rRNA synthesis was increased 2–3.5-fold after the injection of nucleolin antibody (Egyhazi et al. 1988). In the cold-acclimatized carp (*Cyprinus carpio*), there is an up-regulated level of nucleolin accompanied by a concomitant repression of rDNA transcription (Alvarez et al. 2003) and in stage VI *Xenopus* oocytes the level of 40S pre-rRNA was significantly reduced after the injection of an excess of *Xenopus* or hamster nucleolin (Roger et al. 2002). On the contrary, in the chicken cell line DT40, depletion of nucleolin by conditional knockout silences rDNA transcription (Storck et al. 2009). In human cells, nucleolin is also required for rDNA transcription *in vivo*. Knockdown of nucleolin in HeLa and human fibroblast cells decreases polymerase I transcription (Rickards et al. 2007; Ugrinova et al. 2007) whereas an overexpression of nucleolin in HeLa cells leads to an increase of rDNA transcription (Cong and Bouvet, unpublished data).

Nucleolin phosphorylation, which could be triggered by androgens and growth factors (Bonnet et al. 1996; Bouche et al. 1987; Issinger et al. 1988; Suzuki et al. 1991; Tawfic et al. 1994), is accompanied with the active transcription of the rDNA (Issinger et al. 1988). How posttranslational modification of nucleolin affects rDNA transcription is still not completely understood. A link between nucleolin phosphorylation, its proteolysis, and rDNA transcription regulation has been proposed (Bouche et al. 1984). It was speculated that the RBDs of nucleolin associate with the nascent rRNA transcript, whereas the N-terminal domain binds the RNA polymerase I complex, potentially inhibiting rDNA transcription. Once nucleolin is phosphorylated, this could trigger the proteolysis of nucleolin, thus relieving the blockage of rDNA transcription (Bouche et al. 1984). However, recent studies have greatly challenged this model. It has been found that a specific RNA sequence transcribed by the RNA polymerase I was not required to obtained transcription regulation by nucleolin (Roger et al. 2002). Furthermore, all RBDs of nucleolin were not required for cell viability and rDNA transcription regulation, which favors a model where rDNA transcription does not require sequence-specific RNA binding (Storck et al. 2009).

The transcription of rDNA genes is influenced by chromatin accessibility and dynamics (Birch and Zomerdijk 2008; McStay and Grummt 2008). For example, the binding of the nucleolar protein TTF1 (transcription terminator factor 1) to the proximal terminator T0 of rDNA can recruit Tip5 or CSB to establish a silent or active chromatin structure (McStay and Grummt 2008). Several lines of evidences indicate that nucleolin, like TTF1, could regulate rDNA transcription through its ability to regulate rDNA chromatin dynamics. In addition to being associated with rDNA chromatin (Rickards et al. 2007), nucleolin is highly enriched in the promoter and coding regions of the rDNA repeats (Cong and Bouvet, unpublished data).

In vitro, nucleolin is able to assist the deposition of histones on DNA to assemble nucleosomes. The efficiency of nucleolin-mediated histone deposition is similar to that of the other two well-characterized histone chaperones: nucleoplasmin and nucleosome-assembly protein-1 (NAP-1) (Angelov et al. 2006). Nucleolin can also increase the activities of the two well-known chromatin remodelers, SWI/SNF and ACF complex (ATP-dependent chromatin assembly and remodeling factor). Nucleolin promotes the interaction of SWI/SNF with the nucleosome (Angelov et al. 2006) but how the sliding and remodeling activity is activated by nucleolin remains an open question.

It has been also reported that nucleolin can facilitate Pol I transcription of an *in vitro* assembled chromatin template (Rickards et al. 2007). These properties of nucleolin are reminiscent of the FACT complex (*FA*cilitates *C*hromatin *T*ranscription), which is able to facilitate chromatin transcription by RNA polymerase I (Birch et al. 2009). How nucleolin regulates precisely the rDNA transcription remains to be studied. Nevertheless, nucleolin can activate chromatin remodeling (coremodeler activity) and destabilize nucleosomal structure, thus promoting H2A-H2B dimer displacement (Angelov et al. 2006). These different activities of nucleolin might be part of the regulatory mechanism for rDNA transcription.

In addition, the nucleolar structure is dependent on active RNA polymerase I transcription (Grummt 2003). Interestingly, nucleolin depletion leads to a drastic disorganization of the nucleoli (Ma et al. 2007; Ugrinova et al. 2007) similar to that observed on treatment of cells with low amounts of actinomycin D (the polymerase I transcription inhibitor) (Puvion-Dutilleul et al. 1997; Scheer and Benavente 1990), with the injection of anti-UBF antibodies (Rubbi and Milner 2003), or with TIF-IA siRNA (Yuan et al. 2005). After nucleolin depletion, fibrillarin and Ki-67, which are the markers of the spherical structures of the nucleoli or DFCs, and B23 are highly disorganized (Ma et al. 2007). Taken together, these data suggest that the role of nucleolin in nucleolus formation might be related to rDNA transcription.

A recent report using the plant *Arabidopsis thaliana* shed new light on the role of nucleolin in controlling the expression of rRNA genes (Pontvianne et al. 2010). Disruption of AtNUC-L1, the nucleolin-like protein gene from *A. thaliana*, leads to higher transcription levels of rRNA genes. In the genome of *A. thaliana*, there are four major 3' external transcribed spacer (ETS) rRNA gene variants: VAR1, VAR2, VAR3, and VAR4. Intriguingly, VAR1, which is the most highly represented, is not expressed in WT but it is expressed in *Atnuc-L1* mutant plants. The absence of AtNUC-L1 results in an increased amount of RNA Pol I subunit associated with the promoter ETS and coding regions of rRNA genes. It is speculated that the higher level of pre-rRNA might result from the higher loading of RNA Pol I subunits. Interestingly, the levels of pre-rRNA transcripts from rDNA promoter are lower in *Atnuc-L1* than in WT plants, and the higher rRNA transcription level in *Atnuc-L1* mutant plants is probably due to higher Pol I transcription from the IGS. Moreover, the disruption of *AtNUC-L1* gene results in loss of DNA methylation, while the histone modification marks of rRNA genes are not affected in the *Atnuc-L1* plants. This suggests that DNA methylation is required for AtNUC-L1 to regulate Pol I transcription in *A. thaliana*. Collectively, these data provide a novel mechanism by which nucleolin modulates rRNA gene transcription in *Arabidopsis*. However, it seems that nucleolin in plants functions in rRNA gene transcription differently from that in the other experimental systems, but how nucleolin regulates Pol I transcription in mammalian cells remains an open question.

9.4.2 Role in rRNA Maturation and PreRibosome Assembly

Within the nucleolus of eukaryotic cells, pre-rRNA is specifically cleaved to generate the mature rRNA species that will be assembled with imported ribosomal proteins to form the ribosomes that will be then exported in the cytoplasm (Hernandez-Verdun 1991; Reeder 1990; Warner 1989). Nucleolin seems also to be a key player in this pre-rRNA processing and preribosome assembly (Ginisty et al. 1998; Roger et al. 2003).

In yeast, the nucleolin-like proteins NSR1 and GAR2 are required for pre-RNA processing (Gulli et al. 1995a; Kondo and Inouye 1992; Lee et al. 1992). In *Saccharomyces cerevisiae* cells lacking the NSR1 gene, 35S pre-rRNA is blocked and 20S pre-rRNA nearly disappeared (Lee et al. 1992). Cold shock of the *nsr 1* strain leads to drastic impairment of the pre-rRNA processing (Kondo and Inouye 1992). In *S. pombe*, disruption of the GAR2 gene leads to an increase of 35S pre-rRNA and a decrease of mature 18S rRNA (Gulli et al. 1995b). All these data suggest a role of these nucleolin-like proteins in pre-rRNA processing.

Nucleolin plays a role in the first step of mouse ribosomal RNA processing (Ginisty et al. 1998). The interaction of nucleolin with the pre-rRNA substrate is required for the processing of rRNA *in vitro*. It has been found that nucleolin interacts with the UCGA motif present in the ECM, which is located five nucleotides downstream of the first processing site in the 5′ ETS of preribosomal RNA (Ginisty et al. 2000). Moreover, interaction of nucleolin with U3 small nucleolar ribonucleoprotein (snoRNP), which is necessary for the first cleavage in pre-rRNA processing (Kass et al. 1990), is required for pre-rRNA processing. As nucleolin promotes nucleic acid annealing (Hanakahi et al. 2000; Sipos and Olson 1991), it is possible that nucleolin–U3 snoRNP interaction is the basis for the initial recruitment of U3 to pre-rRNA. Then, the U3-rRNA complexes could provide several base pairings important for promoting the pre-rRNA cleavage (Borovjagin and Gerbi 2004).

Apart from its role in pre-rRNA processing, nucleolin may also have a role in preribosome assembly. Nucleolin transiently associates with nascent pre-rRNA and preribosomal particles (Bourbon et al. 1983) and has the ability to promote the secondary structure of ribosomal RNA *in vitro* (Sipos and Olson 1991). The transient interaction of nucleolin with the NRE stem-loop RNA motifs present all along the pre-rRNA (Allain et al. 2000a; Serin et al. 1996) has led to the proposal that nucleolin acts as a RNA chaperone for the correct folding of pre-rRNA during transcription. This proper cotranscriptional folding is required for the orderly interaction with the ribosomal proteins and the formation of the correct folding of preribosomes. This mechanism is supported by experiments in *X. laevis*. Injection of nucleolin in stage VI oocytes leads to incorrect packaging of 40S pre-rRNA (Roger et al. 2003). Interestingly, when rDNA transcription is inhibited by actinomycin D, increasing the amount of nucleolin in the oocyte cannot affect the maturation of the pre-existing 40S particle. These studies indicate that nucleolin is involved in the cotranscriptional packaging of prerRNA, thereby providing a link between RNA polymerase I transcription and preribosome assembly.

Nucleolin is also a shuttling protein that migrates between nucleus and cytoplasm (Borer et al. 1989; Schmidt-Zachmann et al. 1993) suggesting that nucleolin might be involved in the import of cytoplasmic factors required for ribosome assembly in the nucleus, like the ribosomal proteins. Indeed, nucleolin also interacts with some ribosomal proteins through its RGG domain (Bouvet et al. 1998). These nucleolin-associated ribosomal proteins are tightly associated with rRNA and are among the first proteins assembled within the preribosomal particles during transcription (Reboud et al. 1974; Welfle et al. 1976), further supporting the idea that nucleolin functions at an early step of ribosome assembly. However, as nucleolin is not found in mature ribosomes in cytoplasm, this suggests that it is released from the ribosomal complex during the assembly and/or the processing step of preribosomal particles. Taken together, these studies favor a model where during ribosome biogenesis, nucleolin regulates rDNA transcription and could act as an adaptor for the association of different trans-acting factors (ribosomal proteins, snoRNPs, etc.) required for pre-rRNA maturation and preribosome assembly.

9.5 Functions of Nucleolin in Pol II Transcription

In addition to its role in RNA polymerase I transcription regulation, several reports have also implicated nucleolin in RNA polymerase II transcription. Nucleolin is one components of the B cell-specific transcription factor, LR1 (Hanakahi et al. 1997). It was also found that nucleolin interacts with E47, one member of the basic-helix-loop-helix (bHLH) family, which is found in the transcription factors with the potential to form homo- and hetero-dimers mediated by the HLH domain (Dear et al. 1997). Thus, nucleolin might be involved in the transcriptional regulation by LR1 or E47.

Nucleolin is required for the transcription regulation of the Krüppel-like factor 2 (KLF2) (Huddleson et al. 2006). The complex regulation of the *KLF2* gene involves several transacting factors, chromatin modifications, and at least three signaling pathways (Schrick et al. 1999). Nucleolin is bound to the −138/−111 region of the *KLF2* promoter, which is conserved between mouse and human promoters and is critical for *KLF2* expression. Knockdown of nucleolin by siRNAs inhibited the induction of *KLF2* by shear stress (Huddleson et al. 2006).

HPV18-induced cervical carcinogenesis seems to require the expression of nucleolin (Grinstein et al. 2002). Inactivation of nucleolin inhibits E6 and E7 oncogene transcription and selectively decreases cervical cancer cell growth. Nucleolin could regulate HPV18 oncogene transcription through controlling the chromatin structure of the HPV18 enhancer and thus a direct link is provided between nucleolin and HPV18-induced cervical carcinogenesis (Grinstein et al. 2002). Nucleolin can also activate endogenous *CD34* and *Bcl2* gene expression in human CD34-positive hematopoietic cells, thereby enhancing the cell surface CD34 protein expression, which provides insights into processes by which human CD34-positive hematopoietic stem/progenitor cells are maintained (Grinstein et al. 2007).

Nucleolin is involved in the activation of RNA polymerase II gene expression, and in some instances it also seems to be involved in transcription repression. Biochemical and functional studies identified that nucleolin is a transcription repressor for regulation of an acute-phase response gene α-1 acid glycoprotein (AGP) (Yang et al. 1994).

As nucleolin can play a role in both the activation and the repression of transcription, the precise role of nucleolin in the regulation of transcription remains to be elucidated. The action of nucleolin on chromatin could be a key step to explain these observations.

The N-terminal domain of nucleolin possesses acidic stretches which could bind basic proteins like histones; these are characteristic of many proteins with histone chaperone activity (Ginisty et al. 1999; Loyola and Almouzni 2004). As discussed previously for the regulation of polymerase I transcription, this chaperone activity of nucleolin and its ability to modulate the activity of chromatin remodeling complexes could be important also for the regulation of RNA polymerase II transcription (Angelov et al. 2006; Mongelard and Bouvet 2007; Rickards et al. 2007). *In vivo*, FRAP experiments on eGFP-tagged histones (H2B, H4, and macroH2A) in cells depleted of nucleolin by siRNA showed a different behavior of these histones (Gaume et al. 2011). Nuclear histone dynamics was impacted in nucleolin-silenced cells; in particular,

9 The Multiple Properties and Functions of Nucleolin

higher fluorescence recovery kinetics were measured for macroH2A and H2B but not for H4. Interestingly, nucleolin depletion also impacted the dissociation rate constant of H2B and H4. Thus, in live cells, nucleolin also plays a role in chromatin accessibility, probably through its histone chaperone and coremodeling activities.

Therefore, the seemingly contradictory roles of nucleolin in the transcription regulation could be explained if one takes into account that nucleolin can increase the remodeling activity of chromatin remodelers, and these remodelers are involved in either activation or repression of transcription (Tang et al. 2010).

9.6 Role of Nucleolin in Posttranscriptional Regulation

The multiple RNA-binding properties of nucleolin (see above) also make nucleolin a good candidate for being involved in posttranscriptional interactions through direct interaction with RNA. Indeed, nucleolin has been implicated in mRNA stabilization. The JNK MAPK cascade plays an important role in interleukin-2 (IL-2) mRNA stabilization induced by the activation of T cells (Chen et al. 1998). Nucleolin can bind the JNK response element (JRE) and is required for IL-2 mRNA stabilization (Chen et al. 2000). As described above (Sect. 9.2.1), the specific interaction of nucleolin with the ARE in the 3′-UTR of the $Bcl-X_L$ and bcl-2 mRNAs is important to determine the half-life of these mRNAs (Zhang et al. 2008). In addition, nucleolin is one component of the complex that regulates CD154 mRNA turnover (Singh et al. 2004) and is also required for the arsenic-induced stabilization of GADD45α mRNA (Zhang et al. 2006).

Nucleolin could also regulate mRNA translation. For example, nucleolin is able to regulate p53 levels *in vivo*. One group has found that overexpression of nucleolin suppresses p53 translation, whereas nucleolin down-regulation promotes p53 expression (Takagi et al. 2005). The increase of p53 level in nucleolin knockdown cells might be due to the nucleolar stress caused by nucleolin depletion.

Selenoproteins are a small subclass of proteins that contain the essential trace element selenium which plays a variety of important roles in anti-oxidant defense, thyroid hormone metabolism, male reproduction, and development (Hatfield et al. 2006). Using UV crosslinking, nucleolin was identified to bind to a subset of selenoprotein mRNAs with high affinity. Nucleolin depletion did not change the selenoprotein transcript levels, but led to a decrease of the expression of certain selenoproteins, which suggests that nucleolin selectively regulates the expression of some selenoproteins at the translational level (Miniard et al. 2010).

9.7 Functions of Nucleolin in DNA Metabolism

DNA replication, repair, and recombination occur in the nucleus, and they are driven by several regulatory machineries (Boisvert et al. 2007; Gottlieb and Esposito 1989; Hannan et al. 1999). The ability of nucleolin to bind both DNA and proteins involved in DNA metabolism (see Sects. 9.2.1 and 9.2.2) suggests that nucleolin plays an important role in DNA replication, repair, and recombination.

What could be the role of nucleolin in DNA replication? As nucleolin is associated with a DNA synthesome (Applegren et al. 1995), it is suggested that nucleolin might be a member of the DNA replication machinery. It has been proposed that human and pea nucleolin possess DNA helicase activity (Nasirudin et al. 2005; Tuteja et al. 1995), which was attributed to the RGG domain of nucleolin (Nasirudin et al. 2005; Tuteja et al. 1995). Nevertheless, other groups either failed to reproduce these experiments (Ginsty et al. 1999) or found that nucleolin stimulates nucleic acid annealing (Hanakahi et al. 2000; Sapp et al. 1986), which is an antagonistic activity to that of DNA helicase. The reason for these seemingly contradictory data is not clear yet; one possibility might be that according to the purification scheme of nucleolin, it might be copurified with additional factors required for either stimulating nucleolin helicase or annealing activities (Ginsty et al. 1999).

Nucleolin has been shown to redistribute from the nucleolus to the nucleoplasm under some stress conditions such as heat shock (Daniely and Borowiec 2000) or ionizing radiation (IR), or on treatment with the radiomimetic agent camptothecin (CPT) (Daniely et al. 2002). However, the redistribution of nucleolin is stress-selective, as treatment of cells with hydroyurea or UV irradiation does not significantly relocate nucleolin (Daniely et al. 2002). Interestingly, the relocalization of nucleolin is p53-dependent as the mobilization does not occur in p53-null cells, and it was found that the relocalization was stimulated by the physical interaction of nucleolin with p53, but independent of the ability of p53 to activate transcription (Daniely et al. 2002). The mechanism for nucleolin redistribution mediated by specific stresses or p53 needs to be further studied.

Does this stress-dependent relocalization of nucleolin from nucleolus to nucleoplasm have some biological meaning? Genotoxic stress such as heat shock can inhibit DNA replication (Wang et al. 2001). While nucleolin mobilization occurs following some selective stress, it has been proposed that nucleolin could function as a signaling molecule to initiate arrest of DNA replication. Nucleolin forms a complex with RPA to inhibit DNA replication after cell stress (Daniely and Borowiec 2000). RPA can bind the ssDNA, and it plays an important role in DNA metabolism, including DNA replication, nucleotide excision repair, and homologous DNA recombination (Iftode et al. 1999). Intriguingly, heat shock could induce p53-dependent redistribution of nucleolin from the nucleolus to the nucleoplasm, and this relocalization was concomitant with the increase of nucleolin-RPA complex formation. Binding of nucleolin with RPA could prevent the ability of RPA to support DNA replication (Daniely et al. 2002). As nucleolin does not affect the ssDNA binding activity of RPA (Daniely and Borowiec 2000), it is suggested that nucleolin inhibits RPA function by preventing its interaction with other factors.

Previous reports indicate that both the nucleolus and telomeres are related to aging (Johnson et al. 1998). Nucleolin interacts with the telomeric repeat (TTAGGG)n *in vitro* (Ishikawa et al. 1993; Pollice et al. 2000) as well as with telomerase *in vitro* and *in vivo* (Khurts et al. 2004), which suggests that nucleolin might play a specific role in telomere replication and maintenance thus providing a link between telomeres and the nucleolus for understanding aging-related mechanisms. But how nucleolin affects telomere replication remains an open question.

The direct function of nucleolin in DNA replication has been deciphered by *in vitro* biochemical evidence (Seinsoth et al. 2003). Topoisomerase I could activate DNA replication mediated by the simian virus (SV40) large tumor-antigen (T-antigen) hexamer. It has been shown that nucleolin binds the T-antigen hexamer and topoisomerase (Gai et al. 2000) to form a ternary complex (Bullock 1997). Nucleolin could function as a clamp to mediate the cohesion of T-antigen hexamer and topoisomerase I, thus enhancing the bidirectional DNA unwinding.

Recent papers demonstrate the role of nucleolin in viral DNA replication *in vivo* using nucleolin knockdown studies (Calle et al. 2008; Strang et al. 2010). Using siRNA technology, it has been discovered that the efficient replication of herpes simplex virus type 1 (HSV-1) requires nucleolin expression (Calle et al. 2008). Moreover, nucleolin is also necessary for the DNA replication of HCMV (Strang et al. 2010). Nevertheless, that nucleolin can play an opposite role in replication of other viruses has been shown for the tombusvirus, which is inhibited by nucleolin (Jiang et al. 2010b). The molecular details of how nucleolin regulates virus replication needs to be characterized.

Apart from a role in DNA replication, nucleolin has also been implicated in DNA repair and recombination. Nucleolin interacts with some proteins involved in these processes such as with p53 (Daniely et al. 2002), YB-1 (Gaudreault et al. 2004), RPA (Daniely and Borowiec 2000), PCNA (Yang et al. 2009), Rad51 (De et al. 2006), and topoisomerase I (Bharti et al. 1996; Edwards et al. 2000). The *S. cerevisiae* nucleolin NSR1 does not affect the enzymatic activity of topoisomerase I, but it could act in DNA recombination by modulating the cellular localization of yTop1p, the yeast topoisomerase I (Edwards et al. 2000). The LR1 complex can also function in class switch recombination by binding the Ig heavy chain switch region. The presence of nucleolin in this complex suggests that it may take part in this recombination process (Hanakahi et al. 1997).

Altogether, these results show that nucleolin could regulate genome stability either by interacting directly with DNA or through physical interaction with proteins involved in DNA metabolism, thereby modulating the activity of these proteins (Storck et al. 2007; Yang et al. 2002).

9.8 Functions of Nucleolin in Cell Cycle Regulation, Cell Division, and Proliferation

The expression of nucleolin has been correlated with the rate of cell proliferation (Derenzini 2000). In tumors or other actively dividing cells, nucleolin is highly synthesized (Derenzini et al. 1995; Roussel and Hernandez-Verdun 1994), while in nondividing cells the level of nucleolin is very low (Sirri et al. 1995). While the level of nucleolin is low in cells cultured in serum-free medium, cell division induction by pp60v-src induces the expression of nucleolin in mid and late G1, indicating that nucleolin might be required for cell cycle progression into G1 phase (Gillet et al. 1993). Therefore, nucleolin is often used as a marker for cell proliferation (Bates et al. 2009; de Verdugo et al. 1995; Mongelard and Bouvet 2010; Roussel and Hernandez-Verdun 1994).

Controlled proteolysis and posttranslational modifications of nucleolin are also correlated with the regulation of cell proliferation. Nucleolin proteolysis causes T lymphocyte-mediated apoptotic cell death. Granzyme A can bind and cleave nucleolin (Chen et al. 1991; Fang and Yeh 1993; Pasternack et al. 1991). The nucleolin cleavage products could then stimulate autolytic endonucleases, which fragment DNA to induce apoptosis (Arends et al. 1990; Smyth et al. 1994). In nondividing cells, nucleolin could catalyze its own degradation (Chen et al. 1991). *In vitro*, this degradation could be blocked by the addition of nuclear extracts from proliferative cells, suggesting the presence of an inhibitor in rapidly dividing cells that prevents nucleolin degradation (Chen et al. 1991).

Nucleolin phosphorylation is correlated with increased cell proliferation (Geahlen and Harrison 1984; Miranda et al. 1995). In interphase, nucleolin is phosphorylated by CK2 on serine residues, while in mitosis, Cdk1 phosphorylates nucleolin threonine residues (Belenguer et al. 1990; Caizergues-Ferrer et al. 1987). It is speculated that successive phosphorylation of nucleolin by cdk1 and CK2 could be a mechanism for nucleolin to regulate the cell cycle and cell division. In addition, nucleolin phosphorylated by cdk1 was associated with the spindle poles from prometaphase to anaphase (Ma et al. 2007), indicating that nucleolin phosphorylation may be involved in the cell cycle regulation.

However, as ribosome biogenesis is required for cell proliferation, and nucleolin seems indispensible for efficient ribosomal RNA transcription and preribosome assembly; until very recently, it was not really clear if nucleolin had a direct role in cell division and proliferation, or if it was only through the control of ribosome biogenesis. Recent work showing that nucleolin is able to repress p53 expression (Takagi et al. 2005) to mediate the anti-apoptotic effect of heat-shock protein 70(Hsp70) during oxidative stress (Jiang et al. 2010a) and to work as a target of the anti-proliferative G-rich oligonucleotides (Bates et al. 1999; Xu et al. 2001) and that it is required for proper chromosome segregation (Ma et al. 2007) suggested that nucleolin could indeed affect cell division, proliferation, and growth independently of its role on ribosome biogenesis. Interestingly, it has been recently shown that nucleolin can act synergistically with Ras and ErbB1 to facilitate cell growth in soft agar and tumor growth in nude mice (Farin et al. 2011).

In *Arabidopsis*, nucleolin is required for plant development (Kojima et al. 2007; Petricka and Nelson 2007; Pontvianne et al. 2007). Absence of AtNUC-L1 results in severe plant growth and development defects. For example, *Atnuc-L1* plants grew slower than WT plants, with defective vascular patterns and pod development (Kojima et al. 2007; Petricka and Nelson 2007; Pontvianne et al. 2007).

To further explore the role of nucleolin in cell proliferation, siRNA was used to knock down nucleolin in HeLa cells and in human primary fibroblast cells. Nucleolin depletion results in a decrease in cell growth, an increase in apoptosis, and an arrest in G2 phase. Increased multinuclear cells and cells with micronuclei are also observed after nucleolin depletion (Ugrinova et al. 2007). Moreover, inactivation of nucleolin leads to an increased number of centrosomes with a multipolar spindle structure (Ugrinova et al. 2007). Furthermore, nucleolin knockdown also resulted in a prolonged cell cycle with defects in chromosome congression and spindle organization (Ma et al. 2007). In conditional knockout DT40 cells, the expression of different mutants of nucleolin, which lack two or three RBDs, showed that these domains harbor redundant functions and that nucleolin's roles in transcription, rRNA maturation, and nucleolar shape can be partially uncoupled (Storck et al. 2009). This indeed suggests that the different domains of nucleolin probably participate in the formation of multiple protein complexes that participate in many different cellular functions.

9.9 Conclusions

Despite these numerous studies on nucleolin during these past 40 years, there is still a long way to go to fully understand the role of this protein. Although there are no doubts now that nucleolin is required for the organization and function of the nucleolus, and plays an essential role in the regulation of cell cycle and cell proliferation, the detailed mechanisms involved are still missing. The increasing numbers of reports on the different functions of nucleolin outside the nucleolus and even on the cell surface are sometimes very puzzling. What is really needed now are new approaches to have an integrated view of nucleolin *in vivo*. The development of animal models to study the function of nucleolin *in vivo* during development and in different tissues together with proteomic, transcriptomic, and genomic approaches should bring in the future interesting observations to help to understand the multiple functions of nucleolin.

Acknowledgments The author's work is supported by grants from Agence Nationale de la Recherche (ANR-07-BLAN-0062-01), Région Rhône-Alpes MIRA 2007 and 2010, Association pour la Recherche sur le Cancer n° ECL2010R01122, CEFIPRA n° 3803-1, and CNRS.

References

Allain FH, Bouvet P, Dieckmann T, Feigon J (2000a) Molecular basis of sequence-specific recognition of pre-ribosomal RNA by nucleolin. EMBO J 19:6870–6881

Allain FH, Gilbert DE, Bouvet P, Feigon J (2000b) Solution structure of the two N-terminal RNA-binding domains of nucleolin and NMR study of the interaction with its RNA target. J Mol Biol 303:227–241

Alvarez M, Quezada C, Navarro C, Molina A, Bouvet P, Krauskopf M, Vera MI (2003) An increased expression of nucleolin is associated with a physiological nucleolar segregation. Biochem Biophys Res Commun 301:152–158

Angelov D, Bondarenko VA, Almagro S, Menoni H, Mongelard F, Hans F, Mietton F, Studitsky VM, Hamiche A, Dimitrov S, Bouvet P (2006) Nucleolin is a histone chaperone with FACT-like activity and assists remodeling of nucleosomes. EMBO J 25:1669–1679

Applegren N, Hickey RJ, Kleinschmidt AM, Zhou Q, Coll J, Wills P, Swaby R, Wei Y, Quan JY, Lee MY et al (1995) Further characterization of the human cell multiprotein DNA replication complex. J Cell Biochem 59:91–107

Arends MJ, Morris RG, Wyllie AH (1990) Apoptosis. The role of the endonuclease. Am J Pathol 136:593–608

Arumugam S, Miller MC, Maliekal J, Bates PJ, Trent JO, Lane AN (2010) Solution structure of the RBD1,2 domains from human nucleolin. J Biomol NMR 47:79–83

Barrijal S, Perros M, Gu Z, Avalosse BL, Belenguer P, Amalric F, Rommelaere J (1992) Nucleolin forms a specific complex with a fragment of the viral (minus) strand of minute virus of mice DNA. Nucleic Acids Res 20:5053–5060

Bates PJ, Kahlon JB, Thomas SD, Trent JO, Miller DM (1999) Antiproliferative activity of G-rich oligonucleotides correlates with protein binding. J Biol Chem 274:26369–26377

Bates PJ, Laber DA, Miller DM, Thomas SD, Trent JO (2009) Discovery and development of the G-rich oligonucleotide AS1411 as a novel treatment for cancer. Exp Mol Pathol 86:151–164

Belenguer P, Baldin V, Mathieu C, Prats H, Bensaid M, Bouche G, Amalric F (1989) Protein kinase NII and the regulation of rDNA transcription in mammalian cells. Nucleic Acids Res 17:6625–6636

Belenguer P, Caizergues-Ferrer M, Labbe JC, Doree M, Amalric F (1990) Mitosis-specific phosphorylation of nucleolin by p34cdc2 protein kinase. Mol Cell Biol 10:3607–3618

Bharti AK, Olson MO, Kufe DW, Rubin EH (1996) Identification of a nucleolin binding site in human topoisomerase I. J Biol Chem 271:1993–1997

Birch JL, Zomerdijk JC (2008) Structure and function of ribosomal RNA gene chromatin. Biochem Soc Trans 36:619–624

Birch JL, Tan BC, Panov KI, Panova TB, Andersen JS, Owen-Hughes TA, Russell J, Lee SC, Zomerdijk JC (2009) FACT facilitates chromatin transcription by RNA polymerases I and III. EMBO J 28:854–865

Bogre L, Jonak C, Mink M, Meskiene I, Traas J, Ha DT, Swoboda I, Plank C, Wagner E, Heberle-Bors E, Hirt H (1996) Developmental and cell cycle regulation of alfalfa nucMs1, a plant homolog of the yeast Nsr1 and mammalian nucleolin. Plant Cell 8:417–428

Boisvert FM, van Koningsbruggen S, Navascues J, Lamond AI (2007) The multifunctional nucleolus. Nat Rev Mol Cell Biol 8:574–585

Bonnet H, Filhol O, Truchet I, Brethenou P, Cochet C, Amalric F, Bouche G (1996) Fibroblast growth factor-2 binds to the regulatory beta subunit of CK2 and directly stimulates CK2 activity toward nucleolin. J Biol Chem 271:24781–24787

Borer RA, Lehner CF, Eppenberger HM, Nigg EA (1989) Major nucleolar proteins shuttle between nucleus and cytoplasm. Cell 56:379–390

Borovjagin AV, Gerbi SA (2004) Xenopus U3 snoRNA docks on pre-rRNA through a novel base-pairing interaction. RNA 10:942–953

Bouche G, Caizergues-Ferrer M, Bugler B, Amalric F (1984) Interrelations between the maturation of a 100 kDa nucleolar protein and pre rRNA synthesis in CHO cells. Nucleic Acids Res 12:3025–3035

Bouche G, Gas N, Prats H, Baldin V, Tauber JP, Teissie J, Amalric F (1987) Basic fibroblast growth factor enters the nucleolus and stimulates the transcription of ribosomal genes in ABAE cells undergoing G0–G1 transition. Proc Natl Acad Sci U S A 84:6770–6774

Bourbon H, Bugler B, Caizergues-Ferrer M, Amalric F (1983) Role of phosphorylation on the maturation pathways of a 100 kDa nucleolar protein. FEBS Lett 155:218–222

Bourbon HM, Prudhomme M, Amalric F (1988) Sequence and structure of the nucleolin promoter in rodents: characterization of a strikingly conserved CpG island. Gene 68:73–84

Bouvet P, Diaz JJ, Kindbeiter K, Madjar JJ, Amalric F (1998) Nucleolin interacts with several ribosomal proteins through its RGG domain. J Biol Chem 273:19025–19029

Bouvet P, Allain FH, Finger LD, Dieckmann T, Feigon J (2001) Recognition of pre-formed and flexible elements of an RNA stem-loop by nucleolin. J Mol Biol 309:763–775

Brooks TA, Hurley LH (2010) Targeting MYC expression through G-quadruplexes. Genes Cancer 1:641–649

Bugler B, Bourbon H, Lapeyre B, Wallace MO, Chang JH, Amalric F, Olson MO (1987) RNA binding fragments from nucleolin contain the ribonucleoprotein consensus sequence. J Biol Chem 262:10922–10925

Bullock PA (1997) The initiation of simian virus 40 DNA replication in vitro. Crit Rev Biochem Mol Biol 32:503–568

Caizergues-Ferrer M, Belenguer P, Lapeyre B, Amalric F, Wallace MO, Olson MO (1987) Phosphorylation of nucleolin by a nucleolar type NII protein kinase. Biochemistry 26:7876–7883

Caizergues-Ferrer M, Mariottini P, Curie C, Lapeyre B, Gas N, Amalric F, Amaldi F (1989) Nucleolin from *Xenopus laevis*: cDNA cloning and expression during development. Genes Dev 3:324–333

Calle A, Ugrinova I, Epstein AL, Bouvet P, Diaz JJ, Greco A (2008) Nucleolin is required for an efficient herpes simplex virus type 1 infection. J Virol 82:4762–4773

Carpentier M, Morelle W, Coddeville B, Pons A, Masson M, Mazurier J, Legrand D (2005) Nucleolin undergoes partial N- and O-glycosylations in the extranuclear cell compartment. Biochemistry 44:5804–5815

Chang YI, Hsu SC, Chau GY, Huang CY, Sung JS, Hua WK, Lin WJ (2011) Identification of the methylation preference region in heterogeneous nuclear ribonucleoprotein K by protein arginine methyltransferase 1 and its implication in regulating nuclear/cytoplasmic distribution. Biochem Biophys Res Commun 404:865–869

Chathoth KT, Ganesan G, Rao MR (2009) Identification of a novel nucleolin related protein (NRP) gene expressed during rat spermatogenesis. BMC Mol Biol 10:64

Chen CM, Chiang SY, Yeh NH (1991) Increased stability of nucleolin in proliferating cells by inhibition of its self-cleaving activity. J Biol Chem 266:7754–7758

Chen CY, Del Gatto-Konczak F, Wu Z, Karin M (1998) Stabilization of interleukin-2 mRNA by the c-Jun NH2-terminal kinase pathway. Science 280:1945–1949

Chen CY, Gherzi R, Andersen JS, Gaietta G, Jurchott K, Royer HD, Mann M, Karin M (2000) Nucleolin and YB-1 are required for JNK-mediated interleukin-2 mRNA stabilization during T-cell activation. Genes Dev 14:1236–1248

Chen Y, Zhou X, Liu N, Wang C, Zhang L, Mo W, Hu G (2008) Arginine methylation of hnRNP K enhances p53 transcriptional activity. FEBS Lett 582:1761–1765

Choi J, Kim B, Heo K, Kim K, Kim H, Zhan Y, Ranish JA, An W (2007) Purification and characterization of cellular proteins associated with histone H4 tails. J Biol Chem 282:21024–21031

Cimato TR, Tang J, Xu Y, Guarnaccia C, Herschman HR, Pongor S, Aletta JM (2002) Nerve growth factor-mediated increases in protein methylation occur predominantly at type I arginine methylation sites and involve protein arginine methyltransferase 1. J Neurosci Res 67:435–442

Craig N, Kass S, Sollnerwebb B (1991) Sequence organization and RNA structural motifs directing the mouse primary ribosomal-RNA-processing event. Mol Cell Biol 11:458–467

Dambara A, Morinaga T, Fukuda N, Yamakawa Y, Kato T, Enomoto A, Asai N, Murakumo Y, Matsuo S, Takahashi M (2007) Nucleolin modulates the subcellular localization of GDNF-inducible zinc finger protein 1 and its roles in transcription and cell proliferation. Exp Cell Res 313:3755–3766

Daniely Y, Borowiec JA (2000) Formation of a complex between nucleolin and replication protein A after cell stress prevents initiation of DNA replication. J Cell Biol 149:799–810

Daniely Y, Dimitrova DD, Borowiec JA (2002) Stress-dependent nucleolin mobilization mediated by p53-nucleolin complex formation. Mol Cell Biol 22:6014–6022

de Carcer G, Cerdido A, Medina FJ (1997) NopA64, a novel nucleolar phosphoprotein from proliferating onion cells, sharing immunological determinants with mammalian nucleolin. Planta 201:487–495

de Verdugo UR, Selinka HC, Huber M, Kramer B, Kellermann J, Hofschneider PH, Kandolf R (1995) Characterization of a 100-kilodalton binding protein for the six serotypes of coxsackie B viruses. J Virol 69:6751–6757

De A, Donahue SL, Tabah A, Castro NE, Mraz N, Cruise JL, Campbell C (2006) A novel interaction [corrected] of nucleolin with Rad51. Biochem Biophys Res Commun 344:206–213

Dear TN, Hainzl T, Follo M, Nehls M, Wilmore H, Matena K, Boehm T (1997) Identification of interaction partners for the basic-helix-loop-helix protein E47. Oncogene 14:891–898

Derenzini M (2000) The AgNORs. Micron 31:117–120

Derenzini M, Sirri V, Pession A, Trere D, Roussel P, Ochs RL, Hernandez-Verdun D (1995) Quantitative changes of the two major AgNOR proteins, nucleolin and protein B23, related to stimulation of rDNA transcription. Exp Cell Res 219:276–282

Dickinson LA, Kohwi-Shigematsu T (1995) Nucleolin is a matrix attachment region DNA-binding protein that specifically recognizes a region with high base-unpairing potential. Mol Cell Biol 15:456–465

Didier DK, Klee HJ (1992) Identification of an *Arabidopsis* DNA-binding protein with homology to nucleolin. Plant Mol Biol 18:977–979

Dranovsky A, Vincent I, Gregori L, Schwarzman A, Colflesh D, Enghild J, Strittmatter W, Davies P, Goldgaber D (2001) Cdc2 phosphorylation of nucleolin demarcates mitotic stages and Alzheimer's disease pathology. Neurobiol Aging 22:517–528

Edwards TK, Saleem A, Shaman JA, Dennis T, Gerigk C, Oliveros E, Gartenberg MR, Rubin EH (2000) Role for nucleolin/Nsr1 in the cellular localization of topoisomerase I. J Biol Chem 275:36181–36188

Egyhazi E, Pigon A, Chang JH, Ghaffari SH, Dreesen TD, Wellman SE, Case ST, Olson MO (1988) Effects of anti-C23 (nucleolin) antibody on transcription of ribosomal DNA in *Chironomus* salivary gland cells. Exp Cell Res 178:264–272

Erard MS, Belenguer P, Caizergues-Ferrer M, Pantaloni A, Amalric F (1988) A major nucleolar protein, nucleolin, induces chromatin decondensation by binding to histone H1. Eur J Biochem 175:525–530

Fang SH, Yeh NH (1993) The self-cleaving activity of nucleolin determines its molecular dynamics in relation to cell proliferation. Exp Cell Res 208:48–53

Farin K, Di Segni A, Mor A, Pinkas-Kramarski R (2009) Structure-function analysis of nucleolin and ErbB receptors interactions. PLoS One 4:e6128

Farin K, Schokoroy S, Haklai R, Cohen-Or I, Elad-Sfadia G, Reyes-Reyes ME, Bates PJ, Cox AD, Kloog Y, Pinkas-Kramarski R (2011) Oncogenic synergism between ErbB1, nucleolin and mutant Ras. Cancer Res 71(6):2140–2151

Finger LD, Johansson C, Rinaldi B, Bouvet P, Feigon J (2004) Contributions of the RNA-binding and linker domains and RNA structure to the specificity and affinity of the nucleolin RBD12/NRE interaction. Biochemistry 43:6937–6947

Gai D, Roy R, Wu C, Simmons DT (2000) Topoisomerase I associates specifically with simian virus 40 large-T-antigen double hexamer-origin complexes. J Virol 74:5224–5232

Garcia MC, Williams J, Johnson K, Olden K, Roberts JD (2011) Arachidonic acid stimulates formation of a novel complex containing nucleolin and RhoA. FEBS Lett 585:618–622

Gaudreault I, Guay D, Lebel M (2004) YB-1 promotes strand separation in vitro of duplex DNA containing either mispaired bases or cisplatin modifications, exhibits endonucleolytic activities and binds several DNA repair proteins. Nucleic Acids Res 32:316–327

Gaume X, Monier K, Argoul F, Mongelard F, Bouvet P (2011) Biochem Res Int. 2011:187624

Geahlen RL, Harrison ML (1984) Induction of a substrate for casein kinase II during lymphocyte mitogenesis. Biochim Biophys Acta 804:169–175

Ghisolfi L, Kharrat A, Joseph G, Amalric F, Erard M (1992) Concerted activities of the RNA recognition and the glycine-rich C-terminal domains of nucleolin are required for efficient complex formation with pre-ribosomal RNA. Eur J Biochem 209:541–548

Ghisolfi-Nieto L, Joseph G, Puvion-Dutilleul F, Amalric F, Bouvet P (1996) Nucleolin is a sequence-specific RNA-binding protein: characterization of targets on pre-ribosomal RNA. J Mol Biol 260:34–53

Gilchrist JS, Abrenica B, DiMario PJ, Czubryt MP, Pierce GN (2002) Nucleolin is a calcium-binding protein. J Cell Biochem 85:268–278

9 The Multiple Properties and Functions of Nucleolin

Gillet G, Michel D, Crisanti P, Guerin M, Herault Y, Pessac B, Calothy G, Brun G, Volovitch M (1993) Serum factors and v-src control two complementary mitogenic pathways in quail neuroretinal cells in culture. Oncogene 8:565–574

Ginisty H, Amalric F, Bouvet P (1998) Nucleolin functions in the first step of ribosomal RNA processing. EMBO J 17:1476–1486

Ginisty H, Sicard H, Roger B, Bouvet P (1999) Structure and functions of nucleolin. J Cell Sci 112(pt 6):761–772

Ginisty H, Serin G, Ghisolfi-Nieto L, Roger B, Libante V, Amalric F, Bouvet P (2000) Interaction of nucleolin with an evolutionarily conserved pre-ribosomal RNA sequence is required for the assembly of the primary processing complex. J Biol Chem 275:18845–18850

Ginisty H, Amalric F, Bouvet P (2001) Two different combinations of RNA-binding domains determine the RNA binding specificity of nucleolin. J Biol Chem 276:14338–14343

Gonzalez V, Hurley LH (2010) The C-terminus of nucleolin promotes the formation of the c-MYC G-quadruplex and inhibits c-MYC promoter activity. Biochemistry 49:9706–9714

Gottlieb S, Esposito RE (1989) A new role for a yeast transcriptional silencer gene, SIR2, in regulation of recombination in ribosomal DNA. Cell 56:771–776

Grinstein E, Wernet P, Snijders PJ, Rosl F, Weinert I, Jia W, Kraft R, Schewe C, Schwabe M, Hauptmann S, Dietel M, Meijer CJ, Royer HD (2002) Nucleolin as activator of human papillomavirus type 18 oncogene transcription in cervical cancer. J Exp Med 196:1067–1078

Grinstein E, Shan Y, Karawajew L, Snijders PJ, Meijer CJ, Royer HD, Wernet P (2006) Cell cycle-controlled interaction of nucleolin with the retinoblastoma protein and cancerous cell transformation. J Biol Chem 281:22223–22235

Grinstein E, Du Y, Santourlidis S, Christ J, Uhrberg M, Wernet P (2007) Nucleolin regulates gene expression in CD34-positive hematopoietic cells. J Biol Chem 282:12439–12449

Grummt I (2003) Life on a planet of its own: regulation of RNA polymerase I transcription in the nucleolus. Genes Dev 17:1691–1702

Guderian G, Peter C, Wiesner J, Sickmann A, Schulze-Osthoff K, Fischer U, Grimmler M (2011) RioK1, a new interactor of protein arginine methyltransferase 5 (PRMT5), competes with pICln for binding and modulates PRMT5 complex composition and substrate specificity. J Biol Chem 286:1976–1986

Gulli MP, Girard JP, Zabetakis D, Lapeyre B, Melese T, Caizergues-Ferrer M (1995a) gar2 is a nucleolar protein from Schizosaccharomyces pombe required for 18S rRNA and 40S ribosomal subunit accumulation. Nucleic Acids Res 23:1912–1918

Gulli MP, Girard JP, Zabetakis D, Lapeyre B, Melese T, Caizerguesferrer M (1995b) Gar2 Is a nucleolar protein from Schizosaccharomyces pombe required for 18s ribosomal-RNA and 40s ribosomal-subunit accumulation. Nucleic Acids Res 23:1912–1918

Gulli MP, Faubladier M, Sicard H, Caizergues-Ferrer M (1997) Mitosis-specific phosphorylation of gar2, a fission yeast nucleolar protein structurally related to nucleolin. Chromosoma 105:532–541

Hanakahi LA, Dempsey LA, Li MJ, Maizels N (1997) Nucleolin is one component of the B cell-specific transcription factor and switch region binding protein, LR1. Proc Natl Acad Sci U S A 94:3605–3610

Hanakahi LA, Sun H, Maizels N (1999) High affinity interactions of nucleolin with G-G-paired rDNA. J Biol Chem 274:15908–15912

Hanakahi LA, Bu Z, Maizels N (2000) The C-terminal domain of nucleolin accelerates nucleic acid annealing. Biochemistry 39:15493–15499

Hannan RD, Cavanaugh A, Hempel WM, Moss T, Rothblum L (1999) Identification of a mammalian RNA polymerase I holoenzyme containing components of the DNA repair/replication system. Nucleic Acids Res 27:3720–3727

Hatfield DL, Carlson BA, Xu XM, Mix H, Gladyshev VN (2006) Selenocysteine incorporation machinery and the role of selenoproteins in development and health. Prog Nucleic Acid Res Mol Biol 81:97–142

Heo K, Kim B, Kim K, Choi J, Kim H, Zhan Y, Ranish JA, An W (2007) Isolation and characterization of proteins associated with histone H3 tails in vivo. J Biol Chem 282:15476–15483

Hernandez-Verdun D (1991) The nucleolus today. J Cell Sci 99(pt 3):465–471

Herrera AH, Olson MO (1986) Association of protein C23 with rapidly labeled nucleolar RNA. Biochemistry 25:6258–6264

Hirano M, Kaneko S, Yamashita T, Luo H, Qin W, Shirota Y, Nomura T, Kobayashi K, Murakami S (2003) Direct interaction between nucleolin and hepatitis C virus NS5B. J Biol Chem 278:5109–5115

Hovanessian AG (2006) Midkine, a cytokine that inhibits HIV infection by binding to the cell surface expressed nucleolin. Cell Res 16:174–181

Huddleson JP, Ahmad N, Lingrel JB (2006) Up-regulation of the KLF2 transcription factor by fluid shear stress requires nucleolin. J Biol Chem 281:15121–15128

Iftode C, Daniely Y, Borowiec JA (1999) Replication protein A (RPA): the eukaryotic SSB. Crit Rev Biochem Mol Biol 34:141–180

Ishikawa F, Matunis MJ, Dreyfuss G, Cech TR (1993) Nuclear proteins that bind the pre-mRNA 3′ splice site sequence r(UUAG/G) and the human telomeric DNA sequence d(TTAGGG)n. Mol Cell Biol 13:4301–4310

Ishimaru D, Zuraw L, Ramalingam S, Sengupta TK, Bandyopadhyay S, Reuben A, Fernandes DJ, Spicer EK (2010) Mechanism of regulation of bcl-2 mRNA by nucleolin and A+U-rich element-binding factor 1 (AUF1). J Biol Chem 285:27182–27191

Issinger OG, Martin T, Richter WW, Olson M, Fujiki H (1988) Hyperphosphorylation of N-60, a protein structurally and immunologically related to nucleolin after tumour-promoter treatment. EMBO J 7:1621–1626

Jiang Y, Xu XS, Russell JE (2006) A nucleolin-binding 3′ untranslated region element stabilizes beta-globin mRNA in vivo. Mol Cell Biol 26:2419–2429

Jiang B, Zhang B, Liang P, Song J, Deng H, Tu Z, Deng G, Xiao X (2010a) Nucleolin/C23 mediates the antiapoptotic effect of heat shock protein 70 during oxidative stress. FEBS J 277:642–652

Jiang Y, Li Z, Nagy PD (2010b) Nucleolin/Nsr1p binds to the 3′ noncoding region of the tombusvirus RNA and inhibits replication. Virology 396:10–20

Johansson C, Finger LD, Trantirek L, Mueller TD, Kim S, Laird-Offringa IA, Feigon J (2004) Solution structure of the complex formed by the two N-terminal RNA-binding domains of nucleolin and a pre-rRNA target. J Mol Biol 337:799–816

Johansson H, Svensson F, Runnberg R, Simonsson T, Simonsson S (2010) Phosphorylated nucleolin interacts with translationally controlled tumor protein during mitosis and with Oct4 during interphase in ES cells. PLoS One 5:e13678

Johnson FB, Marciniak RA, Guarente L (1998) Telomeres, the nucleolus and aging. Curr Opin Cell Biol 10:332–338

Kass S, Tyc K, Steitz JA, Sollner-Webb B (1990) The U3 small nucleolar ribonucleoprotein functions in the first step of preribosomal RNA processing. Cell 60:897–908

Khurts S, Masutomi K, Delgermaa L, Arai K, Oishi N, Mizuno H, Hayashi N, Hahn WC, Murakami S (2004) Nucleolin interacts with telomerase. J Biol Chem 279:51508–51515

Kojima H, Suzuki T, Kato T, Enomoto K, Sato S, Tabata S, Saez-Vasquez J, Echeverria M, Nakagawa T, Ishiguro S, Nakamura K (2007) Sugar-inducible expression of the nucleolin-1 gene of *Arabidopsis thaliana* and its role in ribosome synthesis, growth and development. Plant J 49:1053–1063

Kondo K, Inouye M (1992) Yeast NSR1 protein that has structural similarity to mammalian nucleolin is involved in pre-rRNA processing. J Biol Chem 267:16252–16258

Lapeyre B, Amalric F (1985) A powerful method for the preparation of cDNA libraries: isolation of cDNA encoding a 100-kDal nucleolar protein. Gene 37:215–220

Lapeyre B, Bourbon H, Amalric F (1987) Nucleolin, the major nucleolar protein of growing eukaryotic cells: an unusual protein structure revealed by the nucleotide sequence. Proc Natl Acad Sci U S A 84:1472–1476

Lee WC, Zabetakis D, Melese T (1992) NSR1 is required for pre-rRNA processing and for the proper maintenance of steady-state levels of ribosomal subunits. Mol Cell Biol 12:3865–3871

Leitinger N, Wesierska-Gadek J (1993) ADP-ribosylation of nucleolar proteins in HeLa tumor cells. J Cell Biochem 52:153–158

Lischwe MA, Smetana K, Olson MOJ, Busch H (1979) Protein-C23 and protein-B23 are the major nucleolar silver staining proteins. Life Sci 25:701–708

Lischwe MA, Richards RL, Busch RK, Busch H (1981) Localization of phosphoprotein C23 to nucleolar structures and to the nucleolus organizer regions. Exp Cell Res 136:101–109

Lischwe MA, Roberts KD, Yeoman LC, Busch H (1982) Nucleolar specific acidic phosphoprotein C23 is highly methylated. J Biol Chem 257:4600–4602

Lischwe MA, Cook RG, Ahn YS, Yeoman LC, Busch H (1985) Clustering of glycine and NG, NG-dimethylarginine in nucleolar protein C23. Biochemistry 24:6025–6028

Liu HT, Yung BY (1999) In vivo interaction of nucleophosmin/B23 and protein C23 during cell cycle progression in HeLa cells. Cancer Lett 144:45–54

Losfeld ME, Khoury DE, Mariot P, Carpentier M, Krust B, Briand JP, Mazurier J, Hovanessian AG, Legrand D (2009) The cell surface expressed nucleolin is a glycoprotein that triggers calcium entry into mammalian cells. Exp Cell Res 315:357–369

Loyola A, Almouzni G (2004) Histone chaperones, a supporting role in the limelight. Biochim Biophys Acta 1677:3–11

Ma N, Matsunaga S, Takata H, Ono-Maniwa R, Uchiyama S, Fukui K (2007) Nucleolin functions in nucleolus formation and chromosome congression. J Cell Sci 120:2091–2105

Mamrack MD, Olson MO, Busch H (1979) Amino acid sequence and sites of phosphorylation in a highly acidic region of nucleolar nonhistone protein C23. Biochemistry 18:3381–3386

Maridor G, Nigg EA (1990) cDNA sequences of chicken nucleolin/C23 and NO38/B23, two major nucleolar proteins. Nucleic Acids Res 18:1286

McGrath KE, Smothers JF, Dadd CA, Madireddi MT, Gorovsky MA, Allis CD (1997) An abundant nucleolar phosphoprotein is associated with ribosomal DNA in *Tetrahymena* macronuclei. Mol Biol Cell 8:97–108

McStay B, Grummt I (2008) The epigenetics of rRNA genes: from molecular to chromosome biology. Annu Rev Cell Dev Biol 24:131–157

Miniard AC, Middleton LM, Budiman ME, Gerber CA, Driscoll DM (2010) Nucleolin binds to a subset of selenoprotein mRNAs and regulates their expression. Nucleic Acids Res 38:4807–4820

Miranda GA, Chokler I, Aguilera RJ (1995) The murine nucleolin protein is an inducible DNA and ATP binding protein which is readily detected in nuclear extracts of lipopolysaccharide-treated splenocytes. Exp Cell Res 217:294–308

Mongelard F, Bouvet P (2007) Nucleolin: a multiFACeTed protein. Trends Cell Biol 17:80–86

Mongelard F, Bouvet P (2010) AS-1411, a guanosine-rich oligonucleotide aptamer targeting nucleolin for the potential treatment of cancer, including acute myeloid leukemia. Curr Opin Mol Ther 12:107–114

Morimoto H, Okamura H, Haneji T (2002) Interaction of protein phosphatase 1 delta with nucleolin in human osteoblastic cells. J Histochem Cytochem 50:1187–1193

Murayama R, Harada Y, Shibata T, Kuroda K, Hayakawa S, Shimizu K, Tanaka T (2007) Influenza A virus non-structural protein 1 (NS1) interacts with cellular multifunctional protein nucleolin during infection. Biochem Biophys Res Commun 362:880–885

Nasirudin KM, Ehtesham NZ, Tuteja R, Sopory SK, Tuteja N (2005) The Gly-Arg-rich C-terminal domain of pea nucleolin is a DNA helicase that catalytically translocates in the 5′- to 3′-direction. Arch Biochem Biophys 434:306–315

Olson MO, Thompson BA (1983) Distribution of proteins among chromatin components of nucleoli. Biochemistry 22:3187–3193

Olson MO, Ezrailson EG, Guetzow K, Busch H (1975) Localization and phosphorylation of nuclear, nucleolar and extranucleolar non-histone proteins of Novikoff hepatoma ascites cells. J Mol Biol 97:611–619

Olson MO, Rivers ZM, Thompson BA, Kao WY, Case ST (1983) Interaction of nucleolar phosphoprotein C23 with cloned segments of rat ribosomal deoxyribonucleic acid. Biochemistry 22:3345–3351

Orrick LR, Olson MO, Busch H (1973) Comparison of nucleolar proteins of normal rat liver and Novikoff hepatoma ascites cells by two-dimensional polyacrylamide gel electrophoresis. Proc Natl Acad Sci U S A 70:1316–1320

Pasternack MS, Bleier KJ, McInerney TN (1991) Granzyme A binding to target cell proteins. Granzyme A binds to and cleaves nucleolin in vitro. J Biol Chem 266:14703–14708

Pellar GJ, DiMario PJ (2003) Deletion and site-specific mutagenesis of nucleolin's carboxy GAR domain. Chromosoma 111:461–469

Peter M, Nakagawa J, Doree M, Labbe JC, Nigg EA (1990) Identification of major nucleolar proteins as candidate mitotic substrates of cdc2 kinase. Cell 60:791–801

Petricka JJ, Nelson TM (2007) *Arabidopsis* nucleolin affects plant development and patterning. Plant Physiol 144:173–186

Pollice A, Zibella MP, Bilaud T, Laroche T, Pulitzer JF, Gilson E (2000) In vitro binding of nucleolin to double-stranded telomeric DNA. Biochem Biophys Res Commun 268:909–915

Pontvianne F, Matia I, Douet J, Tourmente S, Medina FJ, Echeverria M, Saez-Vasquez J (2007) Characterization of AtNUC-L1 reveals a central role of nucleolin in nucleolus organization and silencing of AtNUC-L2 gene in *Arabidopsis*. Mol Biol Cell 18:369–379

Pontvianne F, Abou-Ellail M, Douet J, Cornella P, Matia I, Chandrasekhara C, Debures A, Blevins T, Cooke R, Medina FJ, Tourmente S, Pikaard CS, Saez-Vasquez J (2010) Nucleolin is required for DNA methylation state and the expression of rRNA gene variants in *Arabidopsis thaliana*. PLoS Genet 6:e1001225

Puvion-Dutilleul F, Puvion E, Bachellerie JP (1997) Early stages of pre-rRNA formation within the nucleolar ultrastructure of mouse cells studied by in situ hybridization with a 5′ETS leader probe. Chromosoma 105:496–505

Raman B, Guarnaccia C, Nadassy K, Zakhariev S, Pintar A, Zanuttin F, Frigyes D, Acatrinei C, Vindigni A, Pongor G, Pongor S (2001) N(omega)-arginine dimethylation modulates the interaction between a Gly/Arg-rich peptide from human nucleolin and nucleic acids. Nucleic Acids Res 29:3377–3384

Rankin ML, Heine MA, Xiao S, LeBlanc MD, Nelson JW, DiMario PJ (1993) A complete nucleolin cDNA sequence from *Xenopus laevis*. Nucleic Acids Res 21:169

Rao SV, Mamrack MD, Olson MO (1982) Localization of phosphorylated highly acidic regions in the NH2-terminal half of nucleolar protein C23. J Biol Chem 257:15035–15041

Reboud JP, Reboud AM, Madjar JJ, Buisson M (1974) Study of protein reactivity in rat liver ribosomes. Effect of RNA addition to autodigested ribosomes. Acta Biol Med Ger 33:661–666

Reeder RH (1990) rRNA synthesis in the nucleolus. Trends Genet 6:390–395

Rickards B, Flint SJ, Cole MD, LeRoy G (2007) Nucleolin is required for RNA polymerase I transcription in vivo. Mol Cell Biol 27:937–948

Roger B, Moisand A, Amalric F, Bouvet P (2002) Repression of RNA polymerase I transcription by nucleolin is independent of the RNA sequence that is transcribed. J Biol Chem 277:10209–10219

Roger B, Moisand A, Amalric F, Bouvet P (2003) Nucleolin provides a link between RNA polymerase I transcription and pre-ribosome assembly. Chromosoma 111:399–407

Roussel P, Hernandez-Verdun D (1994) Identification of Ag-NOR proteins, markers of proliferation related to ribosomal gene activity. Exp Cell Res 214:465–472

Roussel P, Belenguer P, Amalric F, Hernandez-Verdun D (1992) Nucleolin is an Ag-NOR protein; this property is determined by its amino-terminal domain independently of its phosphorylation state. Exp Cell Res 203:259–269

Rubbi CP, Milner J (2003) Disruption of the nucleolus mediates stabilization of p53 in response to DNA damage and other stresses. EMBO J 22:6068–6077

Saez-Vasquez J, Caparros-Ruiz D, Barneche F, Echeverria M (2004) A plant snoRNP complex containing snoRNAs, fibrillarin, and nucleolin-like proteins is competent for both rRNA gene binding and pre-rRNA processing in vitro. Mol Cell Biol 24:7284–7297

Said EA, Courty J, Svab J, Delbe J, Krust B, Hovanessian AG (2005) Pleiotrophin inhibits HIV infection by binding the cell surface-expressed nucleolin. FEBS J 272:4646–4659

Salazar R, Brandt R, Kellermann J, Krantz S (2000) Purification and characterization of a 200 kDa fructosyllysine-specific binding protein from cell membranes of U937 cells. Glycoconj J 17:713–716

Sapp M, Knippers R, Richter A (1986) DNA binding properties of a 110 kDa nucleolar protein. Nucleic Acids Res 14:6803–6820

Sato H, Kusumoto-Matsuo R, Ishii Y, Mori S, Nakahara T, Shinkai-Ouchi F, Kawana K, Fujii T, Taketani Y, Kanda T. Kukimoto I (2009) Identification of nucleolin as a protein that binds to human papillomavirus type 16 DNA. Biochem Biophys Res Commun 387:525–530

Scheer U, Benavente R (1990) Functional and dynamic aspects of the mammalian nucleolus. Bioessays 12:14–21

Schmidt-Zachmann MS. Dargemont C, Kuhn LC, Nigg EA (1993) Nuclear export of proteins: the role of nuclear retention. Cell 74:493–504

Schrick JJ, Hughes MJ, Anderson KP, Croyle ML, Lingrel JB (1999) Characterization of the lung Kruppel-like transcription factor gene and upstream regulatory elements. Gene 236:185–195

Schwab MS, Dreyer C (1997) Protein phosphorylation sites regulate the function of the bipartite NLS of nucleolin. Eur J Cell Biol 73:287–297

Seinsoth S, Uhlmann-Schiffler H, Stahl H (2003) Bidirectional DNA unwinding by a ternary complex of T antigen, nucleolin and topoisomerase I. EMBO Rep 4:263–268

Semba S, Mizuuchi E, Yokozaki H (2010) Requirement of phosphatase of regenerating liver-3 for the nucleolar localization of nucleolin during the progression of colorectal carcinoma. Cancer Sci 101:2254–2261

Serin G, Joseph G, Faucher C, Ghisolfi L, Bouche G, Amalric F, Bouvet P (1996) Localization of nucleolin binding sites on human and mouse pre-ribosomal RNA. Biochimie 78:530–538

Serin G, Joseph G, Ghisolfi L, Bauzan M, Erard M, Amalric F, Bouvet P (1997) Two RNA-binding domains determine the RNA-binding specificity of nucleolin. J Biol Chem 272:13109–13116

Shi H, Huang Y, Zhou H, Song X, Yuan S, Fu Y, Luo Y (2007) Nucleolin is a receptor that mediates antiangiogenic and antitumor activity of endostatin. Blood 110:2899–2906

Sicard H, Faubladier M, Noaillac-Depeyre J, Leger-Silvestre I, Gas N, Caizergues-Ferrer M (1998) The role of the *Schizosaccharomyces pombe* gar2 protein in nucleolar structure and function depends on the concerted action of its highly charged N terminus and its RNA-binding domains. Mol Biol Cell 9:2011–2023

Singh K, Laughlin J, Kosinski PA, Covey LR (2004) Nucleolin is a second component of the CD154 mRNA stability complex that regulates mRNA turnover in activated T cells. J Immunol 173:976–985

Sipos K, Olson MO (1991) Nucleolin promotes secondary structure in ribosomal RNA. Biochem Biophys Res Commun 177:673–678

Sirri V, Roussel P, Trere D, Derenzini M, Hernandez-Verdun D (1995) Amount variability of total and individual Ag-NOR proteins in cells stimulated to proliferate. J Histochem Cytochem 43:887–893

Smyth MJ, Browne KA, Thia KY, Apostolidis VA, Kershaw MH, Trapani JA (1994) Hypothesis: cytotoxic lymphocyte granule serine proteases activate target cell endonucleases to trigger apoptosis. Clin Exp Pharmacol Physiol 21:67–70

Srivastava M, Fleming PJ, Pollard HB, Burns AL (1989) Cloning and sequencing of the human nucleolin cDNA. FEBS Lett 250:99–105

Storck S, Shukla M, Dimitrov S, Bouvet P (2007) Functions of the histone chaperone nucleolin in diseases. Subcell Biochem 41:125–144

Storck S, Thiry M, Bouvet P (2009) Conditional knockout of nucleolin in DT40 cells reveals the functional redundancy of its RNA-binding domains. Biol Cell 101:153–167

Strang BL, Boulant S, Coen DM (2010) Nucleolin associates with the human cytomegalovirus DNA polymerase accessory subunit UL44 and is necessary for efficient viral replication. J Virol 84:1771–1784

Sun D, Guo K, Shin YJ (2011) Evidence of the formation of G-quadruplex structures in the promoter region of the human vascular endothelial growth factor gene. Nucleic Acids Res 39(4):1256–1265

Suzuki N, Kobayashi M, Sakata K, Suzuki T, Hosoya T (1991) Synergistic stimulatory effect of glucocorticoid, EGF and insulin on the synthesis of ribosomal RNA and phosphorylation of nucleolin in primary cultured rat hepatocytes. Biochim Biophys Acta 1092:367–375

Takagi M, Absalon MJ, McLure KG, Kastan MB (2005) Regulation of p53 translation and induction after DNA damage by ribosomal protein L26 and nucleolin. Cell 123:49–63

Tang L, Nogales E, Ciferri C (2010) Structure and function of SWI/SNF chromatin remodeling complexes and mechanistic implications for transcription. Prog Biophys Mol Biol 102:122–128

Tawfic S, Goueli SA, Olson MO, Ahmed K (1994) Androgenic regulation of phosphorylation and stability of nucleolar protein nucleolin in rat ventral prostate. Prostate 24:101–106

Tediose T, Kolev M, Sivasankar B, Brennan P, Morgan BP, Donev R (2010) Interplay between REST and nucleolin transcription factors: a key mechanism in the overexpression of genes upon increased phosphorylation. Nucleic Acids Res 38:2799–2812

Teng Y, Girvan AC, Casson LK, Pierce WM Jr, Qian M, Thomas SD, Bates PJ (2007) AS1411 alters the localization of a complex containing protein arginine methyltransferase 5 and nucleolin. Cancer Res 67:10491–10500

Tong CG, Reichler S, Blumenthal S, Balk J, Hsieh HL, Roux SJ (1997) Light regulation of the abundance of mRNA encoding a nucleolin-like protein localized in the nucleoli of pea nuclei. Plant Physiol 114:643–652

Tuteja N, Huang NW, Skopac D, Tuteja R, Hrvatic S, Zhang J, Pongor S, Joseph G, Faucher C, Amalric F et al (1995) Human DNA helicase IV is nucleolin, an RNA helicase modulated by phosphorylation. Gene 160:143–148

Ugrinova I, Monier K, Ivaldi C, Thiry M, Storck S, Mongelard F, Bouvet P (2007) Inactivation of nucleolin leads to nucleolar disruption, cell cycle arrest and defects in centrosome duplication. BMC Mol Biol 8:66

Wang Y, Guan J, Wang H, Leeper D, Iliakis G (2001) Regulation of DNA replication after heat shock by replication protein a-nucleolin interactions. J Biol Chem 276:20579–20588

Warner JR (1989) Synthesis of ribosomes in *Saccharomyces cerevisiae*. Microbiol Rev 53:256–271

Welfle H, Henkel B, Bielka H (1976) Ionic interactions in eukaryotic ribosomes: splitting of the subunits of rat liver ribosomes by treatment with monovalent cations. Acta Biol Med Ger 35:401–411

Xu X, Hamhouyia F, Thomas SD, Burke TJ, Girvan AC, McGregor WG, Trent JO, Miller DM, Bates PJ (2001) Inhibition of DNA replication and induction of S phase cell cycle arrest by G-rich oligonucleotides. J Biol Chem 276:43221–43230

Yanagida M, Shimamoto A, Nishikawa K, Furuichi Y, Isobe T, Takahashi N (2001) Isolation and proteomic characterization of the major proteins of the nucleolin-binding ribonucleoprotein complexes. Proteomics 1:1390–1404

Yang TH, Tsai WH, Lee YM, Lei HY, Lai MY, Chen DS, Yeh NH, Lee SC (1994) Purification and characterization of nucleolin and its identification as a transcription repressor. Mol Cell Biol 14:6068–6074

Yang C, Maiguel DA, Carrier F (2002) Identification of nucleolin and nucleophosmin as genotoxic stress-responsive RNA-binding proteins. Nucleic Acids Res 30:2251–2260

Yang C, Kim MS, Chakravarty D, Indig FE, Carrier F (2009) Nucleolin binds to the proliferating cell nuclear antigen and inhibits nucleotide excision repair. Mol Cell Pharmacol 1:130–137

Yuan X, Zhou Y, Casanova E, Chai M, Kiss E, Grone HJ, Schutz G, Grummt I (2005) Genetic inactivation of the transcription factor TIF-IA leads to nucleolar disruption, cell cycle arrest, and p53-mediated apoptosis. Mol Cell 19:77–87

Zhang Y, Bhatia D, Xia H, Castranova V, Shi X, Chen F (2006) Nucleolin links to arsenic-induced stabilization of GADD45alpha mRNA. Nucleic Acids Res 34:485–495

Zhang J, Tsaprailis G, Bowden GT (2008) Nucleolin stabilizes Bcl-X L messenger RNA in response to UVA irradiation. Cancer Res 68:1046–1054

Zhou G, Seibenhener ML, Wooten MW (1997) Nucleolin is a protein kinase C-zeta substrate. Connection between cell surface signaling and nucleus in PC12 cells. J Biol Chem 272:31130–31137

Chapter 10
The Multifunctional Nucleolar Protein Nucleophosmin/NPM/B23 and the Nucleoplasmin Family of Proteins

Shea Ping Yip, Parco M. Siu, Polly H.M. Leung, Yanxiang Zhao, and Benjamin Y.M. Yung

10.1 Introduction

Nucleophosmin 1 (NPM1) is also known as nucleophosmin (NPM), nucleolar phosphoprotein B23, and numatrin in mammals, and NO38 in amphibians. NPM1 was first discovered as a nucleolar protein in rat liver cells and Novikoff hepatoma ascites cells by two-dimensional polyacrylamide gel electrophoresis (Table 10.1) (Orrick et al. 1973). It was first named as B23 because it was the 23rd protein in region B of the gel slab when the protein spots were numbered in order of decreasing mobility in both electrophoretic dimensions. It serves as a nuclear chaperone and has many other important functions (Grisendi et al. 2006). Nucleophosmin 2 (NPM2), originally called nucleoplasmin, was first identified and purified from the eggs of the African clawed frog *Xenopus laevis* (Laskey et al. 1978), in which it is the most abundant nuclear protein (Mills et al. 1980). It binds histones and transfers them to DNA, and facilitates the assembly of nucleosomes – the basic building block of chromatin. Nucleophosmin 3 (NPM3) was cloned and characterized as a novel gene (*npm3*) in mouse, and found to be very similar to human NPM1 and *Xenopus* npm2 in amino acid sequence, protein features, and exon organization (MacArthur and Shackleford 1997). In the same year, NPM3 was also discovered by sodium dodecyl sulfate polyacrylamide gel electrophoresis as a very acidic protein (NO29) immunolocalized in the nucleoli of *Xenopus* oocytes, and forming complex with NPM1 (Zirwes et al. 1997). It was proposed to be involved in the assembly of preribosomal particles.

B.Y.M. Yung (✉)
Department of Health Technology and Informatics, The Hong Kong Polytechnic University,
Hung Hom, Kowloon, Hong Kong SAR, China
e-mail: htbyung@inet.polyu.edu.hk

M.O.J. Olson (ed.), *The Nucleolus*, Protein Reviews 15,
DOI 10.1007/978-1-4614-0514-6_10, © Springer Science+Business Media, LLC 2011

Table 10.1 Members of the nucleophosmin/nucleoplasmin (NPM) family

	Nucleophosmin 1	Nucleophosmin 2	Nucleophosmin 3
When first discovered	1973 in rat	1978 in frog	1997 in mouse
Expression & subcellular location	Ubiquitous, mainly nucleolar	Only in oocytes & eggs, nuclear	Ubiquitous, mainly nucleolar
Basic features in *Homo sapiens*[a]			
Gene symbol	*NPM1*	*NPM2*	*NPM3*
Chromosome location	5q35	8p21.3	10q24.31
Official full name	Nucleophosmin (nucleolar phosphoprotein B23, numatrin)	Nucleophosmin/ nucleoplasmin 2	Nucleophosmin/ nucleoplasmin 3
Other names	*B23, NPM, MGC104254*	Nucleoplasmin	*PORMIN, TMEM123*
No. of exons	12	9	6
No. of transcripts and accession no.	3	1	1
	Variant 1: NM_002520	NM_182795	NM_006993
	Variant 2: NM_199185		
	Variant 3: NM_001037738		
No. of amino acids	Isoform 1: 294	214	178
	Isoform 2: 265		
	Isoform 3: 259		

[a]Information from NCBI (http://www.ncbi.nlm.nih.gov/)

NPM1, NPM2, NPM3, and invertebrate NPM-like proteins form the nucleophosmin/nucleoplasmin family of nuclear chaperones and are found throughout the animal kingdom (Eirín-López et al. 2006; Frehlick et al. 2007). Nuclear chaperones serve to ensure proper assembly of nucleosomes and proper formation of higher order structures of chromatin. In fact, this family of proteins has such diverse functions in cellular processes such as chromatin remodeling, ribosome biogenesis, genome stability, centrosome replication, and transcriptional regulation. Of the members of this family, NPM1 is the most studied because it is often altered in expression levels in tumors and mutated/translocated in hematological malignancies (Grisendi et al. 2006). It is the main focus of this review. NPM2 and NPM3 are less well characterized, and are also discussed wherever appropriate. Invertebrate NPM-like proteins are not discussed in this review.

10.2 Structure and Expression of *NPM* Genes

The human *NPM1* gene spans a genomic region of about 23 kb at chromosome 5q35 and has 12 exons (Table 10.1). It can be transcribed as three variants. Transcript variant 1 is ubiquitously expressed and is the major and the longest transcript, giving rise to isoform 1 of 294 amino acids. Transcript variant 2 skips an in-frame exon (exon 8) and produces a shorter protein (isoform 2) of 265 amino acids, whose functions and expression pattern are not known. Transcript variant 3 utilizes an alternate

3′ exon (exon 10) and hence produces a 259-amino-acid protein with a C-terminus different from that of isoform 1. The corresponding isoforms 1 and 3 in rat are B23.1 and B23.2, and have different subcellular distribution patterns (Wang et al. 1993). Isoform 1 is localized in the nucleolus (Michalik et al. 1981; Spector et al. 1984) while isoform 3 is found in the nucleoplasm with low expression levels (Dalenc et al. 2002).

The human *NPM2* gene is about 12 kb long at chromosome 8p21.3 and has nine exons (Table 10.1). Its single transcript produces a protein of 214 amino acids. NPM2 has a rather restricted tissue distribution and is found in the nucleus of oocytes and eggs only (Laskey et al. 1978; Mills et al. 1980; Burns et al. 2003). The human *NPM3* gene is only about 2 kb long at chromosome 10q24.31 and has six exons (Table 10.1). Similar to NPM1, NPM3 is ubiquitously expressed, and is localized to the nucleolus (MacArthur and Shackleford 1997; Zirwes et al. 1997; Shackleford et al. 2001).

10.3 Structure–Function Relationship of NPM Proteins

The NPM proteins share strong sequence and structural homology. The sequence of NPM proteins can be divided into several domains with distinct biochemical, structural, and functional properties (Fig. 10.1) (Hingorani et al. 2000; Frehlick et al. 2007; Okuwaki 2008). In addition to these modular domains, the NPM proteins also contain several sequence motifs, including the nuclear localization signal (NLS),

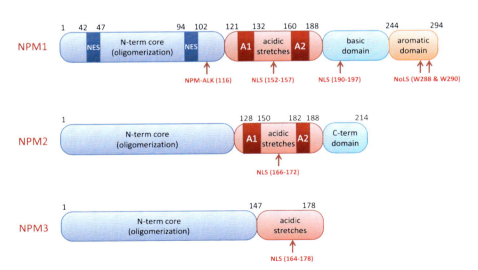

Fig. 10.1 Domain organization of NPM proteins. *NES* nuclear export signal; *NLS* nuclear localization signal; *NoLS* nucleolus localization signal; the NPM-ALK fusion protein that is involved in hematological malignancies

216 S.P. Yip et al.

the nuclear export signal (NES), and the nucleolus localization signal (NoLS) that are critical for the localization of NPM proteins in the nucleolus as well as their nucleo-cytoplasmic shuttling (Grisendi et al. 2006; Frehlick et al. 2007; Okuwaki 2008). In this section, we intend to discuss the structural and biochemical properties of each individual domain and how these properties correlate with its respective function. The 294-amino-acid NPM1 protein is mostly used as the representative example.

10.3.1 The N-Terminal Core Region (Residues 1–120)

The N-terminal region of NPM1 (residues 1–120) is commonly referred to as the "core" because this region is the most conserved among the NPM family of proteins. This region is largely hydrophobic and folds into a distinct structural domain that is protease resistant (Dutta et al. 2001). This core domain is responsible for oligomerization and the molecular chaperone activity of NPM1 by suppressing the misfolding and aggregation of target proteins in the crowded environment in the nucleolus (Szebeni and Olson 1999). This region can also interact with core histone proteins H2A, H2B, H3, and H4 to function as a histone chaperone and facilitate nucleosome assembly (Okuwaki et al. 2001b). A functional role of this domain in ribosomal biogenesis and p53-related tumor suppression has also been implicated (Colombo et al. 2002; Murano et al. 2008).

The X-ray crystal structure of the human NPM1 core region (residues 9–122) reveals a compact domain consisting of two four-stranded β-sheets packed in a β-sandwich topology (Fig. 10.2a) (Lee et al. 2007). The core region forms a tight pentameric assembly through hydrophobic interactions between the monomeric subunits (Fig. 10.2b). Moreover, in the crystallization environment, two pentameric complexes align along their fivefold symmetry axis and associate in head-to-head fashion to form a decameric assembly (Fig. 10.2c). The monomeric and pentameric structures of human NPM1 core domain are highly similar to that of the core region

Fig. 10.2 Structure of the N-terminal core region of NPM1. (**a**) Structure of the core region in monomeric form. The *dotted line* indicates the disordered surface loop. (**b**) Structure of the core region in pentameric form. The *dot in the center* indicates the fivefold axis. (**c**) Structure of the core region in decameric assembly as observed in crystal lattice. The histone-octamer is modeled to contact the lateral surface of NPM1 decameric ring to form the NPM-histone complex. The *dotted lines* indicate highly flexible and disordered loops. (**d**) Comparison of human NPM1 decameric complex (*colored blue*) with that of *Xenopus* npm1 (*colored magenta*). The two structures are aligned by superposing on the top pentamer of the decameric ring (*left*), and the rotational shift for the bottom pentamer as a result of such alignment is illustrated with an arrow (*right*). The structural files for the N-terminal pentameric domain of human NPM1 (PDB ID 2P1B; Lee et al. 2007) and *Xenopus* npm1 (PDB ID 1XE0; Namboodiri et al. 2004) are downloaded from the public database RCSB Protein Data Bank (www.rcsb.org), and the figures are prepared using the program CCP4mg (Potterton et al. 2004)

10 Nucleophosmin/NPM/B23 and the Nucleoplasmin Family of Proteins 217

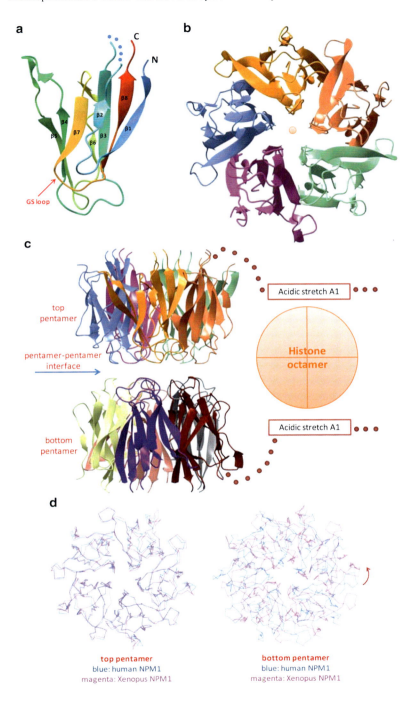

of several other NPM proteins including the *Xenopus* NO38 (i.e., npm1) (Namboodiri et al. 2004), the *Xenopus* nucleoplasmin (i.e., npm2) (Dutta et al. 2001), and the *Drosophila* nucleoplasmin-like protein (i.e., dNPL) (Namboodiri et al. 2003). All these structures share the same β-sandwich fold with root mean square deviation of ~1.0 Å for the Cα atoms (Lee et al. 2007). Such an oligomeric assembly observed in crystal structures agrees with previous findings that NPM proteins have a propensity to oligomerize with the benefit of enhanced thermostability (Umekawa et al. 1993; Herrera et al. 1996). A structure-based model of the NPM-histone complex, mainly based on the structures of the *Xenopus* NO38 and nucleoplasmin core domain, was proposed to consist of the NPM decamer and the histone octamer, with either the H2A-H2B dimer or the H3-H4 tetramer contacting the NPM core on the lateral surface of the decameric ring (Fig. 10.2c) (Dutta et al. 2001; Namboodiri et al. 2004). However, many details of this modeled oligomeric assembly still await further clarification.

A notable difference between the structure of human NPM1 core domain and structures of the other NPM proteins is the plasticity of the pentamer–pentamer interface observed in crystal lattice. When the decameric structure of human NPM1 core region is superimposed onto that of *Xenopus* NO38 by aligning one of the pentamers (referred to as the top one), the other pentamer (referred to as the bottom one) in the decameric assembly shows a large rotational offset (~20°) from that of *Xenopus* npm1 core (Fig. 10.2d) and a relative smaller rotational offset (~10°) from that of *Xenopus* npm2 core (Lee et al. 2007). Such rotational offset does not change the pattern of which monomer of the top pentamer contacts the corresponding monomer of the bottom pentamer, but it does lead to different sets of polar interactions between these monomers to stabilize the pentamer–pentamer interface (Lee et al. 2007). In addition, the molecular composition of the lateral surface for the NPM decamer assembly would be affected by such rotational offset as well. This structural plasticity is likely due to the small interface area between the pentamers (~560 Å^2) and the limited number of residues directly involved in polar interactions at the interface (Lee et al. 2007). The differences in the decameric assembly of human and *Xenopus* NPM proteins could have a significant implication for their respective histone chaperone function because the histones were proposed in the above-mentioned structural model to contact the lateral surface of the decameric assembly. Indeed, some substrate preferences toward the various core histone proteins have been observed for NPM proteins, with *Xenopus* npm1 showing preference for the H3-H4 tetramers and the *Xenopus* npm2 preferring the H2A-H2B dimers (Dutta et al. 2001; Namboodiri et al. 2003).

The stable oligomeric assembly of the N-terminal core region is critical for mediating the interaction between NPM1 and the tumor suppressor ARF, which is the protein translated from the transcript produced by an alternate reading frame of the cyclin-dependent kinase inhibitor 2A (*CDKN2A*) gene, and is also known as p14ARF in humans and p19Arf in mouse on the basis of their molecular weights (Gallagher et al. 2006). The NPM1–ARF interaction can lead to cell growth arrest and tumor suppression in either p53-dependent manner through the p53-Mdm2 pathway or p53-independent manner by inhibiting ribosome biogenesis and cell proliferation (Bertwistle et al. 2004; Brady et al. 2004; Itahana et al. 2003; Korgaonkar et al. 2005).

Several conserved residues critical for NPM1–ARF interaction (Leu102, Gly105, and Gly107) are located on a short GS loop between β7 and β8 strand, in close proximity to the pentamer–pentamer interface of the decameric ring (Fig. 10.2a) (Enomoto et al. 2006). Mutations of these residues (Leu102Ala, Gly105Ala, and Gly107Ala) abolish the NPM1–ARF interaction and lead to loss of the NPM1 oligomeric state, probably by disturbing the pentamer–pentamer interactions and weakening the hydrophobic interactions between the monomer subunits within the pentameric assembly (Enomoto et al. 2006). The monomeric mutated NPM1 is delocalized from the nucleolus to the nucleoplasm and displays increased susceptibility to ubiquitination and proteasomal degradation (Enomoto et al. 2006).

The N-terminal core region domain contains two well-studied NES motifs, both of which are critical for the nucleo-cytoplasmic shuttling of NPM1 (Wang et al. 2005; Maggi et al. 2008). The first motif (residues 42–47) is located on the β3 strand forming part of inter-subunit interface within the pentameric ring (Fig. 10.2a). Mutation of two conserved leucine residues in this sequence motif (Leu42Ala and Leu44Ala) leads to a defective NPM1 that cannot be shuttled out of nucleus (Maggi et al. 2008). Such defect inhibits the nuclear export of both the 40S and 60S ribosomal units, which associate with NPM1 and depend on the shuttling of NPM1 for their export (Maggi et al. 2008). Thus, for this shuttling-defective NPM1 mutant, the available pool of cytoplasmic ribosome units would be reduced and the overall protein synthesis diminished, leading to inhibition of cell proliferation. The second NES motif (residues 94–102) is located on the β7 strand on the outer lateral surface of the pentameric ring. This motif leads to nuclear export of NPM1 by the Ran-Crm1 complex, and enables the association of NPM1 with the centrosome during mitosis to prevent centrosome reduplication (Wang et al. 2005). Removal of this NES motif by deleting residues 94–102 or mutating two conserved leucine residues (Leu100Ala and Leu102Ala) leads to nuclear retention of NPM1 and supernumerary centrosomes (Wang et al. 2005).

10.3.2 The Acidic Stretches A1 (Residues 121–132) and A2 (Residues 160–188)

Following the N-terminal hydrophobic core region, the NPM1 protein contains a long stretch of unstructured segment enriched in clusters of acidic residues, including the acidic stretch A1 (residues 121–132) and A2 (residues 160–188) (Grisendi et al. 2006; Frehlick et al. 2007; Okuwaki 2008). Dictated by the many acidic aspartate and glutamate residues, the distinct electrostatic property of this region suggests that it can possibly be engaged in interaction with basic proteins such as histones and sperm basic proteins to facilitate nucleosome assembly and chromatin remodeling. Indeed, the acidic stretch A1 was found to be critical for the histone chaperone activity of NPM1. While an NPM1 construct encompassing both the pentameric core domain and the acidic stretch A1 shows ~97% chaperone activity, the pentameric domain alone shows only 30% chaperone activity and may be incapable of assembling nucleosomes (Hingorani et al. 2000). In the proposed model of

NPM–histone complex mentioned above, the acidic stretch A1 was included as a critical element to bind to histone proteins and facilitate the assembly of the NPM–histone complex (Fig. 10.2c) (Dutta et al. 2001; Namboodiri et al. 2004).

This acidic region is also found to be critical for NPM1-mediated viral replication (see Sect. 10.6.1). One study showed that both acidic stretches A1 and A2 are critical for promoting in vitro replication of adenovirus DNA complexed with viral basic core proteins by mediating direct interaction between NPM1 and the viral basic proteins to enable them to serve as molecular chaperones and to facilitate the transfer of viral DNA onto these basic proteins (Okuwaki et al. 2001a; Samad et al. 2007). Similarly, another study demonstrated that the human hepatitis delta virus antigen molecules (HDVAgs) interact with NPM1 to up-regulate NPM1 expression and enhance viral replication (Huang et al. 2001). In vitro and in vivo studies mapped this NPM1–HDVAgs interaction to the N-terminal basic region of HDVAgs and the acidic stretch A1, probably because of their electrostatic complementarity (Huang et al. 2001). Lastly, the acidic region in general has been speculated to contribute to the relatively high affinity of NPM1 for peptides containing sequences of NLSs of the SV40 T-antigen type, such as the HIV Rev protein to stimulate its nuclear import and viral replication (Szebeni et al. 1997).

10.3.3 The Basic Domain (Residues 189–243)

This region refers to an unstructured segment (residues 189–243) that is enriched in the basic residues lysine and arginine. This basic domain is important for the nucleic acid binding activity of NPM1, likely because of its favorable electrostatic properties to associate with acidic DNA/RNA molecules (Hingorani et al. 2000). Such binding could be positive for NPM1's function in nucleosome assembly by coordinating the packaging of DNA molecules onto core histone proteins. Moreover, this binding activity would be beneficial to the functional role of NPM1 in ribosomal biogenesis because NPM1 should recognize and associate with preribosomal particles to transport them from nucleus to cytoplasm. In fact, NPM1 has been found to interact with rRNA and several ribonuclear proteins in the 40S and 60S ribosome units (Yu et al. 2006; Maggi et al. 2008). This basic region also mediates the interaction of NPM1 with tumor suppressor p53. The NPM1–p53 interaction leads to enhanced stability of p53 and up-regulation of its tumor suppression function (Colombo et al. 2002).

10.3.4 The C-Terminal Aromatic Domain (Residues 244–294)

The C-terminal region of NPM1 (residue 244–294) is unique to the isoform 1 of NPM1 (corresponding to B23.1 variant in rat) but absent in isoform 3 (B23.2 in rat), which contains only residues 1–259 of NPM1 isoform 1 (corresponding to residues 1–257 of Npm1 protein in rat) (Chang and Olson 1989). This region contains the nucleolar localization signal (NoLS) that is critical for the nucleus–cytosol shuttling

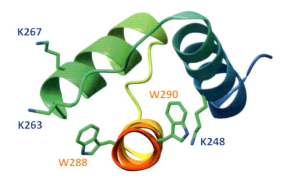

Fig. 10.3 Structure of the C-terminal aromatic domain of NPM1. Residues important for nucleolus localization are shown in stick model and colored in the cpk scheme. The structural file for human NPM1 C-terminal aromatic region (PDB ID 2VXD; Grummitt et al. 2008) is downloaded from the public database RCSB Protein Data Bank (www.rcsb.org), and the figure is prepared using the program CCP4mg (Potterton et al. 2004)

of NPM1 (Hingorani et al. 2000). Furthermore, this region contains several functionally important aromatic residues such as Trp288 and Trp290 that are important for the nucleolar localization of NPM1 (Nishimura et al. 2002) but are frequently mutated in acute myeloid leukemia (AML) (Falini et al. 2006). Such mutated NPM1 proteins are found to be aberrantly localized in the cytoplasm, instead of the nucleus, in leukemic cells (see Sect. 10.5.1.1).

Recently, the 3D structure for the C-terminal aromatic region of NPM1 was solved by nuclear magnetic resonance spectroscopy, which reveals a novel structural folding consisting of three helices packed tightly via a hydrophobic core formed by the conserved aromatic tryptophan and phenylalanine residues (Fig. 10.3) (Grummitt et al. 2008). Several well-conserved aromatic residues, including Trp288 and Trp290, are found to form a hydrophobic core and are critical to the structural folding of this domain. Mutation of these tryptophan residues to alanine results in collapse of the structure and loss of nucleolar localization (Grummitt et al. 2008). This result indicates that the structural integrity of this domain is critical to its nucleolar localization. Several surface lysine residues (Lys248, Lys263, and Lys267) have also been found to be critical to nucleolar localization of NPM1 although they are not critical to the structural integrity of this domain. Thus, these lysine residues could serve a functional, instead of a structural, role for nucleolar localization of NPM1.

10.4 Physiological Functions of NPM Proteins

Elucidating the physiological functions of NPM1 has become an important and exciting research topic, particularly in the field of cancer research. Studies show that NPM1 is a multifunctional protein widely involved in many vital biological processes. The fact that NPM1 can influence both proliferative and growth-suppressive

Table 10.2 NPM1: summary of physiological functions, posttranslational modifications, and alterations in human cancers

Physiological functions
- As a molecular chaperone
- Involvement in ribosome biogenesis
- Regulation of transcription
- Inhibition of apoptosis
- Modulation of tumor suppressors
- Maintenance of genomic stability
- Regulation of cell cycle

Posttranslational modifications
- Phosphorylation
- Dephosphorylation
- Acetylation
- Poly(ADP-ribosyl)ation
- Ubiquitination
- Sumoylation

Alterations in human cancers
- Overexpression in tumors of different origins
- Mutations causing NPMc+ acute myeloid leukemia
- Balanced translocations in lymphomas and leukemias

cellular events brings the scenario to a complex molecular situation that has to be completely resolved by future research efforts. Although the exact biological functions of NPM1 and its involved molecular mechanisms have yet to be fully unraveled, some clues on the physiological functions of NPM1 have been acquired since the discovery of this intriguing protein. This section mainly focuses on the physiological functions of NPM1: molecular chaperone function, ribosome biogenesis, transcriptional regulation, apoptosis inhibition, tumor suppressor modulation, genomic stability maintenance, and cell cycle regulation (Table 10.2). The physiological functions of NPM2 and NPM3 are also briefly described followed by the introduction of the identified posttranslational modifications of NPM proteins.

10.4.1 NPM1 as a Molecular Chaperone

NPM1 has been shown to have the ability to function as molecular chaperone for both proteins and nucleic acids (Okuwaki et al. 2001b; Szebeni and Olson 1999). The observed molecular chaperone activity of NPM1 is thought to prevent protein aggregation and protein misfolding (Szebeni and Olson 1999). Indeed, NPM1 has been shown to suppress the aggregation and misfolding of target proteins through the structural properties of its N-terminal core domain (Hingorani et al. 2000). During cell cycle, NPM1 works as a histone chaperone to control the assembly and disassembly of chromatin formation (Okuwaki et al. 2001b). To function as a molecular chaperone, NPM1 has to possess nucleo-cytoplasmic shuttling activities attributable to the presence of NES and NLS in NPM1 (Dingwall et al. 1987; Hingorani et al. 2000).

10.4.2 Involvement of NPM1 in Ribosome Biogenesis

NPM1 has all the necessary machinery features for the processing and assembly of ribosomes. These features include the abundant NPM1 localization in nucleolus and the ability to shuttle between nucleus and cytoplasm to bind nucleic acids and to transport preribosomal particles (Borer et al. 1989; Yun et al. 2003; Wang et al. 1994; Dumbar et al. 1989; Prestayko et al. 1974; Olson et al. 1986). In fact, NPM1 has been found to have intrinsic ribonuclease activity and be able to process preribosomal RNA in the internal transcribed spacer sequence (Savkur and Olson 1998; Herrera et al. 1995). This ribonuclease-mediated processing of ribosomal RNA is facilitated necessarily by the chaperone property of NPM1, which is important in preventing the aggregation of proteins in nucleolus during the process of ribosomal assembly. This proposition is supported by the finding that knockdown of NPM1 changes the ribosome profile and suppression of NPM1 inhibits the processing of preribosomal RNA (Grisendi et al. 2005; Itahana et al. 2003). One of the common morphological features of tumor cells is the enlarged nucleoli. This observation is conceivably linked to the proposed oncogenic role of NPM1 because the frequently observed aberrantly high NPM1 expression in rapidly proliferating tumor cells is generally consistent with the rapid ribosome biogenesis in maintaining the proliferative potential of tumor cells (Ruggero and Pandolfi 2003).

The physiological function of NPM1 in ribosomal biogenesis has been collectively proposed on the basis of the indirect evidence showing the intimate association of NPM1 with the synthetic machinery of ribosome. Nevertheless, this remains to be fully understood as it has been reported that NPM1-deficient embryos can survive to mid-gestation (Colombo et al. 2005; Grisendi et al. 2005), but embryonic lethality at very early stages has been demonstrated in embryos deficient in pescadillo, another nonribosomal protein participating in ribosome biogenesis (Lerch-Gaggl et al. 2002). These findings illustrate the difference of the survival time between NPM1-deficient and pescadillo-deficient embryos and raise the question whether NPM1 might not be an essential protein in the process of ribosomal biogenesis. Nonetheless, it is still possible that other factors such as ribosomal storage might have compensated for the loss of NPM1 in the NPM1-deficient embryos until the stage of mid-gestation (Grisendi et al. 2006).

10.4.3 Regulation of Gene Transcription by NPM1

NPM1 is also involved in transcriptional regulation and contributes to the cell growth control. The function of transcriptional regulation of NPM1 is related to the interaction of NPM1 with various transcription factors including NFκB, YY1, ARF, and IRF1 (Colombo et al. 2002; Dhar et al. 2004; Inouye and Seto 1994; Kondo et al. 1997; Korgaonkar et al. 2005). NPM1 has been shown to act as a coactivator for NFκB in regulating the expression of antioxidant enzyme MnSOD (Dhar et al. 2004). NPM1 can also alter the transcriptional activity of IRF1 and p53 (Colombo et al. 2002; Kondo et al. 1997). NPM1 is able to form a stable complex with YY1 and, interestingly, the transcriptional repressive function of YY1 can be relieved by the interaction with NPM1 (Inouye and Seto 1994). NPM1 can also modulate the binding of NFκB, E2F1, and pRB for the activation of E2F1 promoter (Lin et al. 2006).

A close interacting relationship is found between NPM1 and AP2α. In retinoic acid-induced differentiation in human leukemia HL-60 cells, a decline in c-myc, NPM1, and its promoter activity is observed (Yung 2004). The transcriptional mechanism underlying the down-regulation of NPM1 during retinoic acid-induced granulocytic differentiation involves dynamic changes in the promoter occupancy of different transcriptional regulators and these include NPM1 and AP2α (Liu et al. 2007a). It is intriguing that NPM1 is demonstrated to be recruited by AP2α to the promoters of some retinoic acid-responsive genes such as b-Myb, HSP60, and p120. These findings illustrate that NPM1 might be able to serve as a negative coregulator during retinoic acid signaling-induced gene expression. All these data indicate the potential important function of NPM1 as a molecular regulator of gene expression at the transcriptional level. By examining the U1 bladder cancer U4 premalignant cells, the intimate relationship of NPM1 with Ras and c-Myc has been further demonstrated in proliferation of cells associated with a high degree of malignancy (Yeh et al. 2006).

10.4.4 Inhibition of Apoptosis by NPM1

Overexpression of NPM1 can promote cell survival and this is accomplished partly through the inhibition of apoptosis (Ye 2005). Several pieces of evidenc support the anti-apoptotic role of NPM1. First of all, the oncogenic role of NPM1 hinges on its ability in inhibiting apoptosis in response to hypoxia (Li et al. 2004). As NPM1 is a transcriptional target of hypoxia inducible factor 1α (HIF1α), it has been reported that aberrantly increased expression of NPM1 might result in inhibition of the activation of tumor suppressor p53 and thus dampen p53-mediated activation of apoptosis during hypoxia-driven tumor progression (Li et al. 2004). Second, overexpression of NPM1 has been shown to induce resistance to ultraviolet irradiation-induced apoptosis, which is mediated by the tumor suppressor interferon regulatory factor-1 (IRF1, a transcription factor involved in DNA damage-induced apoptosis and cell cycle arrest) in NIH-3 T3 fibroblast cells (Wu et al. 2002b; Kondo et al. 1997). Third, as demonstrated in hematopoietic cells, overexpression of NPM1 has been shown to be related to the suppression of the activation of interferon-inducible, double-stranded RNA-dependent protein kinase (PKR), which normally induces apoptosis (Pang et al. 2003; Jagus et al. 1999). This observation has been associated with the aberrant proliferation of tumor cells. Fourth, NPM1 is shown to have a functional role as the receptor for second messenger phosphatidylinositol(3,4,5)-trisphosphate (PIP3) in nucleus and this NPM1–PIP3 complex is a downstream effector of phosphatidylinositol 3 kinase. The formation of nuclear NPM1–PIP3 complex has been shown to suppress apoptosis by inhibiting the DNA fragmentation activity of the pro-apoptotic factor caspase-activated DNase in PC12 cells treated with nerve growth factor (Ahn et al. 2005). Furthermore, the anti-apoptotic function of NPM1 has been related to its inhibitory effect on tumor suppressor p53. NPM1 interacts with p53 in hypoxic cells to hinder the hypoxia-induced activation of p53 phosphorylation (Li et al. 2004). There is evidence that NPM1 might inhibit the phosphorylation of p53 at serine 15 and thus abrogate the induction of p21 (Li et al. 2005). Last but not least, NPM1 interacts with the pro-apoptotic protein GADD45α, which is responsive to genotoxic stress. While GADD45α lacks a classical NLS, NPM1 has been shown to serve as the molecular chaperone for GADD45α in controlling its subcellular distribution. Most importantly, disruption of NPM1–GADD45α complex has been found to impair the cell cycle arrest and apoptotic functions of GADD45α (Gao et al. 2005).

NPM1 is one of the key elements in the down-regulation of nucleolar function for cellular apoptosis, as exemplified by the death of human promyelocytic leukemia HL-60 cells induced by sodium butyrate (BuONa; a cell growth inhibitor) and vanadate (a tyrosine phosphatase inhibitor) (Liu and Yung 1998). NPM1 is decreased via down-regulated transcriptional process during the BuONa/vanadate-induced apoptosis. As no decrease in NPM1 mRNA and the telomerase activities is observed during the growth arrest by serum-starvation, the decrease in NPM1 mRNA expression and telomerase activity in HL-60 cells subsequent to BuONa/vanadate treatment is suggested to be attributed to cellular apoptosis rather than the growth arrest induced by BuONa/vanadate. The anti-apoptotic role of NPM1 is further substantiated by the data that BuONa-induced apoptosis and inhibition of telomerase activity are significantly potentiated after NPM1 antisense oligomer treatment (Liu and Yung 1998).

It is intriguing that NPM1 might also serve as one of the key elements in the down-regulation of nucleolar function for cellular differentiation. The regulatory role of NPM1 in cellular differentiation has been demonstrated in the granulocytic differentiation of HL-60 cells induced by retinoic acid (Hsu and Yung 1998). NPM1 is reduced via transcriptionally mediated down-regulation during the retinoic acid-induced differentiation. Conversely, there is no decline of NPM1 mRNA during the growth arrest as induced by serum-starvation. These findings suggest that the decrease in NPM1 mRNA expression in HL-60 cells subsequent to retinoic acid treatment is attributed to cellular differentiation rather than the growth arrest induced by retinoic acid. The retinoic acid-induced cellular differentiation is shown to be potentiated by NPM1 antisense oligomer treatment (Hsu and Yung 1998).

10.4.5 Modulation of Tumor Suppressors by NPM1

NPM1 is also involved in the regulation of activity and stability of some key tumor suppressors such as ARF and p53. NPM1 is associated with the stabilization of ARF by retarding the turnover of ARF and this stabilization is essential to maintain the biological function of ARF (Kuo et al. 2004, 2008). The protective effect of

NPM1 on ARF turnover involves both proteasome-dependent and -independent degradation. While ARF is known to suppress cell proliferation through both p53-dependent and -independent pathways (Bertwistle et al. 2004; Brady et al. 2004; Itahana et al. 2003; Korgaonkar et al. 2005), the deficiency of NPM1 has been demonstrated to result in acceleration of tumorigenesis and this is probably attributed to the destabilization of ARF (Sherr 2006). Taken together, NPM1 can possibly work with ARF in mediating the response to oncogenic stimulus. ARF is able to suppress cell proliferation by inhibiting the biogenesis of ribosome through the retardation of the production of rRNA (Sugimoto et al. 2003). Thus, the interaction of NPM1 and ARF in nucleolus is another way of controlling the cell proliferative activities.

Tumor suppressor p53 is another protein that is proposed to be modulated by NPM1, and is an important protein responsible for the prevention of cell growth and cell division when genomic stability is not achieved or DNA integrity is severely damaged (Levine 1997). A putative link has been established between the integrity of nucleolus and p53 stability, in which the nucleolus is believed to play a role in sensing the abundance of p53 in proliferating cells (Rubbi and Milner 2003). Indeed, NPM1 has been shown to promote p53 stability when undergoing nucleoplasmic relocalization (Horn and Vousden 2004). It has been demonstrated that NPM1 in the nucleolus can increase the stability of p53 by suppressing the physical binding interaction between MDM2 and p53 in response to ultraviolet irradiation (Kurki et al. 2004a, b). In response to cellular stress stimulus, disruption of nucleolar integrity induces the translocation of nucleolar NPM1 between subcellular compartments and the relocalized NPM1 can participate in the corresponding reaction mediated by p53 (Rubbi and Milner 2003; Horn and Vousden 2004).

10.4.6 Role of NPM1 in the Maintenance of Genomic Stability

NPM1 is implicated in the maintenance of genomic stability through participation in DNA repair process and control of cellular ploidy. In response to DNA double-strand breaks, NPM1 has been shown to act as a chromatin-binding factor (Lee et al. 2005). The early response of NPM1 to DNA damage involves rapid transcriptional up-regulation of NPM1 following ultraviolet irradiation (Wu and Yung 2002; Wu et al. 2002a). Increased DNA repair has been shown to be associated with elevated expression of NPM1 (Wu et al. 2002b). NPM1 also works to maintain the genomic stability during the cell cycle through the regulation of centrosome duplication, which is associated with the proposed physiological function of NPM1 in the regulation of cell cycle.

10.4.7 Regulation of the Cell Cycle by NPM1

The nucleolus undergoes reversible disassembly during mitosis and NPM1 is observed to translocate from nucleolar remnants to the cytoplasm (Hernandez-Verdun and Gautier 1994). NPM1 is found to translocate to the chromosome periphery and the cytoplasmic entities called nucleolus-derived foci (NDF) (Dundr et al. 2000). It then redistributes to the poles of the mitotic spindle to interact with a nuclear matrix protein called NuMA to control the formation of centrosomes in prometaphase and mitotic poles in metaphase (Compton and Cleveland 1994; Zatsepina et al. 1999). The consistent observation that NPM1 is present at the mitotic spindle poles suggests the protective role of NPM1 in preventing hyper-amplification of centrosome to ensure successful progression through G2-M phases (Tokuyama et al. 2001; Zatsepina et al. 1999; Zhang et al. 2004). Indeed, NPM1 is not classified as a centrosomal protein but it has been proposed to be involved in the duplication of centrosomes. This is supported by the data that NPM1 is associated specifically with the CDK2-cyclin E-mediated phosphorylation, which facilitates centrosome duplication (Okuda 2002; Tokuyama et al. 2001; Andersen et al. 2003).

Modification of the phosphorylation state of NPM1 by various protein kinases during cell cycle has been documented (Jiang et al. 2000). For instance, NPM1 has been identified as a substrate of cyclin-dependent kinase (CDK) 2-cyclin E complex in the regulatory process of centrosome duplication (Okuda et al. 2000). NPM1 has also been reported to be phosphorylated by cdc2 kinase during mitosis (Peter et al. 1990) and by nuclear casein kinase 2 during interphase (Chan et al. 1990). NPM1 has been suggested to be the candidate substrate for BRCA1-associated RING domain 1 (BRCA1-BARD1) ubiquitin ligase and the complex of BRCA1-BARD1-NPM1 has been shown to localize at centrosomes during mitosis (Sato et al. 2004). It is also proposed that the ubiquitylational interaction of NPM1 with BRCA1-BARD1 might be an important process in maintaining the integrity of spindle poles and genomic integrity (Grisendi et al. 2006).

10.4.8 Physiological Functions of NPM2 and NPM3

NPM2 can bind to histone proteins and is thus proposed to mediate the assembly of nucleosomes from histones and DNA (Earnshaw et al. 1980; Laskey et al. 1978). NPM2 has also been found to be involved in facilitating the postfertilization decondensation and remodeling of paternal chromosome by its binding activities with sperm nuclear basic proteins (Philpott and Leno 1992).

NPM3 is involved in the biogenesis of ribosomal RNA by interacting with NPM1 (Huang et al. 2005). It has been demonstrated that overexpression of NPM3 decreased the rates of pre-rRNA synthesis and processing, but overexpression of a NPM3 mutant that did not interact with NPM1 did not change the pre-rRNA synthesis and processing (Huang et al. 2005). Moreover, the expression level of NPM3 has been shown to be correlated with the process of decondensation of paternal chromosome after fertilization (McLay and Clarke 2003). Intriguingly, NPM3 has also been found to be associated with the histone tail peptides and serves as a histone-binding protein in mouse embryonic stem cells (Motoi et al. 2008). Thus, NPM3 is believed to be a chromatin-remodeling protein responsible for the unique chromatin structure and replicative capacity of embryonic stem cells. Recently, NPM3 was shown to interact with all the individual core histones and was able to enhance transcription via the modulation of the histone chaperone activities of its interacting partner NPM1 in vitro (Gadad et al. 2010).

10.4.9 Posttranslational Modification of NPM Proteins

Some physiological functions of NPM1 are regulated through posttranslational modification mechanisms including phosphorylation, dephosphorylation, acetylation, poly(ADP-ribosyl)ation, ubiquitination, and sumoylation (Table 10.2). However, the complete profile of posttranslational modifications of NPM1 remains to be fully elucidated.

For phosphorylation of NPM1, several kinases have been identified. Phosphorylation of NPM1 by cyclin E/cdk2 during G1 phase has been documented and this may be related to the initiation of centrosome duplication by dissociating NPM1 from centriole (Tarapore et al. 2006; Tokuyama et al. 2001). The RNA-binding activity of NPM1 is diminished after cdc2-mediated phosphorylation of Thr199 of NPM1 during mitosis, and this is suggested to link to the disassembly of nucleolus by disrupting the RNA-protein binding interaction of NPM1 (Hisaoka et al. 2010; Okuwaki et al. 2002). Phosphorylation of Thr199 has also been implicated in inhibiting GCN5-mediated histone acetylation (Zou et al. 2008). During mitosis, NPM1 is found to be phosphorylated by Polo-like kinase 1 (Plk1), which might be an important event in mediating mitosis (Zhang et al. 2004). During interphase, NPM1 has also been reported to be phosphorylated by casein kinase 2 (CK2) and this is thought to have a role in regulating the nucleolar structure by modulating the dynamic localization of NPM1 between nucleolus and nucleoplasm (Szebeni et al. 2003; Negi and Olson 2006). Dephosphorylation can be another mechanism in regulating the functions of NPM1. A serine/threonine protein phosphatase called PP1β has been shown to dephosphorylate NPM1 in response to DNA damage during ultraviolet irradiation and this process is suggested to facilitate the DNA repair process (Lin et al. 2010).

Acetylation of NPM1 increases its binding affinity to histone and this has been suggested to be involved in the NPM1-mediated regulation of chromatin transcription (Swaminathan et al. 2005). While histone acetyltransferase p300 activates transcription by acetylating the histones and "loosening" the tightly packed chromatin structure, the p300 enzyme also leads to acetylation of NPM1. Such acetylation potentiates the activating effect of p300 on transcription activation by ~fourfold. In vitro experiments have mapped the acetylation sites of NPM1 mostly to the C-terminal region (Lys212, Lys229, Lys230, Lys248 or Lys250, Lys257 and Lys292), which need to be confirmed by further in vivo experiments.

Factors involved in poly(ADP-ribosyl)ation such as poly(ADP-ribose) polymerase 1 (PARP1) and PARP2 have been shown to have association with NPM1 (Meder et al. 2005). Poly(ADP-ribosyl)ation of NPM1 might contribute to the formation of chromatin because PARP1 can serve as a molecular linker that regulates the structure of chromatin (Kim et al. 2004). NPM1 has been shown to be the ubiquitination substrate of BRCA1-BARD1 ubiquitin ligase, but intriguingly the product is not targeted for proteasome-dependent protein degradation unless destabilized by tumor suppressor ARF (Sato et al. 2004; Itahana et al. 2003). Sumoylation is shown to be another mechanism that can modulate the activities of NPM1. In particular, NPM1 has been demonstrated to be sumoylated by ARF and this can increase the stability and modulate the subcellular localization of NPM1 (Liu et al. 2007b; Tago et al. 2005).

The biological activities of NPM2 are also regulated via phosphorylation. The activity of NPM2 that binds and removes sperm basic proteins, and replaces them with histones has been found to depend on the massive hyperphosphorylation of NPM2 that occurs when oocytes mature into eggs (Leno et al. 1996). The hyperphosphorylation of NPM2 is proposed to modulate the rapid changes in chromatin structure that accompany early development in *Xenopus*. The function of NPM2 to exchange the H2A-H2B heterodimers for sperm-specific proteins is shown to be mediated by adding 14–20 phosphates to each NPM2 monomer (Cotten et al. 1986).

10.5 Alteration of NPM1 in Human Cancers

Overexpression of NPM1 in general promotes cell growth and proliferation, particularly through enhancing ribosome biogenesis (see Sect. 10.4.2) (Grisendi et al. 2006). Another main effect of NPM1 overexpression is the inhibition of apoptosis (see Sect. 10.4.4) via several different pathways (Grisendi et al. 2006; Ye 2005). As such, NPM1 has been implicated in tumorigenesis. Indeed, NPM1 is overexpressed in many tumors of different origins (Table 10.2): gastric (Tanaka et al. 1992), colon (Nozawa et al. 1996), liver (Yun et al. 2007), breast (Skaar et al. 1998), ovarian (Shields et al. 1997), prostate (Léotoing et al. 2008; Subong et al. 1999), bladder (Tsui et al. 2004), thyroid (Pianta et al. 2010), brain (Gimenez et al. 2010), and multiple myeloma (Weinhold et al. 2010). In particular, NPM1 overexpression may be correlated with clinical features in some cases. Overexpression of NPM1 in hepatocellular carcinoma was found to be correlated with clinical prognostic parameters such as serum alpha fetal protein level, tumor pathological grading, and liver cirrhosis (Yun et al. 2007) – suggesting the potential of NPM1 overexpression as a marker of hepatocellular carcinoma. NPM1 overexpression was associated with recurrence and progression of bladder cancer (Tsui et al. 2004). Overall, the observation that NPM1 overexpression promotes tumor development tends to suggest its role as a proto-oncogene.

Genetic alteration of the *NPM1* gene was not found in common solid cancers including lung, hepatocellular, breast, colorectal, and gastric carcinomas (Jeong et al. 2007). However, the *NPM1* gene is a common target for genetic alteration in hematological malignancies (lymphomas and leukemias) (Naoe et al. 2006; Falini et al. 2007b; Rau and Brown 2009). The genetic alterations include frameshift mutations, translocations, and deletions (Table 10.2). The main focus of this section is on the *NPM1* gene mutations in humans.

10.5.1 Acute Myeloid Leukemia Carrying Cytoplasmic NPM (NPMc+ AML)

The breakthrough in this field came with the first report by Falini et al. (2005) that aberrant cytoplasmic localization (instead of nucleolar) of the NPM1 protein (NPMc+) in leukemic blast cells was due to frameshift mutations in exon 12 of the *NPM1* gene in patients with AML carrying a normal karyotype. NPMc+ AML accounts for ~30% of all cases of adult AML, or ~60% of all cases of adult AML *with* normal karyotype (Falini et al. 2007b; Rau and Brown 2009). The significance of the finding is that NPMc+ due to *NPM1* frameshift mutations is the single most common somatic mutation in adult AML. Of note is the less frequent occurrence (~7%) of NPMc+ in AML in children (Falini et al. 2007b; Rau and Brown 2009). This difference may reflect the difference in molecular pathogenesis of AML carrying normal karyotype in adults and children.

10.5.1.1 The *NPM1* Mutations Producing NPMc+

A recent compilation documents over 50 reported somatic *NPM1* mutations that include insertions, insertions and deletions (indels), base substitutions, and their combinations (Rau and Brown 2009). About 50% of the mutations are 4-base insertions between the second and the third base of the Trp288 codon (TG^G). About 20% of the mutations are insertions of 4–14 bases between Gln289 and Trp290 codons (CAG^TGG). Another 20% are insertions of 8–12 bases between the first and the second base of Trp290 codon (T^GG) followed by deletions of 2–5 bases. All these mutations produce a shift in the reading frame of the transcript from the point of insertion or deletion. Many more mutations are expected to be discovered. However, the most common mutation is a duplication (a type of insertion) of a 4-base sequence TCTG at positions 956–959 of the reference sequence (NM_002520, Table 10.1), and accounts for 70–80% of adult NPMc+ AML. This was designated as mutation A by Falini et al. (2005). About 15% of adult NPMc+ AML cases are due to mutations B (CATG insertion) or D (CCTG insertion) at the same position. The remaining mutations are all rare. Interestingly, a genome-wide computational analysis indicated that the generation of a new NES motif (see below) by a duplication of the TCTG sequence was unique to the *NPM1* mutation – a genetic event specific to AML (Liso et al. 2008).

Despite such heterogeneity at the DNA sequence level, the mutations produce two alterations at the C-terminus of the mutated NPM1, *both* of which are crucial to the aberrant export of the mutated protein from the nucleolus to the cytoplasm (Falini et al. 2006). The first critical alteration is the *loss* of Trp288 and Trp290 residues or just Trp290 alone, which are essential to the nucleolar localization of the wildtype NPM1 (Nishimura et al. 2002). Loss of these tryptophan residues disrupts the triple helix structure of the NPM1 C-terminal domain (Fig. 10.3; see Sect. 10.3.4) and thus greatly reduces the NPM1 localization in the nucleolus (Grummitt et al. 2008).

The second critical alteration is the *generation* of a new leucine-rich NES motif in the new C-terminus (Nakagawa et al. 2005), in addition to the original two NES motifs (residues 42–47 and 94–102) in the N-terminal core region (see Sect. 10.3.1).

Insertion in between the bases of the Trp288 codon results in the loss of both tryptophan residues while insertion (with or without concomitant deletion) at positions after the Trp288 codon removes the Trp290 residue only. There is a strong correlation between the loss of one or two tryptophan residues and the type of new NES motif generated at the C-terminus. Loss of both tryptophan residues is always found with the common NES motif **Leu**-xxx-**Val**-xx-**Val**-x-**Leu**, where x is any amino acid. On the contrary, loss of Trp290 alone is always found with the much less frequent *variant* NES motif **Leu**-**Trp** xx-**X**-xx-**Val**-x-**Leu**, where X replaces the Val residue and can be leucine, methionine, phenylalanine, or cysteine (Rau and Brown 2009). Experimentally, the *variant* NES motifs were found to provide a much *stronger* force than the common NES motif in driving the Trp288-containing mutated NPM1 from the nucleolus to the cytoplasm (Bolli et al. 2007). Tested with the same experimental system, the physiological N-terminal NES motifs were found to be weak in transporting the wildtype NPM1 protein to the cytoplasm, and this explains the dominant localization of the wildtype NPM1 protein in the nucleolus. The artificial combination of a common weak C-terminal NES motif with a Trp288-containing NPM1 mutant localized the mutated protein mainly in the nucleoplasm and nucleolus with much less export to the cytoplasm. Intriguingly, a weak C-terminal NES motif together the retention of the Trp288 residue has never been detected in any primary AML samples. Therefore, this strongly suggests that cytoplasmic mislocalization of the mutated NPM1 protein is critical to the development of AML (Bolli et al. 2007).

The critical role of NPMc+ in the development of AML carrying normal karyotype is further supported by the report of NPMc+ generated by rare mutations found outside exon 12, namely, a mutation affecting the splicing donor site of exon 9 (Mariano et al. 2006) and two different insertions in exon 11 (Albiero et al. 2007; Pitiot et al. 2007). In all three cases, the mutations produce truncated mutated NPM1 proteins and hence abolish both Trp288 and Trp290 residues, and simultaneously create new functional NES motifs at the new respective C-termini (Mariano et al. 2006; Albiero et al. 2007; Falini et al. 2007a). In other words, these rare mutants utilize the same mechanism of transporting the mutated NPM1 to the cytoplasm as those mutations occurring in exon 12.

In NPMc+ AML, the leukemic blast cells are heterozygous with one mutated *NPM1* allele and one wildtype allele (Falini et al. 2007b). While the mutated NPM1 protein is strictly localized to the cytoplasm, the wildtype NPM1 protein can be detected in both nucleoplasm and cytoplasm (Falini et al. 2006; Bolli et al. 2009). All mutated NPM1 protein retains the N-terminal oligomerizaton domain (see Sect. 10.3.1), and hence can form heterodimers with wildtype NPM1 protein. As such, mutated NPM1 protein can recruit wildtype NPM1 protein from nucleolus to nucleoplasm and cytoplasm. Indeed, in vitro transfection studies demonstrated the coimmunoprecipitation of mutated and wildtype NPM1 proteins.

10.5.1.2 Putative Mechanisms Leading to NPMc+ AML

Cytoplasmic localization of mutated NPM1 protein is believed to play a critical role in leukemogenesis. However, how somatic *NPM1* frameshift mutations lead to NPMc+ AML remains elusive. The putative underlying mechanisms can be explored from two perspectives (Falini et al. 2007c, 2011). First, the remaining single copy of wildtype *NPM1* allele produces wildtype NPM1 protein, which is less than that in normal counterparts and is also dislocated to the nucleoplasm and cytoplasm as a result of forming heterodimers with mutated NPM1 protein. Second, the mutated *NPM1* allele produces mutated NPM1 protein, which is dislocated to the cytoplasm by its very nature, may recruit and hence dislocate other interacting nuclear proteins to the cytoplasm, and may also interact with other new partners in the cytoplasm.

Mutant mice with only one functional *Npm1* gene (*Npm1*[+/−]) showed greater instability in their genome and developed a hematological syndrome analogous the myelodysplastic syndrome in humans (Grisendi et al. 2005). When compared to wildtype mice (*Npm1*[+/+]), *Npm1*[+/−] mice showed a much higher frequency of developing malignancies including hematological malignancies, particularly myeloid malignancies (Sportoletti et al. 2008). In addition, chromosomal abnormalities were also consistently found in these mice. This shows that *Npm1* is a haploinsufficient tumor suppressor in hemopoiesis. Given that NPMc+ AML is mainly found in patients with normal karyotype, factors other than haploinsufficiency must also contribute to the development of AML.

Removal of the critical C-terminal tryptophan residues and generation of a new C-terminal NES motif dictate the localization of the mutated NPM1 protein in the cytoplasm, instead of in the nucleolus (Falini et al. 2006) (see Sect. 10.5.1.1). Intriguingly, mutated NPM1 protein may still be able to interact with other nuclear proteins that interact with the wildtype NPM1 protein in normal cells, and dislocate them to the cytoplasm. Indeed, at least four such nucleolar/nuclear interacting partners have been found to interact with the mutated NPM1 protein, be dislocated to the cytoplasm, and hence have their physiological functions abrogated: mouse p19Arf and human p14ARF (den Besten et al. 2005; Colombo et al. 2006; Bolli et al. 2009), hexamethylene bis-acetamide-inducible protein 1 (HEXIM1) (Gurumurthy et al. 2008), the F-box protein Fbw7γ (Bonetti et al. 2008), and Miz1 (Wanzel et al. 2008). Attenuation of the functions of these interacting proteins are suggested to contribute to the oncogenic effect of NPMc+, as briefly explained below one by one. First, ARF is a well-known tumor suppressor and is stabilized by wildtype NPM1 protein in the nucleolus (Gallagher et al. 2006; Sherr 2006). In vitro experiments have shown that mutated NPM1 protein can interact directly with ARF and shuttle it to the cytoplasm, but cannot protect it from degradation (den Besten et al. 2005; Colombo et al. 2006). Second, HEXIM1 is an inhibitor of the positive transcription elongation factor b (P-TEFb), which is itself an important transcriptional regulator of the enzyme RNA polymerase II (Dey et al. 2007). Wildtype NPM1 negatively regulates HEXIM1 via proteasome-mediated degradation while mutated NPM1 associates with and shuttles HEXIM1 to the cytoplasm and hence promotes P-TEFb-mediated transcription in the nucleus (Gurumurthy et al. 2008).

10 Nucleophosmin/NPM/B23 and the Nucleoplasmin Family of Proteins

Third, the F-box protein Fbw7γ is a component of the E3 ligase complex, which ubiquitinates and degrades the oncoprotein c-Myc via the proteasome pathway (Welcker and Clurman 2008). Wildtype NPM1 protein localizes and stabilizes Fbw7γ in the nucleolus, and hence regulates the turnover of c-Myc (Bonetti et al. 2008). Mutated NPM1 protein interacts with Fbw7γ and dislocates it to the cytoplasm, where it is degraded. As a result, c-Myc is stabilized – a situation that reflects the oncogenic potential of NPMc+ (Bonetti et al. 2008). Fourth, Miz1 is a Myc-associated zinc-finger protein and, when bound to Myc, enables Myc to suppress transcription of the genes encoding the cell cycle inhibitors p15Ink4b and p21Cip1 (Adhikary and Eilers 2005). Wildtype NPM1 localizes Miz1 to the nucleolus and is an essential coactivator of Miz1. However, mutated NPM1 protein re-directs Miz1 to the cytoplasm and exhibits dominant-negative effect on Miz1 (Wanzel et al. 2008). Thus, disruption of Miz1 function may contribute to the transforming potential of NPMc+.

NPMc+ may acquire new function in its new environment – the cytoplasm. This is indeed the case. Mutated NMP1 directly interacts with the active cell-death proteases caspase 6 and caspase 8, and inhibits their activities, and thereby protects the cells from apoptosis (Leong et al. 2010). In addition, mutated NPM1 also suppresses myeloid differentiation mediated by caspase 6 and caspase 8. This new data provide the first evidence for the myeloid-restricted leukemogenic property of NPMc+.

In transgenic zebrafish embryos with forced ubiquitous expression of human NPMc+, primitive early myeloid cells expand in numbers (Bolli et al. 2010). There are also increased numbers of definitive erythromyeloid progenitors in the posterior blood island and hematopoietic stem cells in the aorta ventral wall. In transgenic mice expressing human NPMc+ under the influence of a myeloid-specific promoter, expansion of myeloid cells is noted in bone marrow and spleen (Cheng et al. 2010). However, both transgenic models show no evidence of AML. This suggests that the current animal models are not adequate and may not mimic the condition in the human NPMc+ AML.

10.5.1.3 Cell of Origin in NPMc+ AML

Wildtype NPM1 is predominantly located in the nucleolus while the mutated NPM1 is aberrantly localized in the cytoplasm (NPMc+). Because of their uniqueness, the mutation and the cytoplasmic localization of the mutated protein can be used as clonal markers to study the cell lineage involved in NPMc+ AML. Clonal NPM1 mutations are found in myeloid, monocytic, erythroid, and megakaryocytic cells, but not in fibroblasts and endothelial cells (Pasqualucci et al. 2006). In addition, two or more myeloid hemopoietic cell lineages are affected in about 62% of NPMc+ AML cases while the remaining 38% involve only one myeloid cell lineage (Pasqualucci et al. 2006). On the other hand, B and T lymphoid cells are not part of the mutated clones in NPMc+ AML (Martelli et al. 2008). This indicates that NPMc+ AML arise from a common myeloid or an earlier progenitor that is incapable of differentiating into lymphoid lineages.

The leukemic blast cells are negative for CD34 (i.e., <10% CD34+ cells) in over 90% of NPMc+ AML cases (Falini et al. 2005, 2011). The surface marker CD34 is typically present on hematopoietic stem cells. This raises the question whether the *NPM1* mutation arises in a CD34– multipotent hemotapoietic progenitor (Engelhardt et al. 2002) or whether there exists a small pool of CD34+/CD38– *NPM1*-mutated progenitor. CD34+/CD38– cells usually contain the leukemia stem cells that are capable of propagating and maintaining the leukemia phenotype in immuocompromised mice (Estrov 2010). Indeed, CD34+ cells from NPMc+ AML carry the *NPM1* mutation and, when transplanted into immunocompromised mice, generate a leukemia phenotype that is the same as the original patient's disease in all aspects (Martelli et al. 2010). On the other hand, the evidence for the engraftment capability of CD34– cells from NPMc+ AML is less consistent (Martelli et al. 2010; Taussig et al. 2010). These findings may reflect the heterogeneity of leukemia stem cells in NPMc+ AML.

Leukemic blast cells from NPMc+ AML show characteristic gene expression signature and microRNA signature. In general, gene expression profiling shows down-regulation of *CD34* and up-regulation of several members of the homeodomain-containing family of transcription factors, which include *HOX* genes and *TALE* genes (Alcalay et al. 2005; Andreeff et al. 2008; Becker et al. 2010; Mullighan et al. 2007; Verhaak et al. 2005). Intriguingly, *HOX* and *TALE* genes are known to be important in the maintenance of stem cells – a finding supporting that the cell of origin for NPMc+ AML is a multipotent hematopoietic progenitor. On the other hand, unique microRNA signature includes up-regulation of miR-10a, miR-10b, miR-196a, miR-196b, several members of let-7, and miR-29 families (Debernardi et al. 2007; Garzon et al. 2008; Jongen-Lavrencic et al. 2008), and down-regulation of miR-204 and miR-128a (Garzon et al. 2008). It is of interest to note that miR-10a, -10b, 196a, and -196b are located within the genomic cluster of the *HOX* genes (Jongen-Lavrencic et al. 2008). Moreover, miR-204 has been shown to down-regulate *HOXA10* and *MEISI* genes (Garzon et al. 2008), a finding linking the down-regulation of miR-204 to the up-regulation of *HOXA10* and *MEIS1* in NPMc+ AML.

10.5.1.4 Distinctive Features of NPMc+ AML

Since the first report of NPMc+ AML in 2005, many studies have been done on this group of acute leukemias. It was listed as a new provisional entity (AML with mutated *NPM1*) under the category of "AML with recurrent genetic abnormalities" in the 2008 *World Health Organization (WHO) Classification of Tumours of Haematopoietic and Lymphoid Tissue* (Swerdlow et al. 2008). As a group, NPMc+ AML has many distinctive features. *NPM1* mutations are unique to AML, usually de novo AML, (Falini et al. 2005, 2007c; Liso et al. 2008) and mutually exclusive of other "AML with recurrent genetic abnormalities" listed in the 2008 WHO Classification (Falini et al. 2005, 2008b). They are usually detected in all cells of the leukemic population and are stable over the course of the disease (Chou et al. 2006; Falini et al. 2008a).

As such, monitoring of minimal residual disease can easily be achieved by detection of the *NPM1* mutations with real-time quantitative polymerase chain reaction assay (Wertheim and Bagg 2008). *NPM1* mutations appear to precede other associated mutations like fms-like tyrosine kinase internal tandem duplication (*FLT3*-ITD), which is found in 40% of NPMc+ AML (Thiede et al. 2006; Gale et al. 2008). Moreover, NPMc+ AML cells have distinct gene and microRNA expression profiles (see Sect. 10.5.1.3). Taken together, these features suggest that the *NPM1* mutation is a founder genetic alteration in NPMc+ AML (Falini et al. 2011).

NPMc+ AML is more frequent in adults (~30% of cases) than in children (~7% of cases) (Falini et al. 2005; Cazzaniga et al. 2005; Brown et al. 2007). Interestingly, no NPM1 mutation has been detected in AML patients younger than 3 years old (Brown et al. 2007). The type of NPM1 mutations is also different in adult and childhood AML (Thiede et al. 2007). NPMc+ AML is more common in AML patients with normal karyotype (~85% of cases), and frequently involves multiple lineages (Falini et al. 2005, 2007c). It shows good response to induction therapy, and the prognosis is relatively good in the absence of *FTL3*-ITD mutations (Falini et al. 2005, 2007c, 2011).

10.5.2 Lymphomas and Leukemias Carrying NPM1 Gene Translocations

The *NPM1* gene at chromosome 5q35 is translocated in anaplastic large-cell lymphoma (ALCL) and in rare variants of AML. Translocation produces an oncogenic fusion protein and a reduced level of the wildtype NPM1 protein encoded by the remaining copy of the functional allele (heterozygosity). The fusion protein is made up of the N-terminus of the NPM1 protein and the C-terminus of the partner protein encoded by the other gene involved in the translocation. The role of the NPM1 moiety in these fusion proteins has not been fully elucidated although it may just serve to promote heterodimer formation, and hence shuttle the fusion protein to the nucleus.

ALCL is a T-cell lymphoma characterized by CD30 expression (Falini 2001; Falini et al. 2007b). It accounts for ~3% of adult non-Hodgkin's lymphoma and 10–30% of childhood lymphoma. About 60% of ALCL cases express the tyrosine kinase gene *ALK* (anaplastic lymphoma receptor tyrosine kinase) and are known as ALK+ ALCL. In general, ALK+ ALCL show good response to induction therapy and has good prognosis. The majority (~85%) of such cases carry the t(2;5)(p23;q35) chromosome translocation, which joins the *NPM1* gene on 5q35 with the *ALK* gene on 2p23 to produce the chimeric gene *NPM1-ALK* and hence the fusion protein. The remaining 15% are heterogeneous at the molecular level because the translocations involve other chromosomal partners. The NPM1-ALK fusion protein consists of the N-terminal part of the NPM1 protein (the first 116 amino acids; carrying the oligomerization domain) and the entire cytoplasmic domain of the ALK protein (the last 563 amino acids; carrying the catalytic domain). Through the oligomerization domain, the NPM1-ALK fusion proteins form homodimers with each other, and heterodimers with the wildtype NPM1 protein. Because of the NPM1 promoter, the fusion protein is ectopically expressed in lymphoid cells and the constitutive activation of the tyrosine kinase domain is thought to contribute to the tumor formation. The fusion protein is expressed in the cytoplasm as expected but is also unexpectedly localized in the nucleus because of the shuttling of the heterodimers composed of the fusion protein and the wildtype NPM1, which still possesses the nucleus localization signal and hence imports the heterodimers into the nucleus.

Acute promyelocytic leukemia (APL) is characterized by a maturational block at the promyelocytic stage and classically carries the t(15;17) translocation, which generates the fusion protein PML-RARA (Zelent et al. 2001). Chromosomal translocation t(5;17)(q35;q12) fuses the *NPM1* gene at 5q35 to the retinoic acid receptor alpha (*RARA*) gene at 17q12, and produces the fusion protein NPM1-RARA (Falini et al. 2007b). This translocation is extremely rare and has so far been reported in a few children with APL. Leukemic cells express NPM1-RARA fusion protein, and its reciprocal, the wildtype NPM1 and the wildtype RARA. The fusion protein affects the expression of retinoid-responsive genes, disrupts the retinoic acid signaling pathway, and arrests myeloid differentiation at the promyelocytic stage. Like other APL, APL with t(5;17) shows good response to differentiation therapy with all-trans retinoic acid.

The chromosomal translocation t(3;5)(q25;q35) is found in myelodysplastic syndrome and in <1% of AML (Raimondi et al. 1989; Falini et al. 2007b). It fuses the *NPM1* gene at 5q35 to the myelodysplasia/myeloid leukemia factor 1 (*MLF1*) gene at 3q25. The fusion protein is composed of the N-terminal portion of the NPM1 protein and almost the entire MLF1 protein (only without the first 16 amino acids). Since wildtype MLF1 is not expressed in normal hematopoietic tissues, it is speculated that the NPM1-MLF1 fusion protein promotes malignant transformation via ectopic expression in hematopoietic cells (Hitzler et al. 1999). Like the NPM1-ALK fusion protein, the NPM1-MLF1 fusion protein is expectedly expressed in the cytoplasm, and unexpectedly in the nucleus.

10.6   Interactions Between Viruses and NPM1

Viruses are obligate intracellular parasites with small-sized genomes (in the form of either DNA or RNA) and very limited coding capacities, and hence their replication and metabolic activities rely on the host cellular machineries (Flint 2000). Most DNA viruses, negative-sense RNA viruses with segmented genomes, and retroviruses replicate in the nucleus and frequently interact with nuclear or nucleolar proteins. In contrast, most positive-sense RNA viruses replicate in the cytoplasm and are not much dependent on the nucleus. However, there are growing evidences that this group of viruses also relies on the nucleus because the proteins involved in viral replication and assembly are localized in the nucleolus (Mai et al. 2006; Perkins et al. 1989; Tamini et al. 2005; Tsuda et al. 2006). Interaction between viruses and nucleus requires trafficking of viral products in and out of the nucleus.

10 Nucleophosmin/NPM/B23 and the Nucleoplasmin Family of Proteins

Apart from the diverse functions discussed in Sect. 10.4, NPM1 is also known to affect viral replication and assembly during infection (Hiscox 2007). More importantly, such interaction often results in dislocation and loss of normal functions of nucleolar proteins. This consequently leads to disruption of normal host cell functions. Most of the published studies concerning virus–NPM1 interaction focus on NPM1.1 (or B23.1), while the role of NPM1.3 (or B23.2) in viral activities is not well-understood. This section reviews some examples of virus–NPM1 interaction and discusses the influence of such interaction on the viruses and the host.

10.6.1 Influence of Virus–NPM1 Interaction on Virus Replication Cycle

The role of NPM1 in transporting human immunodeficiency virus type 1 (HIV-1) proteins to nucleolus is well-documented. HIV is the causative agent of acquired immunodeficiency syndrome (AIDS), and possesses two identical copies of positive-sense RNA genomes. Replication of HIV-1 is a complicated process and involves both nucleus and cytoplasm. During replication, the RNA genome is being reverse-transcribed into DNA in the cytoplasm and transported to the nucleus, where the DNA is transcribed into mRNA and the latter then returns to the cytoplasm for subsequent translation. The HIV-1 regulatory protein Rev is involved in the export of partially spliced or unspliced mRNA from the nucleus (Perkins et al. 1989). As Rev is localized in the nucleolus, it was speculated that certain host factors would be involved in the transportation of Rev in and out of the nucleolus. Using affinity chromatography, Fankhauser et al. (1991) confirmed the participation of NPM1 in the transit of Rev between cytoplasm and nucleus, which allowed for further rounds of export of HIV-1 mRNA. Apart from Rev, NPM1 also localizes the HIV Tat protein into nucleolus (Li 1997), where Tat will recruit cellular cofactor (positive transcription elongation factor b or P-TEFb) and transactivate proviral DNA transcription. This process is critical for HIV replication.

Japanese encephalitis virus (JEV) can cause acute encephalitis in humans. It belongs to the same Flaviviridae family as hepatitis C virus (HCV). It is transmitted among mosquitoes and pigs, and transmission to humans may occur when the number of infected mosquito vectors increases to a very high level. The RNA genome encodes the envelope, structural proteins, as well as core and the non-structural proteins. Similar to HCV and other flaviviridae, JEV core protein is localized in the cytoplasm and nucleus (Bulich and Aaskov 1992; Tsuda et al. 2006). According to mutation and animal inoculation studies, localization of core protein in the nucleus is crucial to the replication of JEV (Mori et al. 2005). During JEV infection, amino acids Gly42 and Pro43 of the JEV core protein interact with the N-terminal region of NPM1, and this interaction results in transportation of the viral core protein into the nucleus (Tsuda et al. 2006). Both chaperone and RNA binding activities of NPM1 are important for JEV replication. Besides, dislocation of NPM1 from nucleus to cytoplasm has also been noted. Although the precise mechanism is not well-defined, it is possible that NPM1 is retained in the cytoplasm by JEV-induced cytoplasmic factors or NPM1 is released into the cytoplasm on disruption of nuclear organization (Tsuda et al. 2006).

HDV is a negative-sense RNA virus that requires the presence of hepatitis B virus (HBV) as a helper virus for replication. Patients with chronic HBV infection and superinfected with HDV can suffer from much more severe complications such as fulminant hepatitis, cirrhosis, and hepatocellular carcinoma. The virus possesses an RNA genome and HDVAg in two isoforms; the small form is involved in HDV RNA replication and the large form in virus assembly (Casey 2006). Replication of HDV occurs in the nucleolus, and NPM1 interacts with HDV antigens and modulates viral RNA replication (Huang et al. 2001; Li et al. 2006). During HDV infection, NPM1 is up-regulated and interacts with the small and the large form (to a lesser extent) of HDVAg. The interaction domains of NPM1–HDVAg are within the NLS of HDVAg and NPM1 acts as a shuttle protein for transportation of HDVAg into the nucleus. Apart from this, NPM1 also plays a role in HDV RNA replication. Huang et al. (2001) demonstrated that exogenous NPM1 had stimulatory effect on HDV RNA replication, while deleting the HDV binding site in NPM1 impaired this effect. Besides, they also observed colocalization of the small HDVAg with NPM1 and nucleolin in the nucleolus. As nucleolin can serve as a transcriptional factor (Yang et al. 1994), this may in turn confer a regulatory role for HDVAg on HDV RNA replication.

Adenovirus is a double-stranded DNA virus causing a wide range of human infections including respiratory, ocular, and gastrointestinal tract infections (Lenaerts et al. 2008). Replication of the viral genome relies on three early proteins, namely, the viral polymerase (Adpol); preterminal protein (pTP), which primes DNA synthesis (Liu et al. 2003); and DNA-binding protein (DBP), which initiates DNA replication (de Jong et al. 2003). Two NPM1 isoforms are involved in adenovirus DNA replication (Hindley et al. 2007). During viral genome replication, NPM1.1 and NPM1.3 interact differently with the viral early proteins pTP and DBP, and NPM1.3 is initially localized in the DBP/viral DNA-rich regions. Once the viral pTP expression increases, NPM1.1 is localized in the pTP-rich regions and interacts with pTP. This is followed by recruiting NPM1.3 into the pTP-rich regions and interaction with the pTP/NPM1.1/viral DNA complex. Apart from genome replication, NPM1 also plays a role in viral assembly of adenovirus. During the late stage of viral infection, the viral protein V interacts with NPM1, which acts as a chaperone by transferring the newly synthesized core protein to the viral DNA genome (Matthews and Russell 1998; Samad et al. 2007).

HBV belongs to the Hepadnaviridae family and is associated with cirrhosis and hepatocellular carcinoma (Ganem and Schneider 2001). It possesses a partial double-stranded DNA genome which encodes four viral proteins, namely, the core protein, surface protein, polymerase, and the X protein (Ganem and Schneider 2001). The HBV core protein consists of an assembly domain at the N-terminal and a viral replication regulatory domain at the C-terminal (Kang et al. 2006; Zlotnick et al. 1996). Ning and Shih (2004) have reported the colocalization of HBV core antigen with nucleolin and NPM1 in the nucleolus, and cells with such colocalization exhibited binucleated and apoptotic morphology. Recently, Lee et al. (2009) demonstrated the involvement of NPM1 in the assembly of HBV, in which the N-terminal of the HBV core protein bound to NPM1 during viral encapsidation. Their study also showed that amino acid residues 259–294 of NPM1 were essential for the interaction with HBV core protein.

NPM1 also interacts with Kaposi sarcoma herpes virus (KSHV) to regulate viral latency. KSHV is a DNA virus belonging to family γ-herpesviruses. The virus is capable of cell transformation and is associated with Kaposi's sarcoma and AIDS-related non-Hodgkin lymphoma (Renne et al. 1996; Zhong et al. 1996). KSHV remains in latency stage after infection but the virus can be reactivated by various intra- and extra-cellular factors, including cytokines, hypoxia, and chemical agents (Miller et al. 2007). In addition, interactions of the viral protein with host transcription factors and components of the host cellular signaling pathways may reactivate the virus. The KSHV latent protein v-cyclin and host cellular CDK6 kinase can phosphorylate NPM1 on threonine 199 (Sarek et al. 2010). Phosphorylation of NPM1 facilitates interaction of NPM1 with the latency-associated nuclear antigen, a repressor for viral lytic replication. Depletion of NPM1 causes KSHV reactivation; this demonstrated that NPM1 is a regulator of KSHV latency.

10.6.2 Influence of Virus–NPM1 Interaction on Host Cell Cycle

With an understanding of the interaction between NPM1 and the Rev protein of HIV-1, the clinical implication has also been studied. Miyazaki et al. (1996) reported that an overexpression of Rev altered the nucleolar architecture, and this correlated with the accumulation of NPM1. An elevated level of NPM1 may alter the cell cycle control as the nucleolar protein is involved in ribosome biogenesis (Okuda 2002). In fact, the T-lymphocytes from HIV-1-infected patients had changes in the nucleolar architecture and this correlated with a loss of cell cycle control (Galati et al. 2003).

HCV is associated with posttransfusion hepatitis, which may progress to cirrhosis and hepatocellular carcinoma (Allain 2000; Barazani et al. 2007). The positive-sense RNA genome of HCV encodes the viral envelope, core protein, and several other nonstructural proteins (Choo et al. 1991; Takamizawa et al. 1991). Among various HCV viral proteins, the core protein is the best studied. Instead of being a viral nucleocapsid, the core protein interacts with various host cellular factors and influences various host cell functions including apoptosis, signal transduction, and transcriptional regulation (Fischer et al. 2007; Koike 2007). One of the host cellular factors that interact with HCV is NPM1. The virus interacts with NPM1 through the NLSs located in the core protein amino acid 51–100 (Chen et al. 2003) and is transported into the nucleolus. During HCV infection, NPM1 together with YY1 and P300 forms complex with HCV (Mai et al. 2006). This relieves the suppression effect of YY1 on the NPM1 promoter, thereby up-regulating the expression of NPM1, which in turn activates RNA polymerase I transcription and results in higher rate of ribosome biogenesis. Overall, this promotes cell proliferation during viral infection, and consequently leads to cell transformation and hepatocellular carcinogenesis.

The SARS-coronavirus (SARS-CoV) is the pathogen responsible for severe acute respiratory syndrome (SARS), a human respiratory infection that first identified in Southern China in 2002 (Christian et al. 2004; Ksiazek et al. 2003). The 30-kb positive-sense RNA genome encodes the viral envelope, nucleocapsid, hemagglutinin, and membrane-associated proteins (Wang et al. 2003). The nucleocapsid protein is involved in viral assembly and also regulation of signal transduction. It is mainly localized in the cytoplasm of SARS-CoV-infected cells, but the protein is also present at low level in the nucleus (Tamini et al. 2005). The nucleocapsid protein of SARS-CoV is able to interact with various cellular proteins, including cyclophylin A (Luo et al. 2004), human ubiquitin-conjugating enzyme (Fan et al. 2006), and CDK-cyclin complex proteins (Surjit et al. 2006). NPM1 interacts with nucleocapsid protein of SARS-CoV and, during the interaction, the viral protein competitively inhibits the interaction of NPM1 with CDK2 kinase, and thereby inhibits the phosphorylation of NPM1 (Zeng et al. 2008). A decrease in phosphorylation inhibits the duplication of centromere, which leads to subsequent cell cycle arrest. In addition, interaction of the viral nucleocapsid protein with NPM1 may also cause defects in ribosome synthesis and results in suppression of gene expression, or leads to protein misfolding (Zeng et al. 2008).

Adenovirus mobilizes NPM1 from the nucleolus to nucleoplasm and cytoplasm for genome replication and viral assembly. This causes disruption of the nucleolus, which leads to inhibition of rRNA processing and transportation in the host cell (Castiglia and Flint 1983; Matthews 2001).

To date, much progress has been achieved in understanding the interaction between viruses and nucleolus. The study of viral interactions with the nucleolus enables a deeper understanding of the pathogenic mechanisms of the viral pathogens. The findings also provide new insights into various nucleolar activities and functions. More importantly, the knowledge on the biological pathways involved in virus and nucleolus interactions can be exploited for design of novel therapeutics against viral infections.

10.7   Concluding Remarks

NPM1 is truly a multifunctional protein whose functions have been unraveled in the past few decades through the work of many research groups while NPM2 and NPM3 have been less well studied. Other than the many processes in which NPM1 takes part, the major interest comes from its involvement in human cancers, particularly AML. Its significance stems from the fact that NPMc+ AML accounts for ~30% of all AML cases and usually has good prognosis. Its clinical importance also comes from its involvement in virus replication, particularly in the era of outbreaks of infectious diseases. A lot more remain to be discovered and learned for all three NPM proteins in the years to come.

10 Nucleophosmin/NPM/B23 and the Nucleoplasmin Family of Proteins 241

Acknowledgments Our work on nucleophosmin is currently supported by grants from The Hong Kong Polytechnic University (1-BD03, G-U702 and G-U915), and was also supported by ROC National Science Council Grants and Chang Gung Research Grants.

References

Adhikary S, Eilers M (2005) Transcriptional regulation and transformation by Myc proteins. Nat Rev Mol Cell Biol 6:635–645

Ahn JY, Liu X, Cheng D, Peng J, Chan PK, Wade PA, Ye K (2005) Nucleophosmin/B23, a nuclear PI(3,4,5)P(3) receptor, mediates the antiapoptotic actions of NGF by inhibiting CAD. Mol Cell 18:435–445

Albiero E, Madeo D, Bolli N, Giaretta I, Bona ED, Martelli MF, Nicoletti I, Rodeghiero F, Falini B (2007) Identification and functional characterization of a cytoplasmic nucleophosmin leukaemic mutant generated by a novel exon-11 NPM1 mutation. Leukemia 21:1099–1103

Alcalay M, Tiacci E, Bergomas R, Bigerna B, Venturini E, Minardi SP, Meani N, Diverio D, Bernard L, Tizzoni L, Volorio S, Luzi L, Colombo E, Lo Coco F, Mecucci C, Falini B, Pelicci PG (2005) Acute myeloid leukemia bearing cytoplasmic nucleophosmin (NPMc+ AML) shows a distinct gene expression profile characterized by up-regulation of genes involved in stem-cell maintenance. Blood 106:899–902

Allain JP (2000) Emerging viral infections relevant to transfusion medicine. Blood Rev 14:173–181

Andersen JS, Wilkinson CJ, Mayor T, Mortensen P, Nigg EA, Mann M (2003) Proteomic characterization of the human centrosome by protein correlation profiling. Nature 426:570–574

Andreeff M, Ruvolo V, Gadgil S, Zeng C, Coombes K, Chen W, Kornblau S, Barón AE, Drabkin HA (2008) HOX expression patterns identify a common signature for favorable AML. Leukemia 22:2041–2047

Barazani Y, Hiatt JR, Tong MJ, Busuttil RW (2007) Chronic viral hepatitis and hepatocellular carcinoma. World J Surg 3:1243–1248

Becker H, Marcucci G, Maharry K, Radmacher MD, Mrózek K, Margeson D, Whitman SP, Wu YZ, Schwind S, Paschka P, Powell BL, Carter TH, Kolitz JE, Wetzler M, Carroll AJ, Baer MR, Caligiuri MA, Larson RA, Bloomfield CD (2010) Favorable prognostic impact of NPM1 mutations in older patients with cytogenetically normal de novo acute myeloid leukemia and associated gene- and microRNA-expression signatures: a Cancer and Leukemia Group B study. J Clin Oncol 28:596–604

Bertwistle D, Sugimoto M, Sherr CJ (2004) Physical and functional interactions of the Arf tumor suppressor protein with nucleophosmin/B23. Mol Cell Biol 24:985–996

Bolli N, Nicoletti I, De Marco MF, Bigerna B, Pucciarini A, Mannucci R, Martelli MP, Liso A, Mecucci C, Fabbiano F, Martelli MF, Henderson BR, Falini B (2007) Born to be exported: COOH-terminal nuclear export signals of different strength ensure cytoplasmic accumulation of nucleophosmin leukemic mutants. Cancer Res 67:6230–6237

Bolli N, De Marco MF, Martelli MP, Bigerna B, Pucciarini A, Rossi R, Mannucci R, Manes N, Pettirossi V, Pileri SA, Nicoletti I, Falini B (2009) A dose-dependent tug of war involving the NPM1 leukaemic mutant, nucleophosmin, and ARF. Leukemia 23:501–509

Bolli N, Payne EM, Grabher C, Lee JS, Johnston AB, Falini B, Kanki JP, Look AT (2010) Expression of the cytoplasmic NPM1 mutant (NPMc+) causes the expansion of hematopoietic cells in zebrafish. Blood 115:3329–3340

Bonetti P, Davoli T, Sironi C, Amati B, Pelicci PG, Colombo E (2008) Nucleophosmin and its AML-associated mutant regulate c-Myc turnover through Fbw7 gamma. J Cell Biol 182:19–26

Borer RA, Lehner CF, Eppenberger HM, Nigg EA (1989) Major nucleolar proteins shuttle between nucleus and cytoplasm. Cell 56:379–390

Brady SN, Yu Y, Maggi LB Jr, Weber JD (2004) ARF impedes NPM/B23 shuttling in an Mdm2-sensitive tumor suppressor pathway. Mol Cell Biol 24:9327–9338

Brown P, McIntyre E, Rau R, Meshinchi S, Lacayo N, Dahl G, Alonzo TA, Chang M, Arceci RJ, Small D (2007) The incidence and clinical significance of nucleophosmin mutations in childhood AML. Blood 110:979–985

Bulich R, Aaskov JG (1992) Nuclear localization of dengue 2 virus core protein detected with monoclonal antibodies. J Gen Virol 73:2999–3003

Burns KH, Viveiros MM, Ren Y, Wang P, DeMayo FJ, Frail DE, Eppig JJ, Matzuk MM (2003) Roles of NPM2 in chromatin and nucleolar organization in oocytes and embryos. Science 300:633–666

Casey JL (2006) Hepatitis delta virus. Springer, Berlin

Castiglia CL, Flint SJ (1983) Effects of adenovirus infection on rRNA synthesis and maturation in HeLa cells. Mol Cell Biol 3:662–671

Cazzaniga G, Dell'Oro MG, Mecucci C, Giarin E, Masetti R, Rossi V, Locatelli F, Martelli MF, Basso G, Pession A, Biondi A, Falini B (2005) Nucleophosmin mutations in childhood acute myelogenous leukemia with normal karyotype. Blood 106:1419–1422

Chan PK, Liu QR, Durban E (1990) The major phosphorylation site of nucleophosmin (B23) is phosphorylated by a nuclear kinase II. Biochem J 270:549–552

Chang JH, Olson MO (1989) A single gene codes for two forms of rat nucleolar protein B23 mRNA. J Biol Chem 264:11732–11737

Chen SY, Kao CF, Chen CM, Shih CM, Hsu MJ, Chao CH, Wang SH, You LR, Lee YH (2003) Mechanisms for inhibition of hepatitis B virus gene expression and replication by hepatitis C virus core protein. J Biol Chem 278:591–607

Cheng K, Sportoletti P, Ito K, Clohessy JG, Teruya-Feldstein J, Kutok JL, Pandolfi PP (2010) The cytoplasmic NPM mutant induces myeloproliferation in a transgenic mouse model. Blood 115:3341–3345

Choo QL, Richman KH, Han JH, Berger K, Lee C, Dong C, Gallegos C, Coit D, Medina-Selby R, Barr PJ, Weiner AJ, Bradley DW, Kuo G, Houghton M (1991) Genetic organization and diversity of the hepatitis C virus. Proc Natl Acad Sci USA 88:2451–2455

Chou WC, Tang JL, Lin LI, Yao M, Tsay W, Chen CY, Wu SJ, Huang CF, Chiou RJ, Tseng MH, Lin DT, Lin KH, Chen YC, Tien HF (2006) Nucleophosmin mutations in de novo acute myeloid leukemia: the age-dependent incidences and the stability during disease evolution. Cancer Res 66:3310–3316

Christian MD, Poutanen SM, Loutfy MR, Muller MP, Low DE (2004) Severe acute respiratory syndrome. Clin Infect Dis 38:1420–1427

Colombo E, Marine JC, Danovi D, Falini B, Pelicci PG (2002) Nucleophosmin regulates the stability and transcriptional activity of p53. Nat Cell Biol 4:529–533

Colombo E, Bonetti P, Lazzerini DE, Martinelli P, Zamponi R, Marine JC, Helin K, Falini B, Pelicci PG (2005) Nucleophosmin is required for DNA integrity and p19Arf protein stability. Mol Cell Biol 25:8874–8886

Colombo E, Martinelli P, Zamponi R, Shing DC, Bonetti P, Luzi L, Volorio S, Bernard L, Pruneri G, Alcalay M, Pelicci PG (2006) Delocalization and destabilization of the Arf tumor suppressor by the leukemia-associated NPM mutant. Cancer Res 66:3044–3050

Compton DA, Cleveland DW (1994) NuMA, a nuclear protein involved in mitosis and nuclear reformation. Curr Opin Cell Biol 6:343–346

Cotten M, Sealy L, Chalkley R (1986) Massive phosphorylation distinguishes *Xenopus laevis* nucleoplasmin isolated from oocytes or unfertilized eggs. Biochemistry 25:5063–5069

Dalenc F, Drouet J, Ader I, Delmas C, Rochaix P, Favre G, Cohen-Jonathan E, Toulas C (2002) Increased expression of a COOH-truncated nucleophosmin resulting from alternative splicing is associated with cellular resistance to ionizing radiation in HeLa cells. Int J Cancer 100:662–668

de Jong RN, van der Vliet PC, Brenkman AB (2003) Adenovirus DNA replication: protein priming, jumping back and the role of the DNA binding protein DBP. Curr Top Microbiol Immunol 272:187–211

10 Nucleophosmin/NPM/B23 and the Nucleoplasmin Family of Proteins

Debernardi S, Skoulakis S, Molloy G, Chaplin T, Dixon-McIver A, Young BD (2007) MicroRNA miR-181a correlates with morphological sub-class of acute myeloid leukaemia and the expression of its target genes in global genome-wide analysis. Leukemia 21:912–916

den Besten W, Kuo ML, iams RT, Sherr CJ (2005) Myeloid leukemia-associated nucleophosmin mutants perturb p53-dependent and independent activities of the Arf tumor suppressor protein. Cell Cycle 4:1593–1598

Dey A, Chao SH, Lane DP (2007) HEXIM1 and the control of transcription elongation: from cancer and inflammation to AIDS and cardiac hypertrophy. Cell Cycle 6:1856–1863

Dhar SK, Lynn BC, Daosukho C, St CD (2004) Identification of nucleophosmin as an NF-kappaB co-activator for the induction of the human SOD2 gene. J Biol Chem 279:28209–28219

Dingwall C, Dilworth SM, Black SJ, Kearsey SE, Cox LS, Laskey RA (1987) Nucleoplasmin cDNA sequence reveals polyglutamic acid tracts and a cluster of sequences homologous to putative nuclear localization signals. EMBO J 6:69–74

Dumbar TS, Gentry GA, Olson MO (1989) Interaction of nucleolar phosphoprotein B23 with nucleic acids. Biochemistry 28:9495–9501

Dundr M, Misteli T, Olson MO (2000) The dynamics of postmitotic reassembly of the nucleolus. J Cell Biol 150:433–446

Dutta S, Akey IV, Dingwall C, Hartman KL, Laue T, Nolte RT, Head JF, Akey CW (2001) The crystal structure of nucleoplasmin-core: implications for histone binding and nucleosome assembly. Mol Cell 8:841–853

Earnshaw WC, Honda BM, Laskey RA, Thomas JO (1980) Assembly of nucleosomes: the reaction involving *X. laevis* nucleoplasmin. Cell 21:373–383

Eirín-López JM, Frehlick LJ, Ausió J (2006) Long-term evolution and functional diversification in the members of the nucleophosmin/nucleoplasmin family of nuclear chaperones. Genetics 173:1835–1850

Engelhardt M, Lubbert M, Guo Y (2002) CD34(+) or CD34(−): which is the more primitive? Leukemia 16:1603–1608

Enomoto T, Lindström MS, Jin A, Ke H, Zhang Y (2006) Essential role of the B23/NPM core domain in regulating ARF binding and B23 stability. J Biol Chem 281:18463–18472

Estrov Z (2010) The leukemia stem cell. Cancer Treat Res 145:1–17

Falini B (2001) Anaplastic large cell lymphoma: pathological, molecular and clinical features. Br J Haematol 114:741–760

Falini B, Mecucci C, Tiacci E, Alcalay M, Rosati R, Pasqualucci L, La Starza R, Diverio D, Colombo E, Santucci A, Bigerna B, Pacini R, Pucciarini A, Liso A, Vignetti M, Fazi P, Meani N, Pettirossi V, Saglio G, Mandelli F, Lo-Coco F, Pelicci PG, Martelli MF, GIMEMA Acute Leukemia Working Party (2005) Cytoplasmic nucleophosmin in acute myelogenous leukemia with a normal karyotype. N Engl J Med 352:254–266

Falini B, Bolli N, Shan J, Martelli MP, Liso A, Pucciarini A, Bigerna B, Pasqualucci L, Mannucci R, Rosati R, Gorello P, Diverio D, Roti G, Tiacci E, Cazzaniga G, Biondi A, Schnittger S, Haferlach T, Hiddemann W, Martelli MF, Gu W, Mecucci C, Nicoletti I (2006) Both carboxyterminus NES motif and mutated tryptophan(s) are crucial for aberrant nuclear export of nucleophosmin leukemic mutants in NPMc+ AML. Blood 107:4514–4523

Falini B, Albiero E, Bolli N, De Marco MF, Madeo D, Martelli M, Nicoletti I, Rodeghiero F (2007a) Aberrant cytoplasmic expression of C-terminal-truncated NPM leukaemic mutant is dictated by tryptophans loss and a new NES motif. Leukemia 21:2052–2054

Falini B, Nicoletti I, Bolli N, Martelli MP, Liso A, Gorello P, Mandelli F, Mecucci C, Martelli MF (2007b) Translocations and mutations involving the nucleophosmin (NPM1) gene in lymphomas and leukemias. Haematologica 92:519–532

Falini B, Nicoletti I, Martelli MF, Mecucci C (2007c) Acute myeloid leukemia carrying cytoplasmic/ mutated nucleophosmin (NPMc+ AML): biologic and clinical features. Blood 109:874–885

Falini B, Martelli MP, Mecucci C, Liso A, Bolli N, Bigerna B, Pucciarini A, Pileri S, Meloni G, Martelli MF, Haferlach T, Schnittger S (2008a) Cytoplasmic mutated nucleophosmin is stable in primary leukemic cells and in a xenotransplant model of NPMc+ acute myeloid leukemia in SCID mice. Haematologica 93:775–779

Falini B, Mecucci C, Saglio G, Lo Coco F, Diverio D, Brown P, Pane F, Mancini M, Martelli MP, Pileri S, Haferlach T, Haferlach C, Schnittger S (2008b) NPM1 mutations and cytoplasmic nucleophosmin are mutually exclusive of recurrent genetic abnormalities: a comparative analysis of 2562 patients with acute myeloid leukemia. Haematologica 93:439–442

Falini B, Martelli MP, Bolli N, Sportoletti P, Liso A, Tiacci E, Haferlach T (2011) Acute myeloid leukemia with mutated nucleophosmin (NPM1): is it a distinct entity? Blood 117:1109–1120

Fan Z, Zhuo Y, Tan X, Zhou Z, Yuan J, Qiang B, Yan J, Peng X, Gao GF (2006) SARS-CoV nucleocapsid protein binds to hUbc9, a ubiquitin conjugating enzyme of the sumoylation system. J Med Virol 78:1365–1373

Fankhauser C, Izaurralde E, Adachi Y, Wingfield P, Laemmli UK (1991) Specific complex of human immunodeficiency virus type 1 rev and nucleolar B23 proteins: dissociation by the Rev response element. Mol Cell Biol 11:2567–2575

Fischer R, Baumert T, Blum HE (2007) Hepatitis C virus infection and apoptosis. World J Gastroenterol 3:4865–4872

Flint SJ (2000) Principles of virology: molecular biology, pathogenesis, and control. ASM Press, Washington

Frehlick LJ, Eirín-López JM, Ausió J (2007) New insights into the nucleophosmin/nucleoplasmin family of nuclear chaperones. Bioessays 29:49–59

Gadad SS, Shandilya J, Kishore AH, Kundu TK (2010) NPM3, a member of the nucleophosmin/ nucleoplasmin family, enhances activator-dependent transcription. Biochemistry 49:1355–1357

Galati D, Paiardini M, Cervasi B, Albrecht H, Bocchino M, Costantini A, Montroni M, Magnani M, Piedimonte G, Silvestri G (2003) Specific changes in the posttranslational regulation of nucleolin in lymphocytes from patients infected with human immunodeficiency virus. J Infect Dis 188:1483–1491

Gale RE, Green C, Allen C, Mead AJ, Burnett AK, Hills RK, Linch DC, Medical Research Council Adult Leukaemia Working Party (2008) The impact of FLT3 internal tandem duplication mutant level, number, size and interaction with NPM1 mutations in a large cohort of young adult patients with acute myeloid leukemia. Blood 111:2776–2784

Gallagher SJ, Kefford RF, Rizos H (2006) The ARF tumour suppressor. Int J Biochem Cell Biol 38:1637–1641

Ganem D, Schneider R (2001) Hepadnaviridae: the viruses and their replication. In: Knipe DM, Howley PM, Griffin DE, Lamb RA, Martin MA, Roizman B, Straus SE (eds) Fields virology, 4th edn. Lippincott iams & Wilkins, Philadelphia

Gao H, Jin S, Song Y, Fu M, Wang M, Liu Z, Wu M, Zhan Q (2005) B23 regulates GADD45a nuclear translocation and contributes to GADD45a-induced cell cycle G2-M arrest. J Biol Chem 280:10988–10996

Garzon R, Garofalo M, Martelli MP, Briesewitz R, Wang L, Fernandez-Cymering C, Volinia S, Liu CG, Schnittger S, Haferlach T, Liso A, Diverio D, Mancini M, Meloni G, Foa R, Martelli MF, Mecucci C, Croce CM, Falini B (2008) Distinctive microRNA signature of acute myeloid leukemia bearing cytoplasmic mutated nucleophosmin. Proc Natl Acad Sci USA 105:3945–3950

Gimenez M, Souza VC, Izumi C, Barbieri MR, Chammas R, Oba-Shinjo SM, Uno M, Marie SK, Rosa JC (2010) Proteomic analysis of low- to high-grade astrocytomas reveals an alteration of the expression level of raf kinase inhibitor protein and nucleophosmin. Proteomics 10:2812–2821

Grisendi S, Bernardi R, Rossi M, Cheng K, Khandker L, Manova K, Pandolfi PP (2005) Role of nucleophosmin in embryonic development and tumorigenesis. Nature 437:147–153

Grisendi S, Mecucci C, Falini B, Pandolfi PP (2006) Nucleophosmin and cancer. Nat Rev Cancer 6:493–505

Grummitt CG, Townsley FM, Johnson CM, Warren AJ, Bycroft M (2008) Structural consequences of nucleophosmin mutations in acute myeloid leukemia. J Biol Chem 283:23326–23332

Gurumurthy M, Tan CH, Ng R, Zeiger L, Lau J, Lee J, Dey A, Philp R, Li Q, Lim TM, Price DH, Lane DP, Chao SH (2008) Nucleophosmin interacts with HEXIM1 and regulates RNA polymerase II transcription. J Mol Biol 378:302–317

10 Nucleophosmin/NPM/B23 and the Nucleoplasmin Family of Proteins

Hernandez-Verdun D, Gautier T (1994) The chromosome periphery during mitosis. Bioessays 16:179–185

Herrera JE, Savkur R, Olson MO (1995) The ribonuclease activity of nucleolar protein B23. Nucleic Acids Res 23:3974–3979

Herrera JE, Correia JJ, Jones AE, Olson MO (1996) Sedimentation analyses of the salt- and divalent metal ion-induced oligomerization of nucleolar protein B23. Biochemistry 35: 2668–2673

Hindley CE, Davidson AD, Matthews DA (2007) Relationship between adenovirus DNA replication proteins and nucleolar proteins B23.1 and B23.2. J Gen Virol 88:3244–3248

Hingorani K, Szebeni A, Olson MO (2000) Mapping the functional domains of nucleolar protein B23. J Biol Chem 275:24451–24457

Hisaoka M, Ueshima S, Murano K, Nagata K, Okuwaki M (2010) Regulation of nucleolar chromatin by B23/nucleophosmin jointly depends upon its RNA binding activity and transcription factor UBF. Mol Cell Biol 30:4952–4964

Hiscox JA (2007) RNA viruses: hijacking the dynamic nucleolus. Nat Rev Microbiol 5:119–127

Hitzler JK, Witte DP, Jenkins NA, Copeland NG, Gilbert DJ, Naeve CW, Look AT, Morris SW (1999) cDNA cloning, expression pattern, and chromosomal localization of Mlf1, murine homologue of a gene involved in myelodysplasia and acute myeloid leukemia. Am J Pathol 155:53–59

Horn HF, Vousden KH (2004) Cancer: guarding the guardian? Nature 427:110–111

Hsu CY, Yung BY (1998) Down-regulation of nucleophosmin/B23 during retinoic acid-induced differentiation of human promyelocytic leukemia HL-60 cells. Oncogene 16:915–923

Huang WH, Yung BY, Syu WJ, Lee YH (2001) The nucleolar phosphoprotein B23 interacts with hepatitis delta antigens and modulates the hepatitis delta virus RNA replication. J Biol Chem 276:25166–25175

Huang N, Negi S, Szebeni A, Olson MO (2005) Protein NPM3 interacts with the multifunctional nucleolar protein B23/nucleophosmin and inhibits ribosome biogenesis. J Biol Chem 280:5496–5502

Inouye CJ, Seto E (1994) Relief of YY1-induced transcriptional repression by protein-protein interaction with the nucleolar phosphoprotein B23. J Biol Chem 269:6506–6510

Itahana K, Bhat KP, Jin A, Itahana Y, Hawke D, Kobayashi R, Zhang Y (2003) Tumor suppressor ARF degrades B23, a nucleolar protein involved in ribosome biogenesis and cell proliferation. Mol Cell 12:1151–1164

Jagus R, Joshi B, Barber GN (1999) PKR, apoptosis and cancer. Int J Biochem Cell Biol 31:123–138

Jeong EG, Lee SH, Yoo NJ, Lee SH (2007) Absence of nucleophosmin 1 (NPM1) gene mutations in common solid cancers. APMIS 115:341–234

Jiang PS, Chang JH, Yung BY (2000) Different kinases phosphorylate nucleophosmin/B23 at different sites during G(2) and M phases of the cell cycle. Cancer Lett 153:151–160

Jongen-Lavrencic M, Sun SM, Dijkstra MK, Valk PJ, Löwenberg B (2008) MicroRNA expression profiling in relation to the genetic heterogeneity of acute myeloid leukemia. Blood 111:5078–5085

Kang HY, Lee S, Park SG, Yu J, Yim K, Jung G (2006) Phosphorylation of hepatitis B virus Cp at Ser87 facilitates core assembly. Biochem J 398:311–317

Kim MY, Mauro S, Gevry N, Lis JT, Kraus WL (2004) NAD+-dependent modulation of chromatin structure and transcription by nucleosome binding properties of PARP-1. Cell 119:803–814

Koike K (2007) Hepatitis C virus contributes to hepatocarcinogenesis by modulating metabolic and intracellular signaling pathways. J Gastroenterol Hepatol 22(suppl 1):S108–S111

Kondo T, Minamino N, Nagamura-Inoue T, Matsumoto M, Taniguchi T, Tanaka N (1997) Identification and characterization of nucleophosmin/B23/numatrin which binds the anti-oncogenic transcription factor IRF-1 and manifests oncogenic activity. Oncogene 15:1275–1281

Korgaonkar C, Hagen J, Tompkins V, Frazier AA, Allamargot C, Quelle FW, Quelle DE (2005) Nucleophosmin (B23) targets ARF to nucleoli and inhibits its function. Mol Cell Biol 25:1258–1271

Ksiazek TG, Erdman D, Goldsmith CS, Zaki SR, Peret T, Emery S, Tong S, Urbani C, Comer JA, Lim W, Rollin PE, Dowell SF, Ling AE, Humphrey CD, Shieh WJ, Guarner J, Paddock CD,

Rota P, Fields B, DeRisi J, Yang JY, Cox N, Hughes JM, LeDuc JW, Bellini WJ, Anderson LJ; SARS Working Group (2003) A novel coronavirus associated with severe acute respiratory syndrome. N Engl J Med 348:1953–1966

Kuo ML, den Besten W, Bertwistle D, Roussel MF, Sherr CJ (2004) N-terminal polyubiquitination and degradation of the Arf tumor suppressor. Genes Dev 18:1862–1874

Kuo ML, den Besten W, Thomas MC, Sherr CJ (2008) Arf-induced turnover of the nucleolar nucleophosmin-associated SUMO-2/3 protease Senp3. Cell Cycle 7:3378–3387

Kurki S, Peltonen K, Laiho M (2004a) Nucleophosmin, HDM2 and p53: players in UV damage incited nucleolar stress response. Cell Cycle 3:976–979

Kurki S, Peltonen K, Latonen L, Kiviharju TM, Ojala PM, Meek D, Laiho M (2004b) Nucleolar protein NPM interacts with HDM2 and protects tumor suppressor protein p53 from HDM2-mediated degradation. Cancer Cell 5:465–475

Laskey RA, Honda BM, Mills AD, Finch JT (1978) Nucleosomes are assembled by an acidic protein which binds histones and transfers them to DNA. Nature 275:416–420

Lee SY, Park JH, Kim S, Park EJ, Yun Y, Kwon J (2005) A proteomics approach for the identification of nucleophosmin and heterogeneous nuclear ribonucleoprotein C1/C2 as chromatin-binding proteins in response to DNA double-strand breaks. Biochem J 388:7–15

Lee HH, Kim HS, Kang JY, Lee BI, Ha JY, Yoon HJ, Lim SO, Jung G, Suh SW (2007) Crystal structure of human nucleophosmin-core reveals plasticity of the pentamer-pentamer interface. Proteins 69:672–678

Lee SJ, Shim HY, Hsieh A, Min JY, Jung G (2009) Hepatitis B virus core interacts with the host cell nucleolar protein, nucleophosmin 1. J Microbiol 47:746–752

Lenaerts L, De Clercq E, Naesens L (2008) Clinical features and treatment of adenovirus infections. Rev Med Virol 18:357–374

Leno GH, Mills AD, Philpott A, Laskey RA (1996) Hyperphosphorylation of nucleoplasmin facilitates *Xenopus* sperm decondensation at fertilization. J Biol Chem 271:7253–7256

Leong SM, Tan BX, Bte Ahmad B, Yan T, Chee LY, Ang ST, Tay KG, Koh LP, Yeoh AE, Koay ES, Mok YK, Lim TM (2010) Mutant nucleophosmin deregulates cell death and myeloid differentiation through excessive caspase-6 and -8 inhibition. Blood 116:3286–3296

Léotoing L, Meunier L, Manin M, Mauduit C, Decaussin M, Verrijdt G, Claessens F, Benahmed M, Veyssière G, Morel L, Beaudoin C (2008) Influence of nucleophosmin/B23 on DNA binding and transcriptional activity of the androgen receptor in prostate cancer cell. Oncogene 27:2858–2867

Lerch-Gaggl A, Haque J, Li J, Ning G, Traktman P, Duncan SA (2002) Pescadillo is essential for nucleolar assembly, ribosome biogenesis, and mammalian cell proliferation. J Biol Chem 277:45347–45355

Levine AJ (1997) p53, the cellular gatekeeper for growth and division. Cell 88:323–331

Li YP (1997) Protein B23 is an important human factor of the nucleolar localization of the human immunodeficiency virus protein Tat. J Virol 71:4098–4102

Li J, Zhang X, Sejas DP, Bagby GC, Pang Q (2004) Hypoxia-induced nucleophosmin protects cell death through inhibition of p53. J Biol Chem 279:41275–41279

Li J, Zhang X, Sejas DP, Pang Q (2005) Negative regulation of p53 by nucleophosmin antagonizes stress-induced apoptosis in human normal and malignant hematopoietic cells. Leuk Res 29:1415–1423

Li YJ, Macnaughton T, Gao L, Lai MM (2006) RNA-templated replication of hepatitis delta virus: genomic and antigenomic RNAs associate with different nuclear bodies. J Virol 80:6478–6486

Lin CY, Liang YC, Yung BY (2006) Nucleophosmin/B23 regulates transcriptional activation of E2F1 via modulating the promoter binding of NF-kappaB, E2F1 and pRB. Cell Signal 18:2041–2048

Lin CY, Tan BC, Liu H, Shih CJ, Chien KY, Lin CL, Yung BY (2010) Dephosphorylation of nucleophosmin by PP1β facilitates pRB binding and consequent E2F1-dependent DNA repair. Mol Biol Cell 21:4409–4417

10 Nucleophosmin/NPM/B23 and the Nucleoplasmin Family of Proteins

Liso A, Bogliolo A, Freschi V, Martelli MP, Pileri SA, Santodirocco M, Bolli N, Martelli MF, Falini B (2008) In human genome, generation of a nuclear export signal through duplication appears unique to nucleophosmin (NPM1) mutations and is restricted to AML. Leukemia 22:1285–1289

Liu WH, Yung BY (1998) Mortalization of human promyelocytic leukemia HL-60 cells to be more susceptible to sodium butyrate-induced apoptosis and inhibition of telomerase activity by down-regulation of nucleophosmin/B23. Oncogene 17:3055–3064

Liu H, Naismith JH, Hay RT (2003) Adenovirus DNA replication. Curr Top Microbiol Immunol 272:131–164

Liu H, Tan BC, Tseng KH, Chuang CP, Yeh CW, Chen KD, Lee SC, Yung BY (2007a) Nucleophosmin acts as a novel AP2alpha-binding transcriptional corepressor during cell differentiation. EMBO Rep 8:394–400

Liu X, Liu Z, Jang SW, Ma Z, Shinmura K, Kang S, Dong S, Chen J, Fukasawa K, Ye K (2007b) Sumoylation of nucleophosmin/B23 regulates its subcellular localization, mediating cell proliferation and survival. Proc Natl Acad Sci USA 104:9679–9684

Luo C, Luo H, Zheng S, Gui C, Yue L, Yu C, Sun T, He P, Chen J, Shen J, Luo X, Li Y, Liu H, Bai D, Shen J, Yang Y, Li F, Zuo J, Hilgenfeld R, Pei G, Chen K, Shen X, Jiang H (2004) Nucleocapsid protein of SARS coronavirus tightly binds to human cyclophilin A. Biochem Biophys Res Commun 321:557–565

MacArthur CA, Shackleford GM (1997) Npm3: a novel, widely expressed gene encoding a protein related to the molecular chaperones nucleoplasmin and nucleophosmin. Genomics 42:137–140

Maggi LB Jr, Kuchenruether M, Dadey DY, Schwope RM, Grisendi S, Townsend RR, Pandolfi PP, Weber JD (2008) Nucleophosmin serves as a rate-limiting nuclear export chaperone for the mammalian ribosome. Mol Cell Biol 28:7050–7065

Mai RT, Yeh TS, Kao CF, Sun SK, Huang HH, Wu Lee YH (2006) Hepatitis C virus core protein recruits nucleolar phosphoprotein B23 and coactivator p300 to relieve the repression effect of transcriptional factor YY1 on B23 gene expression. Oncogene 25:448–462

Mariano AR, Colombo E, Luzi L, Martinelli P, Volorio S, Bernard L, Meani N, Bergomas R, Alcalay M, Pelicci PG (2006) Cytoplasmic localization of NPM in myeloid leukemias is dictated by gain-of-function mutations that create a functional nuclear export signal. Oncogene 25:4376–4380

Martelli MP, Manes N, Pettirossi V, Liso A, Pacini R, Mannucci R, Zei T, Bolli N, di Raimondo F, Specchia G, Nicoletti I, Martelli MF, Falini B (2008) Absence of nucleophosmin leukaemic mutants in B and T cells from AML with NPM1 mutations: implications for the cell of origin of NPMc+ AML. Leukemia 22:195–198

Martelli MP, Pettirossi V, Thiede C, Bonifacio E, Mezzasoma F, Cecchini D, Pacini R, Tabarrini A, Ciurnelli R, Gionfriddo I, Manes N, Rossi R, Giunchi L, Oelschlägel U, Brunetti L, Gemei M, Delia M, Specchia G, Liso A, Di Ianni M, Di Raimondo F, Falzetti F, Del Vecchio L, Martelli MF, Falini B (2010) CD34+ cells from AML with mutated NPM1 harbor cytoplasmic mutated nucleophosmin and generate leukemia in immunocompromised mice. Blood 116:3907–3922

Matthews DA (2001) Adenovirus protein V induces redistribution of nucleolin and B23 from nucleolus to cytoplasm. J Virol 75:1031–1038

Matthews DA, Russell WC (1998) Adenovirus core protein V is delivered by the invading virus to the nucleus of the infected cell and later in infection is associated with nucleoli. J Gen Virol 79:1671–1675

McLay DW, Clarke HJ (2003) Remodelling the paternal chromatin at fertilization in mammals. Reproduction 125:625–633

Meder VS, Boeglin M, de Murcia G, Schreiber V (2005) PARP-1 and PARP-2 interact with nucleophosmin/B23 and accumulate in transcriptionally active nucleoli. J Cell Sci 118:211–222

Michalik J, Yeoman LC, Busch H (1981) Nucleolar localization of protein B23 (37/5.1) by immunocytochemical techniques. Life Sci 28:1371–1379

Miller G, El-Guindy A, Countryman J, Ye J, Gradoville L (2007) Lytic cycle switches of oncogenic human gammaherpesviruses. Adv Cancer Res 97:81–109

Mills AD, Laskey RA, Black P, De Robertis EM (1980) An acidic protein which assembles nucleosomes in vitro is the most abundant protein in *Xenopus* oocyte nuclei. J Mol Biol 139:561–568

Miyazaki Y, Nosaka T, Hatanaka M (1996) The posttranscriptional regulator Rev of HIV: implications for its interaction with the nucleolar protein B23. Biochimie 78:1081–1086

Mori Y, Okabayashi T, Yamashita T, Zhao Z, Wakita T, Yasui K, Hasebe F, Tadano M, Konishi E, Moriishi K, Matsuura Y (2005) Nuclear localization of Japanese encephalitis virus core protein enhances viral replication. J Virol 79:3448–3458

Motoi N, Suzuki K, Hirota R, Johnson P, Oofusa K, Kikuchi Y, Yoshizato K (2008) Identification and characterization of nucleoplasmin 3 as a histone-binding protein in embryonic stem cells. Dev Growth Differ 50:307–320

Mullighan CG, Kennedy A, Zhou X, Radtke I, Phillips LA, Shurtleff SA, Downing JR (2007) Pediatric acute myeloid leukemia with NPM1 mutations is characterized by a gene expression profile with dysregulated HOX gene expression distinct from MLL-rearranged leukemias. Leukemia 21:2000–2009

Murano K, Okuwaki M, Hisaoka M, Nagata K (2008) Transcription regulation of the rRNA gene by a multifunctional nucleolar protein, B23/nucleophosmin, through its histone chaperone activity. Mol Cell Biol 28:3114–3126

Nakagawa M, Kameoka Y, Suzuki R (2005) Nucleophosmin in acute myelogenous leukemia. N Engl J Med 352:1819–1820

Namboodiri VM, Dutta S, Akey IV, Head JF, Akey CW (2003) The crystal structure of *Drosophila* NLP-core provides insight into pentamer formation and histone binding. Structure 11:175–186

Namboodiri VM, Akey IV, Schmidt-Zachmann MS, Head JF, Akey CW (2004) The structure and function of *Xenopus* NO38-core, a histone chaperone in the nucleolus. Structure 12:2149–2160

Naoe T, Suzuki T, Kiyoi H, Urano T (2006) Nucleophosmin: a versatile molecule associated with hematological malignancies. Cancer Sci 97:963–969

Negi SS, Olson MO (2006) Effects of interphase and mitotic phosphorylation on the mobility and location of nucleolar protein B23. J Cell Sci 119:3676–3685

Ning B, Shih C (2004) Nucleolar localization of human hepatitis B virus capsid protein. J Virol 78:13653–13668

Nishimura Y, Ohkubo T, Furuichi Y, Umekawa H (2002) Tryptophans 286 and 288 in the C-terminal region of protein B23.1 are important for its nucleolar localization. Biosci Biotechnol Biochem 66:2239–2242

Nozawa Y, Van Belzen N, Van der Made AC, Dinjens WN, Bosman FT (1996) Expression of nucleophosmin/B23 in normal and neoplastic colorectal mucosa. J Pathol 178:48–52

Okuda M (2002) The role of nucleophosmin in centrosome duplication. Oncogene 21:6170–6174

Okuda M, Horn HF, Tarapore P, Tokuyama Y, Smulian AG, Chan PK, Knudsen ES, Hofmann IA, Snyder JD, Bove KE, Fukasawa K (2000) Nucleophosmin/B23 is a target of CDK2/cyclin E in centrosome duplication. Cell 103:127–140

Okuwaki M (2008) The structure and functions of NPM1/nucleophsmin/B23, a multifunctional nucleolar acidic protein. J Biochem 143:441–448

Okuwaki M, Iwamatsu A, Tsujimoto M, Nagata K (2001a) Identification of nucleophosmin/B23, an acidic nucleolar protein, as a stimulatory factor for in vitro replication of adenovirus DNA complexed with viral basic core proteins. J Mol Biol 311:41–55

Okuwaki M, Matsumoto K, Tsujimoto M, Nagata K (2001b) Function of nucleophosmin/B23, a nucleolar acidic protein, as a histone chaperone. FEBS Lett 506:272–276

Okuwaki M, Tsujimoto M, Nagata K (2002) The RNA binding activity of a ribosome biogenesis factor, nucleophosmin/B23, is modulated by phosphorylation with a cell cycle-dependent kinase and by association with its subtype. Mol Biol Cell 13:2016–2030

Olson MO, Wallace MO, Herrera AH, Marshall-Carlson L, Hunt RC (1986) Preribosomal ribonucleoprotein particles are a major component of a nucleolar matrix fraction. Biochemistry 25:484–491

10 Nucleophosmin/NPM/B23 and the Nucleoplasmin Family of Proteins

Orrick LR, Olson MO, Busch H (1973) Comparison of nucleolar proteins of normal rat liver and Novikoff hepatoma ascites cells by two-dimensional polyacrylamide gel electrophoresis. Proc Natl Acad Sci USA 70:1316–1320

Pang Q, Christianson TA, Koretsky T, Carlson H, David L, Keeble W, Faulkner GR, Speckhart A, Bagby GC (2003) Nucleophosmin interacts with and inhibits the catalytic function of eukaryotic initiation factor 2 kinase PKR. J Biol Chem 278:41709–41717

Pasqualucci L, Liso A, Martelli MP, Bolli N, Pacini R, Tabarrini A, Carini M, Bigerna B, Pucciarini A, Mannucci R, Nicoletti I, Tiacci E, Meloni G, Specchia G, Cantore N, Di Raimondo F, Pileri S, Mecucci C, Mandelli F, Martelli MF, Falini B (2006) Mutated nucleophosmin detects clonal multilineage involvement in acute myeloid leukemia: impact on WHO classification. Blood 108:4146–4155

Perkins A, Cochrane AW, Ruben SM, Rosen CA (1989) Structural and functional characterization of the human immunodeficiency virus rev protein. J Acquir Immune Defic Syndr 2:256–263

Peter M, Nakagawa J, Doree M, Labbe JC, Nigg EA (1990) Identification of major nucleolar proteins as candidate mitotic substrates of cdc2 kinase. Cell 60:791–801

Philpott A, Leno GH (1992) Nucleoplasmin remodels sperm chromatin in *Xenopus* egg extracts. Cell 69:759–767

Pianta A, Puppin C, Franzoni A, Fabbro D, Di Loreto C, Bulotta S, Deganuto M, Paron I, Tell G, Puxeddu E, Filetti S, Russo D, Damante G (2010) Nucleophosmin is overexpressed in thyroid tumors. Biochem Biophys Res Commun 397:499–504

Pitiot AS, Santamaría I, García-Suárez O, Centeno I, Astudillo A, Rayón C, Balbín M (2007) A new type of NPM1 gene mutation in AML leading to a C-terminal truncated protein. Leukemia 21:1564–1566

Potterton L, McNicholas S, Krissinel E, Gruber J, Cowtan K, Emsley P, Murshudov GN, Cohen S, Perrakis A, Noble M (2004) Developments in the CCP4 molecular-graphics project. Acta Crystallogr D Biol Crystallogr 60:2288–2294

Prestayko AW, Klomp GR, Schmoll DJ, Busch H (1974) Comparison of proteins of ribosomal subunits and nucleolar preribosomal particles from Novikoff hepatoma ascites cells by two-dimensional polyacrylamide gel electrophoresis. Biochemistry 13:1945–1951

Raimondi SC, Dubé ID, Valentine MB, Mirro J Jr, Watt HJ, Larson RA, Bitter MA, Le Beau MM, Rowley JD (1989) Clinicopathologic manifestations and breakpoints of the t(3;5) in patients with acute nonlymphocytic leukemia. Leukemia 3:42–47

Rau R, Brown P (2009) Nucleophosmin (NPM1) mutations in adult and childhood acute myeloid leukaemia: towards definition of a new leukaemia entity. Hematol Oncol 27:171–181

Renne R, Zhong W, Herndier B, McGrath M, Abbey N, Kedes D, Ganem D (1996) Lytic growth of Kaposi's sarcoma-associated herpesvirus (human herpesvirus 8) in culture. Nat Med 2:342–346

Rubbi CP, Milner J (2003) Disruption of the nucleolus mediates stabilization of p53 in response to DNA damage and other stresses. EMBO J 22:6068–6077

Ruggero D, Pandolfi PP (2003) Does the ribosome translate cancer? Nat Rev Cancer 3:179–192

Samad MA, Okuwaki M, Haruki H, Nagata K (2007) Physical and functional interaction between a nucleolar protein nucleophosmin/B23 and adenovirus basic core proteins. FEBS Lett 581:3283–3288

Sarek G, Järviluoma A, Moore HM, Tojkander S, Vartia S, Biberfeld P, Laiho M, Ojala PM (2010) Nucleophosmin phosphorylation by v-cyclin-CDK6 controls KSHV latency. PLoS Pathog 6:e1000818

Sato K, Hayami R, Wu W, Nishikawa T, Nishikawa H, Okuda Y, Ogata H, Fukuda M, Ohta T (2004) Nucleophosmin/B23 is a candidate substrate for the BRCA1-BARD1 ubiquitin ligase. J Biol Chem 279:30919–30922

Savkur RS, Olson MO (1998) Preferential cleavage in pre-ribosomal RNA by protein B23 endoribonuclease. Nucleic Acids Res 26:4508–4515

Shackleford GM, Ganguly A, MacArthur CA (2001) Cloning, expression and nuclear localization of human NPM3, a member of the nucleophosmin/nucleoplasmin family of nuclear chaperones. BMC Genomics 2:8

Sherr CJ (2006) Divorcing ARF and p53: an unsettled case. Nat Rev Cancer 6:663–673

Shields LB, Gerçel-Taylor C, Yashar CM, Wan TC, Katsanis WA, Spinnato JA, Taylor DD (1997) Induction of immune responses to ovarian tumor antigens by multiparity. J Soc Gynecol Investig 4:298–304

Skaar TC, Prasad SC, Sharareh S, Lippman ME, Brünner N, Clarke R (1998) Two-dimensional gel electrophoresis analyses identify nucleophosmin as an estrogen regulated protein associated with acquired estrogen-independence in human breast cancer cells. J Steroid Biochem Mol Biol 67:391–402

Spector DL, Ochs RL, Busch H (1984) Silver staining, immunofluorescence, and immunoelectron microscopic localization of nucleolar phosphoproteins B23 and C23. Chromosoma 90:139–148

Sportoletti P, Grisendi S, Majid SM, Cheng K, Clohessy JG, Viale A, Teruya-Feldstein J, Pandolfi PP (2008) Npm1 is a haploinsufficient suppressor of myeloid and lymphoid malignancies in the mouse. Blood 111:3859–3862

Subong EN, Shue MJ, Epstein JI, Briggman JV, Chan PK, Partin AW (1999) Monoclonal antibody to prostate cancer nuclear matrix protein (PRO:4-216) recognizes nucleophosmin/B23. Prostate 39:298–304

Sugimoto M, Kuo ML, Roussel MF, Sherr CJ (2003) Nucleolar Arf tumor suppressor inhibits ribosomal RNA processing. Mol Cell 11:415–424

Surjit M, Liu B, Chow VT, Lal SK (2006) The nucleocapsid protein of severe acute respiratory syndrome-coronavirus inhibits the activity of cyclin–cyclin-dependent kinase complex and blocks S phase progression in mammalian cells. J Biol Chem 281:10669–10681

Swaminathan V, Kishore AH, Febitha KK, Kundu TK (2005) Human histone chaperone nucleophosmin enhances acetylation-dependent chromatin transcription. Mol Cell Biol 25:7534–7545

Swerdlow SH, Campo E, Harris NL, Jaffe ES, Pileri SA, Stein H, Thiele J, Vardiman JW (eds) (2008) WHO classification of tumours of haematopoietic and lymphoid tissues. IARC, Lyon

Szebeni A, Olson MO (1999) Nucleolar protein B23 has molecular chaperone activities. Protein Sci 8:905–912

Szebeni A, Mehrotra B, Baumann A, Adam SA, Wingfield PT, Olson MO (1997) Nucleolar protein B23 stimulates nuclear import of the HIV-1 Rev protein and NLS-conjugated albumin. Biochemistry 36:3941–3949

Szebeni A, Hingorani K, Negi S, Olson MO (2003) Role of protein kinase CK2 phosphorylation in the molecular chaperone activity of nucleolar protein b23. J Biol Chem 278:9107–9115

Tago K, Chiocca S, Sherr CJ (2005) Sumoylation induced by the Arf tumor suppressor: a p53-independent function. Proc Natl Acad Sci USA 102:7689–7694

Takamizawa A, Mori C, Fuke I, Manabe S, Murakami S, Fujita J, Onishi E, Andoh T, Yoshida I, Okayama H (1991) Structure and organization of the hepatitis C virus genome isolated from human carriers. J Virol 65:1105–1113

Tanaka M, Sasaki H, Kino I, Sugimura T, Terada M (1992) Genes preferentially expressed in embryo stomach are predominantly expressed in gastric cancer. Cancer Res 52:3372–3377

Tarapore P, Shinmura K, Suzuki H, Tokuyama Y, Kim SH, Mayeda A, Fukasawa K (2006) Thr199 phosphorylation targets nucleophosmin to nuclear speckles and represses pre-mRNA processing. FEBS Lett 580:399–409

Taussig DC, Vargaftig J, Miraki-Moud F, Griessinger E, Sharrock K, Luke T, Lillington D, Oakervee H, Cavenagh J, Agrawal SG, Lister TA, Gribben JG, Bonnet D (2010) Leukemia-initiating cells from some acute myeloid leukemia patients with mutated nucleophosmin reside in the CD34(−) fraction. Blood 115:1976–1984

Thiede C, Koch S, Creutzig E, Steudel C, Illmer T, Schaich M, Ehninger G (2006) Prevalence and prognostic impact of NPM1 mutations in 1485 adult patients with acute myeloid leukemia (AML). Blood 107:4011–4020

Thiede C, Creutzig E, Reinhardt D, Ehninger G, Creutzig U (2007) Different types of NPM1 mutations in children and adults: evidence for an effect of patient age on the prevalence of the TCTG-tandem duplication in NPM1-exon 12. Leukemia 21:366–367

Tamini KA, Liao Q, Ye L, Zeng Y, Liu J, Zheng Y, Ye L, Yang X, Lingbao K, Gao J, Zhu Y (2005) Nuclear/nucleolar localization properties of C-terminal nucleocapsid protein of SARS coronavirus. Virus Res 114:23–34

Tokuyama Y, Horn HF, Kawamura K, Tarapore P, Fukasawa K (2001) Specific phosphorylation of nucleophosmin on Thr(199) by cyclin-dependent kinase 2-cyclin E and its role in centrosome duplication. J Biol Chem 276:21529–21537

Tsuda Y, Mori Y, Abe T, Yamashita T, Okamoto T, Ichimura T, Moriishi K, Matsuura Y (2006) Nucleolar protein B23 interacts with Japanese encephalitis virus core protein and participates in viral replication. Microbiol Immunol 50:225–234

Tsui KH, Cheng AJ, Chang PL, Pan TL, Yung BY (2004) Association of nucleophosmin/B23 mRNA expression with clinical outcome in patients with bladder carcinoma. Urology 64:839–844

Umekawa H, Chang JH, Correia JJ, Wang D, Wingfield PT, Olson MO (1993) Nucleolar protein B23: bacterial expression, purification, oligomerization and secondary structures of two isoforms. Cell Mol Biol Res 39:635–645

Verhaak RG, Goudswaard CS, van Putten W, Bijl MA, Sanders MA, Hugens W, Uitterlinden AG, Erpelinck CA, Delwel R, Löwenberg B, Valk PJ (2005) Mutations in nucleophosmin (NPM1) in acute myeloid leukemia (AML): association with other gene abnormalities and previously established gene expression signatures and their favorable prognostic significance. Blood 106:3747–3754

Wang D, Umekawa H, Olson MO (1993) Expression and subcellular locations of two forms of nucleolar protein B23 in rat tissues and cells. Cell Mol Biol Res 39:33–42

Wang D, Baumann A, Szebeni A, Olson MO (1994) The nucleic acid binding activity of nucleolar protein B23.1 resides in its carboxyl-terminal end. J Biol Chem 269:30994–30998

Wang J, Ji J, Ye J, Zhao X, Wen J, Li W, Hu J, Li D, Sun M, Zeng H, Hu Y, Tian X, Tan X, Xu N, Zeng C, Wang J, Bi S, Yang H (2003) The structure analysis and antigenicity study of the N protein of SARS-CoV. Genomics Proteomics Bioinformatics 1:145–154

Wang W, Budhu A, Forgues M, Wang XW (2005) Temporal and spatial control of nucleophosmin by the Ran-Crm1 complex in centrosome duplication. Nat Cell Biol 7:823–830

Wanzel M, Russ AC, Kleine-Kohlbrecher D, Colombo E, Pelicci PG, Eilers M (2008) A ribosomal protein L23-nucleophosmin circuit coordinates Mizl function with cell growth. Nat Cell Biol 10:1051–1061

Weinhold N, Moreaux J, Raab MS, Hose D, Hielscher T, Benner A, Meissner T, Ehrbrecht E, Brough M, Jauch A, Goldschmidt H, Klein B, Moos M (2010) NPM1 is overexpressed in hyperdiploid multiple myeloma due to a gain of chromosome 5 but is not delocalized to the cytoplasm. Genes Chromosomes Cancer 49:333–341

Welcker M, Clurman BE (2008) FBW7 ubiquitin ligase: a tumour suppressor at the crossroads of cell division, growth and differentiation. Nat Rev Cancer 8:83–93

Wertheim G, Bagg A (2008) Nucleophosmin (NPM1) mutations in acute myeloid leukemia: an ongoing (cytoplasmic) tale of dueling mutations and duality of molecular genetic testing methodologies. J Mol Diagn 10:198–202

Wu MH, Yung BY (2002) UV stimulation of nucleophosmin/B23 expression is an immediate-early gene response induced by damaged DNA. J Biol Chem 277:48234–48240

Wu MH, Chang JH, Chou CC, Yung BY (2002a) Involvement of nucleophosmin/B23 in the response of HeLa cells to UV irradiation. Int J Cancer 97:297–305

Wu MH, Chang JH, Yung BY (2002b) Resistance to UV-induced cell-killing in nucleophosmin/B23 over-expressed NIH 3 T3 fibroblasts: enhancement of DNA repair and up-regulation of PCNA in association with nucleophosmin/B23 over-expression. Carcinogenesis 23:93–100

Yang TH, Tsai WH, Lee YM, Lei HY, Lai MY, Chen DS, Yeh NH, Lee SC (1994) Purification and characterization of nucleolin and its identification as a transcription repressor. Mol Cell Biol 14:6068–6074

Ye K (2005) Nucleophosmin/B23, a multifunctional protein that can regulate apoptosis. Cancer Biol Ther 4:918–923

Yeh CW, Huang SS, Lee RP, Yung BY (2006) Ras-dependent recruitment of c-Myc for transcriptional activation of nucleophosmin/B23 in highly malignant U1 bladder cancer cells. Mol Pharmacol 70:1443–1453

Yu Y, Maggi LB Jr, Brady SN, Apicelli AJ, Dai MS, Lu H, Weber JD (2006) Nucleophosmin is essential for ribosomal protein L5 nuclear export. Mol Cell Biol 26:3798–3809

Yun JP, Chew EC, Liew CT, Chan JY, Jin ML, Ding MX, Fai YH, Li HK, Liang XM, Wu QL (2003) Nucleophosmin/B23 is a proliferate shuttle protein associated with nuclear matrix. J Cell Biochem 90:1140–1148

Yun JP, Miao J, Chen GG, Tian QH, Zhang CQ, Xiang J, Fu J, Lai PB (2007) Increased expression of nucleophosmin/B23 in hepatocellular carcinoma and correlation with clinicopathological parameters. Br J Cancer 96:477–484

Yung BY (2004) c-Myc-mediated expression of nucleophosmin/B23 decreases during retinoic acid-induced differentiation of human leukemia HL-60 cells. FEBS Lett 578:211–216

Zatsepina OV, Rousselet A, Chan PK, Olson MO, Jordan EG, Bornens M (1999) The nucleolar phosphoprotein B23 redistributes in part to the spindle poles during mitosis. J Cell Sci 112:455–466

Zelent A, Guidez F, Melnick A, Waxman S, Licht JD (2001) Translocations of the RARalpha gene in acute promyelocytic leukemia. Oncogene 20:7186–7203

Zeng Y, Ye L, Zhu S, Zheng H, Zhao P, Cai W, Su L, She Y, Wu Z (2008) The nucleocapsid protein of SARS-associated coronavirus inhibits B23 phosphorylation. Biochem Biophys Res Commun 369:287–291

Zhang H, Shi X, Paddon H, Hampong M, Dai W, Pelech S (2004) B23/nucleophosmin serine 4 phosphorylation mediates mitotic functions of polo-like kinase 1. J Biol Chem 279:35726–35734

Zhong W, Wang H, Herndier B, Ganem D (1996) Restricted expression of Kaposi sarcoma-associated herpesvirus (human herpesvirus 8) genes in Kaposi sarcoma. Proc Natl Acad Sci USA 93:6641–6646

Zirwes RF, Schmidt-Zachmann MS, Franke WW (1997) Identification of a small, very acidic constitutive nucleolar protein (NO29) as a member of the nucleoplasmin family. Proc Natl Acad Sci USA 94:11387–11392

Zlotnick A, Cheng N, Conway JF, Booy FP, Steven AC, Stahl SJ, Wingfield PT (1996) Dimorphism of hepatitis B virus capsids is strongly influenced by the C-terminus of the capsid protein. Biochemistry 35:7412–7421

Zou Y, Wu J, Giannone RJ, Boucher L, Du H, Huang Y, Johnson DK, Liu Y, Wang Y (2008) Nucleophosmin/B23 negatively regulates GCN5-dependent histone acetylation and transactivation. J Biol Chem 283:5728–5737

Chapter 11
Structure and Function of Nopp140 and Treacle

Fang He and Patrick DiMario

11.1 Introduction

Nopp140 and treacle are believed to function as molecular chaperones, delivering small nucleolar ribonucleoprotein complexes (snoRNPs) to the nucleolus where preribosomal RNA is synthesized, cleaved, chemically modified, and assembled into large and small ribosomal subunits. Orthologs of Nopp140 have been identified in an evolutionarily wide range of eukaryotes from yeast to human, but treacle appears to be restricted to the vertebrates. Both proteins share an amino terminal Lis1 homology (LisH) motif in their amino termini, and a large central repeat domain consisting of alternating serine-rich acidic motifs and lysine/proline-rich basic motifs. The carboxy termini of Nopp140 and treacle are dissimilar denoting unique associations. Nopp140 associates preferentially with both box H/ACA snoRNPs that guide pseudouridylation of pre-rRNA, and weakly with box C/D snoRNPs that guide 2'-O-methylation, while treacle interacts with box C/D snoRNPs with no detected interaction with box H/ACA snoRNPs. Nopp140 and treacle may be multifunctional: the large central domains of Nopp140 and treacle serve as scaffolds for delivering and positioning snoRNPs within nucleoli. Nopp140 also participates in Pol I and in one case, Pol II transcription; treacle also interacts with Pol I transcription machinery. Nopp140 is required for normal development in *Drosophila*, while treacle is critical for mammalian neural crest cell development. Loss of treacle function causes a nucleolar stress response that initiates p53-mediated apoptosis in embryonic neural epithelial and neural crest cells leading to the Treacher Collins–Franceschetti Syndrome in humans, a collection of craniofacial malformations.

P. DiMario (✉)
Department of Biological Sciences, Louisiana State University, Baton Rouge, LA, USA
e-mail: pdimari@lsu.edu

M.O.J. Olson (ed.), *The Nucleolus*, Protein Reviews 15,
DOI 10.1007/978-1-4614-0514-6_11, © Springer Science+Business Media, LLC 2011

11.2 Nopp140: Orthologs and Early Reports

Clearly, the majority of work on Nopp140 has been performed by U. Thomas Meier, first as a post doctoral fellow in Gunter Blobel's laboratory at Rockefeller University, and then in his own laboratory at Albert Einstein College of Medicine. While looking for proteins responsible for nuclear-cytoplasmic transport, Meier and Blobel (1990) identified p140 (as it was first called) in rat as a protein that could bind the nuclear localization signal (NLS) of the SV-40 large T antigen. Immuno-fluorescence microscopy showed that p140 localized to the nucleolus, but not to the nucleoplasm or cytoplasm. Colocalization with fibrillarin indicated that p140 localized to the fibrillar regions of the nucleolus. Later, Meier and Blobel (1992) deduced the amino acid sequence of Nopp140 as it is now called. They verified that antibodies against Nopp140 strongly labeled nucleoli, but with the better antibody they could detect weaker nucleoplasmic labeling. Injecting fluorescence labeled antibodies into the cytoplasm caused nucleolar accumulation of the antibodies, indicating that Nopp140 shuttles between the nucleus and the cytoplasm where it binds the antibody and then delivers the antibody to the nucleolus upon nuclear import. Nucleo-cytoplasmic shuttling of Nopp140 was later confirmed by Bellini and Gall (1999).

Nopp140 is conserved among eukaryotes, from yeast to humans (Fig. 11.1). It was actually observed prior to 1990 (Pfeifle and Anderer 1984; Schmidt-Zachmann et al. 1984; Pfeifle et al. 1986). Schmidt-Zachmann et al. (1984) initially characterized the *Xenopus* ortholog of Nopp140, which is now called xNopp180 (Cairns and McStay 1995). The protein has an apparent molecular weight of 180 kDa and a pI of ~4.2; it is enriched in nucleoli of oocytes and somatic cells. Immuno-gold

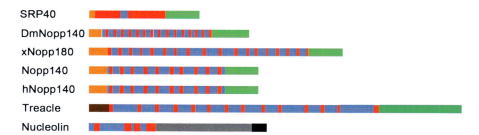

Fig. 11.1 Nopp140 protein compared to structurally related treacle and nucleolin proteins. Bar diagrams of yeast Srp40 (Meier 1996), *Drosophila* DmNopp140-True (Waggener and DiMario 2002), *Xenopus* xNopp140 (Cairns and McStay 1995), rat Nopp140 (Meier and Blobel 1992), human hNopp140 (Pai et al. 1995), human treacle (So et al. 2004), and human nucleolin (Srivastava et al. 1989) were drawn to show their different domains and motifs. Orange represents the conserved N-terminal domain of Nopp140 proteins. *Deep blue* within the amino termini of Nopp140 and treacle represents the LisH motif used as a possible homo-dimerization domain. *Red* represents acidic stretches, and *blue* represents basic stretches within the large central domains of Nopp140 and treacle and within the amino terminal region of nucleolin. *Green* represents the conserved C-terminal domain of Nopp140 proteins. *Gray* and *black* in nucleolin represent the four RNA binding domains and the RGG domain, respectively

labeling showed that the protein localizes to the dense fibrillar component (DFC) of interphase nucleoli, and immuno-fluorescence labeling showed that xNopp180 disperses to the cytoplasm during mitosis (metaphase and anaphase), with no apparent accumulation in the chromosomal nucleolar organizer regions (NORs). In telophase, xNopp180 rapidly reassembles into reforming nucleoli. Its cDNA was isolated from a *Xenopus* oocyte expression library using mAb G1C7 (Cairns and McStay 1995), and the deduced amino acid sequence shows up to 18 alternating acidic and basic stretches in the central domain (Fig. 11.1), while the N- and C-termini show 50 and 59% identities to the respective regions of rat Nopp140.

Pfeifle and Anderer (1984) described a nucleolar protein of 135 kDa in various mouse fibroblast, leukemia, and embryonic cell lines by immuno-fluorescence labeling and western blot analyses. Cross reacting proteins were observed in human cells (128 kDa), chicken cells (130 kDa), and *Drosophila* culture cells (118 kDa). Pulse labeling of enriched nucleoli from cultured mouse cells with [^{32}P] demonstrated preferential labeling of the mouse 135 kDa protein with a phosphoserine/phosphothreonine ratio of 47/1. An interesting point made in this early report is the apparent dependence of pp135 abundance on the cell cycle, with greater abundance of pp135 in rapidly dividing cells versus stationary cells. A subsequent paper (Pfeifle et al. 1986) reported localization of pp135 to NORs during mitosis.

Human Nopp140 was first predicted from cDNA clones that encoded proteins with sequence similarity to rat Nopp140 (Nomura et al. 1994). Pai et al. (1995) discovered human p130 when searching for proteins that fluctuated in abundance between interphase and M-phase. P130 is present in nucleoli of interphase cells, and it is heavily phosphorylated by casein kinase II (CKII) with hyper-phosphorylation occurring during mitosis, presumably by Cdk1. While p130 was undetectable by immuno-fluorescence during metaphase, it localized to prenucleolar bodies in telophase, and eventually to the nucleoli in interphase. Human p130 contains ten alternating acidic and basic repeat stretches (Fig. 11.1), and it shows 74% identity to rat Nopp140 (Pai et al. 1995). Interestingly, an isoform of p130 contains an additional 10 amino acid insertion in the fourth basic region (Pai and Yeh 1996). Both human isoforms (referred to as p130α and p130β) are coexpressed in various cell types, but the transcript encoding the β form (with the extra amino acids) is expressed to a lesser extent than the transcript encoding the α form. Both transcripts show a significant decrease in abundance upon cell cycle arrest.

Sequence similarity in the carboxy terminus of Nopp140 has been the criterion for identifying various orthologs. Srp40 is the ortholog of mammalian Nopp140 in *S. cerevisiae* (Meier 1996); it consists of two acidic clusters separated by one basic stretch. Its carboxy terminal domain is 59% identical to that in mammalian Nopp140. Like rat Nopp140, the acidic domains in Srp40 are rich in Ser residues that are likewise phosphorylated; a GST-Srp40 fusion has a calculated mass at 69 kDa, but it migrates in SDS gels at 110 kDa. Antibodies directed against rat Nopp140 cross react with Srp40, which colocalizes with the yeast fibrillarin ortholog, Nop1. Either the N-terminus or the central domain of Srp40 is sufficient to establish nuclear localization, while the carboxy-terminal domain alone is not (Ikonomova et al. 1997). Deletion of Srp40 retarded yeast growth mildly (it is not an essential protein), and the nucleoli remained morphologically unaffected. The slight growth defects could be rescued by introducing full length rat Nopp140, and the large central repeat domain of rat Nopp140 alone seemed to fully restore the growth defects (Yang and Meier 2003).

Over-expression of Srp40 resulted in growth arrest. Although growth defects were observed when Srp40 was under-expressed or over-expressed, no abnormalities in rRNA transcription/maturation or translocation were detected. These early results suggested that Srp40 is not critical for preribosome assembly or transport. On the other hand, Srp40 is necessary for snoRNA localization to the yeast nuclear body, a structure comparable to the mammalian Cajal body (Verheggen et al. 2001). See below for a more detailed discussion of Nopp140's (Srp40's) role in snoRNP biogenesis.

A cDNA encoding a *Drosophila* Nopp140-like protein was originally isolated by screening a cDNA library prepared from stage 10 egg chambers with a subclone encoding the amino terminal region of *Xenopus* nucleolin (Waggener and DiMario 2002). We originally intended to recover *Drosophila* nucleolin (this was prior to the availability of the *Drosophila* genome). The recovered full length *Drosophila* cDNA, however, encoded a Nopp140-like protein that contained a large central domain of alternating acidic and basic motifs that are quite similar to those in pro-totypical rat Nopp140, but it has a distinctive Arg-Gly–Gly rich (RGG) carboxyl domain. RGG domains are often found in RNA binding proteins such as fibrillarin, vertebrate nucleolin, and many SR and hnRNP-type proteins (Ochs et al. 1985; Lischwe et al. 1985; Nichols et al. 2002). Just as we finished sequencing the *Nopp140-like* cDNA, the *Drosophila* genome became available, and a BLAST search identified conceptual gene *CG7421*, now called *Nopp140*. Two translation products were predicted by alternative splicing: the RGG-containing version that we had in hand and a true Nopp140 ortholog with 64 and 65% sequence identity in its carboxy terminus when compared to the carboxy termini of rat and human Nopp140, respectively. The two *Drosophila* isoforms are now referred to as Nopp140-RGG and Nopp140-True: they share the same amino-terminus and large alternating acidic and basic central domain, but they differ in their carboxy-terminal domains (Waggener and DiMario 2002). Both proteins contain CKII and Cdk1 phosphorylation sites, and they show slower than expected mobility on SDS-gels, migrating at approximately 125–127 kDa. Both proteins localize to nucleoli when expressed in *Drosophila* Schneider II culture cells or in transgenic embryos, larvae, and adults (McCain et al. 2006). Interestingly, with the *Drosophila* genome now well annotated, it appears that *Drosophila* does not encode a close homolog of vertebrate nucleolin, the original target of our cDNA library screen.

Two Nopp140 isoforms were also identified in *Trypanosome brucei* (Kelly et al. 2006). *Tb*Nopp140 and the *Tb*Nopp140-like protein (*Tb*NoLP) share the same central alternating acidic and basic repeat domain. While the C-terminus of *Tb*Nopp140 is similar to the C-terminus of yeast Srp40, the carboxy terminus of *Tb*NoLP also contains a RGG domain, similar to the Nopp140-RGG isoform in *Drosophila*. Both Nopp140 isoforms in *T. brucei* localize to nucleoli, both are phosphorylated, and interestingly, both can be coprecipitated with antibody directed against RNA Pol I (Kelly et al. 2006).

The existence of Nopp140-RGG isoforms in *Drosophila* and *Trypanosome* suggests that these isoforms may have similar interactions and perform similar functions as that of vertebrate nucleolin, at least with respect to its carboxy RGG domain. Similar RGG domains exist in many other RNA-associated proteins, usually near their carboxy termini. These RGG domains are known to bind RNA either directly or indirectly (Kiledjian and Dreyfuss 1992; Godin and Varani 2007). The arginines within the tripeptide repeats are asymmetrically dimethylated (reviewed by McBride and Silver 2001), the domain forms a series of β turns (collectively, a β-spiral) (Ghisolfi et al. 1992), and it likely binds G-quartet RNA structures (e.g., Darnell et al. 2001; Ramos et al. 2003). Precisely what the RGG domain in the Nopp140 *Drosophila* and *Trypanosome* splice variants is doing remains unknown. The analogous carboxy RGG domain in nucleolin has been reported to bind to both RNA and protein (reviewed by Ginisty et al. 1999; see Chap. 9).

11.3 Detailed Molecular Structures of Nopp140 Proteins

Figure 11.1 summarizes the linear domain organization of the various Nopp140 orthologs. The amino terminal portion of Nopp140 contains a LisH (Lis1-homology) motif that is generally accepted to be a dimerization domain (Kim et al. 2004). The LisH motif was originally described in Lis1 from *Mus musculus* (residues 6–39). Lis1 is required for normal neuronal migration during cerebral cortex development; mutations in the human *Lis1* gene lead to Miller-Dieker lissencephaly ("smooth brain"), a defect leading to severe retardation, epilepsy, and eventually death (Reiner et al. 1993; Emes and Ponting 2001). The LisH motif in human Nopp140 consists of amino acid residues 9–42 in which residues K17, K22, and A31 are well conserved in other LisH motifs (Kim et al. 2004). The LisH motif contains two alpha helices necessary for dimerization; two LisH motifs form a homodimer by assembling their alpha helices into a four-helix anti-parallel bundle (Kim et al. 2004; Mateja et al. 2006). Presence of the LisH motif in the amino terminus of Nopp140 suggests homo-dimerization or perhaps hetero-dimerization with another LisH-containing protein.

Mammalian Nopp140 contains a large central domain consisting of 10 repetitive alternating acidic and basic motifs. Nopp140 is heavily phosphorylated (Pfeifle and Anderer 1984), and its pI is quite acidic; for instance, the pI for *Xenopus* xNopp180 is ~4.2 (Schmidt-Zachmann et al. 1984). The deduced rat protein is computationally determined to be 73.4 kDa, but shifts to the apparent weight of 140 kDa in SDS-gels. This shift is likely due to extensive phosphorylation and the resulting high charge density. Dephosphorylation by alkaline phosphatase dramatically shifts Nopp140 downward on SDS-gels compared to the phosphorylated form (Meier and Blobel 1992; Cairns and McStay 1995). Nopp140 expressed in bacteria can be phosphorylated extensively by CKII in vitro (Meier 1996). In fact, rat Nopp140 contains 82 serine residues within the 10 acidic motifs of the central domain; 45 of these serines are recognizable CKII consensus phosphorylation sites

(S/T-X-X-D/E), but once a particular serine is phosphorylated, it serves as the critical acidic residue at the C-terminal side of what becomes another CKII phosphorylation site (Meier and Blobel 1992). Interestingly, Nopp140 forms a stable complex with the β regulatory subunit of CKII in vitro and likely in vivo (Li et al. 1997). Thus the majority of the serine residues within the central domain are probably phosphorylated in vivo. This phosphorylation is required for Nopp140's interaction with snoRNPs (see below).

Besides the acidic motifs, the basic motifs within the large central domain of rat Nopp140 contain a total of 19 protein kinase C (PKC) consensus sites (Meier and Blobel 1992), suggesting regulation by calcium dependent signaling pathways. Most of these PKC sites also form Cdk1/cyclin B phosphorylation sites, and as expected, Cdk1/cyclin B phosphorylation of Nopp140 increases in mitosis, suggesting a link between its M-phase phosphorylation and initial redistribution to the cytoplasm (Pai et al. 1995). Finally, a highly conserved protein kinase A (PKA) site resides in the carboxy terminus of Nopp140 (Meier 1996; Chiu et al. 2002; Kim et al. 2006).

The carboxy terminus of Nopp140 is the most conserved region of the various Nopp140 proteins described (Meier 1996). The terminus actually consists of two identifiable subdomains (NoppCa and NoppCb) encoded by their own exons. This is also the case for the Nopp140-True isoform in *Drosophila* (Waggener and DiMario 2002), indicating exon conservation in *Nopp140* gene organization. The precise functions of these carboxy subdomains remain unknown, but their properties in rat Nopp140 have been well described (Isaac et al. 1998). For instance, the conserved PKA phosphorylation site (Ser_{685} in rat Nopp140 and Ser_{670} in *Drosophila* Nopp140-True) suggests that Nopp140 may be a substrate for signal transduction-mediated phosphorylation cascades that regulate molecular interactions within nucleoli or Cajal bodies (CBs) (Meier 1996).

11.4 Nopp140's Nucleolar Locations and Associations

The location of Nopp140 inside nucleoli has been somewhat controversial: early reports claimed that Nopp140 resides in the DFC of interphase cells (Schmidt-Zachmann et al. 1984; Pfeifle et al. 1986), while later reports (Vandelaer and Thiry 1998; Thiry et al. 2009) using different fixation techniques indicate that Nopp140 can be detected in the fibrillar centers, preferentially on the peripheral edge. There is no significant amount of Nopp140 in the granular regions or in the nucleoplasm.

The location of Nopp140 during mitosis has been equally controversial. While three reports (Pfeifle et al. 1986; Weisenberger and Scheer 1995; Vandelaer and Thiry 1998) describe the association of Nopp140 with M-phase NORs, other reports claim that Nopp140 does not localize to the NORs (Schmidt-Zachmann et al. 1984; Pai et al. 1995; Dundr et al. 1997; Tsai et al. 2008). The most recent study shows that Nopp140 localizes initially to the nucleoplasm in between the chromosomes during prophase, and that it redistributes to perichromosomal regions, to nucleolar derived foci (NDF), and to the cytoplasm, but not to NORs from prometaphase to telophase (Thiry et al. 2009). In telophase, Nopp140 enters reforming nucleoli without detectable association with prenucleolar bodies (Dundr et al. 1997; Thiry et al. 2009).

Similarly, McCain et al. (2006) used GFP-Nopp140 as a marker for nucleologenesis in *Drosophila* embryogenesis, and showed that initial nucleolar formation in stage 13 and 14 blastoderm nuclei occurred without Nopp140's apparent localization to prenucleolar bodies. Interestingly, the first cells to form in the *Drosophila* embryo are the primordial germ cells (pole cells), but these cells lack nucleoli during the blastoderm stages because of repressed DNA transcription (Deshpande et al. 2004). Again, GFP-Nopp140 appeared dispersed within the pole cell nuclei during the blastoderm stages. Pole cells form nucleoli within minutes just as they begin their migration at the start of gastrulation. The dispersed GFP-Nopp140 coalesced rapidly into the forming nucleoli. Contrary to these studies that showed a diffuse distribution of Nopp140 prior to nucleologenesis, Baran et al. (2001) showed that Nopp140 in one-, two-, and four-cell mouse embryos, localizes to peripheral patches of nucleolus precursor bodies (NPBs, the peripheral zones being analogous to prenucleolar bodies) and that Nopp140 shifts to the cortex of the NPB (analogous to the NOR) as rDNA transcription begins in the two-cell embryo.

11.5   Molecular Interactions Indicate Function

Neither deletion of yeast Srp40 (Meier 1996), nor the knock-down of *Drosophila* Nopp140 (Cui and DiMario 2007) seemed to alter nucleolar structure. When over-expressed, however, full length human Nopp140 caused the redistribution of RNA Pol I and largely disrupted nucleolar integrity (Chen et al. 1999). Similar findings were reported for both *Drosophila* Nopp140 isoforms: when the *Drosophila* GFP-Nopp140-True was over-expressed in transgenic larvae, nucleoli within the polyploid cells appeared swollen and disorganized. Over-expression of the GFP-Nopp140-RGG isoform completely disrupted the nucleoli (Cui and DiMario 2007). Taken together, these observations suggest that Nopp140 is not required for nucleolar formation, but its over-expression disrupts nucleolar integrity and function.

Isaac et al. (1998) carefully examined what roles the individual amino terminal, the large central repeat domain, and the carboxy terminal domains have in nucleolar and Cajal body localization and retention. The amino terminal domain (NoppN) was expressed as a fusion (GFP-NLS-HA-NoppN); it localized to the nucleoplasm and the cytoplasm, but it failed to localize to nucleoli or CBs. NoppN contains a putative NES, perhaps explaining its cytoplasmic enrichment. Endogenous Nopp140 was not affected by the over-expression of NoppN.

When over-expressed as a fusion to HA, the conserved C-terminus of Nopp140 (referred to as HA-NoppC) localized to nuclei, but it acted as a dominant negative in that it caused the redistribution of full length Nopp140, the Nopp140-associated protein of 57 kDa (NAP57), and fibrillarin from nucleoli to nucleoplasmic granules. Nucleolin and UBF remained within the nucleoli, which maintained their normal structural integrity. Over-expression of NoppC also caused the disassembly of CBs as judged by the dispersion of p80 coilin to the nucleoplasm. As described above, the carboxy terminal domains of mammalian Nopp140 and *Drosophila* Nopp140-True are encoded by two separate exons. Isaac et al. (1998) showed over-expression of the individual peptides encoded by the individual exons (NoppCa and NoppCb) caused the same dominant-negative phenotypes on nucleoli (redistribution of nucleolar components) and CBs (disruption) as did NoppC itself.

Interestingly, Isaac et al. (1998, 2001) showed that over-expression of the large central domain of Nopp140 (referred to as NoppR) in COS-1 cells caused the formation of phase-dark nuclear rings of 0.5–5 μm diameter. They called these structures R-rings. Over-expression of full-length human Nopp140 can also form R-rings (Kittur et al. 2007). Like NoppC, NoppR caused a dominant-negative effect by redistributing endogenous Nopp140, fibrillarin, NAP57, UBF, and Pol I from nucleoli to the R-rings; however, neither nucleolin nor B23 were affected. Newly synthesized p80 coilin also localized to R-rings, but other Cajal body components (e.g., Sm antigens) failed to redistribute to the R-rings. Subsequent examination of R-rings (Isaac et al. 2001) revealed multilamellar membrane stacks that appear identical to the previously described nucleolar channel system (NCS) found only in postovulation human endometrial cells that are receptive to blastoderm implantation (see references in Isaac et al. 2001).

Kittur et al. (2007) examined the R-rings in even greater detail. The rings apparently form by invagination of the inner nuclear membrane into the nucleoplasm. They used immuno-fluorescence microscopy to show that R-rings form by the accumulation of the highly charged NoppR in patches on the underside of the nuclear envelope. They showed that Nopp140 complexes with calcium in a phosphorylation-dependent manner, and then used electron spectroscopic imaging to show that R-ring formation likely occurs via a calcium-mediated interaction between the multiple phosphates on NoppR and the inner nuclear membrane. The stacked membranes of the R-rings lie within an electron dense matrix that contains Nopp140, its bound calcium, and associated nucleolar components. R-rings are often found in close proximity to the nuclear envelope and nucleoli. Because of their derivation from the nuclear envelope and thus the ER, R-rings contain a mix of rough and smooth ER-associated membrane and luminal proteins. These include calnexin, Sec61, the IP_3 calcium channel, the receptor for the signal recognition particle, BiP, PDI, HMG-CoA reductase, and glucose-6-phosphatase. R-rings, however, are distinct from the nuclear envelope in that they lack the lamin-associated protein, LAP2, nucleoporin p62, and lamin B. R-rings are morphologically indistinguishable from NCSs, and like R-rings, NCSs contain calnexin, BiP, and glucose-6-phosphatase. NCSs, however, contain less Nopp140 and calcium than do the R-rings. What induces NSC formation, what role NSCs play in the receptive phase of the human endometrium, and what molecular relationships they may have with nucleoli remain exciting avenues of exploration.

11.6 Nopp140, a Chaperone for Small Nucleolar Ribonucleoproteins (snoRNPs)

The first indication that Nopp140 interacts with snoRNPs was the discovery that rat NAP57 (dyskerin in humans, Cbf5p in yeast) could coimmunoprecipitate with Nopp140 (Meier and Blobel 1994). Immuno-fluorescence microscopy showed that NAP57, like Nopp140, localized to both nucleoli and CBs, with NAP57 localized primarily in the nucleolar DFCs (Meier and Blobel 1994). NAP57 is a pseudouridy-lase, a component of box H/ACA RNPs, which consist of four proteins (NAP57, GAR1, NHP2, and NOP10) and one of several box H/ACA guide RNAs (Ganot et al. 1997; Henras et al. 1998; Lafontaine et al. 1998; Watkins et al. 1998). Immunoprecipitation of Nopp140 also identified intact snoRNP complexes that contained H/ACA guide RNAs (Yang et al. 2000). Box H/ACA snoRNPs function in site-specific pseudouridylation of pre-rRNA processing, pre-mRNA splicing, and telomere maintenance (reviewed by Meier 2005). In vitro pseudouridylation by box H/ACA snoRNP complexes occurred in an energy and helicase independent reaction *without* the association of Nopp140 (Wang et al. 2002), suggesting that Nopp140 itself is not required for the snoRNP enzymatic reaction. The functional model as proposed by Wang et al. (2002) states that within the nucleolus, Nopp140 acts as a scaffold for multiple snoRNPs as they modify the pre-rRNA. Further, Nopp140's association with at least the box H/ACA snoRNPs is dependent on its extensive CKII phosphorylation, indicating the association between Nopp140 and snoRNPs is electrostatic and reversible.

Besides box H/ACA RNPs, Nopp140 weakly associates with C/D box snoRNPs (Yang et al. 2000) that perform site-specific methylation of the pre-rRNA (Tollervey et al. 1993; Kiss-László et al. 1996; Nicoloso et al. 1996). Components of box C/D RNPs include the four core proteins, NHP2L1/15.5, NAP65, Nop56, and fibrillarin, which is the RNA methyl-transferase. Fibrillarin and NAP65 were found in Nopp140 immunoprecipitates, but under less stringent conditions, suggesting that Nopp140's interactions with C/D box snoRNPs are not as strong as its interactions with box H/ACA snoRNPs (Yang et al. 2000). In a study to determine association between Nopp140 and the specific box U3 C/D box snoRNP complex, Watkins et al. (2004) found Nopp140 associated with both precursor and mature U3-containing snoRNPs in nuclear extracts, but antibodies against U3 snoRNP core proteins (e.g., Nop56, Nop58, fibrillarin) failed to coprecipitate Nopp140 from *nucleolar* extracts. Watkins et al. (2004) concluded that Nopp140, along with two other putative assembly factors, TIP48 and TIP49, participates as a snoRNP biogenesis factor in the nucleoplasmic phase of U3 RNP assembly (perhaps in CBs, see below), but that Nopp140 dissociates from the mature form of the U3 snoRNP once inside the nucleolus.

Determining the role of yeast Srp40 in snoRNP biosynthesis/maintenance has been difficult because of its dependence on the *Shm2* (previously called *LES2*) gene product (Yang et al. 2000; Yang and Meier 2003). *Shm2* encodes a cytosolic serine hydroxymethyltransferase involved in one-carbon metabolism, converting tetrahydrofolate (THF) to 5, 10-methylene THF. Yang et al. (2000) first showed that the single mutations, *srpΔ* or *shm2*, have slight growth defects, and that the double mutant can be rescued by *SRP40* expression from a plasmid either from its own endogenous promoter or from the *GAL10* (conditional) promoter. Depleting Srp40 in the conditional double mutant led to reductions of several box H/ACA RNAs (snR3, snR10, snR11, snR42, and the required snR30), but not the box C/D RNAs U3, U14, and U24 as determined by Northern analyses (Yang et al. 2000; Yang and Meier 2003). The observation indicates that Srp40, like Nopp140, is likely to have a greater role in box H/ACA snoRNP interaction/biosynthesis than it has in that of box C/D snoRNPs. In fact, loss of box H/ACA snoRNAs by depletion of Srp40 is similar to phenotypes observed for the loss of individual H/ACA box proteins, Cbf5, Nhp2p, and Nop10p (Henras et al. 1998; Lafontaine et al. 1998; Watkins et al. 1998).

To further define how *Shm2* might interact with *Srp40*, Yang and Meier (2003) first showed that the triple mutant strain, *srp40Δ shm3 ade3,* is synthetic lethal. Like Shm2, ADE3 is a cytosolic enzyme involved in one-carbon metabolism, producing 5, 10-methylene THF from formate and THF in three steps. Yang and Meier (2003) showed that *SHM2, SRP40,* or *ADE3* could rescue the triple-synthetic lethality when expressed separately from *LEU2* plasmids using their own endogenous promoters. The mechanistic link between Srp40 and the two cytosolic enzymes involved in one-carbon metabolism is perplexing, but Yang and Meier (2003) showed that *catalytic* mutants of Shm2, expressed in the synthetic lethal strain, actually complemented growth. This indicates that loss of one-carbon metabolites is not the reason for lethality, and that Shm2 (and perhaps Ade3) may have secondary, non-catalytic functions related to Srp40. This possibility was strengthened by over-expressing Lsm5 which partially restored growth of the triple mutant. Lsm5 normally resides in yeast Sm-like complexes, and Lsm5 likely has several roles in nuclear RNA (e.g., tRNA) processing. Yang and Meier (2003) showed that ectopic Lsm5 expression provided a partial growth rescue which correlated with partial restorations in box H/ACA snoRNPs snR3 and snR10 abundance. Apparently, Lsm5 interacts with Shm2 as determined by a genome-wide two-hybrid assay, and Yang and Meier (2003) concluded that Lsm5 links the cytosolic enzyme, Shm2, with box H/ACA snoRNPs, and therefore Srp40. The mechanistic details of these interactions, however, remain unknown.

11.7 Nopp140 and Cajal Bodies

Isaac et al. (1998) showed that newly synthesized Nopp140 in transfected culture cells localizes first to nucleoli and then to CBs, suggesting that Nopp140 shuttles snoRNPs from nucleoli and CBs. Using the *Xenopus* oocyte system, Bellini and Gall (1999) showed that shuttling Nopp140 appears simultaneously within nucleoli and CBs as it reenters the nuclei from the cytoplasm, and they reasoned there may be a difference between newly synthesized Nopp140 just arriving to the nucleus for the first time (i.e., Isaac et al. 1998) and mature Nopp140 that shuttles between the nucleolus, CBs, and cytoplasm. What these differences may be remains unknown, but one could easily imagine extensive CKII phosphorylation on the mature Nopp140 versus nascent Nopp140 as a possible determinant in CB localization. Regardless of the differences, it is now well accepted that Nopp140 is the likely chaperone for snoRNPs between CBs and nucleoli.

Most snoRNAs are encoded as introns, and they assemble into snoRNPs without ever leaving the nucleus. Conversely, snoRNAs U3, U8, and U13 are transcribed from their own genes by RNA Pol II and thus have an initial m_7G cap, but they too remain in the nucleus where their 5' caps are trimethylated, and where they too assemble into snoRNPs prior to their delivery to the nucleolus (Narayanan et al. 1999; Verheggen et al. 2002; Boulon et al. 2004). The CB is the nuclear compartment associated with spliceosomal assembly, preassembly of transcription complexes, and the processing of snoRNAs (reviewed by Nizami et al. 2010). Meier and Blobel (1994) first found Nopp140 in CBs by immuno-fluorescence microscopy. This was later confirmed by immuno-electron microscopy with the colocalization of Nopp140 and p80 coilin (Vandelaer and Thiry 1998), a generally accepted marker protein for CBs (see Nizami et al. 2010). Nopp140 can interact with amino terminus of p80 coilin (Isaac et al. 1998); however, the retention time of Nopp140 in CBs is shorter than that of p80 coilin or the survival of motor neuron (SMN) protein; in fact, the transit time for Nopp140 in CBs is similar to the transit times for snoRNP proteins GAR1 (box H/ACA snRNPs) and fibrillarin (box C/D snoRNPs), suggesting an interaction between Nopp140 and these snoRNPs, perhaps for their biogenesis or remodeling while in the CBs (Dundr et al. 2004).

Several studies have examined either the appropriate levels of Nopp140 required for CB integrity, or the localization of Nopp140 to CBs that are depleted of other known constituents. For instance, when over-expressed, the dominant negative NoppC described above disrupts CBs (Isaac et al. 1998). Conversely, depletion of yeast Srp40 disrupts the nucleolar body which may be the yeast complement of the metazoan CB (Isaac et al. 1998; Verheggen et al. 2001).

Lemm et al. (2006) described the RNAi knockdown of SMN or hTGS1 (the methyl-transferase that further methylates m^7G caps to yield 2,2,7-trimethyl G caps on U snRNAs and snoRNAs), and showed that residual coilin-containing nuclear foci maintained snoRNP proteins fibrillarin and Nop58. They reported (but did not show) that Nop56 and Nopp140 were also found in similar residual coilin-containing foci. Lemm et al. (2006) concluded that factors necessary for snoRNP assembly localize to a subclass of coilin-containing nuclear foci that still form in the absence of hTGS1 or SMN.

In a pivotal paper, Renvoisé et al. (2009) showed an inverse correlation between Nopp140 levels in CBs within spinal muscular atrophy (SMA) fibroblasts and the severity of the disease, suggesting the SMN protein is required for Nopp140 localization within CBs. SMA is a neuronal degenerative disease marked by low levels of SMN (for review, see Lorson et al. 2010); SMN is required for snRNP biogenesis both in the cytoplasm and in CBs (Carvalho et al. 1999). As SMN interacts with fibrillarin and GAR1 (Jones et al. 2001; Pellizzoni et al. 2001), it may also function in snoRNP assembly or maturation. Localization of Nopp140 to CBs is significantly reduced in SMA cells, and this reduction is correlated with reduced levels of box

H/ACA snoRNP proteins, GAR1, and NAP57/dyskerin, within CBs. Renvoisé et al. (2009) showed that Nopp140 localizes to CBs in nearly all (96%) COS cells that had been transiently transfected to over-express wild type SMN, while a reduced number (56%) of nontransfected control cells contained Nopp140 within their CBs. Three SMN mutants (SMN472Δ5, SMNexΔ7, and SMNE134K) display progressively severe phenotypes, and they reduce the accumulation of Nopp140 in CBs to correspondingly greater extents (Renvoisé et al. 2009). RNAi knockdown of SMN in control fibroblasts also reduced Nopp140 levels in the CBs, while over-expression of wild-type SMN in primary SMA cells restored Nopp140 levels in the CBs. Although Nopp140 has been shown to interact directly with p80 coilin (Isaac et al. 1998), it is now apparent that wild-type SMN is required for the accumulation of Nopp140 within CBs. The precise function of Nopp140 in CBs remains unknown, but it is becoming increasingly clear that Nopp140 acts with SMN in vital aspects of snoRNP biogenesis or remodeling within the CBs. In the least, Nopp140 can now be used as a CB marker to gauge the severity of SMA (Renvoisé et al. 2009).

11.8 Nopp140 as a Transcription Factor

Intriguing studies indicate that Nopp140 acts as a transcription factor for at least one Pol II gene (Lee et al. 1996; Miau et al. 1997). A C/EBP family member, AGP/EBP, was previously known to induce the acute phase response α_1-acid glycoprotein (*AGP*) gene. In searching for other factors that coactivate the *AGP* gene, the authors identified Nopp140 by coimmunoprecipitation with AGP/EBP followed by LC/MS/MS. Control experiments verified that Nopp140 bound to AGP/EBP in a defined complex rather than to the AGP/EBP antibody. Cotransfection of BHK cells with a reporter construct, *AGP-CAT*, and either *CMV-Nopp140* or *CMV-Nopp140-Reverse* showed enhanced CAT activity only with Nopp140 expression. As there are no known nucleic acid binding domains in vertebrate Nopp140, its coactivation of *AGT-CAT* must be via interaction with identifiable DNA-binding transcription factors.

To verify this possibility, Miau et al. (1997) cotransfected *CMV-Nopp140* and *CMV-AGP/EBP* expression plasmids along with the reporter plasmid and showed that both Nopp140 and AGP/EBP interact synergistically to activate expression of the *AGP-CAT* reporter gene. Functional (CAT) assays using deletions for both Nopp140 and AGP/EBP initially suggested that the carboxy terminal portion of Nopp140 (residues 347–704) is required to interact with the amino-terminal portion of AGP/EBP (residues 21–151). The authors initially concluded that Nopp140 bound to AGP/EBP by way of these identified regions as AGP/EBP bound to its three cognate DNA elements within the *AGP* promoter region. Further work revealed that Nopp140's role in coactivation of *AGP-CAT* is mediated by an additional interaction between Nopp140 and TFIIB. Specifically, the carboxy terminal portion of Nopp140 is critical for its in vitro interaction with TFIIB. The main conclusion of the 1997 paper is that synergistic activation of *AGT-CAT* reporter gene is via a Nopp140-AGP/EBP-TFIIB ternary complex. The one caveat in these experiments is the possible over-expression of Nopp140 and AGP/EBP from strong CMV promoters. A follow-up report (Chiu et al. 2002) found that PKA phosphorylates rat Nopp140 at Ser[113], Ser[627], and Ser[628]. Nopp140 phosphorylated by PKA activates *AGP* gene expression in a synergistic manner with CREB and C/EBPβ, while a mutant version of Nopp140 devoid of the site Ser[627] could not achieve this synergistic activation.

Nopp140 also has a putative role in Pol I transcription. Chen et al. (1999) immunoprecipitated endogenous human Nopp140 from CEM and HeLa cells and showed by SDS-PAGE and mass spectroscopy that the 190 kDa subunit of RNA Pol I coprecipitated. The other protein to coprecipitate was the alpha subunit of CKII, suggesting that Nopp140, CKII, and Pol I form a complex. They reported the same coprecipitation using anti-FLAG to pull down exogenously expressed FLAG-Nopp140. As other reports documented (Schmidt-Zachmann et al. 1984; Pfeifle et al. 1986), Chen et al. (1999) showed Nopp140 colocalizes with Pol I in dot-like structures within the nucleolar DFCs, suggesting a potential interaction. Actinomycin D-mediated segregation of nucleoli maintained similar colocalizations between Nopp140 and Pol I. With low level expression, FLAG-Nopp140 localized to similar dot-like structures within the DFCs, but over-expression of FLAG-Nopp140 clearly disrupted nucleolar morphology, producing large hypertrophied nucleoli. With this over-expression, nucleolin redistributed to the nucleoplasm while Pol I and fibrillarin remained associated with the FLAG-Nopp140 in the enlarged nucleoli, again suggesting possible interactions. Chen et al. (1999) went on to use a Nopp140 deletion series and coimmunoprecipitation to show that the region spanning residues 204–382 (middle portion of the large central domain) interacts with RPA194, the large subunit of Pol I. Exogenous expression of this Nopp140 region (residues 204–383) now tagged with FLAG and an NLS appeared to displace endogenous Pol I in a dominant negative manner. Over-expression of full length Nopp140 or Nopp140 depleted for its carboxy half (Nopp140N382, still containing residues 204–382) resulted in segregation of nucleoli and a block in Pol I transcription as measured by Br-UTP incorporation, similar to the effects of actinomycin D. Chen et al. (1999) concluded that Nopp140N382 competed in a dominant-negative manner with endogenous Nopp140 for Pol I. This was the first description of Nopp140 affecting rRNA transcription.

Yang et al. (2000) then showed that expression of just the conserved carboxy tail of Nopp140 (NoppC) displaced endogenous Nopp140 from nucleoli in a dominant-negative manner, and blocked Pol I transcription as monitored by BrUTP incorporation. Interestingly, Pol I remained in position within these nucleoli. Kelly et al. (2006) also coprecipitated both isoforms of Nopp140 in *Trypanosome* using an antibody against Pol I, adding more validity to the possibility that Nopp140 directly interacts with Pol I as a transcription factor in rDNA transcription. One of the most intriguing hypotheses put forth regarding Nopp140 is that its association with Pol I could provide a molecular link between pre-rRNA transcription and processing, perhaps providing a feedback mechanism to regulate Pol I transcription when ribosome production is perturbed (Chen et al. 1999; Yang et al. 2000).

11.9 Organismal Depletion of Nopp140

We finish our discussion of Nopp140 by describing perturbations in development when it is depleted (Cui and DiMario 2007). RNAi-mediated depletion of *Drosophila Nopp140* mRNAs was measured by RT-PCR, and the loss of Nopp140 was determined by immunofluorescence microscopy. Depletions of *Nopp140* transcripts by 50% or greater caused late larval and pupal lethality; however, a partial depletion of 30% permitted adults to survive, but these adults displayed deformed legs, wings, and cuticle. The defects were reminiscent of craniofacial malformations associated with the Treacher Collins syndrome due to the loss of the related nucleolar protein, treacle (see below). Our initial results suggested that larval diploid precursor cells (imaginal disc cells that generate legs and wings, and histoblasts that generate the adult cuticle) have higher demands for ribosome biogenesis, and are thus more sensitive to ribosome loss. Preliminary results clearly show abundant anti-caspase 3 labeling in wing discs isolated from larvae that express RNAi that depletes Nopp140. Loss of imaginal wing disc cells by apoptosis is thus the most likely explanation for the morphological defects due to loss of Nopp140. Terminally differentiated larval polyploidy cells (i.e., larval midgut cells) appear to respond differently to the loss of Nopp140 by inducing autophagy rather than apoptosis. How different cells respond to the loss of ribosomes may prove to be much more complicated than originally anticipated.

11.10 Treacle and the Treacher Collins–Franceschetti Syndrome 1

Treacle is a nucleolar phosphoprotein structurally related to Nopp140 (Wise et al. 1997; Marsh et al. 1998; Winokur and Shiang 1998; Isaac et al. 2000; Fig. 11.1). Treacle has been studied primarily in mammals (human, mouse, dog) (Dixon et al. 1997; Paznekas et al. 1997, Haworth et al. 2001), but an ortholog exists in *Xenopus* (Gonzales et al. 2005a, b) indicating that treacle is a vertebrate protein. Human treacle is encoded by the *TCOF1* gene at 5q32-q33.1 (Jabs et al. 1991; Dixon et al. 1993; Dixon 1996). *TCOF1* is greater than 20 kbp in length, and it contains 27 exons (see So et al. 2004). Three isoforms of human treacle exist because of alternative splicing; the original human isoform described is 1,411 amino acids in length, but the predominant isoform in terms of abundance is 1,488 residues in length (So et al. 2004). Human treacle has a highly conserved amino terminus of 213 amino acid residues. Like Nopp140, treacle contains a LisH motif (amino acids 5–38) (Emes and Ponting 2001; Kim et al. 2004). The amino terminus is followed by 11 repeating units (10 in the originally described isoform). Each repeat consists of an acidic and a basic motif. Similar to Nopp140, the acidic motifs in treacle are serine-rich with many putative CKII and PKC phosphorylation sites, while the basic motifs are rich in lysine, alanine, and proline. Human treacle

11 Structure and Function of Nopp140 and Treacle   267

expressed in *E. coli* has a predicted size of 144 kDa (Marsh et al. 1998), but the native, highly phosphorylated protein from human cells migrates anomalously at ~220 kDa on SDS-gels presumably because of the extensive phosphorylation and charge density (Isaac et al. 2000).

The carboxy tail of human treacle contains several functional NLSs, and the last 41 residues are necessary for nucleolar retention (Marsh et al. 1998; Winokur and Shiang 1998; Isaac et al. 2000); nonsense mutations yield treacle truncations that fail to translocate into the nucleus or localize within nucleoli (Marsh et al. 1998; Winokur and Shiang 1998). Similar truncations are frequently associated with the Treacher Collins syndrome in humans (Wise et al. 1997).

In mouse and humans, treacle is expressed in a wide variety of embryonic and adult tissues (Dixon et al. 1997; Paznekas et al. 1997), but most significantly, treacle expression is elevated in the embryonic neural folds just prior to neural tube fusion and in the first pharyngeal arch, coincident with primordial tissues known to give rise to craniofacial structures. Mutations in *TCOF1* give rise to the autosomal dominant Treacher Collins–Franceschetti syndrome (TCS; Fazen et al. 1967; Dixon 1996, Trainor et al. 2009). TCS is the most common of congenital craniofacial disorders in humans, afflicting 1 in 50,000 live births (Wise et al. 1997). Defects include hypoplasia of the facial mandible and zygomatic complex, coloboma (lesion) of the lower eyelids, a lack of eye lashes medial to the eye lid defect, downward slanting palpebral fissures, a high incidence of cleft palate, and conductive hearing loss due to malformation of the outer ear and the middle ear ossicles (Dixon 1996). Higher than expected polymorphisms exist within *TCOF1* (Teber et al. 2004), and they may account for the variable expressivity of the TCS.

Disruption of the murine *Tcof1* gene caused severe craniofacial anomalies and perinatal death in *Tcof1*[+/−] mice (Dixon et al. 2000; Dixon et al. 2006). Deletions of 1–40 nucleotides are the most common genetic defects, but insertion-type, splicing, and nonsense mutations also exist (Trainor et al. 2009). The craniofacial defects were traced back to apoptosis in embryonic neural crest cells within the cranial neural folds, specifically a subset of cephalic neural crest cells that display relatively high *Tcof1* expression. To better establish the link between *Tcof1*/treacle and the TCS, Dixon et al. (2000) replaced the first exon in the mouse *Tcof1* gene with the neomycin resistance cassette in embryonic stem cells. Germ line chimeric males were prepared and crossed to wild-type females to produce heterozygous *Tcof1*[+/−] embryos. These heterozygous embryos displayed several major craniofacial deformities beginning at day 8 of development (E8). Whole mount TUNEL assays of E9 *Tcof1*[+/−] embryos showed excessive amounts of apoptosis in the neuro-epithelium of the cranial neural folds and in the neural tube compared to wild-type litter mates. Anti-neurofilament labeling of E10.5 *Tcof1*[+/−] embryos showed a loss of neural crest cell-derived structures such as cranial ganglia, the ophthalmic branch of the trigeminal nerve, the glossopharyngeal ganglia, and an underdevelopment of the dorsal root ganglia. These heterozygous *Tcof1*[+/−] mice died shortly after birth. Interestingly, the particular genetic background of the heterozygous *Tcof1*[+/−] embryos has a significant effect on the penetrance and severity of the cranial defects (Dixon and Dixon 2004). For example, *Tcof1*[+/−] mice with inbred CBA, C57BL6, or C3H genetic backgrounds were lethal displaying severe morphological abnormalities, while the majority of $Tcof1^{+/-}$ mice with DBA/1 and BALB/c backgrounds were normal and viable. This variation is likely due to factors in the different genetic backgrounds, but the identity of these factors remains unknown.

Dixon et al. (2006) showed that treacle is required in a cell-autonomous, spaciotemporal manner for rapidly proliferating cephalic neural crest cells with apparent high demands in protein synthesis. Cells of the neuro-epithelium and neural crest-derived craniofacial mesenchyme in $Tcof1^{+/-}$ E8.75–E9 embryos showed relatively few ribosomes by antibody labeling (mouse monoclonal anti-rRNA antibody, Y10B) compared to wild-type littermates. Induction and migration of cephalic neural crest cells were not affected by the loss of treacle. Rather, treacle was required for ribosome biogenesis, and spatiotemporal haplo-insufficiency of treacle led to apoptosis and loss of these neural epithelial cells finally resulting in TCS. But why these particular neural crest cells are sensitive to the loss of treacle (ribosomes) at this point in development remains uncertain. Malformations associated with TCS are restricted to the head and neck regions, suggesting that other embryonic progenitor cells must have either sufficient amounts of treacle (ribosomes) or lower demands for protein synthesis.

In a seminal study, Jones et al. (2008; see also McKeown and Bronner-Fraser 2008; Sakai and Trainor 2009) showed that the partial loss of treacle in $Tcof1^{+/-}$ mouse embryos (E8.5–E10.5) led to p53 stabilization in the neuroepithelium, and in turn to p53-induced G_1-cell cycle arrest, apoptosis, and ultimately hypoplasia of cranioskeletal structures. Remarkably, however, they were able to rescue this hypoplasia by deleting the $p53$ gene. Rescue occurred in a $p53$ dose-dependent manner; that is they observed complete rescue with $p53^{-/-}$ versus partial rescue with $p53^{+/-}$. Immuno-staining with the Y10 mAb showed that deleting the $p53$ gene had no effect (neither decline, nor restoration) in ribosome biogenesis in the neural crest cells. This indicates that loss of ribosomes was not the direct inducer of apoptosis in these cells, but rather a nucleolar stress response caused by treacle insufficiency somehow triggered p53 stabilization, cell cycle arrest, and finally apoptosis.

The nucleolar stress response is just now coming into focus. While Isaac et al. (2000) showed that the abundance of full length treacle does not vary by more than twofold in fibroblasts derived from both normal individuals and TCS patients, Jones et al. (2008) suggested that the embryonic neural crest cells have a higher threshold requirement for a specific level of treacle due to its high rate of proliferation (Trainor, pers. comm.). A resulting deficiency in ribosome biogenesis thus leads to stress in these neural crest cells. All forms of cell stress disrupt nucleoli to some extent (Rubbi and Milner 2003), and strong evidence now indicates that nucleoli act as stress sensors (Rubbi and Milner 2003; Olson 2004; Ma and Pederson 2008). For example, one hypothesis holds that stress-induced nucleolar disruption in mammalian cells releases p19[ARF] to the nucleoplasm where it blocks the p53-specific ubiquitin ligase, MDM2. Activated p53 acts as a negative regulator of ribosome biogenesis by disrupting normal interactions between RNA Pol I and the upstream binding factor (UBF) and the selectivity factor (SLI) (Zhai and Comai 2000), thereby compounding the loss of nucleolar function. Once stabilized, p53 induces proapoptotic *Bcl* family member genes *Bax* and *Bak* whose protein products facilitate release of cytochrome c from the mitochondria thus inducing a cascade of caspase activity and the initiation of apoptosis.

11.11 Treacle Function

Within mammalian cells, treacle localizes to nucleolar DFCs, but unlike Nopp140, it fails to localize to CBs (Isaac et al. 2000). As far as we know, treacle's only role is in nucleolar ribosome biogenesis (Trainor, pers. comm.). Immunoprecipitations showed a potential association between treacle and the alpha catalytic subunit of CKII, and according to Isaac et al. (2000), there is no apparent interaction between treacle and Nopp140, or between treacle and box H/ACA snoRNP components NAP57 and GAR1. Lin and Yeh (2009), however, did detect an interaction between treacle and Nopp140 by coimmunoprecipitation, specifically between Nopp140 and treacles's carboxy terminus (a robust interaction with residues 962–1488, but a weakened interaction with residues 1294–1488).

Hayano et al. (2003) performed a proteomic analysis of Nop56, a component of nucleolar box C/D small nucleoprotein complexes that direct site specific $2'$-O-methylation of pre-rRNA. Treacle coprecipitated with Nop56-associated pre-rRNP complexes, and its association with Nop56 was independent of RNA, suggesting a protein–protein interaction between treacle and the box C/D snoRNP complexes. Conversely, precipitation of FLAG-tagged treacle-associated complexes identified Nop56. Gonzales et al. (2005b) confirmed a direct interaction between Nop56 (its C-terminal residues 367–594) and treacle, and while no direct interaction was found between fibrillarin and treacle, fibrillarin could be coprecipitated with FLAG-tagged treacle, suggesting an indirect association. The two studies therefore, indicate an interaction between treacle and C/D box snoRNPs mediated by a direct interaction with Nop56.

Gonzales et al. (2005b) further demonstrated that RNAi-mediated depletion of treacle in *Xenopus* oocytes blocked $2'$-O-methylation of nucleotide C_{427} in the 18S region of pre-rRNA, and showed that *Tcof1*[+/−] mouse embryos with either CBA or C57BL/6 genetic backgrounds that are lethal (Dixon and Dixon 2004) were also deficient in pre-rRNA methylation of the corresponding nucleotide, C_{463}. Conversely, *Tcof1*[+/−] mouse embryos with a BALB/c genetic background have no craniofacial malformations (Dixon and Dixon 2004), and they showed normal pre-rRNA methylation. While $2'$-O-methylation was adversely affected in *Tcof1*[+/−] mice with a CBA background, pseudouridylation of U_{1642} in the 18S region was not impaired in these embryos. Hayano et al. (2003) suggested that treacle acts as a chaperone similar to Nopp140, but that treacle and Nopp140 interact with box C/D snoRNPs at different stages during ribosome biogenesis. An equally intriguing possibility is that treacle may preferentially chaperone box C/D snoRNPs while Nopp140 chaperones the box H/ACA snoRNPs preferentially, and box C/D snoRNPs only marginally.

Besides acting as a chaperone for box C/D snoRNPs, treacle may also function in rDNA transcription. Treacle colocalizes with the upstream binding factor (UBF) and Pol I on mitotic NORs, suggesting a role for treacle in the rDNA transcription machinery (Valdez et al. 2004; Lin and Yeh 2009). Treacle also maintains its localization with UBF within the nucleolar caps of actinomycin D-segregated nucleoli (Valdez et al. 2004), and with Pol I in nucleolar condensed spots when UBF is depleted by siRNA (Lin and Yeh 2009). Immunoprecipitation of the FLAG-tagged and nucleolar-localized carboxy-terminal half of treacle successfully pulled down UBF, and yeast two-hybrid confirmed the interaction. Small interfering RNA-mediated depletion of treacle caused a 47% drop in pre-rRNA, suggesting an inhibition in pre-rRNA transcription which was confirmed by RNase protection assays, ^{32}P-metabolic labeling, and BrUTP incorporation (Valdez et al. 2004). Later Gonzales et al. (2005b) used ChIP analysis to show that human treacle binds rDNA within nucleotides −240 to +370, a region that contains the proximal promoter and the 5′ end of the rDNA gene encoding the 5′ ETS of the pre-rRNA.

Lin and Yeh (2009) refined these treacle interactions to −321 to −22 in the HeLa cell rDNA promoter region, and then attributed this interaction to the carboxy-terminal region (residues 1294–1488) of treacle. They showed that siRNA-mediated depletion of treacle redistributed Pol I, UBF, and Nopp140 from nucleoli, even though their overall abundance did not change as assayed by immunoblots. Lin and Yeh (2009) further showed that the central repeat domain of treacle interacts with the Pol I complex in a robust manner but that the carboxy terminus of treacle binds UBF, Nopp140, and rDNA, the last one either directly or indirectly via UBF or Nopp140. Over-expression of this carboxy terminus behaved as a dominant negative in that it caused the redistribution of Pol I, UBF, and Nopp140, resulting in a decline in rDNA transcription as determined by BrU labeling. This dominant-negative behavior of treacle's carboxy terminal domain is reminiscent of the over-expression of NoppC as described above for Nopp140. Lin and Yeh (2009) concluded that central repeat domain of treacle interacts with the Pol I complex to maintain the transcription machinery in the nucleolus, and that the carboxy terminus of treacle is responsible for interacting with the rDNA promoter to help recruit UBF.

The model emerging from the combined observations (Valdez et al. 2004; Gonzales et al. 2005b; Lin and Yeh 2009) indicates that treacle, like Nopp140, links pre-rRNA processing with rDNA transcription. A haplo-insufficiency in vertebrate treacle would therefore disrupt ribosome production at the transcriptional and pre-rRNA processing levels. A resulting loss of functional ribosomes could then stress particularly sensitive embryonic neural crest cells leading to their apoptosis and the loss of critical progenitor cells that normally give rise to adult craniofacial structures. This model explains treacle function in vertebrate cells, but many questions remain unanswered. For example, are there functional overlaps between Nopp140 and treacle in vertebrate cells in rDNA transcription and pre-rRNA processing? Or does Nopp140 perform a preferential function (e.g., the delivery of box H/ACA snoRNPs to the nucleolus) while treacle performs a chaperone function preferentially for box C/D snoRNPs? What replaces treacle function in metazoans that apparently lack treacle (e.g., *Drosophila* and *C. elegans*)? With respect to TCS, why

11 Structure and Function of Nopp140 and Treacle

is *Tcof1*[+/−] haplo-insufficient in some genetic backgrounds but not in others? What are the genetic factors affecting *Tcof1* gene expression, and how do other genes and environmental factors affect TCS severity (i.e., the nucleolar stress that leads to p53 stabilization and apoptosis)? Questions like these will require diversified research approaches. For example, in a recent study, Dauwerse et al. (2011) examined patients who clearly displayed TCS phenotypes, but who did not have mutations in their *Tcof1* genes. Instead, these patients had mutations in their *POLR1D* or *POLR1C* genes that encode RPA40 and RPA16, respectively. RPA40 and RPA16 are shared subunits of RNA polymerases I and III, and are known to interact (Yao et al. 1996). Thus, TCS is a ribosomopathy (Dauwerse et al. 2011) caused by mutations in diverse genes whose protein products are required for ribosome biogenesis.

11.12 Other Related Nucleolar Proteins

The principal common feature between Nopp140 and treacle is the large central domain consisting of alternating acidic and basic domains. Other nucleolar proteins that share related domains include nucleolin (its yeast homolog is NSR1), B23, p100 in *Chironumus*, and NPI46 in yeast. Nucleolin and B23 are reviewed in Chaps. 9 and 10, respectively.

Sun et al. (2002) described a novel nucleolar phosphoprotein p100 in *Chironumus tentans*. It too has large central domain consisting of 12 alternating acidic and basic regions quite similar to those in mammalian Nopp140. Its predicted molecular weight is 63 kDa, but it too has an anomalous migration at 100 kDa on SDS gels. Immediately following the alternating acidic/basic domain is an RGG domain that is about half the size of that found in nucleolin. Interestingly, two C4-type zinc fingers follow the RGG domain in p100, and these domains are followed by a tryptophan-rich carboxy terminus. Because p100 maintains its nucleolar association after RNase treatment, Sun et al. (2002) suggest that p100 is a component of a putative proteinaceous nucleolar framework.

Finally, Shan et al. (1994) described NPI46 in *S. cerevisiae*. It is a non-essential nucleolar protein with a predicted mass of 46.5 kDa, but it migrates at about 70 kDa. Its central domain has three acidic domains reminiscent of those in Nopp140. A basic domain separates acidic domains 1 and 2, and a second basic domain follows the third acidic domain. The stretch of amino acids separating acidic domains 2 and 3 is not considered basic. The acidic domains contain a few CKII phosphorylation sites, but they are not as prevalent as in Nopp140 or treacle. Interestingly, the carboxy terminus (106 amino acids) of NPI46 is homologous to prolyl *cis-trans* isomerases that are classified by their ability to bind FK506, an immunosuppressant drug. Shan et al. (1994) showed that NPI46 has proline isomerase activity, and they suggest that NPI46 functions in ribosome biosynthesis perhaps to fold ribosomal proteins during the assembly process. We should keep in mind, however, that nonribosomal nucleolar proteins such as Nopp140, treacle, nucleolin, etc. have basic regions rich in proline, and prolyl isomerase activity could be invoked in the delivery and release of snoRNPs.

11.13 Future Explorations

Future work on Nopp140 will likely take us into regulatory pathways governing cell growth and cell stress responses (e.g., Ferreira-Cerca and Hurt 2009). For instance, a functional genomics assay identified the *Nopp140* gene in *Drosophila* as one of many genes necessary for ribosome biogenesis that are regulated by dTOR signaling (Guertin et al. 2006). Other *Drosophila* genes identified in the same assay were *Nop60b*, which encodes the H/ACA snoRNP pseudouridylase, and *CG3983* (now called *NS1*), which encodes nucleostemin. Nopp140 itself may be the target of multiple regulatory events as it shuttles from the cytoplasm back into the nucleus, on to CBs where it participates in snoRNP assembly/modification, then to nucleoli where it delivers these snoRNPs and provides a scaffold for their site-directed chemical modification of pre-rRNA, and finally to Pol I where it may regulate rDNA transcription. One can easily imagine differential phosphorylation of Nopp140 by CKII, PKC, PKA, and various phosphatases in response to cell proliferation or cell stress signals. The challenge before us is to determine what regulates post-translational modifications of Nopp140. Like Nopp140, differential phosphorylation of treacle could potentially regulate box C/D snoRNP function and Pol I transcription. The putative roles that Nopp140 and treacle may have in feedback regulation of Pol I transcription remain a fascinating avenue of future investigation. Can Nopp140 and treacle respond to perturbations in pre-rRNA processing or the loss of cytoplasmic ribosomes by regulating Pol I transcription? If so, would they act as transcription activators or repressors? Do Nopp140 and treacle participate in nucleolar stress response, which is only now coming into focus? Just as the nucleolus is considered "plurifunctional" (Pederson 1998), its constituent proteins are likely to be multifunctional as well.

Acknowledgments Preliminary work on *Drosophila* Nopp140 reported here was supported by the National Science Foundation, grant MCB-0919709. We thank Renford Cindass, Jr. (LSU), Dana Meier (Southeastern University of Louisiana), Stephanie Terhoeve (St. Joseph's Academy, Baton Rouge), Helya Ghaffari (University of Maryland, Baltimore County), and Courtney Mumphrey (LSU) for generating the preliminary data on apoptosis in *Drosophila* cells lacking Nopp140.

References

Baran V, Brochard V, Renard JP, Flechon JE (2001) Nopp 140 involvement in nucleologenesis of mouse preimplantation embryos. Mol Reprod Dev 59:277–284

Bellini M, Gall JG (1999) Coilin shuttles between the nucleus and cytoplasm in *Xenopus* oocytes. Mol Biol Cell 10:3425–3434

Boulon S, Verheggen C, Jady BA, Girard C, Pescia C, Paul C, Ospina JK, Kiss T, Matera AG, Bordonné R, Bertrand E (2004) PHAX and CRM1 are required sequentially to transport U3 snoRNA to nucleoli. Mol Cell 16:777–787. doi:10.1016/j.molcel.2004.11.013 DOI:dx.doi.org

Cairns C, McStay B (1995) Identification and cDNA cloning of a Xenopus nucleolar phosphoprotein, xNopp 180, that is the homolog of the rat nucleolar protein Nopp140. J Cell Sci 108: 3339–3347

11 Structure and Function of Nopp140 and Treacle

Carvalho T, Almeida F, Calapez A, Lafarga M, Berciano MT, Carmo-Fonseca M (1999) The spinal muscular atrophy disease gene product, SMN: a link between snRNP biogenesis and the Cajal (coiled) body. J Cell Biol 147:715–728. doi:10.1083/jcb.147.4.715 DOI:dx.doi.org

Chen H-K, Pai C-Y, Huang J-Y, Yeh N-S (1999) Human Nopp 140, which interacts with RNA polymerase I: Implications for rRNA gene transcription and nucleolar structural organization. Mol Cell Biol 19:8536–8546

Chiu C-M, Tsay Y-G, Chang C-J, Lee S-C (2002) Nopp 140 is a mediator of the protein kinase signaling pathway that activates the acute phase response α1-acid glycoprotein gene. J Biol Chem 277:39102–39111

Cui Z, DiMario PJ (2007) RNAi knockdown of Nopp 140 induces Minute-like phenotypes in Drosophila. Mol Biol Cell 18:2179–2191. doi:10.1091/mbc.E07-01-0074 DOI:dx.doi.org

Darnell JC, Jensen KB, Jin P, Brown V, Warren ST, Darnell RB (2001) Fragile X mental retardation protein targets G quartet mRNAs important for neuronal function. Cell 107:489–499. doi:10.1016/S0092-8674(01)00566-9 DOI:dx.doi.org

Dauwerse JG, Dixon J, Seland S, Ruivenkamp CAL, van Haeringen A, Hoefsloot LH, Peters DJM, Clement-de Boers A, Daumer-Haas C, Maiwald R, Zweier C, Kerr B, Cobo AM, Toral JF, Hoogeboom AJM, Lohmann DR, Hehr U, Dixon MJ, Bruening MH, Wieczorek D (2011) Mutations in genes encoding subunits of RNA polymerases I and III cause Treacher Collins syndrome. Nat Genet 43:20–22. doi:10.1038/ng.724 DOI:dx.doi.org

Deshpande G, Calhoun G, Schedl P (2004) Overlapping mechanisms function to establish transcriptional quiescence in the embryonic *Drosophila* germline. Development 131:1247–1257. doi:10.1242/dev.01004 DOI:dx.doi.org

Dixon MJ (1996) Treacher Collins syndrome. Hum Mol Genet 5:1391–1396. doi:10.1111/j.1601-6343.2007.00388.x DOI:dx.doi.org

Dixon J, Dixon MJ (2004) Genetic background has a major effect on the penetrance and severity of craniofacial defects in mice heterozygous for the gene encoding the nucleolar protein treacle. Dev Dyn 229:907–914. doi:10.1002/dvdy.20004 DOI:dx.doi.org

Dixon MJ, Dixon J, Houseal T, Bhatt M, Ward DC, Klinger K, Landes GM (1993) Narrowing the position of the Treacher Collins syndrome locus to a small interval between three new microsatellite markers at 5q32-q33.1. Am J Hum Genet 52:907–914

Dixon J, Hovanes K, Shiang R, Dixon MJ (1997) Sequence analysis, identification of evolutionary conserved motifs and expression analysis of murine *tcof1* provide further evidence for a potential function for the gene and its human homologue, *TCOF1*. Hum Mol Genet 6:727–737. doi:10.1093/hmg/6.5.727 DOI:dx.doi.org

Dixon J, Brakebusch C, Fässler R, Dixon MJ (2000) Increased levels of apoptosis in the prefusion neural folds underlie the craniofacial disorder, Treacher Collins syndrome. Hum Mol Genet 10:1473–1480. doi:10.1093/hmg/9.10.1473 DOI:dx.doi.org

Dixon J, Jones NC, Sandell LL, Jayasinghe SM, Crane J, Rey J-P, Dixon MJ, Trainor PA (2006) *Tcof1*/treacle is required for neural crest cell formation and proliferation deficiencies that cause craniofacial abnormalities. Proc Natl Acad Sci USA 103:13403–13408. doi:10.1073/pnas.0603730103 DOI:dx.doi.org

Dundr M, Meier UT, Lewis N, Rekosh D, Hammarskjöld M-L, Olson MOJ (1997) A class of non-ribosomal nucleolar components is located in chromosome periphery and in nucleolus-derived foci during anaphase and telophase. Chromosoma 105:407–417. doi:10.1007/BF02510477 DOI:dx.doi.org

Dundr M, Hebert MD, Karpova TS, Stanek D, Xu H, Shpargel KB, Meier UT, Neugebaueur KM, Matera AG, Misteli T (2004) In vivo kinetics of Cajal body components. J Cell Biol 164:831–842. doi:10.1083/jcb.200311121 DOI:dx.doi.org

Emes RD, Ponting CP (2001) A new motif linking lissencephaly, Treacher Collins and oral-facial-digital type 1 syndromes, microtubule dynamics and cell migration. Hum Mol Genet 24:2813–2820. doi:10.1093/hmg/10.24.2813 DOI:dx.doi.org

Fazen LE, Elmore J, Nadler HL (1967) Mandibulo-facial dysostosis (Treacher Collins syndrome). Am J Dis Child 113:405–410

Ferreira-Cerca S, Hurt E (2009) Arrest by ribosome. Nature 459:46–47. doi:10.1038/459046a DOI:dx.doi.org

Ganot PM, Caizergues-Ferrer M, Kiss T (1997) The family of ACA small nucleolar RNAs is defined by an evolutionary conserved secondary structure and ubiquitous sequence elements essential for RNA accumulation. Gene Dev 11:941–956. doi:10.1101/gad.11.7.941 DOI:dx.doi.org

Ghisolfi L, Joseph G, Amalric F, Erard M (1992) The glycine-rich domain of nucleolin has an unusual supersecondary structure responsible for its RNA-helix-destabilizing properties. J Biol Chem 267:2955–2959

Ginisty H, Sicard H, Roger B, Bouvet P (1999) Structure and functions of nucleolin. J Cell Sci 112:761–772

Godin KS, Varani G (2007) How arginine-rich domains coordinate mRNA maturation events. RNA Biol 4:69–75

Gonzales B, Yang H, Henning D, Valdez BC (2005a) Cloning and functional characterization of the *Xenopus* orthologue of the Treacher Collins syndrome (*TCOF1*) gene product. Gene 359:73–80

Gonzales B, Henning D, So RB, Dixon J, Dixon ML, Valdez BC (2005b) The Treacher Collins syndrome (*TCOF1*) gene product is involved in pre-rRNA methylation. Hum Mol Genet 14:2035–2043. doi:10.1093/hmg/ddi208 DOI:dx.doi.org

Guertin DA, Guntur KVP, Bell GW, Thoreen CC, Sabatini DM (2006) Functional genomics identifies TOR-regulated genes that control growth and division. Curr Biol 16:958–970. doi:10.1016/j.cub.2006.03.084 DOI:dx.doi.org

Haworth KE, Islam I, Breen M, Putt W, Makrinou E, Binns M, Hopkinson D, Edwards Y (2001) Canine *TCOF1*; cloning, chromosome assignment and genetic analysis in dogs with different head types. Mamm Genome 12:622–629. doi:10.1007/s00335-001-3011-0 DOI:dx.doi.org

Hayano T, Yanagida M, Yamauchi Y, Shinkawa T, Isobe T, Takahashi N (2003) Proteomic analysis of human Nop56p-associated pre-ribosomal ribonucleoproteins complexes. J Biol Chem 278: 34309–34319. doi:10.1074/jbc.M304304200 DOI:dx.doi.org

Henras A, Henry Y, Bousquet-Antonelli C, Noaillac-Depeyre J, Gelugne JP, Caizergues-Ferrer M (1998) Nhp2p and Nop10p are essential for the function of H/ACA snoRNPs. EMBO J 17:7078–7090. doi:10.1093/emboj/17.23.7078 DOI:dx.doi.org

Ikonomova R, Sommer T, Kepes F (1997) The Srp40 protein plays a dose-sensitive role in preribosome assembly or transport and depends on its carboxy-terminal domain for proper localization to the yeast nucleoskelton. DNA Cell Biol 16:1161–1173. doi:10.1089/dna.1997.16.1161 DOI:dx.doi.org

Isaac C, Yang Y, Meier UT (1998) Nopp 140 functions as a molecular link between the nucleolus and the coiled bodies. J Cell Biol 142:319–329. doi:10.1083/jcb.142.2.319 DOI:dx.doi.org

Isaac C, Marsh KL, Paznekas WA, Dixon J, Dixon MJ, Jabs EW, Meier UT (2000) Characterization of the nucleolar gene product, treacle, in Treacher Collins syndrome. Mol Biol Cell 11: 3061–3071

Isaac C, Pollard JW, Meier UT (2001) Intranuclear endoplasmic reticulum induced by Nopp 140 mimics the nucleolar channel system of human endometrium. J Cell Sci 114:4253–4264

Jabs EW, Li X, Coss CA, Taylor EW, Meyers DA, Weber JL (1991) Mapping the Treacher Collins syndrome locus to 5q31.3-q33.3. Genomics 11:193–198

Jones KW, Gorzynski K, Hales CM, Fischer U, Badbanchi F, Terns RM, Terns MP (2001) Direct interaction of the spinal muscular atrophy disease protein SMN with the small nucleolar RNA-associated protein fibrillarin. J Biol Chem 276:38645–38651. doi:10.1074/jbc.M106161200 DOI:dx.doi.org

Jones NC, Lynn ML, Gaudenz K, Sakai D, Aoto K, Rey J-P, Glynn EF, Ellington L, Du C, Dixon J, Dixon MJ, Trainor PA (2008) Prevention of the neurocristopathy Treacher Collins syndrome through inhibition of p53 function. Nat Med 14:125–133. doi:10.1038/nm1725 DOI:dx.doi.org

Kelly S, Singleton W, Wickstead B, Ersfeld K, Gull K (2006) Characterization and differential nuclear localization of Nopp 140 and a novel Nopp140-like protein in Trypanosomes. Eukaryot Cell 5:876–879. doi:10.1128/EC.5.5.876-879.2006 DOI:dx.doi.org

Kiledjian M, Dreyfuss G (1992) Primary structure and binding activity of the hnRNP U protein: binding RNA through RGG box. EMBO J 11:2655–2664

11 Structure and Function of Nopp140 and Treacle

Kim MH, Cooper DR, Oleksy A, Devedjiev Y, Derewenda U, Reiner O, Otlewski J, Derewenda ZS (2004) The structure of the N-terminal domain of the product of the lissencephaly gene Lis1 and its functional implications. Structure 12:987–998. doi:10.1016/j.str.2004.03.024 DOI:dx.doi.org

Kim YK, Lee WK, Jin Y, Lee KJ, Jeon H, Yu YG (2006) Doxobrubicin binds to un-phosphorylated form of hNopp140 and reduces protein kinase CK2-dependent phosphorylation of hNopp140. J Biochem Mol Biol 39:774–781

Kiss-László Z, Henry Y, Bachellerie J-P, Caizergues-Ferrer M, Kiss T (1996) Site-specific ribose methylation or preribosomal RNA: a novel function for small nucleolar RNAs. Cell 85: 1077–1088

Kittur N, Zapantis G, Aubuchon M, Santoro N, Bazett-Jones DP, Meier UT (2007) The nucleolar channel system of human endometrium is related to endoplasmic reticulum and R-rings. Mol Biol Cell 18:2296–2304. doi:10.1091/mbc.E07-02-0154 DOI:dx.doi.org

Lafontaine DLJ, Bousquet-Antonelli C, Henry Y, Caizergues-Ferrer M, Tollervey D (1998) The box H+ACA snoRNAs carry Cbf5p, the putative rRNA pseudouridine synthase. Genes Dev 12:527–537. doi:10.1101/gad.12.4.527 DOI:dx.doi.org

Lee YM, Miau LH, Chang CJ, Lee SC (1996) Transcriptional induction of the alpha-1 acid glycoprotein (AGP) gene by synergistic interaction of two alternative activator forms of AGP/enhancer binding protein (C/EBPβ) and NF- B or Nopp 140. Mol Cell Biol 16:4257–4263

Lemm I, Girard C, Kuhn AN, Watkins NJ, Schneider M, Bordonné R, Lührmann R (2006) Ongoing U snRNP biogenesis is required for the integrity of Cajal bodies. Mol Biol Cell 17:3221–3231. doi:10.1091/mbc.E06-03-0247 DOI:dx.doi.org

Li D, Meier UT, Dobrowolska G, Krebs EG (1997) Specific interaction between casein kinase 2 and nucleolar protein Nopp 140. J Biol Chem 272:3773–3779

Lin C-I, Yeh N-H (2009) Treacle recruits RNA polymerase I complex to the nucleolus that is independent of UBF. Biochem Biophys Res Comm 386:396–401. doi:10.1016/j.bbrc.2009.06.050 DOI:dx.doi.org

Lischwe MA, Cook RG, Ahn YS, Yeoman LC, Busch H (1985) Clustering of glycine and NG, NG-dimethylarginine in nucleolar protein C23. Biochemistry 24:6025–6602. doi:10.1021/bi00343a001 DOI:dx.doi.org

Lorson CL, Rindt H, Shababi M (2010) Spinal muscular atrophy: mechanisms and therapeutic strategies. Hum Mol Genet 19(R1):R111–R118 doi:10.1093/hmg/ddq147 DOI:dx.doi.org

Ma H, Pederson T (2008) Nucleostemin: a multiplex regulator of cell-cycle progression. Trends Cell Biol 18:575–579. doi:10.1016/j.tcb.2008.09.003 DOI:dx.doi.org

Marsh KL, Dixon J, Dixon MJ (1998) Mutations in the Treacher Collins syndrome gene lead to mislocalization of the nucleolar protein treacle. Hum Mol Genet 7:1795–1800. doi:10.1093/hmg/7.11.1795 DOI:dx.doi.org

Mateja A, Ciepicki T, Paduch M, Derewenda ZS, Otlewski J (2006) The dimerization mechanism of LIS1 and its implication for proteins containing the LisH motif. J Mol Biol 357:621–631. doi:10.1016/j.jmb.2006.01.002 DOI:dx.doi.org

McBride AE, Silver PA (2001) State of the Arg: protein methylation at arginine comes of age. Cell 106:5–8

McCain J, Danzy L, Hamdi A, Dellafosse O, DiMario PJ (2006) Tracking nucleolar dynamics with GFP-Nopp 140 during Drosophila oogenesis and embryogenesis. Cell Tissue Res 323: 105–115. doi:10.1007/s00441-005-0044-9 DOI:dx.doi.org

McKeown SJ, Bronner-Fraser M (2008) Saving face: rescuing craniofacial birth defects. Nat Med 14:115–116. doi:10.1038/nm0208-115 DOI:dx.doi.org

Meier UT (1996) Comparison of the rat nucleolar protein Nopp 140 with its yeast homolog SRP40. J Biol Chem 271:19376–19384

Meier UT (2005) The many facets of H/ACA ribonucleoproteins. Chromosoma 114:1–14. doi:10.1007/s00412-005-0333-9 DOI:dx.doi.org

Meier UT, Blobel G (1990) A nuclear localization signal binding protein in the nucleolus. J Cell Biol 111:2235–2245. doi:10.1083/jcb.111.6.2235 DOI:dx.doi.org

Meier UT, Blobel G (1992) Nopp 140 shuttles on tracks between nucleolus and cytoplasm. Cell 70:127–138

Meier UT, Blobel G (1994) NAP57, a mammalian nucleolar protein with a putative homolog in yeast and bacteria. J Cell Biol 127:1505–1514. doi:10.1083/jcb.127.6.1505 DOI:dx.doi.org

Miau L-H, Chang C-J, Tsai WH, Lee S-C (1997) Identification and characterization of a nucleolar phosphoprotein, Nopp 140, as a transcription factor. Mol Cell Biol 17:230–239

Narayanan A, Lukowiak A, Jády BE, Dragon F, Kiss T, Terns RM, Terns MP (1999) Nucleolar localization signals of box H/ACA small nucleolar RNAs. EMBO J 15:5120–5130. doi:10.1093/emboj/18.18.5120 DOI:dx.doi.org

Nichols RC, Wang XW, Tang J, Hamilton BJ, High FA, Herschman HR, Rigby WF (2002) The RGG domain in hnRNP A2 affects subcellular localization. Exp Cell Res 256:522–532. doi:10.1006/excr.2000.4827 DOI:dx.doi.org

Nicoloso M, Qu L-H, Michot B, Bachellerie J-P (1996) Intron-encoded, antisense small nucleolar RNAs: the characterization of nine novel species points to their direct role as guides for the 2'-O-ribose methylation of rRNAs. J Mol Biol 260:178–195. doi:10.1006/jmbi.1996.0391 DOI:dx.doi.org

Nizami ZF, Deryusheva S, Gall JG (2010) Cajal bodies and histone locus bodies. Cold Spring Harb Perspect Biol 2:a000653. doi:10.1101/cshperspect.a000653 DOI:dx.doi.org

Nomura N, Miyajima N, Sazuka T, Tanaka A, Kawarabayasi Y, Sato S, Nagase T, Sake N, Ishikawa A, Tabata S (1994) Prediction of the coding sequences of unidentified human genes, I the coding sequences of 40 new genes (KIAA0001-KIAA0040) deduced by analysis of randomly sampled cDNA clones form human immature myeloid cell line KG-1. DNA Res 1:27–35. doi:10.1093/dnares/1.1.27 DOI:dx.doi.org

Ochs RL, Lischwe MA, Spohn WH, Busch H (1985) Fibrillarin: a new protein of the nucleolus identified by autoimmune sera. Biol Cell 54:123–134

Olson MOJ (2004) Sensing cellular stress: another new function for the nucleolus? Sci STKE 2004:pe10. doi:10.1126/stke.2242004pe10 DOI:dx.doi.org

Pai CY, Yeh NH (1996) Cell proliferation-dependent expression of two isoforms of the nucleolar phosphoprotein p130. Biochem Biophys Res Comm 221:581–587. doi:10.1006/bbrc.1996.0639 DOI:dx.doi.org

Pai CY, Chen HK, Sheu HL, Yeh NH (1995) Cell cycle-dependent alterations of a highly phosphorylated nucleolar protein p130 are associated with nucleologenesis. J Cell Sci 108: 1911–1920

Paznekas WA, Zhang N, Gridley T, Jabs EW (1997) Mouse *TCOF1* is expressed widely, has motifs conserved in nucleolar phosphoproteins, and maps to chromosome 18. Biochem Biophys Res Commun 238:1–6. doi:10.1006/bbrc.1997.7229 DOI:dx.doi.org

Pederson T (1998) The plurifunctional nucleolus. Nucleic Acids Res 26:3871–3876. doi:10.1093/nar/26.17.3871 DOI:dx.doi.org

Pellizzoni L, Baccon J, Charroux B, Dreyfuss G (2001) The survival of motor neurons (SMN) protein interacts with the snoRNP proteins fibrillarin and GAR1. Curr Biol 11:1079–1088. doi:10.1016/S0960-9822(01)00316-5 DOI:dx.doi.org

Pfeifle J, Anderer FA (1984) Isolation and localization of phosphoprotein pp 135 in the nucleoli of various cell lines. Eur J Biochem 139:417–424

Pfeifle J, Boller K, Anderer FA (1986) Phosphoprotein pp 135 is an essential component of the nucleolus organizer region (NOR). Exp Cell Res 162:11–22. doi:10.1016/0014-4827(86)90422-2 DOI:dx.doi.org

Ramos A, Hollingworth D, Pastore A (2003) G-quartet-dependent recognition between the FMRP RGG box and RNA. RNA 9:1198–1207. doi:10.1261/rna.5960503 DOI:dx.doi.org

Reiner O, Carrozzo R, Shen Y, Whenert M, Faustinella F, Dobyns WB (1993) Isolation of a Miller-Dieker lissencephaly gene containing G protein β-subunit-like repeats. Nature 364:717–721. doi:10.1038/364717a0 DOI:dx.doi.org

Renvoisé B, Colasse S, Burlet P, Viollet L, Meier UT, Lefebvre S (2009) The loss of the snoRNP chaperone Nopp140 from Cajal bodies of patient fibroblasts correlates with the severity of spinal muscular atrophy. Hum Mol Genet 18:1181–1189

Rubbi CP, Milner J (2003) Disruption of the nucleolus mediates stabilization of p53 in response to DNA damage and other stresses. EMBO J 22:6068–6077. doi:10.1093/emboj/cdg579 DOI:dx.doi.org

Sakai D, Trainor PA (2009) Treacher Collins syndrome: unmasking the role of Tcof1/treacle. Int J Biochem Cell Biol 41:1229–1232. doi:10.1016/j.biocel.2008.10.026 DOI:dx.doi.org

Schmidt-Zachmann MS, Hügle B, Scheer U, Franke WW (1984) Identification and localization of a novel nucleolar protein of high molecular weight by a monoclonal antibody. Exp Cell Res 153:327–346

Shan X, Xue Z, Mélèse T (1994) Yeast *NPI46* encodes a novel prolyl *cis-trans* isomerase that is located in the nucleolus. J Cell Biol 126:853–862. doi:10.1083/jcb.126.4.853 DOI:dx.doi.org

So RB, Gonzales B, Henning D, Dixon J, Dixon MJ, Valdez BC (2004) Another face of the Treacher Collins syndrome (TCOF1) gene: identification of additional exons. Gene 328:49–57. doi:10.1016/j.gene.2003.11.027 DOI:dx.doi.org

Srivastava M, Fleming PJ, Pollard HB, Burns AL (1989) Cloning and sequencing of the human nucleolin cDNA. FEBS Lett 250:99–105. doi:10.1016/0014-5793(89)80692-1 DOI:dx.doi.org

Sun X, Zhao J, Jin S, Palka K, Visa N, Aissouni Y, Danehoh B, Alzhanova-Ericsson AT (2002) A novel protein localized to the fibrillar compartment of the nucleolus and to the brush border of a secretory cell. Eur J Cell Biol 81:125–137. doi:10.1078/0171-9335-00231 DOI:dx.doi.org

Teber ÖA, Gillessen-Kaesbach G, Fischer S, Böhringer S (2004) Genotyping in 46 patients with tentative diagnosis of Treacher Collins syndrome revealed unexpected phenotypic variation. Eur J Hum Genet 12:879–890

Thiry M, Cheutin T, Lamaye F, Thelen N, Meier UT, O'Donohue M-F, Ploton D (2009) Localization of Nopp 140 within mammalian cells during interphase and mitosis. Histochem Cell Biol 132:129–140. doi:10.1007/s00418-009-0599-8 DOI:dx.doi.org

Tollervey D, Lehtonen H, Jansen R, Kern H, Hurt EC (1993) Temperature-sensitive mutations demonstrate roles for yeast fibrillarin in pre-rRNA processing, pre-rRNA methylation, and ribosome assembly. Cell 72:443–457. doi:10.1016/0092-8674(93)90120-F DOI:dx.doi.org

Trainor PA, Dixon J, Dixon MJ (2009) Treacher Collins syndrome: etiology, pathogenesis and prevention. Eur J Hum Genet 17:275–283. doi:10.1038/ejhg.2008.221 DOI:dx.doi.org

Tsai Y-T, Lin C-I, Chen H-K, LeeK-M HC-Y, Yang S-J, Yeh N-H (2008) Chromatin tethering effects of hNopp 140 are involved in the spatial organization of the nucleus and the rRNA gene transcription. J Biomed Sci 15:471–486. doi:10.1007/s11373-007-9226-7 DOI:dx.doi.org

Valdez BC, Henning D, So RB, Dixon J, Dixon MJ (2004) The Treacher Collins syndrome (*TCOF1*) gene product is involved in ribosomal DNA gene transcription by interacting with upstream binding factor. Proc Natl Acad Sci USA 101:10709–10714. doi:10.1073/pnas.0402492101 DOI:dx.doi.org

Vandelaer M, Thiry M (1998) The phosphoprotein pp 135 is an essential constituent of the fibrillar components of nucleoli and of coiled bodies. Histochem Cell Biol 110:169–177. doi:10.1007/s004180050278 DOI:dx.doi.org

Verheggen C, Mouaikel J, Thiry M, Blanchard J, Tollervey D, Bordonné R, Lafontaine DLJ, Bertrand E (2001) Box C/D small nucleolar RNA trafficking involves small nucleolar RNP proteins, nucleolar factors and a novel nuclear domain. EMBO J 20:5480–5490. doi:10.1093/emboj/20.19.5480 DOI:dx.doi.org

Verheggen C, Lafontaine DLJ, Samasky D, Mouaikel J, Blanchard J-M, Bordonné R, Bertrand E (2002) Mammalian and yeast U3 snoRNPs are matured in specific and related nuclear compartments. EMBO J 21:2736–2745. doi:10.1093/emboj/21.11.2736 DOI:dx.doi.org

Waggener J, DiMario PJ (2002) Two splicing variants of Nopp 140 in *Drosophila melanogaster*. Mol Biol Cell 13:362–381. doi:10.1091/mbc.01-04-0162 DOI:dx.doi.org

Wang C, Query CC, Meier UT (2002) Immunopurified small nucleolar ribonucleoprotein particles pseudouridylate rRNA independently of their association with phosphorylated Nopp 140. Mol Cell Biol 22:8457–8466. doi:10.1128/MCB.22.24.8457-8466.2002 DOI:dx.doi.org

Watkins NJ, Gottschalk A, Neubauer G, Kastner B, Fabrizio P, Mann M, Lührmann R (1998) Cbf5p, a potential pseudouridine synthase, and Nhp2p, a putative RNA-binding protein, are present together with Gar1p in all box H/ACA-motif snoRNPs and constitute a common bipartite structure. RNA 4:1549–1568. doi:10.1017/S1355838298980761 DOI:dx.doi.org

Watkins NJ, Lemm I, Ingelfinger D, Scneider C, Hoβbach M, Urlaub H, Lührmann R (2004) Assembly and maturation of the U3 snoRNP in the nucleoplasm in a large dynamic multiprotein complex. Mol Cell 16:789–798. doi:10.1016/j.molcel.2004.11.012 DOI:dx.doi.org

Weisenberger D, Scheer U (1995) A possible mechanism for the inhibition of ribosomal RNA gene transcription during mitosis. J Cell Biol 129:561–575. doi:10.1083/jcb.129.3.561 DOI:dx.doi.org

Winokur ST, Shiang R (1998) The Treacher Collins syndrome (*TCOF1*) gene product, treacle, is targeted to the nucleolus by signals in its C-terminus. Hum Mol Genet 7:1947–1952. doi:10.1093/hmg/7.12.1947 DOI:dx.doi.org

Wise CA, Chiang LC, Paznekas WA, Sharma M, Musy MM, Ashley JA, Lovett M, Jabs EW (1997) *TCOF1* gene encodes a putative nucleolar phosphoprotein that exhibits mutations in Treacher Collins Syndrome throughout its coding region. Proc Natl Acad Sci USA 94: 3110–3115. doi:10.1073/pnas.94.7.3110 DOI:dx.doi.org

Yang Y, Meier UT (2003) Genetic interaction between a chaperone of small nucleolar ribonucleoprotein particles and cytosolic serine hydroxymethyltransferase. J Biol Chem 278: 23553–23560. doi:10.1074/jbc.M300695200 DOI:dx.doi.org

Yang Y, Isaac C, Wang C, Dragon F, Pogačić V, Meier UT (2000) Conserved composition of mammalian box H/ACA and box C/D small nucleolar ribonucleoprotein particles and their interaction with the common factor Nopp 140. Mol Biol Cell 11:567–577

Yao Y, Yamamoto K, Nishi Y, Nogi Y, Muramatsu M (1996) Mouse RNA polymerase I 16-kDa subunit able to associate with 40-kDa subunit is a homolog of yeast AC19 subunit of RNA polymerase I and III. J Biol Chem 271:32881–32885

Zhai W, Comai L (2000) Repression of RNA polymerase I transcription by the tumor suppressor p53. Mol Cell Biol 20:5930–5938. doi:10.1128/MCB.20.16.5930-5938.2000 DOI:dx.doi.org

Part III
Novel Functions of the Nucleolus

Chapter 12
The Role of the Nucleolus in the Stress Response

Laura A. Tollini, Rebecca A. Frum, and Yanping Zhang

12.1 Introduction

Growing evidence has demonstrated that the p53 tumor suppressor is the major cellular stress sensor. In response to stressors such as DNA damage, oncogene activation, or perturbations to the ribosome, the *TP53* gene product, p53, transactivates genes such as p21, Noxa, Bax, Puma, and GADD45, leading to cell cycle arrest, apoptosis, or senescence. Proper regulation of p53 allows for it to function to repair or eliminate damaged cells after exposure to cellular stress but otherwise remain inactive to permit development and growth; p53 is considered to be a critical and overarching tumor suppressor and is found to be mutated in approximately 50% of all human cancers, suggestive of the importance of p53 expression and function. In contrast, unchecked p53 activity can be detrimental to cells. For example, in mice, deletion of p53's major negative regulator murine double minute 2 (Mdm2), leading to excessive levels of p53 activity, causes embryonic lethality. This phenotype can be rescued by concomitant deletion of p53, highlighting the importance of proper p53 regulation during embryonic development. Mdm2 is considered to be the primary negative regulator of p53, largely through its role as an E3 ubiquitin ligase. Mdm2 and p53 exist in a regulatory feedback loop in which the ability of Mdm2 to ubiquitinate p53 is critical for effective degradation of p53, whereas the *Mdm2* gene itself is a transcriptional target of p53.

Y. Zhang (✉)
Departments of Radiation Oncology and Pharmacology,
University of North Carolina at Chapel Hill,
Chapel Hill, NC, USA

Lineberger Comprehensive Cancer Center, University of North Carolina at Chapel Hill,
Chapel Hill, NC, USA
e-mail: ypzhang@med.unc.edu

M.O.J. Olson (ed.), *The Nucleolus*, Protein Reviews 15,
DOI 10.1007/978-1-4614-0514-6_12, © Springer Science+Business Media, LLC 2011

While important, the Mdm2-p53 regulatory loop does not exist in isolation, with growing evidence indicating that the nucleolus plays a central role in the regulation of these primarily nucleoplasmic proteins. The tumor suppressor ARF (p14[ARF] in humans and p19[ARF] in mice, ARF hereafter), a nucleolar protein, is capable of antagonizing Mdm2's ability to downregulate p53 via several proposed mechanisms, including inhibition of Mdm2 E3 ubiquitin ligase activity and sequestration of Mdm2 in the nucleolus, both of which result in stabilization and activation of p53. The nucleolar protein B23, important in ribosomal biogenesis, has also been shown to function in the cellular stress response. The ability of B23 to sequester ARF in the nucleolus implicates B23 as both an oncogene and a tumor suppressor gene. It is likely that the multiple functions of B23, in both ARF-dependent and -independent pathways, are critical in maintaining normal cell function during different cellular contexts, and that dysfunctional B23 regulation may have deleterious effects on the cell.

In addition to its functions in responding to DNA damage, p53 has also been shown to respond to disturbances of ribosome biogenesis in the nucleolus by inducing cell cycle arrest. Perturbation of any of the steps of ribosome biogenesis, including the coordinated expression of ribosomal RNA (rRNA) and ribosomal proteins (RPs), processing of rRNA, and assembly of the 40S and 60S ribosomal subunits, is thought to lead to nucleolar stress. Upon exposure to nucleolar stress, several RPs such as RPL11, RPL23, and RPL5 have been shown to interact with Mdm2 and inhibit its E3 ubiquitin ligase activity toward p53, resulting in stabilization of p53 and consequently linking disturbances in cell growth with inhibition of cell cycle progression.

The function of ARF, B23, and RPs in regulating p53 activity through the nucleolus places the nucleolus in the pivotal position of linking cell cycle progression with protein synthesis. The ability of p53 to act as a common regulator may serve to coordinate these different cellular processes, allowing for optimized protection of the genetic integrity of the cell in the event of an adverse cellular context. Here, we discuss the current understanding of the role of the nucleolus and nucleolar protein pathways, in terms of how they modulate the cellular stress response and how the integration of these pathways may serve to protect the genetic integrity of the cell.

12.2 ARF Acts as a Tumor Suppressor by Activating the p53 Pathway

The gene for ARF is localized to human chromosome 9p21, a locus shared by the ARF tumor suppressor and cell cycle regulator p16INK4a. Through alternative splicing, the *ARF-INK4a* gene locus can be translated into two distinct proteins, with the ARF protein found to be the result of exon 2 being read in an alternate reading frame (Quelle et al. 1995). Both ARF and p16INKa are frequently deleted, mutated, or methylated at their promoters, leading to loss of their expression in many human tumors. Although loss of either ARF or p16INK4a increases susceptibility to tumors, it appears that loss of ARF is more critical in the development of tumorigenesis (Gazzeri et al. 1998). In mouse models, tumor formation was found to be more likely to occur within the first year of life in ARF-null mice than in mice

12 The Role of the Nucleolus in the Stress Response 283

harboring deletion of p16INK4a (Sherr 2001). Furthermore, mice lacking both p16INK4a and ARF show a similar phenotype to mice lacking ARF alone, suggesting that ARF is playing a more significant role in the prevention of tumorigenesis (Sherr 2001).

The ability of ARF to act as a tumor suppressor involves its activation of the p53 pathway via inactivation of Mdm2's p53-inhibitory functions; p53 has been shown to be induced in response to oncogenic challenges, resulting in p53 mediated transactivation of genes involved in cell cycle arrest, apoptosis, or senescence. Under normal, unstressed conditions, Mdm2 maintains p53 at a very low level. During an oncogenic challenge, however, it is desirable for the cell to inactivate Mdm2's E3 ubiquitin ligase function in order to allow for the p53 pathway to be activated and to respond to the stress. As one of p53's target genes, *Mdm2* is subsequently transactivated as part of the p53 stress response, enabling Mdm2 to once again ubiquitinate and target p53 for degradation following the oncogenic challenge. One way ARF has been proposed to act as a tumor suppressor is by directly binding Mdm2 and sequestering it in the nucleolus, as ARF itself is normally a nucleolar protein, effectively preventing Mdm2 from acting as an ubiquitin ligase. The nucleolar sequestration of Mdm2 has been suggested to allow for the activation of p53 in the nucleoplasm (Tao and Levine 1999; Weber et al. 1999); however, subsequent observations suggest that the ARF-mediated activation of p53 is more complex than simple changes in Mdm2 localization. In contrast, other studies demonstrate that the interaction of ARF with Mdm2 is capable of stabilizing and activating p53 in the nucleoplasm without causing Mdm2 to become localized to the nucleolus (Llanos et al. 2001; Korgaonkar et al. 2002). These observations make the role of the nucleolus in ARF-mediated activation of p53 unclear.

In addition to forming a complex with Mdm2, ARF has been shown to form a ternary complex with Mdm2 and p53 as well (Pomerantz et al. 1998; Zhang et al. 1998; Stott et al. 1998; Kamijo et al. 1998). It has been suggested that this ternary complex blocks the nuclear export of Mdm2 and p53, which would stabilize and activate p53 (Zhang and Xiong 1999). The role of ARF as a tumor suppressor is made apparent by its induction following increased and sustained expression of multiple oncogenes, including Myc (Zindy et al. 1998), E1A (de Stanchina et al. 1998), E2F1(Dimri et al. 2000), and Ras (Palmero et al. 1998), as well as after "culture shock" in mouse embryonic fibroblasts (MEFs) (Kamijo et al. 1997). The induction of ARF enables the p53 response to induce cell cycle arrest, apoptosis, or senescence, and it is through these mechanisms that ARF contributes to protecting cells from oncogenesis. Although the characterization of ARF as a tumor suppressor is largely on the basis of its ability to activate p53, mounting evidence suggests that ARF also possesses a p53-independent role in cell cycle regulation. For example, in the absence of p53, overexpression of p19[ARF] has been demonstrated to induce G1 arrest (Carnero et al. 2000; Weber et al. 2000). ARF-mediated nucleolar sequestration of the oncogenes Myc and E2F1, antagonizing their transcriptional activity, has been proposed as a possible mechanism for ARF's p53-independent tumor suppression function (Sherr 2006). Through its p53-dependent and -independent roles, both of which are mediated largely by the nucleolus, ARF serves a significant role in the stress response.

12.3 ARF-Dependent and -Independent roles of B23 in the Oncogenic Stress Response

Another protein that plays a role in the cellular response to oncogenic stress is B23, also known as NPM, nucleophosmin, NO38, or numatrin. B23 has a well-known role in ribosomal biogenesis, in large part because of its nucleic acid binding property (Wang et al. 1994), RNase activity (Herrera et al. 1995), and ability to act as an endoribonuclease in the nucleolus by cleaving the second internal transcribed spacer (ITS2) in the maturing rRNA transcript (Savkur and Olson 1998). Although B23 is primarily localized to the nucleolus, it is also detected in the nucleoplasm (Spector et al. 1984). Because of its affinity for peptides containing an SV40 nuclear localization signal, B23 is also able to act as a carrier for shuttling some basic proteins into the nucleolus (Szebeni et al. 1995, 1997). As an example of this, basic proteins such as Rev (Fankhauser et al. 1991), Rex (Adachi et al. 1993), Tat (Li 1997), and p120 (Valdez et al. 1994) demonstrate increased nucleolar localization when bound by B23. In addition to facilitating translocation of basic proteins into the nucleolus, B23 has also been suggested to play a role in the transport of ribosomal components from the nucleolus to the cytoplasm (Borer et al. 1989).

Mouse models have indicated that loss of B23 leads to defects in ribosome biogenesis, and that the absence of B23 results in unrestricted centrosome duplication and genomic instability (Grisendi et al. 2005). B23-null mice die between embryonic day E11.5 and E16.5 because of organ development failure and severe anemia. Unlike Mdm2-null and MdmX (a homologue of Mdm2)-null mice, the lethality observed in B23 null mice could not be rescued via concomitant deletion of p53, indicating that this lethality is not due exclusively to dysregulation of p53 (Colombo et al. 2005). Although simultaneous deletion of p53 did not rescue the in vivo phenotype, the proliferative defects observed in $p53^{+/+};B23^{-/-}$ compound MEFs appeared to be partially rescued in $p53^{-/-};B23^{-/-}$ MEFs (Colombo et al. 2005). The same study has shown that B23 was not rate limiting for rRNA processing in the absence of p53 (Colombo et al. 2005). B23 has also been shown to be involved in centrosome duplication; specifically, phosphorylation of B23 on residue Thr199 by cyclin E/CDK2 results in the release of B23's association with the unduplicated centrosome (Okuda et al. 2000; Tokuyama et al. 2001). Despite the reported role of B23 in centrosome duplication, proteomic profiling suggests that B23 is not a component of the centrosome itself (Andersen et al. 2003). Additionally, B23 has been demonstrated to have a role in apoptosis, with its effects on apoptosis occurring in a dose-dependent manner. After ultraviolet (UV)-induced apoptosis, increased levels of B23 have been shown to protect cells from apoptosis, while decreased levels of B23 are associated with sensitization of cells to apoptosis (Wu et al. 2002a, b). Following treatment with a low dose of actinomycin D, UV irradiation, or other inhibitors of rRNA synthesis/processing, B23 is rapidly translocated to the nucleoplasm from the nucleolus (Yung et al. 1985a, b; Kurki et al. 2004).

Multiple mechanisms have been suggested to explain the role of B23 in the cellular stress response, including both ARF-dependent and -independent pathways.

B23 has been implicated in the structure, replication, and repair of DNA through its ability to bind to chromatin following DNA damage (Lee et al. 2005), as well as through its interaction with proteins such as pRb (Takemura et al. 1999), PARP1/2 (Meder et al. 2005), and GADD45 (Gao et al. 2005). Overexpression of B23 is also suggested to stabilize p53 through direct binding (Colombo et al. 2002), although this direct interaction has been disputed (Itahana et al. 2003). In addition, B23 was found to form a transient interaction with Mdm2 following UV irradiation, preventing Mdm2 from being able to degrade p53 (Kurki et al. 2004). Through its interactions, B23 appears to be an additional variable in the p53 response to oncogenic stress.

B23 expression can act to both inhibit cell proliferation and promote oncogenesis. As an example of the former, B23 has been shown to induce senescence after being overexpressed in normal fibroblasts (Colombo et al. 2002). In addition to inducing senescence, overexpression of B23 has been shown to inhibit the cell cycle via induction of G1 cell cycle arrest in cells with wild-type p53, while it increases the percentage of cells in S phase in p53 null cells (Itahana et al. 2003). Overexpression of B23 is also linked to cancer and actively growing cells, with higher expression of B23 present in cancer tissue compared to normal tissue (Zhang 2004). Similarly, in cultured tumor cells, B23 is expressed at higher levels compared to normal cells, and between various cancer cell lines those that grow at a faster rate were shown to possess higher levels of B23 than those growing at a slower rate. In accordance with these findings, neoplastic growth has been shown to be related to high levels of B23 in several cell types (Feuerstein and Mond 1987; Feuerstein et al. 1988; Nozawa et al. 1996). The oncogenic protein c-Myc has also been shown to induce B23 mRNA via binding at its promoter, further supporting the proposed oncogenic potential of B23 (Zeller et al. 2001). The ability of B23 to both suppress and encourage cell cycle progression in various cellular contexts appears to be dependent on additional factors in the cell.

ARF has been suggested as one particular factor that may determine the role of B23 in cell cycle progression. B23 has been shown by several methods to bind to the N-terminal region of ARF; these methods include large-scale coimmunoprecipitation of ARF followed by mass spectrometry (Itahana et al. 2003), tandem affinity purification followed by mass spectrometry (Bertwistle et al. 2004), matrix-assisted laser desorption ionization time-of-flight (MALDI-TOF) analysis of proteins associated with ARF in isolated and purified nucleoli (Brady et al. 2004), and the isolation and identification of ARF-associated phosphoproteins by mass spectrometry (Korgaonkar et al. 2005). It is proposed that ARF binds to the central acidic domain of B23 together with its oligomerization domain (Bertwistle et al. 2004), as well as to the oligomerization domain of B23 alone (Itahana et al. 2003). Additional studies have proposed the heterodimerization domain of B23 to also be required to bind ARF (Korgaonkar et al. 2005). The presence of the acidic poly Glu-Asp residue regions A2 and A3 on the B23 protein has been shown to further enhance and stabilize the ARF–B23 interaction (Itahana et al. 2003). In transfected cells, the ARF mutants that do not bind to B23 were found to be unstable (Kuo et al. 2004), and therefore, the interaction of B23 with ARF is suggested to promote ARF stability in the nucleolus.

The ARF-dependent B23 response to stress has been shown to possess the potential to implicate B23 as either an oncogene or a tumor suppressor gene. The interaction of B23 with ARF has been proposed to have several different outcomes following an oncogenic challenge and the subsequent stress response. The interaction of B23 with ARF serves to localize ARF to the nucleolus, rendering it unable to activate the p53 response, which may suggest a role for B23 as an oncogene (Llanos et al. 2001; Gjerset 2006; Korgaonkar et al. 2005). As further support of this notion, the overexpression of B23 has been shown to interfere with the activation of the p53 pathway by ARF (Korgaonkar et al. 2005). Alternatively, the sequestration of ARF in the nucleolus by B23 could potentially denote B23 as a tumor suppressor. In this sense, the B23-mediated stabilization of ARF is a mechanism through which a stable pool of ARF is formed and maintained in the nucleolus, which can then be released and activated in the event of an oncogenic challenge. This is supported by findings that in the absence of B23, ARF is unstable and cannot localize to the nucleoli (Colombo et al. 2005). In addition to the effect of B23 on ARF stabilization, ARF has been shown to promote the degradation of B23 (Itahana et al. 2003) and prevent its nucleoplasmic shuttling (Brady et al. 2004). It is likely that these various functions of the B23–ARF interaction are critical to maintain normal cell function during different cellular contexts, and that the dysfunction of proper B23 regulation has deleterious effects on the cell.

12.4 The Ribosomal Protein-Mdm2-p53 Pathway

The nucleolus mainly functions in the production of ribosomes, the cellular component necessary for the synthesis of proteins from amino acids. When ribosome biogenesis is disrupted, several RPs have the potential to bind to Mdm2, preventing its interaction with p53 and effectively activating the p53 response. The ability of RPs to influence the p53 pathway serves as a link between the regulation of cell growth or protein synthesis and cell cycle arrest. The knowledge that these two important processes are connected to each other leaves us to determine how the coordination of these processes may be essential to protect the genetic or functional integrity of the cell.

In unstressed conditions, small and large RPs are assembled in the nucleolus and then transported to the cytoplasm for protein synthesis. During nucleolar stress however, ribosome biogenesis is disrupted and the structure of the nucleolus is broken down, which leads to components of the nucleolus being released into the nucleoplasm (Fig. 12.1). One example of nucleolar stress is the disruption of de novo precursor rRNA synthesis, which is required for the assembly of new ribosomes (Lempiainen and Shore 2009). Disruption of proteins necessary for maintaining a supply of rRNA, including Bop1 (Pestov et al. 2001), WDR12 (Holzel et al. 2005), Rbm19 (Zhang et al. 2008), and Wrd36 (Skarie and Link 2008) results in activation of the p53 response and inhibition of the cell cycle (Pestov et al. 2001; Strezoska et al. 2000; Zhang et al. 2008; Holzel et al. 2005; Skarie and Link 2008). Disruption

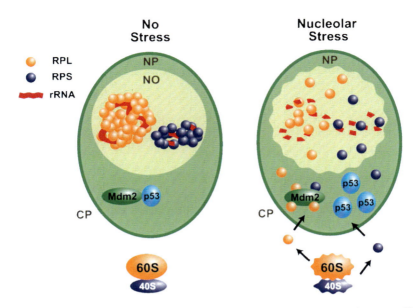

Fig. 12.1 Schematic diagram of RP-Mdm2-p53 pathway regulation by nucleolar stress. Under normal growth conditions (no stress), small (S, 40S), and large (L, 60S) RPs are assembled in the nucleolus (NO) and transported to the cytoplasm (CP) for protein synthesis. Under nucleolar stress, ribosomal biogenesis is inhibited and ribosome-free forms of RPs enter the nucleoplasm (NP) to interact with and inhibit the function of Mdm2, resulting in p53 stabilization and activation. Similarly, RPs, either released from the breaking down (indicated by *wavy edges*) of cytoplasmic ribosomes or overproduction in the cytoplasm, can enter the nucleoplasm to interact with Mdm2.

of components of the ribosome itself, such as RPS9 (Lindstrom and Zhang 2008) and HIP/RPL29 (Liu et al. 2006) also induces p53-mediated cell cycle arrest.

In addition to the dysfunction of nucleolar proteins involved in ribosome biogenesis, other events that induce nucleolar stress and signal to p53 include serum starvation and nucleotide depletion (Zhang and Lu 2009). In the laboratory setting, drugs such as actinomycin D (Perry and Kelley 1970; Sobell 1985), 5-fluorouracil (Longley et al. 2003), or mycophenolic acid (Huang et al. 2008) can be used to induce nucleolar stress by interfering with the production and the availability of rRNA. Several studies have used actinomycin D, a drug that has been used in cancer treatments, to analyze nucleolar stress (da Rocha et al. 2001). Actinomycin D functions by inhibiting transcription by RNA polymerases I, II, and III at concentrations greater than 30 nM, and inhibiting only RNA Pol I-dependent transcription/rRNA production and ribosomal biogenesis at concentrations less than 10 nM (Perry and Kelley 1970; da Rocha et al. 2001; Iapalucci-Espinoza and Franze-Fernandez 1979). Treatment of cells with factors such as a low dose of actinomycin D or serum starvation may function to inhibit the assembly of RPs, resulting in the release of ribosome-free forms of RPs to the nucleoplasm (Scheer and Hock 1999). RPL5, RPL11, and RPL23 are examples of such ribosome-free forms of RPs, which demonstrate increased binding to Mdm2 after nucleolar stress is induced through a multitude of treatments including actinomycin D (Dai and Lu 2004; Dai et al. 2004; Jin et al. 2004), 5-fluorouracil (Sun et al. 2007; Gilkes et al. 2006), serum depletion, contact inhibition (Bhat et al. 2004), mycophenolic acid-mediated depletion of GTP (Sun et al. 2008), and nucleostemin-mediated interference of nucleolar function (Dai et al. 2008).

The RPs that mediate the p53 response to nucleolar stress act primarily through inhibiting the function of p53's primary negative regulator, Mdm2. While Mdm2 is critical in maintaining basal, low levels of p53, as one of p53's target genes *Mdm2* is induced by the p53 stress response, in order to return p53 to low levels after the assault. Mdm2 has been thought to inactivate p53 primarily through two mechanisms: (1) the N-terminus of Mdm2 has been shown to bind to the transactivation domain of p53 and this interaction is thought to physically block or "mask" p53 from having access to the basal transcriptional machinery (Oliner et al. 1993), and (2) the RING domain of Mdm2 has been shown to act as an E3 ubiquitin ligase toward p53, leading to its proteasomal degradation (Haupt et al. 1997; Honda et al. 1997; Kubbutat et al. 1997). Knockout of Mdm2 in mice is embryonic lethal, and this lethality can be rescued by simultaneous deletion of p53, indicating how critical Mdm2 is in proper regulation of p53 (de Oca et al. 1995; Jones et al. 1995). Through the generation of a mouse model with a single amino acid residue substitution (C462A) in the Mdm2 RING finger domain, the region which confers Mdm2 E3 ubiquitin ligase activity, the mechanism through which Mdm2 regulates p53 could be further elucidated (Itahana et al. 2007). The Mdm2[C462A] mutation does not interfere with the interaction of Mdm2 with p53, or RPs, but ablates its E3 ubiquitin ligase activity (Itahana et al. 2007). Despite being able to interact with p53, the Mdm2[C462A] mutation results in embryonic lethality, which can be rescued by concomitant deletion of p53 (Itahana et al. 2007), similar to what is observed in Mdm2 null mice. While the physical interaction of Mdm2 with p53 is still thought to be important in p53 regulation, the Mdm2[C462A] mouse demonstrates that this physical interaction alone is not sufficient for the level of p53 regulation necessary for proper embryonic development (Itahana et al. 2007).

The feedback loop between Mdm2 and p53 may appear counterintuitive; p53 needs to be stabilized and activated in order to transactivate its downstream targets and induce senescence, apoptosis, or cell cycle arrest, but this same stabilization and activation is also responsible for the induction of its major negative regulator, Mdm2. The ability of p53 to regulate itself by playing a role in the induction of its major negative regulator allows for the cell to react to a stressor, and then recover from the stress response itself. In this pathway, the timing and function of Mdm2 and its degradation of p53 are likely critical. Despite Mdm2 being labeled an "oncoprotein," further analysis suggests that its role is hardly so simple. Overexpression of Mdm2 has been shown to induce G1 arrest in normal cells dependent on p53, suggesting that Mdm2 is also capable of acting as a growth suppressor to enable a cellular response and ultimately, cell survival (Frum et al. 2009).

Contrary to its presumed role, preventing Mdm2 from degrading p53 does not directly activate p53, but rather, it more indirectly prevents p53 inactivation.

One mechanism through which exposure to genotoxic chemicals and ionizing or UV irradiation leads to the induction of DNA damage is the activation of the ATM-Chk2 or ATR-Chk1 cascades. The ATM-Chk2 and ATR-Chk1 pathways promote the phosphorylation of both p53 and Mdm2, a modification that prevents the proteins from interacting with each other (Appella and Anderson 2001; Prives 1998). Phosphorylation of p53 on serine 20, prevents Mdm2 from being able to interact with and degrade p53, resulting in increased stability of p53 upon exposure to DNA damage (Hirao et al. 2000). Moreover, different types of stress have been shown to trigger a number of additional posttranslational modifications to p53 and/or Mdm2, including acetylation, sumoylation, methylation, and neddylation, which result in the activation of p53 (Brooks and Gu 2003; Dai et al. 2006a, b; Huang and Berger 2008; Melchior and Hengst 2002; Prives and Manley 2001).

The number of ribosomal proteins involved in p53-Mdm2 regulation is a continuously growing list, with an increasing number of functions. The first RP found to bind to Mdm2 was RPL5, although the functional significance was not clear in early studies (Marechal et al. 1994). The first RP found to bind and inhibit Mdm2 and activate p53 was RPL11 (Lohrum et al. 2003; Zhang et al. 2003). Soon after, RPL5 (Dai and Lu 2004; Marechal et al. 1994) and RPL23 (Dai et al. 2004; Jin et al. 2004) were also shown to interact with Mdm2 and, similarly to RPL11, inhibit its function as an E3 ubiquitin ligase, thus enabling p53 activation. Other RPs such as RPS7 (Chen et al. 2007; Zhu et al. 2009) and RPL26 (Ofir-Rosenfeld et al. 2008) were later found to interact with Mdm2. The interaction of RPS7 with Mdm2 is unique in that RPS7 not only suppresses Mdm2 activity but is itself a substrate for Mdm2 ubiquitination, leading to the proposition that RPS7 acts as both regulator and substrate of Mdm2 (Zhu et al. 2009). RPL26 was found to increase the rate of translation of p53 mRNA by binding to its 5' untranslated region, and in this case, Mdm2 functions to ubiquitinate and degrade RPL26 and inhibit p53 translation (Takagi et al. 2005). The RPS27-like protein (RPS27L), found to interact with Mdm2 through yet another mechanism, is transactivated by p53 directly, which illustrates the capacity for p53 to directly induce an RP and promote apoptosis (He and Sun 2007). Many of the RPs identified thus far, including RPL5, RPL11, RPL23, RPS7, and RPL26, are known bind Mdm2 at its central acidic domain, making this region critical for understanding the RP–Mdm2 interaction.

The Mdm2 protein consists of three highly conserved regions: an N-terminal p53 binding domain, a central acidic region containing a C4 zinc finger domain, and a C-terminal RING finger domain conferring E3 ubiquitin ligase activity (Fig. 12.2). The N-terminal p53 binding domain of Mdm2 is critical not only to bind p53 and mask its transactivation but also, along with the central region of Mdm2, to properly ubiquitinate p53 (Yu et al. 2006a, b).

The C4 zinc finger in Mdm2's central domain shares significant similarity with the zinc ribbon domains found in many ribosomal proteins, and possesses a region matching the X(4)-W-X-C-X(2-4)-C-X(3)-N-X(6)-C-X(2)-C-X(5) (where X is any amino acid) consensus sequence found in RanBP2/NZF C4 zinc fingers (Yu et al. 2006a, b). This group of zinc fingers is known to mediate a variety of protein–protein, protein–DNA, and protein–RNA interactions (Yu et al. 2006a, b).

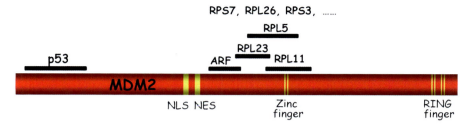

Fig. 12.2 A diagram of the Mdm2 protein and approximate binding domains for p53, ARF, and several ribosomal proteins. The central acidic region including the C4 zinc finger, C-terminal RING finger domain, nuclear localization signal sequence (NLS) and nuclear export signal sequence (NES) of Mdm2 are indicated

Missense mutations in Mdm2's C4 zinc finger domain were found to be associated with follicular lymphomas and liposarcomas; furthermore, cells from these tumors demonstrate accumulation of Mdm2 and a lack of nuclear p53, implicating the zinc finger in the negative regulation of Mdm2 (Schlott et al. 1997; Tamborini et al. 2001). In follicular lymphomas, a cysteine (C) to phenylalanine (F) conversion at amino acid residue 305 of Mdm2 was observed; both in vitro and in vivo studies have shown the Mdm2^{C305F} mutant protein to lack the ability to bind to RPL5 and RPL11 (Lindstrom et al. 2007; Macias et al. 2010). While the Mdm2^{C305F} mutation inhibits Mdm2 from binding to RPL5 and RPL11, it maintains its ability to bind normally to p53 and RPL23 (Macias et al. 2010). The inability of Mdm2^{C305F} to bind to RPL11 and RPL5 prevents these RPs from inhibiting Mdm2, providing an explanation for the high levels of Mdm2 and low levels of p53 observed in the tumor cells. Alternatively, mutating the Mdm2 305 residue to serine results in an Mdm2^{C305S} mutant protein capable of binding to RPL5 and RPL23, but not RPL11(Gilkes et al. 2006). It remains unclear why Mdm2's C4 zinc finger is critical for RPL11 binding, and specifically, whether RPL11 is binding directly to the C4 zinc finger, or if the zinc finger is only necessary to provide the structural stability required for Mdm2 to bind RPL11. Exploration of other RanBP2/NZP C4 zinc fingers has provided some insight into possible mechanisms to explain this dependence, including the concept that Mdm2 and RPL11 may compete to bind to rRNA, mRNA, or the nuclear export receptor protein Crm-1 (Zhang and Lu 2009). Interestingly, although the central acidic domain is highly conserved between Mdm2 and MdmX, these Mdm2-binding RPs have not been shown to interact with MdmX (Gilkes et al. 2006; Jin et al. 2006).

Characterization of Mdm2's C4 zinc finger domain has allowed for identification of binding sites for multiple RPs, but has not served to explain the purpose of having many RPs capable of interacting with Mdm2. Of the more characterized RPs, including RPL5, RPL11, and RPL23, all are found to bind to distinct domains within the Mdm2 central acidic domain in a similar, yet nonidentical manner. Because of the multiple mechanisms through which RPs function, different RPs may influence the p53 response to nucleolar stress differently. Following nucleolar stress, the p53 response may be activated by RPs that bind to and inactivate Mdm2.

The presence of RPs, including RPL11 (Bhat et al. 2004), RPL5 (Dai and Lu 2004), RPS7 (Zhu et al. 2009), and RPS3 (Yadavilli et al. 2009), is important for mediation of cell cycle arrest via the p53 response pathway. In the case of other RPs such as RPL23 (Jin et al. 2004) and RPS6 (Panic et al. 2006), the absence of these proteins leads to activation of the p53 response, resulting in inhibition of the cell cycle in the absence of RPL23, and apoptosis in the absence of RPS6. Depletion of endogenous RPL37 is also shown to increase the level of p53, as well as its downstream targets p21 and Mdm2 (Llanos and Serrano 2010). The variety of effects observed from increased and decreased levels of different RPs suggests that the relative balance of RPs may influence the p53 response to nucleolar stress. For example, knockdown of L37 is shown to increase the level of Mdm2/L11 complexes, promoting inactivation of Mdm2 by L11 and enabling activation of the p53 response (Llanos and Serrano 2010). In this sense, the p53-Mdm2 pathway responds not only to DNA damage but also to ribosomal damage, bringing to light the importance of proper protein synthesis regulation in protecting the cell from abnormal proliferation.

One explanation for the large number of RPs able to bind and inhibit Mdm2 following nucleolar stress is to allow for compensation in the case that one RP is lost. While this explanation is feasible, it appears as if RPs are functioning nonredundantly, as deletion or functional inhibition of a single RP can influence the Mdm2-p53 pathway. Additionally, it has been proposed that multiple RPs exist to serve this seemingly identical function, because under various cellular contexts different RPs, or combinations of RPs, contribute uniquely to determine the response. Nevertheless, mutations in several RPs have been found to be associated with cancer, making it clear that RP binding to Mdm2, to allow for activation of p53 activity, is important for the prevention of tumorigenesis (Draptchinskaia et al. 1999; Ebert et al. 2008; Cmejla et al. 2009; Gazda et al. 2008; Lai et al. 2009).

12.5 Nucleolar Stress Versus Oncogenic Stress: Possible Outcomes in the Cell

Although the cellular response to both nucleolar and oncogenic stressors is mediated by p53, the two pathways appear to be distinct in their function, mechanism, and outcome. One example of this can be observed in knockin mutant Mdm2^{C305F} mice, in which the inability of the mutant Mdm2 to bind RPL5 and RPL11 results in attenuation of the p53 response to nucleolar stress, although it maintains a normal p53 response to DNA damage (Macias et al. 2010). The ability of the DNA damage response to function normally, while simultaneously the nucleolar stress response is largely lost, suggests that these two pathways are functioning in a mutually independent manner. The observed disruption of the Mdm2–p53 interaction, as a result of exposure to both oncogenic and growth stressors, serves to connect the regulation of protein synthesis and cellular division. While serum starvation, nucleotide depletion, actinomycin D treatment, and UV irradiation are all deleterious to a cell's

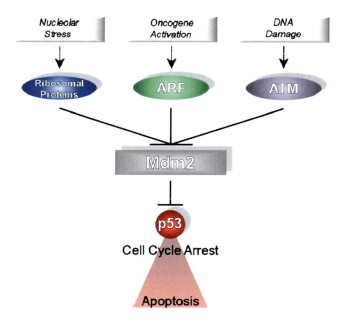

Fig. 12.3 Multiple stress signals are transduced through Mdm2 to p53. Nucleolar stress triggers ribosomal protein-mediated inhibition of Mdm2, hyperproliferative signals induce ARF expression to inhibit Mdm2, and DNA damage can activate ATM kinase to phosphorylate Mdm2. Inhibition of the E3 ligase function of Mdm2 promotes p53 stability, transactivation of target genes, and induction of cell cycle arrest or apoptosis

productivity, the p53 response to each of these stressors determines the cellular outcome. DNA damage evokes p53 activity through sensors such as ATM or ATR, phosphorylating p53 and Mdm2, inhibiting their interaction, and resulting in cell cycle arrest or apoptosis. On the other hand, following nucleolar stress, the nucleolus may contribute to either the activation or inactivation of p53, allowing the nucleolar stress response to potentially promote either cell cycle inhibition or proliferation. Numerous studies have demonstrated the role of the nucleolus in p53 activation via the ability of RPs to bind to Mdm2, inhibiting its degradation of p53; however, the nucleolus may also play a role in the inactivation of p53 via *trans*-nucleolar export of the Mdm2–p53 complex, enabling the degradation and inactivation of p53 (Sherr and Weber 2000).

While oncogenic challenges threaten the fundamental genetic integrity of the cell, nucleolar stress challenges normal cell growth and activity. Upon exposure to an oncogenic challenge, the cell may induce cell cycle arrest to allow for repair or induce apoptosis if the damage is irreparable. In the case of nucleolar stress, the RP-mediated cellular solutions vary depending on whether Mdm2 is bound to RP(s) or p53. Regardless of the type of stress, oncogenic or nucleolar, it is evident that degradation of the integrity of any of the cellular systemic functions, including DNA replication and ribosome synthesis, results in the evocation of p53 (Fig. 12.3). The role of p53 in addressing cellular stress, and the bridge that p53 serves between protein synthesis and cell cycle progression, further emphasizes its role in overall cellular regulation. In the case that a cell is unable to correctly remedy the damage inflicted by oncogenic or nucleolar stressors, and p53 is unable to function to induce senescence or apoptosis, uncontrolled growth, or cancer may result, ultimately causing the destruction of the entire system.

12.6 Conclusions

The nucleolus has long been known as the cellular component critical for the production of ribosomes and enabling protein synthesis, but the studies presented in this chapter demonstrate the additional importance of the nucleolus in the stress response. In addition to mediating cell growth, many examples have suggested that the nucleolus may serve an important role in cell division. Through the sequestration and release of proteins, including ARF and Mdm2, the nucleolus has been demonstrated to be an important facet of cell cycle regulation. While disruption of the nucleolus and the release of RPs into the nucleoplasm have obvious effects on protein synthesis, the ability of RPs to indirectly activate p53 implicates nucleolar disruption as a player in cell cycle arrest as well. Although RPs have been demonstrated to have multiple, and sometimes contrasting, effects on the cell, it is important to note that the variety of these functions may exist to allow for the cell to sense and respond to different forms of stress. There is growing evidence implicating the nucleolus and nucleolar proteins as critical components of the p53 stress response, and with that, carcinogenesis. Overexpression of Mdm2 is observed in a number of cancers, and thus far, drug targeting efforts have focused on the N-terminal p53 binding domain (Issaeva et al. 2004; Shangary et al. 2008; Vassilev et al. 2004) and the C-terminal RING finger E3 ligase domain (Yang et al. 2005). The identification of RP–Mdm2 interactions and the role these interactions play in the p53 response have brought to light the possibility of utilizing the central acidic region of Mdm2 as an anticancer drug target. While much remains unknown regarding the functions of nucleolar proteins, including ARF, B23, and RPs, as well as the nucleolus itself, it is undeniable that these components play an integral role in the p53 stress response.

Acknowledgements We thank Koji Itahana and Chad Deisenroth for figure art. We apologize for not being able to cite all of the relevant papers due to limited space. The work is supported in part by grants from The Leukemia & Lymphoma Society, The American Cancer Society, and The National Institute of Health.

References

Adachi Y, Copeland TD, Hatanaka M, Oroszlan S (1993) Nucleolar targeting signal of Rex protein of human T-cell leukemia virus type I specifically binds to nucleolar shuttle protein B-23. J Biol Chem 268(19):13930–13934

Andersen JS, Wilkinson CJ, Mayor T, Mortensen P, Nigg EA, Mann M (2003) Proteomic characterization of the human centrosome by protein correlation profiling. Nature 426(6966):570–574

Appella E, Anderson CW (2001) Post-translational modifications and activation of p53 by genotoxic stresses. Eur J Biochem 268(10):2764–2772

Bertwistle D, Sugimoto M, Sherr CJ (2004) Physical and functional interactions of the Arf tumor suppressor protein with nucleophosmin/B23. Mol Cell Biol 24(3):985–996

Bhat KP, Itahana K, Jin A, Zhang Y (2004) Essential role of ribosomal protein L11 in mediating growth inhibition-induced p53 activation. Embo J 23(12):2402–2412

Borer RA, Lehner CF, Eppenberger HM, Nigg EA (1989) Major nucleolar proteins shuttle between nucleus and cytoplasm. Cell 56(3):379–390

Brady SN, Yu Y, Maggi LB Jr, Weber JD (2004) ARF impedes NPM/B23 shuttling in an Mdm2-sensitive tumor suppressor pathway. Mol Cell Biol 24(21):9327–9338

Brooks CL, Gu W (2003) Ubiquitination, phosphorylation and acetylation: the molecular basis for p53 regulation. Curr Opin Cell Biol 15(2):164–171

Carnero A, Hudson JD, Price CM, Beach DH (2000) p16^{INK4a} and p19^{ARF} act overlapping pathways in cellular immortalization. Nat Cell Biol 2:148–155

Chen D, Zhang Z, Li M, Wang W, Li Y, Rayburn ER, Hill DL, Wang H, Zhang R (2007) Ribosomal protein S7 as a novel modulator of p53-MDM2 interaction: binding to MDM2, stabilization of p53 protein, and activation of p53 function. Oncogene 26:5029–5037

Cmejla R, Cmejlova J, Handrkova H, Petrak J, Petrtylova K, Mihal V, Stary J, Cerna Z, Jabali Y, Pospisilova D (2009) Identification of mutations in the ribosomal protein L5 (RPL5) and ribosomal protein L11 (RPL11) genes in Czech patients with Diamond-Blackfan anemia. Hum Mutat 30(3):321–327

Colombo E, Marine JC, Danovi D, Falini B, Pelicci PG (2002) Nucleophosmin regulates the stability and transcriptional activity of p53. Nat Cell Biol 4(7):529–533

Colombo E, Bonetti P, Lazzerini Denchi E, Martinelli P, Zamponi R, Marine JC, Helin K, Falini B, Pelicci PG (2005) Nucleophosmin is required for DNA integrity and p19Arf protein stability. Mol Cell Biol 25(20):8874–8886

da Rocha AB, Lopes RM, Schwartsmann G (2001) Natural products in anticancer therapy. Curr Opin Pharmacol 1(4):364–369

Dai MS, Lu H (2004) Inhibition of MDM2-mediated p53 ubiquitination and degradation by ribosomal protein L5. J Biol Chem 279(43):44475–44482

Dai MS, Zeng SX, Jin Y, Sun XX, David L, Lu H (2004) Ribosomal protein L23 activates p53 by inhibiting MDM2 function in response to ribosomal perturbation but not to translation inhibition. Mol Cell Biol 24(17):7654–7668

Dai MS, Jin Y, Gallegos JR, Lu H (2006a) Balance of Yin and Yang: ubiquitylation-mediated regulation of p53 and c-Myc. Neoplasia 8(8):630–644

Dai MS, Shi D, Jin Y, Sun XX, Zhang Y, Grossman SR, Lu H (2006b) Regulation of the MDM2-p53 pathway by ribosomal protein L11 involves a post-ubiquitination mechanism. J Biol Chem 281(34):24304–24313

Dai MS, Sun XX, Lu H (2008) Aberrant expression of nucleostemin activates p53 and induces cell cycle arrest via inhibition of MDM2. Mol Cell Biol 28(13):4365–4376

de Oca M, Luna R, Wagner DS, Lozano G (1995) Rescue of early embryonic lethality in mdm2-deficient mice by deletion of p53. Nature 378(6553):203–206

de Stanchina E, McCurrach ME, Zindy F, Shieh S-Y, Ferbeyre G, Samuelson AV, Prives C, Roussel MF, Sherr CJ, Lowe SW (1998) E1A signaling to p53 involves the p19^{ARF} tumor suppressor. Genes Dev 12:2434–2442

Dimri GP, Itahana K, Acosta M, Campisi J (2000) Regulation of a senescence checkpoint response by the E2F1 transcription factor and p14(ARF) tumor suppressor. Mol Cell Biol 20(1):273–285

Draptchinskaia N, Gustavsson P, Andersson B, Pettersson M, Willig TN, Dianzani I, Ball S, Tchernia G, Klar J, Matsson H, Tentler D, Mohandas N, Carlsson B, Dahl N (1999) The gene encoding ribosomal protein S19 is mutated in Diamond-Blackfan anaemia. Nat Genet 21(2):169–175

Ebert BL, Pretz J, Bosco J, Chang CY, Tamayo P, Galili N, Raza A, Root DE, Attar E, Ellis SR, Golub TR (2008) Identification of RPS14 as a 5q- syndrome gene by RNA interference screen. Nature 451(7176):335–339

12 The Role of the Nucleolus in the Stress Response

Fankhauser C, Izaurralde E, Adachi Y, Wingfield P, Laemmli UK (1991) Specific complex of human immunodeficiency virus type 1 rev and nucleolar B23 proteins: dissociation by the Rev response element. Mol Cell Biol 11(5):2567–2575

Feuerstein N, Mond JJ (1987) "Numatrin," a nuclear matrix protein associated with induction of proliferation in B lymphocytes. J Biol Chem 262(23):11389–11397

Feuerstein N, Spiegel S, Mond JJ (1988) The nuclear matrix protein, numatrin (B23), is associated with growth factor-induced mitogenesis in Swiss 3T3 fibroblasts and with T lymphocyte proliferation stimulated by lectins and anti-T cell antigen receptor antibody. J Cell Biol 107(5):1629–1642

Frum R, Ramamoorthy M, Mohanraj L, Deb S, Deb SP (2009) MDM2 controls the timely expression of cyclin A to regulate the cell cycle. Mol Cancer Res 7(8):1253–1267

Gao H, Jin S, Song Y, Fu M, Wang M, Liu Z, Wu M, Zhan Q (2005) B23 regulates GADD45a nuclear translocation and contributes to GADD45a-induced cell cycle G2-M arrest. J Biol Chem 280(12):10988–10996

Gazda HT, Sheen MR, Vlachos A, Choesmel V, O'Donohue MF, Schneider H, Darras N, Hasman C, Sieff CA, Newburger PE, Ball SE, Niewiadomska E, Matysiak M, Zaucha JM, Glader B, Niemeyer C, Meerpohl JJ, Atsidaftos E, Lipton JM, Gleizes PE, Beggs AH (2008) Ribosomal protein L5 and L11 mutations are associated with cleft palate and abnormal thumbs in Diamond-Blackfan anemia patients. Am J Hum Genet 83(6):769–780

Gazzeri S, Della Valle V, Chaussade L, Brambilla C, Larsen CJ, Brambilla E (1998) The human p19ARF protein encoded by the beta transcript of the p16INK4a gene is frequently lost in small cell lung cancer. Cancer Res 58(17):3926–3931

Gilkes DM, Chen L, Chen J (2006) MDMX regulation of p53 response to ribosomal stress. EMBO J 25(23):5614–5625

Gjerset RA (2006) DNA damage, p14ARF, nucleophosmin (NPM/B23), and cancer. J Mol Histol 37(5–7):239–251

Grisendi S, Bernardi R, Rossi M, Cheng K, Khandker L, Manova K, Pandolfi PP (2005) Role of nucleophosmin in embryonic development and tumorigenesis. Nature 437(7055):147–153

Haupt Y, Maya R, Kazaz A, Oren M (1997) Mdm2 promotes the rapid degradation of p53. Nature 387:296–299

He H, Sun Y (2007) Ribosomal protein S27L is a direct p53 target that regulates apoptosis. Oncogene 26(19):2707–2716

Herrera JE, Savkur R, Olson MO (1995) The ribonuclease activity of nucleolar protein B23. Nucleic Acids Res 23(19):3974–3979

Hirao A, Kong YY, Matsuoka S, Wakeham A, Ruland J, Yoshida H, Liu D, Elledge SJ, Mak TW (2000) DNA damage-induced activation of p53 by the checkpoint kinase Chk2. Science 287(5459):1824–1827

Holzel M, Rohrmoser M, Schlee M, Grimm T, Harasim T, Malamoussi A, Gruber-Eber A, Kremmer E, Hiddemann W, Bornkamm GW, Eick D (2005) Mammalian WDR12 is a novel member of the Pes1-Bop1 complex and is required for ribosome biogenesis and cell proliferation. J Cell Biol 170(3):367–378

Honda R, Tanaka H, Yasuda H (1997) Oncoprotein MDM2 is a ubiquitin ligase E3 for tumor suppressor p53. FEBS Lett 420:25–27

Huang J, Berger SL (2008) The emerging field of dynamic lysine methylation of non-histone proteins. Curr Opin Genet Dev 18(2):152–158

Huang M, Ji Y, Itahana K, Zhang Y, Mitchell B (2008) Guanine nucleotide depletion inhibits preribosomal RNA synthesis and causes nucleolar disruption. Leuk Res 32(1):131–141

Iapalucci-Espinoza S, Franze-Fernandez MT (1979) Effect of protein synthesis inhibitors and low concentrations of actinomycin D on ribosomal RNA synthesis. FEBS Lett 107(2):281–284

Issaeva N, Bozko P, Enge M, Protopopova M, Verhoef LG, Masucci M, Pramanik A, Selivanova G (2004) Small molecule RITA binds to p53, blocks p53-HDM-2 interaction and activates p53 function in tumors. Nat Med 10(12):1321–1328

Itahana K, Bhat KP, Jin A, Itahana Y, Hawke D, Kobayashi R, Zhang Y (2003) Tumor suppressor ARF degrades B23, a nucleolar protein Involved in ribosome biogenesis and cell proliferation. Mol Cell 12(5):1151–1164

Itahana K, Mao H, Jin A, Itahana Y, Clegg HV, Lindstrom MS, Bhat KP, Godfrey VL, Evan GI, Zhang Y (2007) Targeted inactivation of Mdm2 RING finger E3 ubiquitin ligase activity in the mouse reveals mechanistic insights into p53 regulation. Cancer Cell 12(4):355–366

Jin A, Itahana K, O'Keefe K, Zhang Y (2004) Inhibition of HDM2 and activation of p53 by ribosomal protein L23. Mol Cell Biol 24(17):7669–7680

Jin Y, Dai MS, Lu SZ, Xu Y, Luo Z, Zhao Y, Lu H (2006) 14-3-3gamma binds to MDMX that is phosphorylated by UV-activated Chk1, resulting in p53 activation. EMBO J 25(6):1207–1218

Jones SN, Roe AE, Donehower LA, Bradley A (1995) Rescue of embryonic lethality in Mdm2-deficient mice by absence of p53. Nature 378:206–208

Kamijo T, Zindy F, Roussel MF, Quelle DE, Downing JR, Ashmun RA, Grosveld G, Sherr CJ (1997) Tumor suppression at the mouse INK4a locus mediated by the alternative reading frame product p19*ARF*. Cell 91:649–659

Kamijo T, Weber JD, Zambetti G, Zindy F, Roussel M, Sherr CJ (1998) Functional and physical interaction of the ARF tumor suppressor with p53 and MDM2. Proc Natl Acad Sci USA 95:8292–8297

Korgaonkar C, Zhao L, Modestou M, Quelle DE (2002) ARF function does not require p53 stabilization or Mdm2 relocalization. Mol Cell Biol 22(1):196–206

Korgaonkar C, Hagen J, Tompkins V, Frazier AA, Allamargot C, Quelle FW, Quelle DE (2005) Nucleophosmin (B23) targets ARF to nucleoli and inhibits its function. Mol Cell Biol 25(4):1258–1271

Kubbutat MHG, Jones SN, Vousden KH (1997) Regulation of p53 stability by Mdm2. Nature 387:299–303

Kuo ML, den Besten W, Bertwistle D, Roussel MF, Sherr CJ (2004) N-terminal polyubiquitination and degradation of the Arf tumor suppressor. Genes Dev 18(15):1862–1874

Kurki S, Peltonen K, Latonen L, Kiviharju TM, Ojala PM, Meek D, Laiho M (2004) Nucleolar protein NPM interacts with HDM2 and protects tumor suppressor protein p53 from HDM2-mediated degradation. Cancer Cell 5(5):465–475

Lai K, Amsterdam A, Farrington S, Bronson RT, Hopkins N, Lees JA (2009) Many ribosomal protein mutations are associated with growth impairment and tumor predisposition in zebrafish. Dev Dyn 238(1):76–85

Lee SY, Park JH, Kim S, Park EJ, Yun Y, Kwon J (2005) A proteomics approach for the identification of nucleophosmin and heterogeneous nuclear ribonucleoprotein C1/C2 as chromatin-binding proteins in response to DNA double-strand breaks. Biochem J 388(Pt 1):7–15

Lempiainen H, Shore D (2009) Growth control and ribosome biogenesis. Curr Opin Cell Biol 21(6):855–863

Li YP (1997) Protein B23 is an important human factor for the nucleolar localization of the human immunodeficiency virus protein Tat. J Virol 71(5):4098–4102

Lindstrom MS, Zhang Y (2008) Ribosomal protein S9 is a novel B23/NPM-binding protein required for normal cell proliferation. J Biol Chem 283(23):15568–15576

Lindstrom MS, Jin A, Deisenroth C, White Wolf G, Zhang Y (2007) Cancer-Associated Mutations in the MDM2 Zinc Finger Domain Disrupt Ribosomal Protein Interaction and Attenuate MDM2-Induced p53 Degradation. Mol Cell Biol 27(3):1056–1068

Liu JJ, Huang BH, Zhang J, Carson DD, Hooi SC (2006) Repression of HIP/RPL29 expression induces differentiation in colon cancer cells. J Cell Physiol 207(2):287–292

Llanos S, Serrano M (2010) Depletion of ribosomal protein L37 occurs in response to DNA damage and activates p53 through the L11/MDM2 pathway. Cell Cycle 9(19):4005–4012

Llanos S, Clark PA, Rowe J, Peters G (2001) Stabilization of p53 by p14ARF without relocation of MDM2 to the nucleolus. Nat Cell Biol 3(5):445–452

Lohrum MA, Ludwig RL, Kubbutat MH, Hanlon M, Vousden KH (2003) Regulation of HDM2 activity by the ribosomal protein L11. Cancer Cell 3(6):577–587

Longley DB, Harkin DP, Johnston PG (2003) 5-fluorouracil: mechanisms of action and clinical strategies. Nat Rev Cancer 3(5):330–338

Macias E, Jin A, Deisenroth C, Bhat K, Mao H, Lindstrom MS, Zhang Y (2010) An ARF-independent c-MYC-activated tumor suppression pathway mediated by ribosomal protein-Mdm2 Interaction. Cancer Cell 18(3):231–243

Marechal V, Elenbaas B, Piette J, Nicolas J-C, Levine AJ (1994) The ribosomal protein L5 is associated with mdm-2 and mdm2-p53 complexes. MolCell Biol 14:7414–7420

Meder VS, Boeglin M, de Murcia G, Schreiber V (2005) PARP-1 and PARP-2 interact with nucleophosmin/B23 and accumulate in transcriptionally active nucleoli. J Cell Sci 118(Pt 1):211–222

Melchior F, Hengst L (2002) SUMO-1 and p53. Cell Cycle 1(4):245–249

Nozawa Y, Van Belzen N, Van der Made AC, Dinjens WN, Bosman FT (1996) Expression of nucleophosmin/B23 in normal and neoplastic colorectal mucosa. J Pathol 178(1):48–52

Ofir-Rosenfeld Y, Boggs K, Michael D, Kastan MB, Oren M (2008) Mdm2 regulates p53 mRNA translation through inhibitory interactions with ribosomal protein L26. Mol Cell 32(2):180–189

Okuda M, Horn HF, Tarapore P, Tokuyama Y, Smulian AG, Chan PK, Knudsen ES, Hofmann IA, Snyder JD, Bove KE, Fukasawa K (2000) Nucleophosmin/B23 is a target of CDK2/cyclin E in centrosome duplication. Cell 103(1):127–140

Oliner JD, Pietenpol JA, Thiagalingam S, Gyuris J, Kinzler KW, Vogelstein B (1993) Oncoprotein MDM2 conceals the activation domain of tumor suppressor p53. Nature 362:857–860

Palmero I, Pantoja C, Serrano M (1998) p19^{ARF} links the tumor suppressor p53 and Ras. Nature 395:125–126

Panic L, Tamarut S, Sticker-Jantscheff M, Barkic M, Solter D, Uzelac M, Grabusic K, Volarevic S (2006) Ribosomal protein S6 gene haploinsufficiency is associated with activation of a p53-dependent checkpoint during gastrulation. Mol Cell Biol 26(23):8880–8891

Perry RP, Kelley DE (1970) Inhibition of RNA synthesis by actinomycin D: characteristic dose-response of different RNA species. J Cell Physiol 76(2):127–139

Pestov DG, Strezoska Z, Lau LF (2001) Evidence of p53-dependent cross-talk between ribosome biogenesis and the cell cycle: effects of nucleolar protein Bop1 on G(1)/S transition. Mol Cell Biol 21(13):4246–4255

Pomerantz J, Schreiber-Agus N, Liegeois NJ, Silverman A, Alland L, Chin L, Potes J, Chen K, Orlow I, DePinho RA (1998) The INK4a tumor suppressor gene product, p19Arf, interacts with MDM2 and neutralizes MDM2's inhibition of p53. Cell 92:713–723

Prives C (1998) Signaling to p53: breaking the MDM2-p53 circuit. Cell 95:5–8

Prives C, Manley JL (2001) Why is p53 acetylated? Cell 107(7):815–818

Quelle DE, Zindy F, Ashmun R, Sherr CJ (1995) Alterative reading frames of the INK4a tumor suppressor gene encode two unrelated proteins capable of inducing cell cycle arrest. Cell 83:993–1000

Savkur RS, Olson MO (1998) Preferential cleavage in pre-ribosomal RNA byprotein B23 endoribonuclease. Nucleic Acids Res 26(19):4508–4515

Scheer U, Hock R (1999) Structure and function of the nucleolus. Curr Opin Cell Biol 11(3):385–390

Schlott T, Reimer S, Jahns A, Ohlenbusch A, Ruschenburg I, Nagel H, Droese M (1997) Point mutations and nucleotide insertions in the MDM2 zinc finger structure of human tumors. JPathol 182:54–61

Shangary S, Qin D, McEachern D, Liu M, Miller RS, Qiu S, Nikolovska-Coleska Z, Ding K, Wang G, Chen J, Bernard D, Zhang J, Lu Y, Gu Q, Shah RB, Pienta KJ, Ling X, Kang S, Guo M, Sun Y, Yang D, Wang S (2008) Temporal activation of p53 by a specific MDM2 inhibitor is selectively toxic to tumors and leads to complete tumor growth inhibition. Proc Natl Acad Sci U S A 105(10):3933–3938

Sherr CJ (2001) Parsing Ink4a/Arf: "pure" p16-null mice. Cell 106(5):531–534

Sherr CJ (2006) Divorcing ARF and p53: an unsettled case. Nat Rev Cancer 6(9):663–673

Sherr CJ, Weber JD (2000) The ARF-p53 pathway. Curr Opin Genet Dev 10:94–99

Skarie JM, Link BA (2008) The primary open-angle glaucoma gene WDR36 functions in ribosomal RNA processing and interacts with the p53 stress-response pathway. Hum Mol Genet 17(16):2474–2485

Sobell HM (1985) Actinomycin and DNA transcription. Proc Natl Acad Sci U S A 82(16): 5328–5331

Spector DL, Ochs RL, Busch H (1984) Silver staining, immunofluorescence, and immunoelectron microscopic localization of nucleolar phosphoproteins B23 and C23. Chromosoma 90(2): 139–148

Stott FJ, Bates S, James MC, McConnell BB, Starborg M, Brookes S, Palmero I, Ryan K, Hara E, Vousden KH, Peters G (1998) The alternative product from the human CDK2A locus, p14[ARF], participates in a regulatory feedback loop with p53 and MDM2. EMBO J 17:5001–5014

Strezoska Z, Pestov DG, Lau LF (2000) Bop1 is a mouse WD40 repeat nucleolar protein involved in 28S and 5. 8S RRNA processing and 60S ribosome biogenesis. Mol Cell Biol 20(15): 5516–5528

Sun XX, Dai MS, Lu H (2007) 5-fluorouracil activation of p53 involves an MDM2-ribosomal protein interaction. J Biol Chem 282(11):8052–8059

Sun XX, Dai MS, Lu H (2008) Mycophenolic acid activation of p53 requires ribosomal proteins L5 and L11. J Biol Chem 283(18):12387–12392

Szebeni A, Herrera JE, Olson MO (1995) Interaction of nucleolar protein B23 with peptides related to nuclear localization signals. Biochemistry 34(25):8037–8042

Szebeni A, Mehrotra B, Baumann A, Adam SA, Wingfield PT, Olson MO (1997) Nucleolar protein B23 stimulates nuclear import of the HIV-1 Rev protein and NLS-conjugated albumin. Biochemistry 36(13):3941–3949

Takagi M, Absalon MJ, McLure KG, Kastan MB (2005) Regulation of p53 translation and induction after DNA damage by ribosomal protein L26 and nucleolin. Cell 123(1):49–63

Takemura M, Sato K, Nishio M, Akiyama T, Umekawa H, Yoshida S (1999) Nucleolar protein B23.1 binds to retinoblastoma protein and synergistically stimulates DNA polymerase alpha activity. J Biochem 125(5):904–909

Tamborini E, Della Torre G, Lavarino C. Azzarelli A, Carpinelli P, Pierotti MA, Pilotti S (2001) Analysis of the molecular species generated by MDM2 gene amplification in liposarcomas. Int J Cancer 92(6):790–796

Tao W, Levine AJ (1999) Nucleocytoplasmic shuttling of oncoprotein Hdm2 is required for Hdm2-mediated degradation of p53. Proc Natl Acad Sci USA 96:3077–3080

Tokuyama Y, Horn HF, Kawamura K, Tarapore P, Fukasawa K (2001) Specific phosphorylation of nucleophosmin on Thr(199) by cyclin-dependent kinase 2-cyclin E and its role in centrosome duplication. J Biol Chem 276(24):21529–21537

Valdez BC, Perlaky L, Henning D, Saijo Y, Chan PK, Busch H (1994) Identification of the nuclear and nucleolar localization signals of the protein p120. Interaction with translocation protein B23. J Biol Chem 269(38):23776–23783

Vassilev LT, Vu BT, Graves B, Carvajal D, Podlaski F, Filipovic Z, Kong N, Kammlott U, Lukacs C, Klein C, Fotouhi N, Liu EA (2004) In vivo activation of the p53 pathway by small-molecule antagonists of MDM2. Science 303(5659):844–848

Wang D, Baumann A, Szebeni A, Olson MO (1994) The nucleic acid binding activity of nucleolar protein B23.1 resides in its carboxyl-terminal end. J Biol Chem 269(49):30994–30998

Weber JD, Taylor LJ, Roussel MF, Sherr CJ, Bar-Sagi D (1999) Nucleolar Arf sequesters Mdm2 and activates p53. Nat Cell Biol 1:20–26

Weber JD, Jeffers JR, Rehg JE, Randle DH, Lozano G, Roussel MF, Sherr CJ, Zambetti GP (2000) p53-independent functions of the p19(ARF) tumor suppressor. Genes Dev 14(18):2358–2365

Wu MH, Chang JH, Yung BY (2002a) Resistance to UV-induced cell-killing in nucleophosmin/B23 over-expressed NIH 3T3 fibroblasts: enhancement of DNA repair and up-regulation of PCNA in association with nucleophosmin/B23 over-expression. Carcinogenesis 23(1):93–100

Wu MH, Chang JH, Chou CC, Yung BY (2002b) Involvement of nucleophosmin/B23 in the response of HeLa cells to UV irradiation. Int J Cancer 97(3):297–305

Yadavilli S, Mayo LD, Higgins M, Lain S, Hegde V, Deutsch WA (2009) Ribosomal protein S3: A multi-functional protein that interacts with both p53 and MDM2 through its KH domain. DNA Repair (Amst) 8(10):1215–1224

Yang Y, Ludwig RL, Jensen JP, Pierre SA, Medaglia MV, Davydov IV, Safiran YJ, Oberoi P, Kenten JH, Phillips AC, Weissman AM, Vousden KH (2005) Small molecule inhibitors of HDM2 ubiquitin ligase activity stabilize and activate p53 in cells. Cancer Cell 7(6):547–559

Yu GW, Allen MD, Andreeva A, Fersht AR, Bycroft M (2006a) Solution structure of the C4 zinc finger domain of HDM2. Protein Sci 15(2):384–389

Yu GW, Rudiger S, Veprintsev D, Freund S, Fernandez-Fernandez MR, Fersht AR (2006b) The central region of HDM2 provides a second binding site for p53. Proc Natl Acad Sci U S A 103(5):1227–1232

Yung BY, Busch H, Chan PK (1985a) Translocation of nucleolar phosphoprotein B23 (37 kDa/pI 5.1) induced by selective inhibitors of ribosome synthesis. Biochim Biophys Acta 826(4):167–173

Yung BY, Busch RK, Busch H, Mauger AB, Chan PK (1985b) Effects of actinomycin D analogs on nucleolar phosphoprotein B23 (37,000 daltons/pI 5.1). Biochem Pharmacol 34(22):4059–4063

Zeller KI, Haggerty TJ, Barrett JF, Guo Q, Wonsey DR, Dang CV (2001) Characterization of nucleophosmin (B23) as a Myc target by scanning chromatin immunoprecipitation. J Biol Chem 276(51):48285–48291

Zhang Y (2004) The ARF-B23 connection: implications for growth control and cancer treatment. Cell Cycle 3(3):259–262

Zhang Y, Lu H (2009) Signaling to p53: ribosomal proteins find their way. Cancer Cell 16(5):369–377

Zhang Y, Xiong Y (1999) Mutation in human ARF exon 2 disrupt its nucleolar localization and impair its ability to block nuclear export of MDM2 and p53. Mol Cell 3:579–591

Zhang Y, Xiong Y, Yarbrough WG (1998) ARF promotes MDM2 degradation and stabilizes p53: ARF-INK4a locus deletion impairs both the Rb and p53 tumor suppression pathways. Cell 92:725–734

Zhang Y, Wolf GW, Bhat K, Jin A, Allio T, Burkhart WA, Xiong Y (2003) Ribosomal protein L11 negatively regulates oncoprotein MDM2 and mediates a p53-dependent ribosomal-stress checkpoint pathway. Mol Cell Biol 23(23):8902–8912

Zhang J, Tomasini AJ, Mayer AN (2008) RBM19 is essential for preimplantation development in the mouse. BMC Dev Biol 8:115

Zhu Y, Poyurovsky MV, Li Y, Biderman L, Stahl J, Jacq X, Prives C (2009) Ribosomal protein S7 is both a regulator and a substrate of MDM2. Mol Cell 35(3):316–326

Zindy F, Eischen CM, Randle DH, Kamijo T, Cleveland JL, Sherr CJ, Roussel MF (1998) Myc signaling via the ARF tumor suppressor regulates p53-dependent apoptosis and immortalization. Genes Dev 12:2424–2433

Chapter 13
New Frontiers in Nucleolar Research: Nucleostemin and Related Proteins

Robert Y. L. Tsai

13.1 Introduction

A proteomic catalog of proteins that are constantly or transiently associated with the nucleolus has expanded our current view of this subnuclear organelle to include both the ribosomal and nonribosomal functions (Pederson and Tsai 2009) (also see Chap.2 in this book). One of the nucleolar proteins capable of exercising nonribosomal activities is nucleostemin. The cDNA sequence of nucleostemin first appeared in the database of genes whose expression was increased by 17β-estradiol treatment of MCF-7 human breast cancer cells (Charpentier et al. 2000). It was not until 2002 that it was discovered again through a PCR-based subtractive screen that searched for genes preferentially expressed by the undifferentiated neural stem cells compared to their differentiated progeny, at which time it began to be understood biologically (Tsai and McKay 2002). Because of its enriched expression in neural stem cells and embryonic stem (ES) cells and nucleolar distribution, it was given the name, nucleostemin. The first study also revealed nucleostemin's function in maintaining the continuous proliferation of stem cells and cancer cells. Following this initial report on nucleostemin, a series of studies came out that investigated its expression characteristics, biological importance, regulatory pathway, mechanism of action, and potential therapeutic value with regard to tumorigenesis and tissue injury. More recently, a number of reports began to describe the functions of GNL3 in the invertebrate species as well as GNL3L in mammals. We now know that some of the key cell biological activities, including the cell cycle control, telomere and genome maintenance, ribosome biosynthesis, and possibly others, are dynamically monitored by this group of essential genes. Even though they have just begun to show their presence in the scientific arena, one may expect to see their rapid soar to stardom in the integrated world of cell biology.

R.Y.L. Tsai (✉)
Center for Cancer and Stem Cell Biology, Alkek Institute of Biosciences and Technology,
Texas A&M Health Science Center, Houston, TX, USA
e-mail: rtsai@ibt.tamhsc.edu

M.O.J. Olson (ed.), *The Nucleolus*, Protein Reviews 15,
DOI 10.1007/978-1-4614-0514-6_13, © Springer Science+Business Media, LLC 2011

13.2 A Phylogenetic View of Nucleostemin Family Proteins

All proteins containing the same signature MMR1_HSR1 motif as nucleostemin belong to the YlqF/YawG GTPase family, which can be found from prokaryotes to eukaryotes (Figs. 13.1a and 13.2) (Leipe et al. 2002). In bacteria and yeast, there are five distinct subclasses of MMR1_HSR1 proteins, that is, YjeQ (*E. coli*), MJ1464 (*Methanococcus jannaschii*), YqeH (*B. subtilis*), YlqF (*B. subtilis*), and YawG (*Schizosaccharomyces pombe*). The YqeH, YlqF, and YawG genes bear the highest homology with the vertebrate nucleostemin/GNL3L/Ngp-1, Mtg1, and Lsg1/Gnl1 gene families, respectively. The eukaryotic members of the MMR1_HSR1 family proteins can be found throughout the cytoplasm, mitochondria, nucleus, and nucleolus (Reynaud et al. 2005). Among them, nucleostemin, GNL3L, and Ngp-1 are the only MMR1_HSR1 proteins showing a nucleolar predominant distribution and are most closely related to each other (Fig. 13.1). On the basis of these criteria, they are classified as a distinct subfamily of proteins. The existence and transition of the nucleostemin family genes throughout phylogeny have been analyzed by a genome-wide comparison of all sequences that share homology with the human nucleostemin (AAH01024), GNL3L (AAH11720), and Ngp-1 (AAH09250) proteins (Table 13.1) (Tsai and Meng 2009). The search result reveals that nucleostemin, GNL3L, and Ngp-1 exist as different genes in all vertebrate species. Several human nucleostemin protein variants were found that differed by four amino acids or less

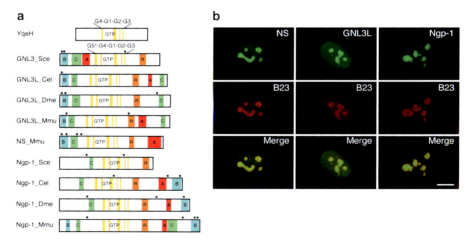

Fig. 13.1 Protein structures and subcellular distributions of nucleostemin family proteins. (**a**) Schematic diagrams of nucleostemin (NS), GNL3L, and Ngp-1 proteins of different species. *YqeH Bacilis subtilis*; *Sce Saccharomyces cerevisiae*; *Cel Caenorhabditis elegans*; *Dme Drosophila melanogaster*; *Mmu Mus musculus*; *B* basic; *C* coiled-coil; *GTP* GTP-binding motifs (G5*, G4, G1, G2, and G3); *R* RNA-binding; *A* acidic domain. (**b**) Subcellular distributions of nucleostemin, GNL3L and Ngp-1 in U2OS cells were shown using a C-terminally fused green fluorescent protein (GFP) (*green*) and counterstained with anti-B23 immunofluorescence (*red*). Scale bar: 10 μm

13 New Frontiers in Nucleolar Research...

				G5 DARXP	G4 NKXL	G1 GXPNVGKSS	G2 GXT	G3 DXPG
Consensus				**DARXP**	**NKXL**	**GXPNVGKSS**	**GXT**	**DXPG**
NS/GNL Family	NS	Human		DARDP	NKSDL	GFPNVGKSS	GLT	DSPS
		Cow		DARDP	NKSDL	GFPNVGKSS	GLT	DSPC
		Rat		DARDP	NKSDL	GFPNVGKSS	GLT	DSPC
		Mouse		DARDP	NKSDL	GFPNVGKSS	GLT	DSPC
		Chicken		DARDP	NKIDL	GFPNVGKSS	GVT	DSPS
		Frog		DARDP	NKSDL	GFANVGKSS	GTT	DSPA
		Fish		DARDP	NKIDL	GFPNVGKSS	GLT	DSPG
	Gnl3l	Human		DARDP	NKIDL	GLPNVGKSS	GIT	DAPG
		Cow		DSRDP	NKIDL	GLPNVGKSS	GVT	DAPG
		Rat						
		Mouse		DARDP	NKIDL	GLPNVGKSS	GVT	DAPG
		Chicken						
		Frog		DARDP	NKIDL	GFPNVGKSS	GVT	DCPG
		Fish		DARDP	NKIDL	GFPNVGKSS	GVT	DCPG
		Fly		DARDP	NKADL	GIPNVGKSS	GST	DCPG
		Worm		DARDP	NKIDL	GFPNVGKSS	GIT	DSPG
		Yeast		DARDP	NKVDL	GYPNVGKSS	GVT	DSPG
	Ngp1	Human		DARDP	NKCDL	GYPNVGKSS	GET	DCPG
		Cow		DARDP	NKCDL	GYPNVGKSS	GET	DCPG
		Rat		DARDP	NKCDL	GYPNVGKSS	GET	DCPG
		Mouse		DARDP	NKCDL	GYPNVGKSS	GET	DCPG
		Chicken		DARDP	NKCDL	GYPNVGKSS	GET	DCPG
		Frog		DSRDP	NKCDL	GYPNVGKSS	GET	DCPG
		Fish		DARDP	NKCDL	GYPNVGKSS	GET	DCPG
		Fly		DARDP	NKVDL	GYPNVGKSS	GET	DCPG
		Worm		DARDP	NKVDL	GYPNVGKSS	GET	DSPG
		Yeast		DARDP	NKCDL	GYPNTGKSS	GET	DCPG
LSG1	Lsg1	Human		DARNP	NKADL	GYPNVGKSS	GHT	DCPG
		Rat		DARNP	NKADL	GYPNVGKSS	GHT	DCPG
		Mouse		DARNP	NKADL	GYPNVGKSS	GHT	DCPG
		Yeast		DARNP	NKADL	GYPNVGKSS	GKT	DCPG
	Gnl1	Human			NKVDL	GFPNVGKSS	GHT	DCPG
		Mouse			NKVDL	GFPNVGKSS	GHT	DCPG
MTG1	Mtg1	Human		DARIP	NKMDL	GVPNVGKSS	GIT	
		Mouse		DARIP	NKMDL	GVPNVGKSS	GIT	DTPG
		Yeast		DIRAP	NKTDV	GMPNVGKST	GVT	DTPG

Fig. 13.2 Nucleostemin family proteins contain a signature MMR1_HSR1 domain of five circularly permuted GTP-binding motifs. Sequence comparison of the five circularly permuted GTP-binding motifs of nucleostemin (NS), GNL3L, Ngp-1, Lsg1, and Mtg1 proteins of different species reveals the consensus sequences as well as the subfamily-specific residues (marked in *color*)

Table 13.1 Homology comparison between proteins in the nucleostemin family

| Species | Accession No. | Gene | Protein Identity (%) | | | |
			Human NS	Human GNL3L	Human Ngp-1	Chicken NS
B. taurus	XP_874221	NS	78.22	30.22	17.98	39.74
	NP_001029479	GNL3L	28.52	91.58	18.01	24.79
	NP_001039432	Ngp-1	18.01	19.97	87.06	19.02
M. musculus	NP_705775	NS	71.22	28.86	17.49	38.63
	NP_932778	GNL3L	27.75	86.82	18.52	25.28
	CAM20897	Ngp-1	16.64	18.09	85.36	17.48
R. norvegicus	NP_783170	NS	74.13	28.86	16.15	38.34
	NP_001075427	GNL3L	27.92	87.84	18.12	25.42
	NP_001020907	Ngp-1	12.04	18.87	83.45	18.6
G. gallus	XP_414249	NS	41.16	24.93	18.8	100
	N/A	GNL3L				
	XP_417761	Ngp-1	15.66	17.61	68.29	18.13
X. laevis	NP_001080648	NS	45.45	27.67	19.15	33.14
	NP_001088820	GNL3L	36.1	50.43	19.21	28.53
	NP_001080513	Ngp-1	17.53	19.15	69.03	18.73
C. pyrrhogaster	BAF31324	NS	45.33	28.98	15.8	38.89
	N/A	GNL3L				
	N/A	Ngp-1				
D. rerio	NP_001002297	NS	44.21	26.86	15.87	33.62
	NP_001002875	GNL3L	35.28	45.63	18.54	29.84
	NP_998389	Ngp-1	18.67	20.37	65.55	18.89
S. purpuratus	XP_783153	GNL3	33.33	33.9	18.01	
	XP_001192782	Ngp-1	15.43	15.3	40.3	
D. melanogaster	AAM49824	GNL3	31.15	31.83	20.03	
	NP_611232	Ngp-1	19.51	20.06	44.31	
A. gambiae	XP_313813	GNL3	30.62	30.05	16.58	
	XP_309340	Ngp-1	18.69	19.94	47.56	
C. elegans	NP_495749	GNL3	30.43	32.45	17.53	
	NP_492275	Ngp-1	18.55	21.47	40.05	
A. thaliana	NP_187361	GNL3	31.69	32.03	19.47	
	NP_175706	Ngp-1	19.43	18.98	36.19	
S. cerevisiae	EDN62972	GNL3	22.5	19.59	14	
	NP_014451	Ngp-1	11.11	18.03	34.14	
S. pombe	NP_596651	GNL3	21.17	23.03	13.2	
	NP_593896	Ngp-1	21.43	18.92	34.51	

and were therefore deemed the results of polymorphism or sequencing errors. A testicular protein (AAB09043) that shows 94.8% protein identity to Ngp-1 was reported in mice but not in other mammalian species. The biological significance of this testicular protein is unknown. Even though *G. gallus* (chicken) contains only one nucleostemin gene that shows 41.1%, 24.9%, and 18.8% protein identity to human nucleostemin, GNL3L, and Ngp-1, respectively, there is a good reason to believe that an unidentified avian GNL3L may exist.

Table 13.2 Homology comparison between nucleostemin superfamily proteins in *H. sapiens*

Accession No.	Gene	Protein identity (%)					
		NS	GNL3L	Ngp-1	Lsg1	GNL1	Mtg1
AAH01024	NS	100	28.69	18.04	12.77	12.03	11.29
AAH11720	GNL3L	28.69	100	19.47	10.62	9.23	13.75
AAH09250	Ngp-1	18.04	19.47	100	11.08	12.86	8.07
NP_060855	Lsg1	12.77	10.62	11.08	100	23.34	6.83
NP_005266	GNL1	12.03	9.23	12.86	23.34	100	10.71
AAH26039	Mtg1	11.29	13.75	8.07	6.83	10.71	100

Interestingly, only one ortholog, GNL3, was found in the invertebrate genome for both nucleostemin and GNL3L. GNL3 is found in organisms with complete genome information, including *D. melanogaster* (CG3983), *C. elegans* (K01C8.9), *S. cerevisiae* (Nug1), and *S. pombi* (NM_001022573, Grn1). These GNL3 proteins bear similar homology to human nucleostemin and GNL3L. In contrast, Ngp-1 exists as a single gene and is highly conserved from yeast to humans. A search of the *S. purpuratus* (sea urchin), *A. gambiae* (mosquito), and *A. thaliana* (thale cress) database identified only one GNL3 and one Ngp-1 in each species. The sea urchin, mosquito, and thale cress GNL3 proteins exhibit equal homology to both human nucleostemin and GNL3L. Although GNL3 or Ngp-1 cannot be found at the present time in *H. magnipapillata* (Hydra), *S. japonicum* (Trematoda), *S. mediterranea* (Turbellaria), and *N. vectensis* (anemone), they are expected to show up when the genomes of those invertebrate species are completely sequenced. Therefore, all present evidence strongly supports the idea that nucleostemin and GNL3L may have been created from a common invertebrate ancestor at or around the beginning of vertebrate evolution.

The next kindred of the nucleostemin subfamily are *Lsg1*, *Gnl1*, and *Mtg1* (Table 13.2). Like nucleostemin and GNL3L, *Lsg1* and *Gnl1* are represented by one gene in the lower organisms (e.g., *C. elegans*) and appear as separate genes only in the organisms such as *D. melanogaster* or higher (Reynaud et al. 2005). Many of the YlqF/YawG proteins, including nucleostemin, GNL3L, and Ngp-1 were shown to have an ability to bind GTP molecules (Tsai and McKay 2005; Meng et al. 2006). However, few have been experimentally confirmed to possess intrinsic GTPase activity; notable exceptions are YjeQ (Daigle et al. 2002), Lsg1 (Reynaud et al. 2005), and drosophila GNL3 (Rosby et al. 2009).

13.3 Nucleostemin

13.3.1 Nucleostemin Gene and Protein Structures

The gene encoding nucleostemin is found on human chromosome 3 (52,720,785-52,728,291) and mouse chromosome 14 (31,825,738-31,831,052), spans approximately 6.6 kilobases (kb), and contains 15 exons. The nucleostemin locus is located in the human 3p21 chromosomal region, which is a risk locus for human major mood disorders, such as bipolar and major depressive disorders (McMahon et al. 2010). Deletion of the 3p21 region is also a frequent and early event in the formation of lung, breast, kidney, and other cancers. A hallmark of the nucleostemin protein is the MMR1_HSR1 GTP-binding (G) domain, as defined by the Pfam database. The distinguishing characteristic of the MMR1_HSR1 domain is the five G motifs arranged in a circularly permuted order, where the highly conserved G5 variant (G5*) motif (DARxP) and the G4 motif (NKxDL) are positioned N-terminally to the G1 (GxPNVGKSS), G2 (GxT), and G3 (DxPG) motifs (Vernet et al. 1994) (Figs. 13.1a and 13.2). Besides the G motifs, most nucleostemin family proteins contain a basic domain at their N-terminus, a coiled-coil domain, a RNA-binding motif, and an acidic domain (Fig. 13.1a).

13.3.2 Expression Profile of Nucleostemin

A major interest in nucleostemin is inspired by its abundant expression in cells capable of continuous (or self-renewing) proliferation. Besides neural stem cells, nucleostemin has been found highly expressed by multiple additional stem cell types, including embryonic stem (ES) cells, mesenchymal stem cells, primordial germ cells, and putative cardiac stem cells (Tsai and McKay 2002; Baddoo et al. 2003; Liu et al. 2004; Kafienah et al. 2006; Ohmura et al. 2008; Siddiqi et al. 2008). In normal adult tissues, high levels of nucleostemin can be spotted only in the testis, where it is expressed by the spermatogonia and primary spermatocytes (Tsai and McKay 2002; Ohmura et al. 2008). Notably, the expression of nucleostemin can be reactivated during the process of tissue regeneration in some adult cells. It has been reported that the expression of nucleostemin can be turned on in the regenerating lens and limbs in newt (*Cynops pyrrhogaster*) (Maki et al. 2007). Unlike the process seen in higher vertebrates, regeneration in newt involves reversal (dedifferentiation) and plasticity (transdifferentiation) of the differentiated cells at the injury site. The molecular mechanism underlying the dedifferentiation event is not entirely clear, but is associated with upregulation of several genes, one of them being nucleostemin. Nucleostemin expression is switched on in the originally differentiated pigmented epithelial cells after lentectomy and in the muscle cells after limb amputation. The appearance of nucleostemin in those cells represents an early molecular event in the dedifferentiation process that precedes their S-phase reentry or expression of FGF2, Pax6, Sox2, and MafB (Maki et al. 2007). The connection between nucleostemin and tissue regeneration is also supported by another observation made in the adult mouse myocardium. Nucleostemin expression was found to be markedly increased by acute myocardiac infarction in some cardiomyocytes located in the border zone adjacent to the infarct regions or by chronic heart failure throughout the myocardium (Siddiqi et al. 2008). The most common cells where one can find high levels of nucleostemin expression are human tumors and cancer cell lines (Tsai and McKay 2002). Of particular interest is the enhanced expression of nucleostemin in the tumor-initiating cells (TIC) or cancer stem cells (CSC) of mammary and brain tumors (Tamase et al. 2009; Lin et al. 2010). In mammary tumors, nucleostemin is expressed more abundantly in the TIC, the basal subtype of human breast cancers, and the late-stage mouse mammary tumors, compared to the non-TIC tumor cell, the luminal subtype of human breast cancers, and the early-stage mouse mammary tumors (Lin et al. 2010). A key feature of nucleostemin expression in dividing cells is that it disappears before these cells exit the cell cycle. During the development of the mouse embryonic cortex, the level of nucleostemin decreases significantly between E10.5 and E12.5, 2–4 days before the drops in nestin (a marker for neuroepithelial progenitors) and proliferative cell nuclear antigen (Tsai and McKay 2002). The delay of cell cycle exit following the disappearance of nucleostemin was also observed in neural stem cells and mesenchymal stem cells undergoing induced differentiation in culture (Tsai and McKay 2002; Yaghoobi et al. 2005), and may work in the reverse direction during the cell cycle reentry of regenerating cells (Maki et al. 2007; Siddiqi et al. 2008). Finally, it should be noted that the preferential expression of nucleostemin in continuously dividing cells cannot be interpreted as a stem cell-specific expression, because cell differentiation does not always preclude self-renewal and a low level of nucleostemin may still linger in differentiated cells with some degrees of proliferative potential. Therefore, a more fitting and inclusive view of the nucleostemin pathway is one used by self-renewing or long-term dividing cells to stay in the cell cycle or by differentiated cells to reenter the cell cycle. Changing the level of nucleostemin above or below a threshold level may be the leading event that signals the cell cycle reentry during regeneration or the cell cycle exit during differentiation.

13.3.3 Nucleolar Distribution of Nucleostemin

It has become clear that most, if not all, nucleolar proteins shuttle between the nucleolus and the nucleoplasm at a fast pace, but how these proteins move so rapidly between the nucleolar-bound and unbound states and why they behave in such a way remain a mystery. To address these issues at the molecular level, a combined strategy of photobleaching and mutagenesis was used to determine the mechanism that drives the static and dynamic nucleolar distribution of nucleostemin (Tsai and McKay 2005; Meng et al. 2006). Somewhat unexpectedly, the nucleolar localization signal (NoLS) of nucleostemin, as defined by the ability to mediate the nucleolar accumulation of tagged epitope or green fluorescent protein (GFP), was found in two regions of nucleostemin – its N-terminal basic domain and its GTP-binding domain. As the story unfolds, it turns out that the N-terminal basic domain, which contains a stretch of basic residues that are usually found in the NoLS of other nucleolar proteins, can only confer a static nucleolar distribution. It alone, however, cannot recapitulate the dynamic properties of the full-length nucleostemin, including a longer nucleolar retention and a GTP-regulated distribution. These additional dynamic features require the GTP-binding domain of nucleostemin, which does not

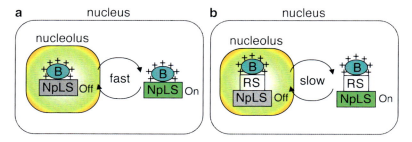

Fig. 13.3 A mechanistic model for the active cycling of nucleostemin family proteins. The nucleolar localization of nucleostemin involves a positively charged, basic nucleolus-targeting domain (B), a nucleolar retention signal (RS), and a nucleoplasmic localization signal coupled with a regulatory module (NpLS). Compared to the fast cycling proteins (**a**), the nucleolar retention signal slows down the exchange rate of the cycling proteins (**b**). In the case of nucleostemin family proteins, the ability of the nucleoplasmic localization signal (NpLS) to detain proteins in the nucleoplasm is regulated by their GTP-binding state

contain the conventional basic residue-rich sequence. Furthermore, the steady-state accumulation, and dynamic cycling to and from the nucleolus, is controlled by another domain that favors its nucleoplasmic localization and acts essentially as a nucleoplasmic localization signal (NpLS) (Tsai and McKay 2005; Meng et al. 2007). A key signal that controls the nucleolar trafficking of nucleostemin is its GTP binding, which functions as a molecular switch for turning off the nucleoplasmic anchoring activity of the adjacent NpLS (Meng et al. 2007) (Fig. 13.3). Interestingly, it has been observed that inhibiting the 26S proteosome-mediated protein degradation pathway by MG132 can stabilize nucleostemin in the face of guanine nucleotide depletion and reverse the nucleoplasmic distribution of non-GTP-bound nucleostemin mutants back to the nucleolus (Huang et al. 2009). The latter finding may suggest that the degradation of nucleostemin protein is coupled to its nucleoplasmic distribution through GTP binding.

13.3.4 Nucleostemin Functions

The biological importance of nucleostemin is best exemplified by the early embryonic lethal phenotype of nucleostemin-null mice (Beekman et al. 2006; Zhu et al. 2006). Almost all nucleostemin-null embryos die around the preimplantation blastocyst stage. The few embryos remaining alive at E3.5 show decreased proliferation without a clear increase of cell death. A reduced growth and G2/M arrest phenotypes were seen in the haploinsufficient mouse embryonic fibroblast (MEF) cells isolated from E13.5 embryonic mesenchyme (Zhu et al. 2006). The role of nucleostemin in maintaining the continuous proliferation of cancer cells and stem cells in culture is supported by the majority of studies published to date (Tsai and McKay

2002; Baddoo et al. 2003; Ma and Pederson 2007; Dai et al. 2008; Meng et al. 2008; Ohmura et al. 2008). Depletion of nucleostemin has been shown to perturb the cell cycle progression either at the G1/S or G2/M transition, depending on the tumor cell types tested as well as the groups that conducted the experiments. On one hand, increased G1 and reduced S phase cell percentages have been described in the nucleostemin-depleted U2OS cells (Dai et al. 2008) and bone marrow-derived mesenchymal stem cells (Jafarnejad et al. 2008). On the other hand, nucleostemin depletion leads to increased G2/M and reduced G1 phase cell percentage in the MEF cells, HEK293 cells, HCT116 (p53-wild-type and null) cells, and U2OS cells (Zhu et al. 2006; Meng et al. 2008; Meng et al. 2010). One group has reported a G1/S arrest effect in SW1710 bladder carcinoma cells and a G2/M arrest phenotype in 5,637 bladder carcinoma cells following nucleostemin knockdown (Nikpour et al. 2009). These somewhat conflicting data may suggest that the nucleostemin activity does not act directly on the cell cycle checkpoint control, and, therefore, the cell cycle phenotypes of nucleostemin-knockdown cells may be determined by other yet unknown factors that vary among different cell types. Conversely, overexpression of nucleostemin can either promote cell proliferation if expressed at a low-to-moderate level (Zhu et al. 2006) or perturb cell cycle progression at the G1/S transition if expressed at a high level (Dai et al. 2008). This paradoxical decrease in cell proliferation by high levels of nucleostemin overexpression may result from a disruption of the physiological conformation of the endogenous nucleostemin complex, and appear to trigger cell cycle arrest through a pathway different from that of nucleostemin knockdown. It has also been reported that mammalian nucleostemin may play a role in the ribosomal biosynthesis, as knockdown of nucleostemin showed an effect in delaying the transition of 32S pre-rRNA to 28S rRNA (Romanova et al. 2009). On the basis of this finding, one may expect nucleostemin to coreside with rRNAs in the nucleolus. On the contrary, confocal and electron spectroscopic analyses showed that nucleostemin is excluded from the fibrillar center and dense fibrillar component, and is localized in a subnucleolar compartment deficient of nascent rRNAs and 28S RNA-containing ribosomes (Politz et al. 2005). This subnucleolar distribution of nucleostemin argues against a direct role of nucleostemin in the processing of pre-rRNAs.

13.3.5 Nucleostemin-Interacting Proteins and Their Mechanisms of Action

The molecular mechanisms underlying the observed activities of nucleostemin have been investigated in the context of the proteins that interact with it. The first protein reported to bind nucleostemin is p53 – a key cell cycle regulator and tumor suppressor. It was shown that p53 is required for the cell death phenotype related to the overexpression of GTP-binding domain mutants of nucleostemin (Tsai and McKay 2002). To date, the p53 dependence of nucleostemin-regulated cell cycle function is still under debate, as the cell cycle arrest phenotype of nucleostemin depletion has

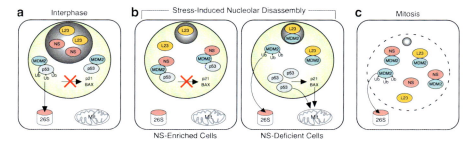

Fig. 13.4 Nucleoplasmic mobilization of nucleostemin stabilizes MDM2 and inhibits the p53 activity. (**a**) In dividing interphase cells, nucleostemin (NS) is localized in the nucleolus (*gray circle*), while MDM2 resides in the nucleoplasm (*yellow circle*) and blocks the activities of p53 by ubiquitylation (Ub) and transcriptional inhibition. (**b**) The nucleoli are disassembled under nucleolar stress conditions. In the nucleostemin-enriched cells (*left panel*), nucleoplasmic translocation of nucleostemin inhibits the p53 activity by stabilizing MDM2 and by competing against L23 for MDM2 binding. In the nucleostemin-deficient cells (*right panel*), MDM2 is either sequestered in the nucleolus by L23 or degraded, leading to G2/M arrest and cell death. (**c**) Mitosis-induced nucleolar disassembly releases nucleostemin into the nucleoplasm/cytoplasm, which stabilizes MDM2 and inactivates p53 functions. Mit, mitochondria; 26S, 26S-proteasome. Reproduced with permission of the Company of Biologists

been shown to be either p53-dependent (Ma and Pederson 2007) or p53-independent (Beekman et al. 2006; Nikpour et al. 2009). It was recently discovered by two groups that the apparent association between nucleostemin and p53 is actually mediated by a p53 regulator, mouse double minute 2 (MDM2) (Dai et al. 2008; Meng et al. 2008). Nucleoplasmic mobilization of nucleostemin stabilizes MDM2 proteins and inhibits the transcriptional activity of p53 (Meng et al. 2008) (Fig. 13.4). These experimental results inspire the idea that the nucleolus may operate as a counting device, which tallies the number of cell divisions by the loss of MDM2 proteins through each round of mitosis and signals cell cycle exit when MDM2 protein falls below a threshold level (Meng et al. 2008). Here, nucleostemin plays a role in inactivating this counting mechanism and safeguarding the proliferative potential of continuously dividing cells. Another study reported that nucleostemin depletion can also enhance MDM2 interaction with L5 and L11 and it induces p53 activation (Dai et al. 2008). While it is clear that nucleostemin depletion can induce the transcriptional activation of p53 (Dai et al. 2008; Meng et al. 2008), minor discrepancies can still be found in different studies. Some reports show an increased p53 level by nucleostemin knockdown (Ma and Pederson 2007; Dai et al. 2008). Others show no change in the p53 protein level or degradation rate in response to nucleostemin knockdown (Meng et al. 2008). Reciprocally, MDM2 is responsible for the degradation of nucleostemin protein induced by guanine nucleotide depletion (Huang et al. 2009). This result somewhat contradicts the finding that overexpression of ARF decreases the protein level of nucleostemin (Ma and Pederson

2007), as ARF is known to neutralize the MDM2 effect on degrading p53 and thereby stabilize the p53 protein.

Nucleostemin has been shown to interact with TRF1 (telomeric repeat binding factor 1) (Zhu et al. 2006) and RSL1D1 (ribosomal L1 domain containing 1) (Meng et al. 2006). Overexpression of nucleostemin can increase the degradation of TRF1 proteins in the cyclohexamide protein stability assay (Zhu et al. 2006), which is opposite to the TRF1 stabilization effect of GNL3L (Zhu et al. 2009). The other nucleostemin-interacting protein, RSL1D1, has been shown to delay replicative senescence of human fibroblast cells by regulating PTEN (Ma et al. 2008). Whether RSL1D1 serves as the primary nucleolar target of the antisenescence activity of nucleostemin, as seen in mouse embryonic fibroblast cells (Zhu et al. 2006), awaits further investigation. Interestingly, RSL1D1 shows a higher binding affinity with nucleostemin and Ngp-1 than with GNL3L. The lack of interaction between GNL3L and RSL1D1 may provide a molecular explanation for the much shorter nucleolar residence time of GNL3L than that of nucleostemin and Ngp-1 (Meng et al. 2006). Two other proteins, ARF and B23/nucleophosmin, were reported to coimmunoprecipitate with nucleostemin (Ma and Pederson 2007, 2008). So far, the functional consequences of these interactions remain unclear.

One notable feature of all nucleostemin-interacting proteins identified so far (except for RSL1D1) is that they are all localized outside the nucleolus. The association between nucleostemin and these nonnucleolar proteins in living cells can be envisioned in several ways. First, nucleostemin shuttles between the nucleolus and nucleoplasm on a GTP-driven cycle, in which the GTP-bound nucleostemin is localized in the nucleolus and the non-GTP-bound nucleostemin is distributed in the nucleoplasm. This cycling behavior provides individual nucleostemin proteins the opportunity to interact with proteins residing in the nucleoplasm (Tsai and McKay 2005). Second, those nucleoplasmic proteins may transit through the nucleolus during a specific phase of their life cycle. For example, despite the nucleoplasmic predominant distribution of p53, it has also been spotted at the active transcriptional site within the nucleolus (Rubbi and Milner 2000). Third, nucleostemin can be relocated to the nucleoplasm upon nucleolar disassembly during mitosis or induced by drugs that block the RNA polymerase activity or *de novo* GTP synthesis. Finally, the interaction between nucleostemin and these proteins may be indirectly mediated by other unidentified factors.

13.4 GNL3L

13.4.1 *Expression Profile and Subcellular Distribution*

The gene encoding GNL3L is located on the human X chromosome (54,556,644-54,587,504) and mouse X chromosome (147,420,572-147,451,811), spans approximately 30 kb, and contains 16 exons. In addition to the MMR1-HSR1 domain, the

GNL3L protein contains the regular basic, coiled-coil, RNA-binding, and acidic domains, plus one extra coiled-coil domain at the C-terminus (Fig. 13.1a). Unlike nucleostemin, the GNL3L expression is lower in the undifferentiated neural stem cells and higher in their differentiated progeny, and is continuously maintained throughout neural development. In adult animals, GNL3L is expressed most highly in the neural tissues, including the brain and eye, and can also be detected in the muscle and kidney at low levels (Yasumoto et al. 2007). Within the cell, the GNL3L protein displays a higher nucleoplasmic intensity and a much shorter nucleolar retention time than nucleostemin does (Meng et al. 2007).

13.4.2 GNL3L Functions on MDM2, TERT, TRF1, and ERRγ

13.4.2.1 GNL3L Provides a Constitutive Mechanism to Stabilize MDM2 Protein

GNL3L, like its vertebrate paralog nucleostemin, can function as a MDM2 regulator (Meng et al. 2010). GNL3L and nucleostemin share many similar characteristics in their binding and regulation of MDM2. The MDM2 binding of both proteins occurs in the nucleoplasm, requires the central domain of MDM2, and can mediate their association with p53. The MDM2 interaction of GNL3L requires its GTP-binding- or intermediate-domain, which is different from the MDM2-interactive coiled-coil- and acidic-domains of nucleostemin. Functionally, GNL3L exhibits the same activity as nucleostemin in stabilizing MDM2 protein and preventing its ubiquitylation, and is capable of rescuing the MDM2 ubiquitylation effect caused by nucleostemin depletion. Similar to the nucleostemin activity, MDM2 destabilization by GNL3L knockdown affects neither the ubiquitylation nor the stability of p53 proteins, and may therefore inhibit the transcriptional activation of p53 through a direct binding mechanism without affecting the p53 protein level. Despite their similar effects on MDM2 stabilization, GNL3L and nucleostemin are designed to operate under distinct biological contexts. The binding and ubiquitylation activity of GNL3L on MDM2 is constitutively active and does not so much depend on the nucleolar release mechanism. In contrast, the MDM2-stabilizing effect of nucleostemin is kept inactive in normal dividing interphase cells by the nucleolar sequestration mechanism and becomes active only when the protein is released from the nucleolus. As a result, the nucleostemin-dependent MDM2 stabilization pathway operates mainly for the purpose of cell cycle counting or under stress conditions. Such a difference in the GNL3L and nucleostemin-mediated MDM2 regulation may be due to their differential nucleolar-nucleoplasmic partitioning dynamics (Meng et al. 2007). Consistent with its MDM2-stabilizing ability, GNL3L knockdown triggers G2/M arrest and upregulates specific p53 downstream targets, that is, Bax, 14-3-3σ, and p21, more so in the p53-wild-type than in the p53-null HCT116 cells. In accordance with the upregulation, a significant percentage of human colorectal and gastric cancers expresses high levels of GNL3L and low levels of 14-3-3σ and p21, suggesting that GNL3L may have a potential role as a tumor-promoting factor via MDM2 stabilization in human cancers.

13.4.2.2   GNL3L Stabilizes TRF1 and Promotes the G2/M Transition

The connection between GNL3L and telomere maintenance was first shown by a study that analyzed the protein components of the telomerase complex, which identified GNL3L as one of the proteins that interact with telomerase reverse transcriptase (TERT) (Fu and Collins 2007). GNL3L negatively regulates the telomere length without affecting the telomerase activity, suggesting that GNL3L binding may reduce the bioavailability of TERT at its active site – the telomere. The telomere regulatory function of GNL3L can also be mediated by its ability to bind TRF1 in living cells (Zhu et al. 2006). The interaction between GNL3L and TRF1 occurs in the nucleoplasm and results in an increased homodimerization and telomeric association of TRF1. Such an effect will have a negative impact on telomere elongation, as TRF1 was shown to negatively regulate the telomere length (van Steensel and de Lange 1997). GNL3L binding also triggers other effects on TRF1, including a reduced PML-body recruitment of telomere-bound TRF1, a process known as alternative lengthening of telomere (ALT)-associated PML body or APB, and an increased stability of TRF1 proteins through inhibition of TRF1 ubiquitylation and binding to FBX4, an E3 ubiquitin ligase for TRF1. At the cellular level, the TRF1 protein-stabilizing activity of GNL3L mediates the mitotic increase of TRF1 protein and promotes the metaphase-to-anaphase transition. The TRF1 modulatory function of GNL3L not only reveals new aspects of telomere regulation but also raises the possibility that nucleostemin and GNL3L may oppositely regulate the same pathway in some biological arenas.

13.4.2.3   GNL3L Inhibits the Transcriptional Activation
of Estrogen Receptor-Related Proteins

The estrogen receptor-related protein-γ (ERRγ) is the first published GNL3L-interacting protein, identified from a yeast two-hybrid screen using GNL3L as the bait (Yasumoto et al. 2007). GNL3L and ERRγ are coexpressed in the eye, kidney, and muscle, and coreside in the nucleoplasm. The interaction between GNL3L and ERRγ requires the intermediate domain of GNL3L and the AF2-domain of ERRγ. Gain- and loss-of-function experiments demonstrate that GNL3L inhibits the transcriptional activities of ERR family proteins by reducing their steroid receptor coactivator (SRC) binding and SRC-mediated transcriptional coactivation. This ERRγ-inhibitory activity of GNL3L does not require its nucleolar localization, which is consistent with the idea that the location of this interaction is in the nucleoplasm. Notably, only GNL3L but not nucleostemin or Ngp-1 has the ability to bind ERRγ, indicating that GNL3L may have a broader functional spectrum than nucleostemin does.

13.4.2.4 Nucleostemin Depletion may Trigger a Compensatory Upregulation of GNL3L

The expression of GNL3L is linked to the level of nucleostemin by the data showing that nucleostemin depletion is accompanied by a compensatory upregulation of GNL3L at the mRNA and protein levels. This new discovery was first made in the siRNA knockdown experiment, and later extended to two physiological events known to reduce the expression of nucleostemin, that is, cell differentiation (Tsai and McKay 2002) and guanine nucleotide depletion (Huang et al. 2009). Indeed, the decrease of nucleostemin expression during neural differentiation is accompanied by a reciprocal increase of GNL3L expression. In addition, guanine nucleotide depletion can trigger nucleostemin decrease and GNL3L increase simultaneously, and this increase of GNL3L can be reversed by MG132 treatment but not by restoring the level of nucleostemin protein. The latter finding argues against a direct control over GNL3L expression by the nucleostemin protein. It should be noted that this phenomenon does not occur in all cancer cell lines. It appears in HCT116 and U2OS cells, but not in HeLa or HEK293 cells. Considering that HCT116 and U2OS cells are Rb1-wild-type and p16-inactive, HeLa cells are Rb1-inactive and p16-wild-type, and HEK293 cells are Rb1-wild-type and p16-wild-type, this cell type-dependent upregulation of GNL3L by nucleostemin knockdown should not involve the p53 and Rb1 pathways and may explain some of the cell type-specific phenotypes of nucleostemin knockdown (Nikpour et al. 2009). One inference of these results is that some cells may survive the consequence of nucleostemin deletion better than others in the cases where nucleostemin and GNL3L share redundant functions, such as the regulation of MDM2 proteins.

13.5 GNL3: The Invertebrate Ortholog of Nucleostemin and GNL3L

Nucleostemin and GNL3L share a common invertebrate ortholog (i.e., GNL3) and become separate genes only in the vertebrate lineages. A critical question that needs to be addressed is how much the mammalian nucleostemin resembles GNL3L and GNL3 in their biological functions. The best way to understand the functional conservation and divergence of nucleostemin, GNL3L, and GNL3 is through the rescue experiment. So far, functional complementation between nucleostemin and GNL3L in mammalian cells has only been examined in the context of MDM2 regulation, where they show redundant activities in stabilizing MDM2 proteins (Meng et al. 2010). Functional rescue of invertebrate GNL3 by human nucleostemin and GNL3L have been tried in *S. pombe* (Du et al. 2006). Deletion of GNL3 in *S. pombe* (*a.k.a.* Grn1) results in a slow-growth phenotype and defects in 35S rRNA processing and nucleolar export of the 60S-complex. Interestingly, the *Grn1*-null phenotype can be rescued only by human GNL3L and not by nucleostemin. A similar experiment was performed in *C. elegans* (Kudron and Reinke 2008), which showed that murine nucleostemin fails to rescue the GNL3/nst-1-deficient growth phenotype in *C. elegans*. The nematode GNL3/nst-1 (K01C8.9) is expressed by both the proliferating and differentiated cells, and is required for both larval growth and germline stem cell division. One notable feature of the nematode GNL3 is its relatively high nucleoplasmic distribution, which resembles more the localization of mammalian GNL3L than that of nucleostemin. Although information on whether mammalian GNL3L can rescue the nematode GNL3/nst-1-deficient phenotype is lacking, this study still supports that nucleostemin may have developed its own unique function that distinguishes itself from GNL3L and GNL3.

A common feature of all invertebrate GNL3 proteins is their functional connection with the synthesis of the 60S ribosome complex. In *S. cerevisiae*, GNL3 (or Nug1p) is required for its viability and 60S ribosome export, but has little effect on the pre-rRNA processing (Bassler et al. 2001). In *S. pombi*, GNL3 (or Grn1p) is involved in both 35S rRNA processing and 60S-complex export (Du et al. 2006). In the nst-1 mutant *C. elegans*, rRNA analyses revealed a decrease of 18S and 26S rRNAs. In *Drosophila melanogaster*, GNL3 is an essential gene for the larval and pupal development, and depletion of the drosophila GNL3 blocks the nucleolar release of the large ribosomal subunit and leads to a loss of cytoplasmic ribosomes (Rosby et al. 2009). Considering that invertebrate GNL3 functionally resembles mammalian GNL3L more than it does nucleostemin, one may infer that during the gene divergence of nucleostemin and GNL3L in the vertebrate lineage, the ribosome biosynthetic activity of invertebrate GNL3 may have been passed on mainly to GNL3L, thereby freeing nucleostemin to perform other cell type-specific functions.

13.6 Other Nucleostemin-Related Proteins

13.6.1 Ngp-1

The third member of the nucleostemin family and by far the most neglected one is Ngp-1. The gene encoding Ngp-1 is found on the human chromosome 1 (38,032,417-38,061,496) and mouse chromosome 4 (124,707,282-124,732,616), spans approximately 29 kb, and contains 16 exons. It has been shown that yeast Ngp-1 protein, Nog2p, is associated with the large pre-60S complex and that Nog2p deletion causes defects in the processing of 27S rRNA precursor to 5.8S and 25S rRNA (Saveanu et al. 2001). In *Drosophila melanogaster*, deletion of Ngp-1 perturbed development and caused lethality at the first instar larval stage (Matsuo et al. 2010). Even though the vertebrate Ngp-1 was identified 6 years ahead of nucleostemin, little is known about its biological activity to this date, except that it is preferentially expressed in the adult testis (Racevskis et al. 1996). As Ngp-1 represents a single gene subfamily highly conserved from yeast to human and as all invertebrate Ngp-1 proteins are connected to the pre-60S complex, one may reasonably expect the vertebrate Ngp-1 to exercise similar functions in ribosome biosynthesis.

13.6.2 Lsg1 and Mtg1

Unlike the nucleostemin family, these two proteins are not localized in the nucleolus, and will not therefore be a focus in this book. Lsg1p distributes in cytoplasmic punctates. Yeast Lsg1p binds to cytoplasmic free 60S subunits and is involved in the final step of 60S biogenesis and export (Kallstrom et al. 2003). In *D. melanogaster*, the serotonergic neuron expression of Lsg1p (CG14788) is required for controlling the body size (Kaplan et al. 2008). Although referred to as nucleostemin 3 (or NS3) in that study, this gene is, in fact, the *bona fide* ortholog of human Lsg1. On the other hand, NS1, NS2, and NS4 correspond to GNL3 (CG3983), Ngp-1 (or GNL2, CG6501), and GNL1 (CG9320), respectively. Mtg1 is localized to the mitochondrial inner membrane. Its target of action is not well understood but likely to be at the 21S rRNA or the large subunit of mitochondrial ribosomes (Barrientos et al. 2003).

13.7 Perspectives

13.7.1 The Dynamic Nucleolus

It has long been known that the nucleolar compartment can be disassembled under stress conditions or during mitosis. Stress-induced nucleolar disassembly provides a mechanism that mobilizes a number of nucleolar proteins capable of activating or inhibiting the transcriptional activity of p53 via their interaction with MDM2 (Tao and Levine 1999; Zhang et al. 2003; Bernardi et al. 2004; Dai et al. 2004, 2008; Jin et al. 2004; Kurki et al. 2004; Meng et al. 2008). The nucleolus has also been proposed as a stress sensor that stabilizes p53 under a variety of cellular stresses (Rubbi and Milner 2003) (also see Chap. 12 in this book). From this perspective, the nucleolus can be viewed as a stress response organelle. During the mitotic window, most nucleolar proteins in the granular component are also temporarily released into the prophase nucleoplasm and mitotic cytoplasm, which may allow the activities of some proteins to be modified as a function of cell division. More recently, it has become clear that most nucleolus-concentrated proteins undergo dynamic shuttling between the nucleolus and the nucleoplasm (Phair and Misteli 2000; Chen and Huang 2001; Pederson 2001; Misteli 2005; Tsai and McKay 2005). The dynamic movement of nucleolar proteins can be controlled by defined intracellular signals, such as the GTP molecule or hydrogen ion. For nucleostemin, it is the GTP-binding domain that mediates its nucleolar retention and the GTP-bound state that releases the nucleoplasmic anchoring activity of the neighboring NpLS (Tsai and McKay 2005; Meng et al. 2006). For VHL (von Hippel-Lindau) tumor suppressor protein, it is the hydrogen ion (pH) that promotes its nucleolar sequestration, which elicits a transient and reversible inhibition that eventually stabilizes HIF (hypoxia-inducible factor) (Mekhail et al. 2004).

13.7.2 The Evolving Nucleostemin

From the evolutionary standpoint, gene divergence of nucleostemin and GNL3L is a recent event that coincides with the evolution of the vertebrate species. Functional comparison of mammalian nucleostemin and GNL3L shows that they possess distinct expression patterns, cell biological properties, and mostly nonredundant functions, and that only human GNL3L but not nucleostemin is capable of compensating for the loss-of-function phenotypes of their common invertebrate ortholog, GNL3 or Grn1p, in *S. pombe*. If true, it will be interesting to learn which new functions have been introduced as a result of the birth of nucleostemin and how invertebrates live without these newly created functions. As one key function of nucleostemin is to maintain continuous cell proliferation during development and tissue regeneration, a difference between the nucleostemin-present vertebrate cells and the nucleostemin-absent invertebrate cells may have to do with the mechanisms that they use to maintain self-renewal and to repair injured tissues. From this perspective, nucleostemin may be used to define novel attributes of tissue regeneration in the vertebrate lineages.

References

Baddoo M, Hill K, Wilkinson R, Gaupp D, Hughes C, Kopen GC, Phinney DG (2003) Characterization of mesenchymal stem cells isolated from murine bone marrow by negative selection. J Cell Biochem 89:1235–1249

Barrientos A, Korr D, Barwell KJ, Sjulsen C, Gajewski CD, Manfredi G, Ackerman S, Tzagoloff A (2003) MTG1 codes for a conserved protein required for mitochondrial translation. Mol Biol Cell 14:2292–2302

Bassler J, Grandi P, Gadal O, Lessmann T, Petfalski E, Tollervey D, Lechner J, Hurt E (2001) Identification of a 60S preribosomal particle that is closely linked to nuclear export. Mol Cell 8:517–529

Beekman C, Nichane M, De Clercq S, Maetens M, Floss T, Wurst W, Bellefroid E, Marine JC (2006) Evolutionarily Conserved Role of Nucleostemin: Controlling Proliferation of Stem/Progenitor Cells during Early Vertebrate Development. Mol Cell Biol 26:9291–9301

Bernardi R, Scaglioni PP, Bergmann S, Horn HF, Vousden KH, Pandolfi PP (2004) PML regulates p53 stability by sequestering Mdm2 to the nucleolus. Nat Cell Biol 6:665–672

Charpentier AH, Bednarek AK, Daniel RL, Hawkins KA, Laflin KJ, Gaddis S, MacLeod MC, Aldaz CM (2000) Effects of estrogen on global gene expression: identification of novel targets of estrogen action. Cancer Res 60:5977–5983

Chen D, Huang S (2001) Nucleolar components involved in ribosome biogenesis cycle between the nucleolus and nucleoplasm in interphase cells. J Cell Biol 153:169–176

Dai MS, Zeng SX, Jin Y, Sun XX, David L, Lu H (2004) Ribosomal protein L23 activates p53 by inhibiting MDM2 function in response to ribosomal perturbation but not to translation inhibition. Mol Cell Biol 24:7654–7668

Dai MS, Sun XX, Lu H (2008) Aberrant expression of nucleostemin activates p53 and induces cell cycle arrest via inhibition of MDM2. Mol Cell Biol 28:4365–4376

Daigle DM, Rossi L, Berghuis AM, Aravind L, Koonin EV, Brown ED (2002) YjeQ, an essential, conserved, uncharacterized protein from Escherichia coli, is an unusual GTPase with circularly permuted G-motifs and marked burst kinetics. Biochemistry 41:11109–11117

Du X, Rao MR, Chen XQ, Wu W, Mahalingam S, Balasundaram D (2006) The homologous putative GTPases Grn1p from fission yeast and the human GNL3L are required for growth and play a role in processing of nucleolar pre-rRNA. Mol Biol Cell 17:460–474

Fu D, Collins K (2007) Purification of human telomerase complexes identifies factors involved in telomerase biogenesis and telomere length regulation. Mol Cell 28:773–785

Huang M, Itahana K, Zhang Y, Mitchell BS (2009) Depletion of guanine nucleotides leads to the Mdm2-dependent proteasomal degradation of nucleostemin. Cancer Res 69:3004–3012

Jafarnejad SM, Mowla SJ, Matin MM (2008) Knocking-down the expression of nucleostemin significantly decreases rate of proliferation of rat bone marrow stromal stem cells in an apparently p53-independent manner. Cell Prolif 41:28–35

Jin A, Itahana K, O'Keefe K, Zhang Y (2004) Inhibition of HDM2 and activation of p53 by ribosomal protein L23. Mol Cell Biol 24:7669–7680

Kafienah W, Mistry S, Williams C, Hollander AP (2006) Nucleostemin is a marker of proliferating stromal stem cells in adult human bone marrow. Stem Cells 24:1113–1120

Kallstrom G, Hedges J, Johnson A (2003) The putative GTPases Nog1p and Lsg1p are required for 60S ribosomal subunit biogenesis and are localized to the nucleus and cytoplasm, respectively. Mol Cell Biol 23:4344–4355

Kaplan DD, Zimmermann G, Suyama K, Meyer T, Scott MP (2008) A nucleostemin family GTPase, NS3, acts in serotonergic neurons to regulate insulin signaling and control body size. Genes Dev 22:1877–1893

Kudron MM, Reinke V (2008) C. elegans nucleostemin is required for larval growth and germline stem cell division. PLoS Genet 4:e1000181

Kurki S, Peltonen K, Latonen L, Kiviharju TM, Ojala PM, Meek D, Laiho M (2004) Nucleolar protein NPM interacts with HDM2 and protects tumor suppressor protein p53 from HDM2-mediated degradation. Cancer Cell 5:465–475

Leipe DD, Wolf YI, Koonin EV, Aravind L (2002) Classification and evolution of P-loop GTPases and related ATPases. J Mol Biol 317:41–72

Lin T, Meng L, Li Y, Tsai RY (2010) Tumor-initiating function of nucleostemin-enriched mammary tumor cells. Cancer Res 70:9444–9452

Liu SJ, Cai ZW, Liu YJ, Dong MY, Sun LQ, Hu GF, Wei YY, Lao WD (2004) Role of nucleostemin in growth regulation of gastric cancer, liver cancer and other malignancies. World J Gastroenterol 10:1246–1249

Ma H, Pederson T (2007) Depletion of the nucleolar protein nucleostemin causes G1 cell cycle arrest via the p53 pathway. Mol Biol Cell 18:2630–2635

Ma H, Pederson T (2008) Nucleophosmin is a binding partner of nucleostemin in human osteosarcoma cells. Mol Biol Cell 19:2870–2875

Ma L, Chang N, Guo S, Li Q, Zhang Z, Wang W, Tong T (2008) CSIG inhibits PTEN translation in replicative senescence. Mol Cell Biol 28:6290–6301

Maki N, Takechi K, Sano S, Tarui H, Sasai Y, Agata K (2007) Rapid accumulation of nucleostemin in nucleolus during newt regeneration. Dev Dyn 236:941–950

Matsuo E, Kanno S, Matsumoto S, Tsuneizumi K (2010) Drosophila nucleostemin 2 proved essential for early eye development and cell survival. Biosci Biotechnol Biochem 74:2120–2123

McMahon FJ, Akula N, Schulze TG, Muglia P, Tozzi F, Detera-Wadleigh SD, Steele CJ, Breuer R, Strohmaier J, Wendland JR, Mattheisen M, Muhleisen TW, Maier W, Nothen MM, Cichon S, Farmer A, Vincent JB, Holsboer F, Preisig M, Rietschel M (2010) Meta-analysis of genome-wide association data identifies a risk locus for major mood disorders on 3p21.1. Nat Genet 42:128–131

Mekhail K, Gunaratnam L, Bonicalzi ME, Lee S (2004) HIF activation by pH-dependent nucleolar sequestration of VHL. Nat Cell Biol 6:642–647

Meng L, Yasumoto H, Tsai RY (2006) Multiple controls regulate nucleostemin partitioning between nucleolus and nucleoplasm. J Cell Sci 119:5124–5136

Meng L, Zhu Q, Tsai RY (2007) Nucleolar trafficking of nucleostemin family proteins: common versus protein-specific mechanisms. Mol Cell Biol 27:8670–8682

Meng L, Lin T, Tsai RY (2008) Nucloplasmic mobilization of nucleostemin stabilizes MDM2 and promotes G2-M progression and cell survival. J Cell Sci 121:4037–4046

Meng L, Hsu JK, Tsai RY (2010) NL3L depletion destabilizes MDM2 and induces p53-dependent G2/M arrest. Oncogene 30:1716–1726

Misteli T (2005) Going in GTP cycles in the nucleolus. J Cell Biol 168:177–178

Nikpour P, Mowla SJ, Jafarnejad SM, Fischer U, Schulz WA (2009) Differential effects of Nucleostemin suppression on cell cycle arrest and apoptosis in the bladder cancer cell lines 5637 and SW1710. Cell Prolif 42:762–769

Ohmura M, Naka K, Hoshii T, Muraguchi T, Shugo H, Tamase A, Uema N, Ooshio T, Arai F, Takubo K, Nagamatsu G, Hamaguchi I, Takagi M, Ishihara M, Sakurada K, Miyaji H, Suda T, Hirao A (2008) Identification of stem cells during prepubertal spermatogenesis via monitoring of nucleostemin promoter activity. Stem Cells 26:3237–3246

Pederson T (2001) Protein mobility within the nucleus–what are the right moves? Cell 104: 635–638

Pederson T, Tsai RY (2009) In search of non-ribosomal nucleolar protein function and regulation. J Cell Biol 184:771–776

Phair RD, Misteli T (2000) High mobility of proteins in the mammalian cell nucleus. Nature 404:604–609

Politz JC, Polena I, Trask I, Bazett-Jones DP, Pederson T (2005) A nonribosomal landscape in the nucleolus revealed by the stem cell protein nucleostemin. Mol Biol Cell 16:3401–3410

Racevskis J, Dill A, Stockert R, Fineberg SA (1996) Cloning of a novel nucleolar guanosine 5'-triphosphate binding protein autoantigen from a breast tumor. Cell Growth Differ 7:271–280

Reynaud EG, Andrade MA, Bonneau F, Ly TB, Knop M, Scheffzek K, Pepperkok R (2005) Human Lsg1 defines a family of essential GTPases that correlates with the evolution of compartmentalization. BMC Biol 3:21

Romanova L, Grand A, Zhang L, Rayner S, Katoku-Kikyo N, Kellner S, Kikyo N (2009) Critical role of nucleostemin in pre-rRNA processing. J Biol Chem 284:4968–4977

Rosby R, Cui Z, Rogers E, Delivron MA, Robinson VL, Dimario PJ (2009) Knockdown of the Drosophila GTPase, nucleostemin 1, impairs large ribosomal subunit biogenesis, cell growth, and midgut precursor cell maintenance. Mol Biol Cell 20:4424–4434

Rubbi CP, Milner J (2000) Non-activated p53 co-localizes with sites of transcription within both the nucleoplasm and the nucleolus. Oncogene 19:85–96

Rubbi CP, Milner J (2003) Disruption of the nucleolus mediates stabilization of p53 in response to DNA damage and other stresses. Embo J 22:6068–6077

Saveanu C, Bienvenu D, Namane A, Gleizes PE, Gas N, Jacquier A, Fromont-Racine M (2001) Nog2p, a putative GTPase associated with pre-60S subunits and required for late 60S maturation steps. Embo J 20:6475–6484

Siddiqi S, Gude N, Hosoda T, Muraski J, Rubio M, Emmanuel G, Fransioli J, Vitale S, Parolin C, D'Amario D, Schaefer E, Kajstura J, Leri A, Anversa P, Sussman MA (2008) Myocardial induction of nucleostemin in response to postnatal growth and pathological challenge. Circ Res 103:89–97

Tamase A, Muraguchi T, Naka K, Tanaka S, Kinoshita M, Hoshii T, Ohmura M, Shugo H, Ooshio T, Nakada M, Sawamoto K, Onodera M, Matsumoto K, Oshima M, Asano M, Saya H, Okano H, Suda T, Hamada J, Hirao A (2009) Identification of tumor-initiating cells in a highly aggressive brain tumor using promoter activity of nucleostemin. Proc Natl Acad Sci USA 106:17163–17168

Tao W, Levine AJ (1999) P19(ARF) stabilizes p53 by blocking nucleo-cytoplasmic shuttling of Mdm2. Proc Natl Acad Sci USA 96:6937–6941

Tsai RY, McKay RD (2002) A nucleolar mechanism controlling cell proliferation in stem cells and cancer cells. Genes Dev 16:2991–3003

Tsai RY, McKay RD (2005) A multistep, GTP-driven mechanism controlling the dynamic cycling of nucleostemin. J Cell Biol 168:179–184

Tsai RY, Meng L (2009) Nucleostemin: A latecomer with new tricks. Int J Biochem Cell Biol 41:2122–2124 van Steensel B, de Lange T (1997) Control of telomere length by the human telomeric protein TRF1. Nature 385:740–743

Vernet C, Ribouchon MT, Chimini G, Pontarotti P (1994) Structure and evolution of a member of a new subfamily of GTP-binding proteins mapping to the human MHC class I region. Mamm Genome 5:100–105

Yaghoobi MM, Mowla SJ, Tiraihi T (2005) Nucleostemin, a coordinator of self-renewal, is expressed in rat marrow stromal cells and turns off after induction of neural differentiation. Neurosci Lett 390:81–86

Yasumoto H, Meng L, Lin T, Zhu Q, Tsai RY (2007) GNL3L inhibits activity of estrogen-related receptor gamma by competing for coactivator binding. J Cell Sci 120:2532–2543

Zhang Y, Wolf GW, Bhat K, Jin A, Allio T, Burkhart WA, Xiong Y (2003) Ribosomal protein L11 negatively regulates oncoprotein MDM2 and mediates a p53-dependent ribosomal-stress checkpoint pathway. Mol Cell Biol 23:8902–8912

Zhu Q, Yasumoto H, Tsai RY (2006) Nucleostemin delays cellular senescence and negatively regulates TRF1 protein stability. Mol Cell Biol 26:9279–9290

Zhu Q, Meng L, Hsu JK, Lin T, Teishima J, Tsai RY (2009) GNL3L stabilizes the TRF1 complex and promotes mitotic transition. J Cell Biol 185:827–839

Chapter 14
Viruses and the Nucleolus

David Matthews, Edward Emmott, and Julian Hiscox

14.1 Introduction

The nucleolus is a dynamic sub-nuclear structure with roles in ribosome subunit biogenesis, mediation of cell stress responses and regulation of cell growth (Boulon et al. 2010). The proteome and structure of the nucleolus are constantly changing in response to metabolic conditions, and virus infection represents one of the major challenges to nucleolar function (Greco 2009; Hiscox 2002, 2003, 2007; Hiscox et al. 2010). Viruses are obligate intracellular parasites and rely on the host cell for genome replication, protein expression and assembly of new virus particles. During infection there is a constant war between viruses trying to subvert the host cell and host-mediated anti-viral activity and interaction with the nucleolus is likely to be a key stage in this.

Interaction with the nucleolus is a pan-virus phenomenon and evidence suggests that proteins from many different types of viruses, such as those with DNA, RNA or RNA/DNA (e.g. retroviruses) genomes, encode proteins that can localise to the nucleolus during infection (Table 14.1). These examples include viruses with DNA genomes including the poxviruses, which replicate in the cytoplasm, as well as the herpes and adenoviruses, which replicate in the nucleus. HIV-1, perhaps the classic example of a retrovirus, undergoes an initial replication event in the cytoplasm and then further activity in the nucleus. RNA viruses encompass genomes of single-stranded positive and negative polarity and also double-stranded RNA. Established dogma suggests that positive strand-RNA viral genome synthesis and transcription occur in the cytoplasm. Examples of negative strand RNA viruses can be found, which replicate in the cytoplasm (most of the Mononegavirales) and the nucleus (e.g. influenza viruses).

J. Hiscox (✉)
Institute of Molecular and Cellular Biology, Faculty of Biological Sciences, and Astbury Centre for Structural Molecular Biology, University of Leeds, Leeds, UK
e-mail: j.a.hiscox@leeds.ac.uk

M.O.J. Olson (ed.), *The Nucleolus*, Protein Reviews 15,
DOI 10.1007/978-1-4614-0514-6_14, © Springer Science+Business Media, LLC 2011

Table 14.1 Representative examples of viral proteins from different classes of viruses that localise to the nucleolus

Family	Subfamily	Genus	Virus	Protein/nucleic acid	Reference
Baltimore class I: double-strand DNA					
Adenoviridae	n/a	Mastadenovirus	Human adenovirus type 5 (Ad5)	E4orf4, preMu, pIVa2, preVII, Mu, UXP, protein V	Miron et al. (2009); Lee et al. (2004); Lee et al. (2003); Lutz and Kedinger (1996); Tollefson et al. (2007); Matthews and Russell (1998)
Asfaviridae	n/a	Asfivirus	African swine fever virus (ASFV)	I14L	Goatley et al. (1999)
Herpesviridae	n/a	Mardivirus	Mareks disease virus/ Gallid herpesvirus 2 (MDV)	MEQ	Liu et al. (1997)
Herpesviridae	Alphaherpesvirinae	Simplexvirus	Herpes simplex virus type 1 (HSV-1)	EAP, gamma(1)34.5, ICP0, ICP27, RL1, US11, UL12, UL24, UL27.5, VP22	Leopardi and Roizman (1996); Cheng et al. (2002); Morency et al. (2005); Mears et al. (1995); Salsman et al. (2008); Maclean et al. (1987); Sagou et al. (2010); Lymberopoulos and Pearson (2007); Salsman et al. (2008); Harms et al. (2000)
Herpesviridae	Alphaherpesvirinae	Simplexvirus	Herpes simplex virus type 2 (HSV-2)	UL24, UL3, UL31	Hong-Yan et al. (2001); Yamada et al. (1999); Zhu et al. (1999)
Herpesviridae	Alphaherpesvirinae	Varicellovirus	Bovine herpesvirus 1 (BHV-1)	BICP27	Guo et al. (2009)

Herpesviridae	Betaherpesvirinae	Cytomegalovirus	Human cytomegalovirus (HCMV)	TLR5, TLR7, TLR9, US33, UL29, UL31, UL44, UL76, UL83 (pp65), UL108, UL123 (IEp72)	Salsman et al. (2008); Strang et al. (2010); Arcangeletti et al. (2003)
Herpesviridae	Gammaherpesvirinae	Lymphocryptovirus	Epstein-Barr virus (EBV/HHV-4)	EBNA-5, v-snoRNA1	Szekely et al. (1995); Hutzinger et al. (2009)
Herpesviridae	Gammaherpesvirinae	Rhadinovirus	Kaposi's sarcoma-associated herpesvirus (KSHV)	KS-Bcl02, ORF57	Kalt et al. (2010); Boyne and Whitehouse (2009)
Herpesviridae	Gammaherpesvirinae	Rhadinovirus	Herpesvirus saimiri (HVS)	ORF57	Boyne and Whitehouse (2006)
Poxviridae	Chordopoxvirinae	Leporipoxvirus	Myxoma virus	M184R	Blanie et al. (2009)
Baltimore class II: single-strand DNA					
Circoviridae	n/a	Cirovirus	Porcine circovirus type 1 (PCV1)	Cap	Finsterbusch et al. (2005)
Geminiviridae	n/a	Begomovirus	Tomato leaf curl Java virus (TYLCV)	CP	Sharma and Ikegami (2009)
Geminiviridae	n/a	Begomovirus	Tomato yellow leaf curl virus (TYLCV)	CP	Rojas et al. (2001)
Parvoviridae	Parvovirinae	Dependovirus	Adeno-associated virus	AAP, Rep, Capsid	Sonntag et al. (2010); Wistuba et al. (1997)
Parvoviridae	Parvovirinae	Parvovirus	Minute virus of mice (MVM)	Viral DNA	Walton et al. (1989)
Baltimore class III: double-strand RNA					
Reoviridae	n/a	Orthoreovirus	Avian reovirus	sigmaA	Vazquez-Iglesia et al. (2009)
Baltimore class IV: positive-sense, single-strand RNA					
n/a		Pomovirus	Potato mop top virus (PMTV)	TGB1	Wright et al. (2010)

(continued)

Table 14.1 (continued)

Family	Subfamily	Genus	Virus	Protein/nucleic acid	Reference
n/a	n/a	Umbravirus	Groundnut rosette virus (GRV)	ORF3	Ryabov et al. (1998)
Arteriviridae	n/a	Arterivirus	Lactate dehydrogenase-elevating virus (LDV)	N	Mohammadi et al. (2009)
Arteriviridae	n/a	Arterivirus	Porcine respiratory and reproductive virus (PRRSV)	N	Rowland et al. (1999)
Bromoviridae	n/a	Cucumovirus	Cucumber mosaic virus (CMV)	2b, Capsid	Gonzalez et al. (2010); Lin et al. (1996)
Coronaviridae	Coronavirinae	Alphacoronavirus	Alphacoronavirus 1/ Transmissible grastroenteritis virus (TGEV)	N	Wurm et al. (2001)
Coronaviridae	Coronavirinae	Betacoronavirus	Murine coronavirus/ Mouse hepatitis virus (MHV)	N	Wurm et al. (2001)
Coronaviridae	Coronavirinae	Betacoronavirus	Severe acute respiratory syndrome coronavirus (SARS-CoV)	3b, N	You et al. (2005); Yuan et al. (2005)
Coronaviridae	Coronavirinae	Gammacoronavirus	Avian coronavirus/ Infectious bronchitis virus (IBV)	N	Hiscox et al. (2001)
Flaviviridae	n/a	Flavivirus	Dengue virus (DEN)	Core protein	Wang et al. (2002)
Flaviviridae	n/a	Flavivirus	Hepatitis C virus (HCV)	Core protein (wild-type and mutant), NS5B mutant	Realdon et al. (2004); Falcon et al. (2003); Hirano et al. (2003)

Family	Subfamily	Genus	Virus	Protein	Reference
Flaviviridae	n/a	Flavivirus	Japanese encephalitis virus (JEV)	Core protein	Mori et al. (2005)
Flaviviridae	n/a	Flavivirus	West nile virus (WNV)	Core protein	Westaway et al. (1997)
Luteoviridae	n/a	Polerovirus	Potato leaf roll virus (PLRV)	CP, P5	Haupt et al. (2005)
Nodaviridae	n/a	Betanodavirus	Greasy grouper nervous necrosis virus (GGNNV)	Alpha	Guo et al. (2003)
Picornaviridae	n/a	Cardiovirus	Encephalomyocarditis virus (ECMV)	2a	Aminev et al. (2003a)
Potyviridae	n/a	Potyvirus	Potato virus A (PVA)	NIa	Rajamaki and Valkonen (2009)
Potyviridae	n/a	Potyvirus	Tabacco etch virus (TEV)	P3, NIa, Nib	Langenberg and Zhang (1997); Baunoch et al. (1991)
Potyviridae	n/a	Potyvirus	Turnip mosaic virus (TuMV)	VPg-Pro (Potential)	Beauchemin et al. (2007)
Togaviridae	n/a	Alphavirus	Semliki forest virus (SFV)	Core (C), nsP2	Michel et al. (1990); Rikkonen et al. (1992)
Baltimore class V: negative/ambisense, single-strand RNA					
Bornaviridae	n/a	Bornavirus	Borna disease virus (BDV)	Genome/anti-genome	Pyper et al. (1998)
Orthomyxoviridae	n/a	Influenzavirus A	Influenza A virus (IAV)	NS1, NP, M1, PB2	Davey et al. (1985); Emmott et al. (2010a)
Orthomyxoviridae	n/a	Isavirus	Infectious Salmon anaemia virus (ISAV)	Genome/anti-genome	Goic et al. (2008)
Paramyxoviridae	Paramyxovirinae	Avulavirus	Newcastle disease virus (NDV)	Matrix protein (M)	Peeples et al. (1992)
Rhabdoviridae	n/a	Nucleorhabdovirus	Maize fine streak virus (MFSV)	N, P	Tsai et al. (2005)

(continued)

Table 14.1 (continued)

Family	Subfamily	Genus	Virus	Protein/nucleic acid	Reference
Baltimore class VI: single-strand RNA – reverse transcription					
Retroviridae	Orthoretrovirinae	Betaretrovirus	Jaagsiekte sheep retrovirus (JSRV)	Env-SP/JSE-SP	Caporale et al. (2009)
Retroviridae	Orthoretrovirinae	Betaretrovirus	Mouse mammary tumour virus (MMTV)	Env-SP/p14, Rem	Hoch-Marchaim et al. (1998); Indik et al. (2005)
Retroviridae	Orthoretrovirinae	Deltaretrovirus	Human T cell lymphotropic virus 1 (HTLV-1)	HBZ-SP1, Rex, p21x (a natural Rex truncation), pX, p30, p30II, Unspliced viral mRNA	Hivin et al. (2007); Kubota et al. (1989); Kubota et al. (1996); Siomi et al. (1988); Ghorbel et al. (2006); Bartoe et al. (2000); Nosaka et al. (1989)
Retroviridae	Orthoretrovirinae	Deltaretrovirus	Human T cell lymphotropic virus 2 (HTLV-2)	Rex	Narayan et al. (2003)
Retroviridae	Orthoretrovirinae	Gammaretrovirus	Maloney murine leukaemia virus (MuLV/MMLV)	Nucleocapsid (NC), Integrase (IN)	Risco et al. (1995)
Retroviridae	Orthoretrovirinae	Lentivirus	Bovine immunodeficiency virus (BIV)	Tat, Rev	Gomez Corredor and Archambault (2009); Fong et al. (1997)
Retroviridae	Orthoretrovirinae	Lentivirus	Carpine arthritis encephalitis virus (CAEV)	Rev-C	Saltarelli et al. (1994)
Retroviridae	Orthoretrovirinae	Lentivirus	Human immunodeficiency virus 1 (HIV-1)	Tat, Rev, unspliced viral mRNA	Cochrane et al. (1990); Ruben et al. (1989)
Retroviridae	Orthoretrovirinae	Lentivirus	Human immunodeficiency virus 2 (HIV-2)	Tat, Rev	Orsini and Debouck (1996); Dillon et al. (1991)

Retroviridae	Orthoretrovirinae	Lentivirus	Visna virus	Rev-V	Schoberg and Clements (1994)
Baltimore class VII: double-strand DNA – reverse transcription					
Hepadnaviridae	n/a	Orthohepadnavirus	Hepatitis B virus (HBV)	Core (various mutants)	Ning and Shih (2004)
Subviral agents: satellite viruses – circular, negative-sense, single-strand RNA					
n/a	n/a	Deltavirus	Hepatitis D virus (HDV)	deltaAg (S and L), anti-genome, RNA	Macnaughton et al. (1990); Li et al. (2006); Kojima et al. (1989)
Subviral agents: Satellite viruses – linear, positive-sense, single-strand RNA					
n/a	n/a	n/a	Satellite panicum mosaic virus (SPMV)	CP	Qi et al. (2008)
Subviral agents: Viroids					
Pospiviroidae	n/a	Pospiviroid	Potato spindle tuber viroid	Genome (positive sense)	Harders et al. (1989)
Endogenous retroviruses					
n/a	n/a	n/a	Human endogenous retrovirus-K (HERV-K)	Env-SP, Np9, cORF	Ruggieri et al. (2009); Armbruester et al. (2004); Lower et al. (1995)

Many of these proteins have been shown to interact with nucleolar proteins

The reason why RNA viruses, and positive-strand RNA viruses in particular, interact with the nucleolus when the site of genome replication is in the cytoplasm is less intuitive. In this latter case, viral proteins that are normally required in the cytoplasm must transit through the nuclear pore complex both to and from the nucleus. This process is crucial for virus biology because if the viral proteins that are required for cytoplasmic functions such as RNA synthesis and encapsidation are sequestered in the nucleolus or nucleus, then progeny virus production will be affected as has been revealed by inhibitor and genetic studies (Lee et al. 2006; Tijms et al. 2002). Viruses may interact with the nucleolus to usurp host cell functions and recruit nucleolar proteins to facilitate virus replication. Investigating the interactions between viruses and the nucleolus may facilitate the design of novel anti-viral therapies both in terms of recombinant vaccines (Pei et al. 2008) and molecular intervention (Rossi et al. 2007), and also contribute to a more detailed understanding of the cell biology of the nucleolus.

For many years our understanding of the interaction of viruses and the nucleolus was phenomenological and focused on identifying viral proteins that localised to this structure, their mechanisms of trafficking and potential interaction with nucleolar proteins (e.g. see Table 14.1). However, recent research capitalising on advances in proteomics, viral genetics and cellular imaging techniques are beginning to increase our understanding of the mechanisms viruses use to subvert host cell nucleoli and facilitate virus biology (Hiscox et al. 2010).

New data are now emerging that support the view that many viruses interact with the nucleus and nucleolus, particularly to facilitate virus replication. One of the best-studied viruses in terms of viral interactions with the nucleolus is HIV-1 and is described in detail in Chap. 17. Although HIV has clearly defined cytoplasmic and nuclear replication strategies, the virus has a positive-sense RNA genome in the sense that the viral capsid contains two copies of positive-sense RNA, but these are reverse transcribed in the cytoplasm and then trafficked to the nucleus, where ultimately the new genome is transcribed and trafficked back to the cytoplasm. Part of the reasoning for the interaction of HIV-1 with the nucleolus is the trafficking of intronless mRNA from the nucleus into the cytoplasm (Michienzi et al. 2000). This is a property shared with herpes viruses and indicated that different viruses have evolved similar strategies involving subversion of nucleolar function for the benefit of virus biology (Boyne and Whitehouse 2006). In the case of HIV-1, this knowledge has also led to the design and implementation of effective genetic therapies against the virus (Unwalla et al. 2008).

14.2   DNA Virus Interactions with the Nucleolus

A large number of viruses with DNA genomes have been shown to interact with nucleolus, and this perhaps is not surprising as most DNA viruses replicate in the nucleus. A genome-wide screen of three distinct herpesviruses, herpes simplex virus 1 (HSV-1), cytomegalovirus (CMV) and Epstein–Barr virus (EBV), has shown that at least 12 herpesvirus-encoded proteins specifically localise to the nucleolus (Salsman et al. 2008), which are implicated in many aspects of the herpesvirus life cycle. Therefore, a number of proteomic studies are currently being undertaken to study changes, in a global context, within the nucleolar proteome during virus infections, and are discussed later (Lam et al. 2010). Several different herpes virus proteins have been shown to cause the redistribution of nucleolar proteins and hence disruption of the nucleolus. These include herpes simplex virus 1, the major tegument structural protein VP22 (Lopez et al. 2008), and the US11 (Xing et al. 2010) and UL24 proteins (Bertrand and Pearson 2008; Lymberopoulos and Pearson 2007). Such disruption in many cases may have a direct effect on nucleolar function.

A significant area of virus biology that has been investigated is the role of viral proteins that traffic through the nucleolus. For example, a number of HIV proteins that traffic through the nucleolus have been implicated in virus mRNA processing (Dundr et al. 1995). Similar observations have also been made in herpesviruses (Boyne and Whitehouse 2006, 2009; Leenadevi and Dalziel 2009). Initial studies utilising the prototype γ-2 herpesvirus, herpes virus saimiri (HVS), demonstrated that the HVS nucleolar trafficking ORF57 protein induces nucleolar redistribution of the host cell human TREX proteins, which are involved in mRNA nuclear export (Boyne and Whitehouse 2006). Intriguingly, ablating ORF57 nucleolar trafficking led to a failure of ORF57-mediated viral mRNA nuclear export (Boyne and Whitehouse 2006). The precise role of this nucleolar sequestration is yet to be determined, but possible effects on viral mRNA/protein processing and viral ribonucleoprotein particle assembly are currently being investigated.

This property may also be conserved in other ORF57 homologues as recent analysis has shown that the ORF57 protein from Kaposi's sarcoma associated herpesvirus (KSHV) also dynamically traffics through the nucleolus (Boyne et al. 2008b). Moreover, on the rapid disorganisation of the nucleolus a reduction is observed in virus mRNA nuclear export (Boyne and Whitehouse 2009). The formation of an ORF57-mediated export competent ribonucleoprotein particle within the nucleolus may also have implications for the translation of viral mRNAs. For example, it has recently been demonstrated that the cellular nucleo-cytoplasmic shuttle protein, PYM, which is involved in translation enhancement, is redistributed to the nucleolus in the presence of the KSHV ORF57 protein (Boyne et al. 2010). This interaction effectively enhances the translation of the predominantly intronless transcripts made by KSHV, and draws parallels with potential translation enhancement of positive strand RNA virus genomes through their interaction with the nucleolus (discussed later).

A second area of virus replication where nucleolar proteins are sequestered involves the replication of the virus DNA genome. For example, we (Matthews) and others have observed that nucleolar antigens upstream binding factor (UBF) and nucleophosmin (B23.1) are both sequestered into adenovirus DNA replication centres where they promote viral DNA replication (Hindley et al. 2007; Lawrence et al. 2006; Okuwaki et al. 2001). Similarly, in HSV-1 infected cells, a number of nucleolar proteins including nucleolin and UBF are recruited into viral DNA replication centres (Lymberopoulos and Pearson 2010). These are specific sites where replication and encapsidation of the HSV-1 genome occurs. Evidence suggests that sequestration of UBF is essential for viral DNA replication as overexpression of tagged version of UBF acts in a dominant-negative manner inhibiting virus DNA replication (Stow et al. 2009). Moreover, depletion of nucleolin results in reduced virus gene expression and infectious virion production (Calle et al. 2008; Sagou et al. 2010).

In addition to enhancing virus replication, nucleolar proteins are redistributed to alter cellular pathways during infection. For example, the nucleolar targeted HSV-1 US11 protein has been shown to interact with homeodomain-interacting protein kinase 2 (HIPK2), which plays a role in p53-mediated cellular apoptosis and hypoxic response (Calzado et al. 2009) and also participates in the regulation of the cell cycle (Calzado et al. 2007). This interaction alters the sub-cellular localisation of HIPK2 and protects against HIPK2-mediated cell cycle arrest (Giraud et al. 2004). In contrast, the cellular protein, protein interacting with the carboxyl terminus-1 (PICT-1), can sequester the virally encoded apoptosis suppressor protein, KS-Bcl-2 protein, from the mitochondria into the nucleolus to down-regulate its anti-apoptotic activity (Kalt et al. 2010). This is a potential interesting interplay between two sub-cellular structures involved in the viral stress response (Olson 2009), and maybe more common and widespread. For example, bacterial infection has been shown to disrupt the nucleolus through regulating mitochondrial dysfunction (Dean et al. 2010).

14.3 Interactions of RNA Viruses with the Nucleolus

Although many RNA virus proteins have been shown to localise to the nucleolus, most attention has focused on viral capsid proteins. These are proteins that associate with the viral genome for encapsidation and assembly of new virus particles. These proteins may also modulate replication (and transcription, where appropriate) of the viral genome. Increasingly, capsid proteins have also been shown to have a number of roles in modulating host cell signalling pathways and functions. These capsid proteins are referred to as capsid, nucleoproteins or nucleocapsid proteins, depending on the virus. In many cases, they are phosphorylated (Chen et al. 2005), which can modulate activity (Spencer et al. 2008).

Many examples of these proteins have been shown to localise to the nucleolus both when over-expressed and also in infected cells. These include proteins from positive-strand animal and plant RNA viruses, including the coronavirus nucleocapsid protein (Chen et al. 2002; Hiscox et al. 2001; Wurm et al. 2001), the arterivirus nucleocapsid protein (Rowland et al. 1999), the alphavirus capsid protein (Jakob 1994) and non-structural protein nsP2 (Rikkonen et al. 1992, 1994) and the umbravirus ORF3 protein (Ryabov et al. 2004). Capsid proteins from negative-strand RNA viruses also localise to the nucleolus. These have strain dependent localisation of a number of different influenza virus proteins (Emmott et al. 2010c; Han et al. 2010; Melen et al. 2007; Volmer et al. 2010).

For many years this has followed a phenomenological pattern and viral capsid and RNA-binding proteins might simply localise to the nucleolus because they diffuse through the nuclear pore complex and associate with compartments in the nucleus that have high RNA contents – the nucleolus in particular because it is transcriptionally active. In this case, sub-cellular localisation to the nucleolus would have no physiological consequence for the virus or the cell. However, RNA virus replication is error prone and selection pressure might apply to such a fortuitous localisation (given the ~4,500+ nucleolar proteins and their diverse roles (Ahmad et al. 2009)), with the concomitant effect that the virus could select for changes that ultimately disrupt nucleolar function and/or recruit nucleolar proteins to aid virus replication.

There is a potential correlation between the nucleolar localisation of a viral protein and the loss of an essential nucleolar function. The molecular mechanisms responsible for this effect are unknown, but the displacement and re-localisation of nucleolar proteins by viral proteins could increase or decrease the nucleolar, nuclear and/or cytoplasmic pool of these proteins. Certainly, the accumulation of viral proteins in the nucleolus could potentially cause volume exclusion or crowding effects, which have been proposed to play a fundamental role in the formation of nuclear compartments including the nucleolus, and can be addressed by proteomic strategies. Therefore, disruption of nucleolar architecture and function might be common in virus-infected cells if viral proteins target the nucleolus or a stage of the virus lifecycle disrupts nucleolar proteins. For example, poliovirus infection results in the selective redistribution of nucleolin from the nucleolus to the cytoplasm (Waggoner and Sarnow 1998) and inactivation of UBF, which shuts off RNA polymerase I transcription in the host cell. The infection of cells with IBV has been shown to disrupt nucleolar architecture (Dove et al. 2006b) and cause arrest of the cell cycle in the G2/M phase and failure of cytokinesis (Dove et al. 2006a). The IBV and arterivirus nucleocapsid proteins associate with nucleolin and fibrillarin, respectively. Similarly, the HIV-1 Rev protein has been shown to localise to the DFC and GC and over-expression of Rev protein alters the nucleolar architecture and is associated with the accumulation of nucleophosmin (Dundr et al. 1995).

14.4 Trafficking of Virus Proteins to the Nucleolus

Many different virus proteins localise to the nucleolus (Table 14.1). However, predicting viral (and cellular) nucleolar targeting signals has historically been problematic and only recently has bioinformatic software been developed to fascilitate this (Scott et al. 2011). Nucleolar trafficking might be mediated by virtue of the fact that viral proteins that are trafficked to the nucleolus contain motifs that resemble host nucleolar targeting signals, that is, a form of molecular mimicry is used (Rowland and Yoo 2003). The discovery of specific nucleolar trafficking signals in viral proteins has indicated a functional mechanism behind this observed localisation (Lee et al. 2003; Reed et al. 2006; Rowland et al. 2003). Analysis of the different nucleolar trafficking signals identified in viral proteins using dynamic live-cell imaging has certainly demonstrated that different proteins can confer differential trafficking rates and localisation patterns (Emmott et al. 2008). This is very similar to cellular nucleolar proteins (Lechertier et al. 2007).

In some virus proteins, both NLSs and nucleolar targeting signals act in concert to direct a protein to the nucleolus. The arterivirus porcine reproductive and respiratory syndrome virus (PRRSV) nucleocapsid protein localises to the nucleolus and has been shown to contain two potential NLSs, a pat4 and a downstream pat7 motif (Rowland et al. 1999, 2003). Analysis revealed that a 31 amino acid sequence incorporating the pat7 motif could direct the nucleocapsid protein to both the nucleus and nucleolus. The protein also contains a predicted NES, presumably to allow the protein to traffic back into the cytoplasm to contribute to viral function in this compartment. This is common with other similar related proteins. For example, in the avian coronavirus nucleocapsid protein an eight amino acid sequence is necessary and sufficient to target the protein to the nucleolus (Reed et al. 2006) and contains an NES (Reed et al. 2007). Intriguingly, genetic analysis (Lee et al. 2006), dynamic live-cell imaging (You et al. 2008) and use of trafficking inhibitors (Tijms et al. 2002) paint a picture of the requirement of these positive sense RNA virus capsid proteins localising to the nucleolus as soon as they are translated, prior to their involvement in virus replication or assembly. This may be related to subversion of host cell function, protein modification (e.g. phosphorylation) or recruitment of nucleolar proteins.

Viral proteins might also traffic to the nucleolus through association with cellular nucleolar proteins (Yoo et al. 2003). For example, the hepatitis delta antigen has been shown to contain a nucleolar targeting signal that also corresponded to a site that promoted binding to nucleolin (Lee et al. 1998). Mutating this region prevented nucleolin binding to the delta antigen and nucleolar trafficking. By implication, this relates nucleolin binding to nucleolar trafficking (Lee et al. 1998). Certainly, interaction with nucleophosmin and hepatitis delta antigens can modulate viral replication (Huang et al. 2001) and more recently combined proteomic-RNAi screens have revealed many other nucleolar proteins that can be associated with this viral protein (Cao et al. 2009). Trafficking and accumulation of viral proteins to and from the nucleolus, similar to cellular proteins, may also be cell cycle related. For example, the coronavirus nucleocapsid protein localises preferentially to the nucleolus in the G2 phase of the cell cycle (Cawood et al. 2007), as does the human cytomegalovirus protein UL83 in the G1 phase (Arcangeletti et al. 2011). Again these trafficking profiles may be related to the interaction with cellular nucleolar proteins (Emmott and Hiscox 2009).

14.5 Functional Relevance of Nucleolar Interactions to the Viral Life Cycle

Many different examples now exist to show that the disruption of nuclear or nucleolar trafficking of viral proteins affects viral pathogenesis, and argues against nucleolar localisation as a purely phenomenological observation. For example, the

Semliki Forest virus non-structural protein nsP2 can localise to the nucleolus (Peranen et al. 1990; Rikkonen et al. 1992, 1994) and disruption of this localisation through a single amino acid change results in a reduction in neurovirulence (Fazakerley et al. 2002). Such in vitro data has also been backed up by persuasive in vivo data. Mutation of the arterivirus nucleocapsid protein pat7 NLS motif in the context of a full-length clone revealed that this sequence could have a key role in virus pathogenesis in vivo, as animals infected with mutant viruses had shorter viraemia than wild-type viruses (Lee et al. 2006; Pei et al. 2008). Interestingly, reversions occurred in the mutated nucleocapsid gene sequence and although the amino acid sequence of the pat7 motif was altered, its function was not; this new signal was defined as a pat8 motif (Lee et al. 2006). The clear implications of this groundbreaking work is that disruption of nucleolar trafficking of a viral protein proves functional relevance and illustrates the potential of exploiting this knowledge for the generation of growth attenuated recombinant vaccines (Pei et al. 2008; Reed et al. 2006, 2007).

Similarly, point mutations in the Japanese encephalitis virus (JEV) core protein that abolished nuclear and nucleolar localisation resulted in recombinant viruses with impaired replication in mammalian cells, compared to wild type virus (Mori et al. 2005; Tsuda et al. 2006). Curiously, replication of recombinant viruses was not impaired in insect cells, illustrating this could potentially be related to differences in nucleolar architecture and proteomes between these cell types (Thiry and Lafontaine 2005). The JEV core protein has been shown to interact with nucleophosmin and is translocated from the nucleolus to the cytoplasm.

Flaviviruses in general (JEV, Dengue virus and West Nile virus) appear to have a part-nuclear stage to the synthesis of viral RNA and several components of the viral replicase together with newly synthesised RNA have been found in the nucleus of infected cells (Uchil et al. 2006). One intriguing question that has yet to be elucidated is how such viral RNA traffics from the nucleus to the cytoplasm. Most cellular mRNAs are spliced and it is part of the splicing process that signals nuclear export. Certain DNA viruses, such as herpesvirus saimiri, produce intron-less mRNA and these viruses have evolved specific viral proteins (such as herpesvirus saimiri ORF57 (Boyne et al. 2008a)), which interact with the cellular mRNA export machinery (e.g. the mRNA processing and export factor ALY) to traffic viral mRNA from the nucleus to the cytoplasm (Boyne et al. 2008b, 2010; Boyne and Whitehouse 2006) and a similar process might be required by RNA viruses. For example, tomato bushy stunt virus (TBSV) redistributes ALY from the nucleus to the cytoplasm, and this might be a way the virus mediates host cell protein synthesis (Uhrig et al. 2004). In plants RNA silencing, a host defence mechanism targets virus RNAs for degradation in a sequence-specific manner and viruses use several mechanisms to counteract this system (Canto et al. 2006). TBSV encodes a protein, P19, which interferes with this pathway. However, ALY might transport P19 from the cytoplasm to the nucleus or nucleolus and disrupt its silencing suppression activity. Nucleolin has also been shown to be involved in the trafficking of herpes simplex virus type 1 nucleocapsids from the nucleus to the cytoplasm (Sagou et al. 2010), drawing parallels with the involvement of nucleolar proteins in the movement of plant viruses

(Kim et al. 2007a, b). Different plant virus proteins involved in long-distance phloem-associated movement of virus particles or with roles in binding to the RNA virus genomes localise to the nucleolus and other sub-nuclear structures (Kim et al. 2007b; Ryabov et al. 2004). This may be mediated by association with nuclear proteins, as is the case with fibrillarin and the ORF3 protein of plant umbraviruses (Kim et al. 2007a).

Hijacking the nucleolus is not exclusive to plant viruses and may also occur with mammalian viruses. Similar to the plant rhabdovirus maize fine streak virus (MFSV), whose nucleocapsid and phosphoproteins localise to the nucleolus (Tsai et al. 2005), the animal negative-stranded RNA virus Borna disease virus has been reported to use the nucleolus as a site for genome replication, and its RNA-binding protein has the appropriate trafficking signals for import to and export from the cytoplasm to the nucleus (Pyper et al. 1998). The hepatitis delta virus genome also has differential synthesis in the nucleus with RNA being transcribed in the nucleolus (Huang et al. 2001); this is similar to the potato spindle tuber viroid where RNAs of opposite polarity are sequestered in different nuclear compartments, with the positive-sense RNA being transported to the nucleolus. Again localisation to different sub-nuclear strcutures may have different roles in the virus life cycle (Li et al. 2006). An intriguing recent discovery has been made showing that adeno-associated virus (AAV) encodes an additional protein called assembly-activating protein (AAP) that localises to the nucleolus and promotes assembly of the viral capsid (Sonntag et al. 2010).

As a result of their limited genomes and coding capacities, recruitment of cellular proteins with defined functions in RNA metabolism would be a logical step to facilitate RNA virus infection. As nucleolar proteins have many crucial functions in cellular RNA biosynthesis, processing and translation, it comes as no surprise that nucleolar proteins are incorporated into the replication and/or translation complexes formed by RNA viruses. Given that some nucleolar proteins have many different functions, the same nucleolar protein might be used by a virus for different aspects of the replication pathway. Studies suggest that the human rhinovirus 3 C protease (3Cpro) pre-cursors, 3CD' and/or 3CD, localise in the nucleoli of infected cells early in infection and inhibit cellular RNA transcription via proteolytic mechanisms (Amineva et al. 2004). This general property is not restricted to human rhinovirus and in terms of the inhibition of cellular translation has also been described for encephalomyocarditis virus (Aminev et al. 2003a, b), again suggesting roles in translational regulation.

14.6 Applying Quantitative Proteomics to Study Viral Interactions with the Nucleolar Proteome

Given the many roles of the nucleolus in the life cycle of the cell, including as stress sensor (Boulon et al. 2010; Mayer and Grummt 2005), it would seem reasonable that comprehensive unbiased analysis of the nucleolar proteome would yield interesting data, particularly, with providing clues as to what cellular nucleolar functions may

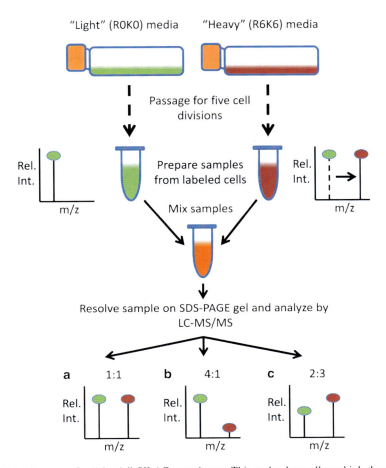

Fig. 14.1 Diagram of a "classic" SILAC experiment. This technology allows high-throughput quantitative proteomics and has been readily applied to the nucleolus, especially when coupled with dynamic live-cell imaging (Andersen et al. 2005). The ability to simultaneously compare up to three different conditions through selection of the appropriate isotope label has enabled the recent studies of how the nucleolar proteome changes in virus-infected cells (Emmott et al. 2010a; Emmott et al. 2010b; Emmott et al. 2010c; Hiscox et al. 2010; Lam et al. 2010)

be altered by virus infection and what mechanisms the nucleolus may use to respond to this. How the nucleolar proteome changes in response to virus-infection has been investigated using stable isotope labelling with amino acids in cell culture (SILAC) coupled to LC-MS/MS and bioinformatics (Fig. 14.1). These studies, led by our laboratories, have analysed purified nucleoli and the nucleus, and have directly stemmed from the pioneering work of the Lamond laboratory in analysing purified nucleoli using quantitative proteomics (Andersen et al. 2005). Viruses investigated so far have included human adenovirus (Lam et al. 2010), avian coronavirus (Emmott et al. 2010a, b), different strains of influenza virus (Emmott et al. 2010c) and human respiratory syncytial virus (Munday et al. 2010). Overall, our data indicates that only a small proportion of nucleolar proteins change in abundance in virus-infected cells, and these tend to be virus-specific. For example, in adenovirus infected cells just 7% of proteins identified show a twofold or greater change compared to almost a third of nucleolar antigens showing a greater than twofold change when cells are treated with ActD which inhibits rRNA synthesis (Lam et al. 2010). What is notable is that direct comparison between the adenovirus data set and the ActD dataset shows no clear correlation (Hiscox et al. 2010; Lam et al. 2010), further supporting the case that adenovirus induces effects on the nucleolus distinct from that of a generalised, non-specific shut down of nucleolar function. This fits well with a previous observation that adenovirus infection does not affect rRNA synthesis even 36 h post-infection (Lawrence et al. 2006). These results were initially surprising given the number of different viral proteins that can localise to this structure and how they interact with nucleolar proteins. This suggests that the nucleolar proteome and architecture is resilient during early stages of infection but may become disrupted as more and more damage accumulates inside cells because of virus activity, as clearly evidenced in live-cell imaging experiments (Bertrand and Pearson 2008; Dove et al. 2006b; Lymberopoulos et al. 2010).

14.7 Future Research Directions

Coupling quantitative proteomic analysis of the nucleolus and deep sequencing throughout infection in time-course experiments of lytic, latent, acute and persistent viruses would reveal valuable insights into the response of the nucleolus to virus infection. Likewise, being able to move from studying cell culture-adapted laboratory strains into clinical isolates replicating in primary cells would yield more biologically relevant information, particularly with regard to the severity of disease and nucleolar changes. These technologies could also be applied to large-scale analysis of viral proteins that traffic to the nucleolus and the cellular nucleolar proteins that they associate with (e.g. using SILAC and EGFP-traps (Trinkle-Mulcahy et al. 2008)), thus generating and integrating interactome networks with the nucleolar proteome during infection.

Acknowledgements DAM and JAH would like to acknowledge their co-workers and collaborators over the years for developing viral interactions with the nucleolus. DAM's research on the nucleolus is supported by the Wellcome Trust and JAH's by the BBSRC and a Leverhulme Trust Research Fellowship. EE is supported by a BBSRC Astbury DTG studentship.

References

Ahmad Y, Boisvert FM, Gregor P, Cobley A, Lamond AI (2009) NOPdb: nucleolar proteome database–2008 update. Nucleic Acids Res 37:D181–D184
Aminev AG, Amineva SP, Palmenberg AC (2003a) Encephalomyocarditis viral protein 2A localizes to nucleoli and inhibits cap-dependent mRNA translation. Virus Research 95:45–57

14 Viruses and the Nucleolus

Aminev AG, Amineva SP, Palmenberg AC (2003b) Encephalomyocarditis virus (EMCV) proteins 2A and 3BCD localize to nuclei and inhibit cellular mRNA transcription but not rRNA transcription. Vir Res 95:59–73

Amineva SP, Aminev AG, Palmenberg AC, Gern JE (2004) Rhinovirus 3 C protease precursors 3CD and 3CD' localize to the nuclei of infected cells. J Gen Virol 85:2969–2979

Andersen JS, Lam YW, Leung AK, Ong SE, Lyon CE, Lamond AI, Mann M (2005) Nucleolar proteome dynamics. Nature 433:77–83

Arcangeletti MC, De Conto F, Ferraglia F, Pinardi F, Gatti R, Orlandini G, Calderaro A, Motta F, Medici MC, Martinelli M, Valcavi P, Razin SV, Chezzi C, Dettori G (2003) Human cytomegalovirus proteins PP65 and IEP72 are targeted to distinct compartments in nuclei and nuclear matrices of infected human embryo fibroblasts. J Cell Biochem 90:1056–1067

Arcangeletti MC, Rodighiero I, Mirandola P, De Conto F, Covan S, Germini D, Razin S, Dettori G, Chezzi C (2011) Cell-cycle-dependent localization of human cytomegalovirus UL83 phosphoprotein in the nucleolus and modulation of viral gene expression in human embryo fibroblasts in vitro. J Cell Biochem 112:307–317

Armbruester V, Sauter M, Roemer K, Best B, Hahn S, Nty A, Schmid A, Philipp S, Mueller A, Mueller-Lantzsch N (2004) Np9 protein of human endogenous retrovirus K interacts with ligand of numb protein X. J Virol 78:10310–10319

Bartoe JT, Albrecht B, Collins ND, Robek MD, Ratner L, Green PL, Lairmore MD (2000) Functional role of pX open reading frame II of human T-lymphotropic virus type 1 in maintenance of viral loads in vivo. J Virol 74:1094–1100

Baunoch DA, Das P, Browning ME, Hari V (1991) A temporal study of the expression of the capsid, cytoplasmic inclusion and nuclear inclusion proteins of tobacco etch potyvirus in infected plants. J Gen Virol 72:487–492

Beauchemin C, Boutet N, Laliberte JF (2007) Visualization of the interaction between the precursors of VPg, the viral protein linked to the genome of turnip mosaic virus, and the translation eukaryotic initiation factor iso 4E in Planta. J Virol 81:775–782

Bertrand L, Pearson A (2008) The conserved N-terminal domain of herpes simplex virus 1 UL24 protein is sufficient to induce the spatial redistribution of nucleolin. J Gen Virol 89:1142–1151

Blanie S, Mortier J, Delverdier M, Bertagnoli S, Camus-Bouclainville C (2009) M148R and M149R are two virulence factors for myxoma virus pathogenesis in the European rabbit. Vet Res 40:11

Boulon S, Westman BJ, Hutten S, Boisvert FM, Lamond AI (2010) The nucleolus under stress. Mol Cell 40:216–227

Boyne JR, Whitehouse A (2006) Nucleolar trafficking is essential for nuclear export of intronless herpesvirus mRNA. Proc Natl Acad Sci USA 103:15190–15195

Boyne JR, Whitehouse A (2009) Nucleolar disruption impairs Kaposi's sarcoma-associated herpesvirus ORF57-mediated nuclear export of intronless viral mRNAs. FEBS Lett 583: 3549–3556

Boyne JR, Colgan KJ, Whitehouse A (2008a) Herpesvirus saimiri ORF57: a post-transcriptional regulatory protein. Front Biosci 13:2928–2938

Boyne JR, Colgan KJ, Whitehouse A (2008b) Recruitment of the complete hTREX complex is required for Kaposi's sarcoma-associated herpesvirus intronless mRNA nuclear export and virus replication. PLoS Pathog 4:e1000194

Boyne JR, Jackson BR, Taylor A, Macnab SA, Whitehouse A (2010) Kaposi's sarcoma-associated herpesvirus ORF57 protein interacts with PYM to enhance translation of viral intronless mRNAs. EMBO J 29:1851–1864

Calle A, Ugrinova I, Epstein AL, Bouvet P, Diaz JJ, Greco A (2008) Nucleolin is required for an efficient herpes simplex virus type 1 infection. J Virol 82:4762–4773

Calzado MA, Renner F, Roscic A, Schmitz ML (2007) HIPK2: a versatile switchboard regulating the transcription machinery and cell death. Cell Cycle 6:139–143

Calzado MA, De La Vega L, Munoz E, Schmitz ML (2009) From top to bottom: the two faces of HIPK2 for regulation of the hypoxic response. Cell Cycle 8:1659–1664

Canto T, Uhrig JF, Swanson M, Wright KM, MacFarlane SA (2006) Translocation of Tomato bushy stunt virus P19 protein into the nucleus by ALY proteins compromises its silencing suppressor activity. J Virol 80:9064–9072

Cao D, Haussecker D, Huang Y, Kay MA (2009) Combined proteomic-RNAi screen for host factors involved in human hepatitis delta virus replication. RNA 15:1971–1979

Caporale M, Arnaud F, Mura M, Golder M, Murgia C, Palmarini M (2009) The signal peptide of a simple retrovirus envelope functions as a posttranscriptional regulator of viral gene expression. J Virol 83:4591–4604

Cawood R, Harrison SM, Dove BK, Reed ML, Hiscox JA (2007) Cell cycle dependent nucleolar localization of the coronavirus nucleocapsid protein. Cell Cycle 6:863–867

Chen H, Wurm T, Britton P, Brooks G, Hiscox JA (2002) Interaction of the coronavirus nucleoprotein with nucleolar antigens and the host cell. J Virol 76:5233–5250

Chen H, Gill A, Dove BK, Emmett SR, Kemp FC, Ritchie MA, Dee M, Hiscox JA (2005) Mass spectroscopic characterisation of the coronavirus infectious bronchitis virus nucleoprotein and elucidation of the role of phosphorylation in RNA binding using surface plasmon resonance. J Virol 79:1164–1179

Cheng G, Brett ME, He B (2002) Signals that dictate nuclear, nucleolar, and cytoplasmic shuttling of the gamma(1)34.5 protein of herpes simplex virus type 1. J Virol 76:9434–9445

Cochrane AW, Perkins A, Rosen CA (1990) Identification of sequences important in the nucleolar localization of human immunodeficiency virus Rev: relevance of nucleolar localization to function. J Virol 64:881–885

Davey J, Colman A, Dimmock NJ (1985) Location of influenza virus M, NP and NS1 proteins in microinjected cells. J Gen Virol 66:2319–2334

Dean P, Scott JA, Knox AA, Quitard S, Watkins NJ, Kenny B (2010) The enteropathogenic E. coli effector EspF targets and disrupts the nucleolus by a process regulated by mitochondrial dysfunction. PLoS Pathog 6:e1000961

Dillon PJ, Nelbock P, Perkins A, Rosen CA (1991) Structural and functional analysis of the human immunodeficiency virus type 2 Rev protein. J Virol 65:445–449

Dove B, Brooks G, Bicknell K, Wurm T, Hiscox JA (2006a) Cell cycle perturbations induced by infection with the coronavirus infectious bronchitis virus and their effect on virus replication. J Virol 80:4147–4156

Dove BK, You JH, Reed ML, Emmett SR, Brooks G, Hiscox JA (2006b) Changes in nucleolar morphology and proteins during infection with the coronavirus infectious bronchitis virus. Cell Microbiol 8:1147–1157

Dundr M, Leno GH, Hammarskjold ML, Rekosh D, Helga-Maria C, Olson MO (1995) The roles of nucleolar structure and function in the subcellular location of the HIV-1 Rev protein. J Cell Sci 108:2811–2823

Emmott E, Hiscox JA (2009) Nucleolar targetting: the hub of the matter. EMBO Reports 10:231–238

Emmott E, Dove BK, Howell G, Chappell LA, Reed ML, Boyne JR, You JH, Brooks G, Whitehouse A, Hiscox JA (2008) Viral nucleolar localisation signals determine dynamic trafficking within the nucleolus. Virology 380:191–202

Emmott E, Rodgers MA, Macdonald A, McCrory S, Ajuh P, Hiscox JA (2010a) Quantitative proteomics using stable isotope labeling with amino acids in cell culture reveals changes in the cytoplasmic, nuclear, and nucleolar proteomes in vero cells infected with the coronavirus infectious bronchitis virus. Mol Cell Proteomics 9:1920–1936

Emmott E, Smith C, Emmett SR, Dove BK, Hiscox JA (2010b) Elucidation of the avian nucleolar proteome by quantitative proteomics using stable isotope labeling with amino acids in cell culture (SILAC) and alteration in the coronavirus infectious bronchitis virus infected cells. Proteomics 10:3558–3562

Emmott E, Wise H, Loucaides EM, Matthews DA, Digard P, Hiscox JA (2010c) Quantitative proteomics using SILAC coupled to LC-MS/MS reveals changes in the nucleolar proteome in influenza A virus infected cells. J Proteome Res 9:5335–5345

Falcon V, Acosta-Rivero N, Chinea G, de la Rosa MC, Menendez I, Duenas-Carrera S, Gra B, Rodriguez A, Tsutsumi V, Shibayama M, Luna-Munoz J, Miranda-Sanchez MM, Morales-Grillo J, Kouri J (2003) Nuclear localization of nucleocapsid-like particles and HCV core protein in hepatocytes of a chronically HCV-infected patient. Biochem Biophys Res Commun 310:54–58

Fazakerley JK, Boyd A, Mikkola ML, Kaariainen L (2002) A single amino acid change in the nuclear localization sequence of the nsP2 protein affects the neurovirulence of semliki forest virus. J Virol 76:392–396

Finsterbusch T, Steinfeldt T, Caliskan R, Mankertz A (2005) Analysis of the subcellular localization of the proteins Rep, Rep' and Cap of porcine circovirus type 1. Virology 343:36–46

Fong SE, Greenwood JD, Williamson JC, Derse D, Pallansch LA, Copeland T, Rasmussen L, Mentzer A, Nagashima K, Tobin G, Gonda MA (1997) Bovine immunodeficiency virus tat gene: cloning of two distinct cDNAs and identification, characterization, and immunolocalization of the tat gene products. Virology 233:339–357

Ghorbel S, Sinha-Datta U, Dundr M, Brown M, Franchini G, Nicot C (2006) Human T-cell leukemia virus type I p30 nuclear/nucleolar retention is mediated through interactions with RNA and a constituent of the 60 S ribosomal subunit. J Biol Chem 281:37150–37158

Giraud S, Diaz-Latoud C, Hacot S, Textoris J, Bourette RP, Diaz JJ (2004) US11 of herpes simplex virus type 1 interacts with HIPK2 and antagonizes HIPK2-induced cell growth arrest. J Virol 78:2984–2993

Goatley LC, Marron MB, Jacobs SC, Hammond JM, Miskin JE, Abrams CC, Smith GL, Dixon LK (1999) Nuclear and nucleolar localization of an African swine fever virus protein, I14L, that is similar to the herpes simplex virus-encoded virulence factor ICP34.5. J Gen Virol 80:525–535

Goic B, Bustamante J, Miquel A, Alvarez M, Vera MI, Valenzuela PD, Burzio LO (2008) The nucleoprotein and the viral RNA of infectious salmon anemia virus (ISAV) are localized in the nucleolus of infected cells. Virology 379:55–63

Gomez Corredor A, Archambault D (2009) The bovine immunodeficiency virus rev protein: identification of a novel lentiviral bipartite nuclear localization signal harboring an atypical spacer sequence. J Virol 83:12842–12853

Gonzalez I, Martinez L, Rakitina DV, Lewsey MG, Atencio FA, Llave C, Kalinina NO, Carr JP, Palukaitis P, Canto T (2010) Cucumber mosaic virus 2b protein subcellular targets and interactions: their significance to RNA silencing suppressor activity. Mol Plant Microbe Interact 23:294–303

Greco A (2009) Involvement of the nucleolus in replication of human viruses. Rev Med Virol 19:201–214

Guo YX, Dallmann K, Kwang J (2003) Identification of nucleolus localization signal of betanodavirus GGNNV protein alpha. Virology 306:225–235

Guo H, Ding Q, Lin F, Pan W, Lin J, Zheng AC (2009) Characterization of the nuclear and nucleolar localization signals of bovine herpesvirus-1 infected cell protein 27. Virus Res 145:312–320

Han H, Cui ZQ, Wang W, Zhang ZP, Wei HP, Zhou YF, Zhang XE (2010) New regulatory mechanisms for the intracellular localization and trafficking of influenza A virus NS1 protein revealed by comparative analysis of A/PR/8/34 and A/Sydney/5/97. J Gen Virol 91:2907–2917

Harders J, Lukacs N, Robert-Nicoud M, Jovin TM, Riesner D (1989) Imaging of viroids in nuclei from tomato leaf tissue by in situ hybridization and confocal laser scanning microscopy. EMBO J 8:3941–3949

Harms JS, Ren X, Oliveira SC, Splitter GA (2000) Distinctions between bovine herpesvirus 1 and herpes simplex virus type 1 VP22 tegument protein subcellular associations. J Virol 74:3301–3312

Haupt S, Stroganova T, Ryabov E, Kim SH, Fraser G, Duncan G, Mayo MA, Barker H, Taliansky M (2005) Nucleolar localization of potato leafroll virus capsid proteins. J Gen Virol 86: 2891–2896

Hindley CE, Davidson AD, Matthews DA (2007) Relationship between adenovirus DNA replication proteins and nucleolar proteins B23.1 and B23.2. J Gen Virol 88:3244–3248

Hirano M, Kaneko S, Yamashita T, Luo H, Qin W, Shirota Y, Nomura T, Kobayashi K, Murakami S (2003) Direct interaction between nucleolin and hepatitis C virus NS5B. J Biol Chem 278:5109–5115

Hiscox JA (2002) Brief review: the nucleolus - a gateway to viral infection? Arch Virol 147:1077–1089

Hiscox JA (2003) The interaction of animal cytoplasmic RNA viruses with the nucleus to facilitate replication. Vir Res 95:13–22

Hiscox JA (2007) RNA viruses: hijacking the dynamic nucleolus. Nat Rev Microbiol 5:119–127

Hiscox JA, Wurm T, Wilson L, Cavanagh D, Britton P, Brooks G (2001) The coronavirus infectious bronchitis virus nucleoprotein localizes to the nucleolus. J Virol 75:506–512

Hiscox JA, Whitehouse A, Matthews DA (2010) Nucleolar proteomics and viral infection. Proteomics. doi:10.1002/pmic.201000251

Hivin P, Basbous J, Raymond F, Henaff D, Arpin-Andre C, Robert-Hebmann V, Barbeau B, Mesnard JM (2007) The HBZ-SP1 isoform of human T-cell leukemia virus type I represses JunB activity by sequestration into nuclear bodies. Retrovirology 4:14

Hoch-Marchaim H, Hasson T, Rorman E, Cohen S, Hochman J (1998) Nucleolar localization of mouse mammary tumor virus proteins in T-cell lymphomas. Virology 242:246–254

Hong-Yan Z, Murata T, Goshima F, Takakuwa H, Koshizuka T, Yamauchi Y, Nishiyama Y (2001) Identification and characterization of the UL24 gene product of herpes simplex virus type 2. Virus Genes 22:321–327

Huang WH, Yung BY, Syu WJ, Lee YH (2001) The nucleolar phosphoprotein B23 interacts with hepatitis delta antigens and modulates the hepatitis delta virus RNA replication. J Biol Chem 276:25166–25175

Hutzinger R, Feederle R, Mrazek J, Schiefermeier N, Balwierz PJ, Zavolan M, Polacek N, Delecluse HJ, Huttenhofer A (2009) Expression and processing of a small nucleolar RNA from the Epstein-Barr virus genome. PLoS Pathog 5:e1000547

Indik S, Gunzburg WH, Salmons B, Rouault F (2005) A novel, mouse mammary tumor virus encoded protein with Rev-like properties. Virology 337:1–6

Jakob R (1994) Nucleolar accumulation of Semliki Forest virus nucleocapsid C protein: influence of metabolic status, cytoskeleton and receptors. J Med Microbiol 40:389–392

Kalt I, Borodianskiy-Shteinberg T, Schachor A, Sarid R (2010) GLTSCR2/PICT-1, a putative tumor suppressor gene product, induces the nucleolar targeting of the Kaposi's sarcoma-associated herpesvirus KS-Bcl-2 protein. J Virol 84:2935–2945

Kim SH, Macfarlane S, Kalinina NO, Rakitina DV, Ryabov EV, Gillespie T, Haupt S, Brown JW, Taliansky M (2007a) Interaction of a plant virus-encoded protein with the major nucleolar protein fibrillarin is required for systemic virus infection. Proc Natl Acad Sci USA 104:11115–11120

Kim SH, Ryabov EV, Kalinina NO, Rakitina DV, Gillespie T, MacFarlane S, Haupt S, Brown JW, Taliansky M (2007b) Cajal bodies and the nucleolus are required for a plant virus systemic infection. EMBO J 26:2169–2179

Kojima T, Callea F, Desmyter J, Shikata T, Desmet VJ (1989) Electron microscopy of ribonucleic acid in nuclear particulate aggregates of hepatitis D using nuclease-gold complexes. J Med Virol 28:183–188

Kubota S, Siomi H, Satoh T, Endo S, Maki M, Hatanaka M (1989) Functional similarity of HIV-I rev and HTLV-I rex proteins: identification of a new nucleolar-targeting signal in rev protein. Biochem Biophys Res Commun 162:963–970

Kubota S, Hatanaka M, Pomerantz RJ (1996) Nucleo-cytoplasmic redistribution of the HTLV-I Rex protein: alterations by coexpression of the HTLV-I p21x protein. Virology 220:502–507

Lam YW, Evans VC, Heesom KJ, Lamond AI, Matthews DA (2010) Proteomics analysis of the nucleolus in adenovirus-infected cells. Mol Cell Proteomics 9:117–130

Langenberg WG, Zhang L (1997) Immunocytology shows the presence of tobacco etch virus P3 protein in nuclear inclusions. J Struct Biol 118:243–247

14 Viruses and the Nucleolus 341

Lawrence FJ, McStay B, Matthews DA (2006) Nucleolar protein upstream binding factor is sequestered into adenovirus DNA replication centres during infection without affecting RNA polymerase I location or ablating rRNA synthesis. J Cell Sci 119:2621–2631

Lechertier T, Sirri V, Hernandez-Verdun D, Roussel P (2007) A B23-interacting sequence as a tool to visualize protein interactions in a cellular context. J Cell Sci 120:265–275

Lee CH, Chang SC, Chen CJ, Chang MF (1998) The nucleolin binding activity of hepatitis delta antigen is associated with nucleolus targeting. J Biol Chem 273:7650–7656

Lee TW, Blair GE, Matthews DA (2003) Adenovirus core protein VII contains distinct sequences that mediate targeting to the nucleus and nucleolus, and colocalization with human chromosomes. J Gen Virol 84:3423–3428

Lee TW, Lawrence FJ, Dauksaite V, Akusjarvi G, Blair GE, Matthews DA (2004) Precursor of human adenovirus core polypeptide Mu targets the nucleolus and modulates the expression of E2 proteins. J Gen Virol 85:185–196

Lee C, Hodgins D, Calvert JG, Welch SK, Jolie R, Yoo D (2006) Mutations within the nuclear localization signal of the porcine reproductive and respiratory syndrome virus nucleocapsid protein attenuate virus replication. Virology 346:238–250

Leenadevi T, Dalziel RG (2009) The alcelaphine herpesvirus-1 ORF 57 encodes a nuclear shuttling protein. Vet Res Commun 33:409–419

Leopardi R, Roizman B (1996) Functional interaction and colocalization of the herpes simplex virus 1 major regulatory protein ICP4 with EAP, a nucleolar-ribosomal protein. Proc Natl Acad Sci USA 93:4572–4576

Li YJ, Macnaughton T, Gao L, Lai MM (2006) RNA-templated replication of hepatitis delta virus: genomic and antigenomic RNAs associate with different nuclear bodies. J Virol 80:6478–6486

Lin NS, Hsieh CE, Hsu YH (1996) Capsid protein of cucumber mosaic virus accumulates in the nuclei and at the periphery of the nucleoli in infected cells. Arch Virol 141:727–732

Liu JL, Lee LF, Ye Y, Qian Z, Kung HJ (1997) Nucleolar and nuclear localization properties of a herpesvirus bZIP oncoprotein, MEQ. J Virol 71:3188–3196

Lopez MR, Schlegel EF, Wintersteller S, Blaho JA (2008) The major tegument structural protein VP22 targets areas of dispersed nucleolin and marginalized chromatin during productive herpes simplex virus 1 infection. Virus Res 136:175–188

Lower R, Tonjes RR, Korbmacher C, Kurth R, Lower J (1995) Identification of a Rev-related protein by analysis of spliced transcripts of the human endogenous retroviruses HTDV/HERV-K. J Virol 69:141–149

Lutz P, Kedinger C (1996) Properties of the adenovirus IVa2 gene product, an effector of late-phase-dependent activation of the major late promoter. J Virol 70:1396–1405

Lymberopoulos MH, Pearson A (2007) Involvement of UL24 in herpes-simplex-virus-1-induced dispersal of nucleolin. Virology 363:397–409

Lymberopoulos MH, Pearson A (2010) Relocalization of upstream binding factor to viral replication compartments is UL24 independent and follows the onset of herpes simplex virus 1 DNA synthesis. J Virol 84:4810–4815

Lymberopoulos MH, Bourget A, Abdeljelil NB, Pearson A (2010) Involvement of the UL24 protein in herpes simplex virus 1-induced dispersal of B23 and in nuclear egress. Virology 412:341–348

MacLean CA, Rixon FJ, Marsden HS (1987) The products of gene US11 of herpes simplex virus type 1 are DNA-binding and localize to the nucleoli of infected cells. J Gen Virol 68:1921–1937

Macnaughton TB, Gowans EJ, Reinboth B, Jilbert AR, Burrell CJ (1990) Stable expression of hepatitis delta virus antigen in a eukaryotic cell line. J Gen Virol 71:1339–1345

Matthews DA, Russell WC (1998) Adenovirus core protein V is delivered by the invading virus to the nucleus of the infected cell and later in infection is associated with nucleoli. J Gen Virol 79:1671–1675

Mayer C, Grummt I (2005) Cellular stress and nucleolar function. Cell Cycle 4:1036–1038

Mears WE, Lam V, Rice SA (1995) Identification of nuclear and nucleolar localization signals in the herpes simplex virus regulatory protein ICP27. J Virol 69:935–947

Melen K, Kinnunen L, Fagerlund R, Ikonen N, Twu KY, Krug RM, Julkunen I (2007) Nuclear and nucleolar targeting of influenza A virus NS1 protein: striking differences between different virus subtypes. J Virol 81:5995–6006

Michel MR, Elgizoli M, Dai Y, Jakob R, Koblet H, Arrigo AP (1990) Karyophilic properties of Semliki Forest virus nucleocapsid protein. J Virol 64:5123–5131

Michienzi A, Cagnon L, Bahner I, Rossi JJ (2000) Ribozyme-mediated inhibition of HIV 1 suggests nucleolar trafficking of HIV-1 RNA. Proc Natl Acad Sci USA 97:8955–8960

Miron MJ, Blanchette P, Groitl P, Dallaire F, Teodoro JG, Li S, Dobner T, Branton PE (2009) Localization and importance of the adenovirus E4orf4 protein during lytic infection. J Virol 83:1689–1699

Mohammadi H, Sharif S, Rowland RR, Yoo D (2009) The lactate dehydrogenase-elevating virus capsid protein is a nuclear-cytoplasmic protein. Arch Virol 154:1071–1080

Morency E, Coute Y, Thomas J, Texier P, Lomonte P (2005) The protein ICP0 of herpes simplex virus type 1 is targeted to nucleoli of infected cells. Brief report. Arch Virol 150:2387–2395

Mori Y, Okabayashi T, Yamashita T, Zhao Z, Wakita T, Yasui K, Hasebe F, Tadano M, Konishi E, Moriishi K, Matsuura Y (2005) Nuclear localization of Japanese encephalitis virus core protein enhances viral replication. J Virol 79:3448–3458

Munday DC, Emmott E, Surtees R, Lardeau CH, Wu W, Duprex WP, Dove BK, Barr JN, Hiscox JA (2010) Quantitative proteomic analysis of A549 cells infected with human respiratory syncytial virus. Mol Cell Proteomics 9:2438–2459

Narayan M, Younis I, D'Agostino DM, Green PL (2003) Functional domain structure of human T-cell leukemia virus type 2 rex. J Virol 77:12829–12840

Ning B, Shih C (2004) Nucleolar localization of human hepatitis B virus capsid protein. J Virol 78:13653–13668

Nosaka T, Siomi H, Adachi Y, Ishibashi M, Kubota S, Maki M, Hatanaka M (1989) Nucleolar targeting signal of human T-cell leukemia virus type I rex-encoded protein is essential for cytoplasmic accumulation of unspliced viral mRNA. Proc Natl Acad Sci USA 86:9798–9802

Okuwaki M, Iwamatsu A, Tsujimoto M, Nagata K (2001) Identification of nucleophosmin/B23, an acidic nucleolar protein, as a stimulatory factor for in vitro replication of adenovirus DNA complexed with viral basic core proteins. J Mol Biol 311:41–55

Olson MO (2009) Induction of apoptosis by viruses: what role does the nucleolus play? Cell Cycle 8:3452–3453

Orsini MJ, Debouck CM (1996) Inhibition of human immunodeficiency virus type 1 and type 2 Tat function by transdominant Tat protein localized to both the nucleus and cytoplasm. J Virol 70:8055–8063

Peeples ME, Wang C, Gupta KC, Coleman N (1992) Nuclear entry and nucleolar localization of the Newcastle disease virus (NDV) matrix protein occur early in infection and do not require other NDV proteins. J Virol 66:3263–3269

Pei Y, Hodgins DC, Lee C, Calvert JG, Welch SK, Jolie R, Keith M, Yoo D (2008) Functional mapping of the porcine reproductive and respiratory syndrome virus capsid protein nuclear localization signal and its pathogenic association. Virus Res 135:107–114

Peranen J, Rikkonen M, Liljestrom P, Kaariainen L (1990) Nuclear localization of Semliki Forest virus-specific nonstructural protein nsP2. J Virol 64:1888–1896

Pyper JM, Clements JE, Zink MC (1998) The nucleolus is the site of Borna disease virus RNA transcription and replication. J Virol 72:7697–7702

Qi D, Omarov RT, Scholthof KB (2008) The complex subcellular distribution of satellite panicum mosaic virus capsid protein reflects its multifunctional role during infection. Virology 376:154–164

Rajamaki ML, Valkonen JP (2009) Control of nuclear and nucleolar localization of nuclear inclusion protein a of picorna-like Potato virus A in Nicotiana species. Plant Cell 21:2485–2502

Realdon S, Gerotto M, Dal Pero F, Marin O, Granato A, Basso G, Muraca M, Alberti A (2004) Proapoptotic effect of hepatitis C virus CORE protein in transiently transfected cells is enhanced by nuclear localization and is dependent on PKR activation. J Hepatol 40:77–85

Reed ML, Dove BK, Jackson RM, Collins R, Brooks G, Hiscox JA (2006) Delineation and modelling of a nucleolar retention signal in the coronavirus nucleocapsid protein. Traffic 7:833–848

Reed ML, Howell G, Harrison SM, Spencer KA, Hiscox JA (2007) Characterization of the nuclear export signal in the coronavirus infectious bronchitis virus nucleocapsid protein. J Virol 81:4298–4304

Rikkonen M, Peranen J, Kaariainen L (1992) Nuclear and nucleolar targeting signals of Semliki Forest virus nonstructural protein nsP2. Virology 189:462–473

Rikkonen M, Peranen J, Kaariainen L (1994) Nuclear targeting of Semliki Forest virus nsP2. Arch Virol Suppl 9:369–377

Risco C, Menendez-Arias L, Copeland TD, Pinto da Silva P, Oroszlan S (1995) Intracellular transport of the murine leukemia virus during acute infection of NIH 3T3 cells: nuclear import of nucleocapsid protein and integrase. J Cell Sci 108:3039–3050

Rojas MR, Jiang H, Salati R, Xoconostle-Cazares B, Sudarshana MR, Lucas WJ, Gilbertson RL, (2001) Functional analysis of proteins involved in movement of the monopartite begomovirus, Tomato yellow leaf curl virus. Virology 291:110–125

Rossi JJ, June CH, Kohn DB (2007) Genetic therapies against HIV. Nat Biotechnol 25: 1444–1454

Rowland RRR, Yoo D (2003) Nucleolar-cytoplasmic shuttling of PRRSV nucleocapsid protein: a simple case of molecular mimicry or the complex regulation by nuclear import, nucleolar localization and nuclear export signal sequences. Vir Res 95:23–33

Rowland RR, Kerwin R, Kuckleburg C, Sperlich A, Benfield DA (1999) The localisation of porcine reproductive and respiratory syndrome virus nucleocapsid protein to the nucleolus of infected cells and identification of a potential nucleolar localization signal sequence. Vir Res 64:1–12

Rowland RRR, Schneider P, Fang Y, Wootton S, Yoo D, Benfield DA (2003) Peptide domains involved in the localization of the porcine reproductive and respiratory syndrome virus nucleocapsid protein to the nucleolus. Virology 316:135–145

Ruben S, Perkins A, Purcell R, Joung K, Sia R, Burghoff R, Haseltine WA, Rosen CA (1989) Structural and functional characterization of human immunodeficiency virus tat protein. J Virol 63:1–8

Ruggieri A, Maldener E, Sauter M, Mueller-Lantzsch N, Meese E, Fackler OT, Mayer J (2009) Human endogenous retrovirus HERV-K(HML-2) encodes a stable signal peptide with biological properties distinct from Rec. Retrovirology 6:17

Ryabov EV, Oparka KJ, Santa Cruz S, Robinson DJ, Taliansky ME (1998) Intracellular location of two groundnut rosette umbravirus proteins delivered by PVX and TMV vectors. Virology 242, 303–313

Ryabov EV, Kim SH, Taliansky M (2004) Identification of a nuclear localization signal and nuclear export signal of the umbraviral long-distance RNA movement protein. J Gen Virol 85:1329–1333

Sagou K, Uema M, Kawaguchi Y (2010) Nucleolin is required for efficient nuclear egress of herpes simplex virus type 1 nucleocapsids. J Virol 84:2110–2121

Salsman J, Zimmerman N, Chen T, Domagala M, Frappier L (2008) Genome-wide screen of three herpesviruses for protein subcellular localization and alteration of PML nuclear bodies. PLoS Pathog 4:e1000100

Saltarelli MJ, Schoborg R, Pavlakis GN, Clements JE (1994) Identification of the caprine arthritis encephalitis virus Rev protein and its cis-acting Rev-responsive element. Virology 199:47–55

Schoborg RV, Clements JE (1994) The Rev protein of visna virus is localized to the nucleus of infected cells. Virology 202:485–490

Scott MS, Boisvert FM, Lamond AI, Barton GJ (2011) PNAC: a protein nucleolar association classifier. BMC Genomics 12:74

Sharma P, Ikegami M (2009) Characterization of signals that dictate nuclear/nucleolar and cytoplasmic shuttling of the capsid protein of Tomato leaf curl Java virus associated with DNA beta satellite. Virus Res 144:145–153

Siomi H, Shida H, Nam SH, Nosaka T, Maki M, Hatanaka M (1988) Sequence requirements for nucleolar localization of human T cell leukemia virus type I pX protein, which regulates viral RNA processing. Cell 55:197–209

Sonntag F, Schmidt K, Kleinschmidt JA (2010) A viral assembly factor promotes AAV2 capsid formation in the nucleolus. Proc Natl Acad Sci USA 107:10220–10225

Spencer KA, Dee M, Britton P, Hiscox JA (2008) Role of phosphorylation clusters in the biology of the coronavirus infectious bronchitis virus nucleocapsid protein. Virology 370:373–381

Stow ND, Evans VC, Matthews DA (2009) Upstream-binding factor is sequestered into herpes simplex virus type 1 replication compartments. J Gen Virol 90:69–73

Strang BL, Boulant S, Coen DM (2010) Nucleolin associates with the human cytomegalovirus DNA polymerase accessory subunit UL44 and is necessary for efficient viral replication. J Virol 84:1771–1784

Szekely L, Jiang WQ, Pokrovskaja K, Wiman KG, Klein G, Ringertz N (1995) Reversible nucleolar translocation of Epstein-Barr virus-encoded EBNA-5 and hsp70 proteins after exposure to heat shock or cell density congestion. J Gen Virol 76:2423–2432

Thiry M, Lafontaine DL (2005) Birth of a nucleolus: the evolution of nucleolar compartments. Trends Cell Biol 15:194–199

Tijms MA, van der Meer Y, Snijder EJ (2002) Nuclear localization of non-structural protein 1 and nucleocapsid protein of equine arteritis virus. J Gen Virol 83:795–800

Tollefson AE, Ying B, Doronin K, Sidor PD, Wold WS (2007) Identification of a new human adenovirus protein encoded by a novel late 1-strand transcription unit. J Virol 81: 12918–12926

Trinkle-Mulcahy L, Boulon S, Lam YW, Urcia R, Boisvert FM, Vandermoere F, Morrice NA, Swift S, Rothbauer U, Leonhardt H, Lamond A (2008) Identifying specific protein interaction partners using quantitative mass spectrometry and bead proteomes. J Cell Biol 183:223–239

Tsai CW, Redinbaugh MG, Willie KJ, Reed S, Goodin M, Hogenhout SA (2005) Complete genome sequence and in planta subcellular localization of maize fine streak virus proteins. J Virol 79:5304–5314

Tsuda Y, Mori Y, Abe T, Yamashita T, Okamoto T, Ichimura T, Moriishi K, Matsuura Y (2006) Nucleolar protein B23 interacts with Japanese encephalitis virus core protein and participates in viral replication. Microbiol Immunol 50:225–234

Uchil PD, Kumar AV, Satchidanandam V (2006) Nuclear localization of flavivirus RNA synthesis in infected cells. J Virol 80:5451–5464

Uhrig JF, Canto T, Marshall D, MacFarlane SA (2004) Relocalization of nuclear ALY proteins to the cytoplasm by the tomato bushy stunt virus P19 pathogenicity protein. Plant Physiol 135:2411–2423

Unwalla HJ, Li H, Li SY, Abad D, Rossi JJ (2008) Use of a U16 snoRNA-containing ribozyme library to identify ribozyme targets in HIV-1. Mol Ther 16:1113–1119

Vazquez-Iglesias L, Lostale-Seijo I, Martinez-Costas J, Benavente J (2009) Avian reovirus sigmaA localizes to the nucleolus and enters the nucleus by a nonclassical energy- and carrier-independent pathway. J Virol 83:10163–10175

Volmer R, Mazel-Sanchez B, Volmer C, Soubies SM, Guerin JL (2010) Nucleolar localization of influenza A NS1: striking differences between mammalian and avian cells. Virol J 7:63

Waggoner S, Sarnow P (1998) Viral ribonucleoprotein complex formation and nucleolar-cytoplasmic relocalization of nucleolin in poliovirus-infected cells. J Virol 72:6699–6709

Walton TH, Moen PT, Jr, Fox E, Bodnar JW (1989) Interactions of minute virus of mice and adenovirus with host nucleoli. J Virol 63:3651–3660

Wang SH, Syu WJ, Huang KJ, Lei HY, Yao CW, King CC, Hu ST (2002) Intracellular localization and determination of a nuclear localization signal of the core protein of dengue virus. J Gen Virol 83:3093–3102

14 Viruses and the Nucleolus

Westaway EG, Khromykh AA, Kenney MT, Mackenzie JM, Jones MK (1997) Proteins C and NS4B of the flavivirus Kunjin translocate independently into the nucleus. Virology 234:31–41

Wistuba A, Kern A, Weger S, Grimm D, Kleinschmidt JA (1997) Subcellular compartmentalization of adeno-associated virus type 2 assembly. J Virol 71:1341–1352

Wright KM, Cowan GH, Lukhovitskaya NI, Tilsner J, Roberts AG, Savenkov EI, Torrance L (2010) The N-terminal domain of PMTV TGB1 movement protein is required for nucleolar localization, microtubule association, and long-distance movement. Mol Plant Microbe Interact 23:1486–1497

Wurm T, Chen H, Britton P, Brooks G, Hiscox JA (2001) Localisation to the nucleolus is a common feature of coronavirus nucleoproteins and the protein may disrupt host cell division. J Virol 75:9345–9356

Xing J, Wu F, Pan W, Zheng C (2010) Molecular anatomy of subcellular localization of HSV-1 tegument protein US11 in living cells. Virus Res 153:71–81

Yamada H, Jiang YM, Zhu HY, Inagaki-Ohara K, Nishiyama Y (1999) Nucleolar localization of the UL3 protein of herpes simplex virus type 2. J Gen Virol 80:2157–2164

Yoo D, Wootton SK, Li G, Song C, Rowland RR (2003) Colocalization and interaction of the porcine arterivirus nucleocapsid protein with the small nucleolar RNA-associated protein fibrillarin. J Virol 77:12173–12183

You J, Dove BK, Enjuanes L, DeDiego ML, Alvarez E, Howell G, Heinen P, Zambon M, Hiscox JA (2005) Subcellular localization of the severe acute respiratory syndrome coronavirus nucleocapsid protein. J Gen Virol 86:3303–3310

You JH, Howell G, Pattnaik AK, Osorio FA, Hiscox JA (2008) A model for the dynamic nuclear/nucleolar/cytoplasmic trafficking of the porcine reproductive and respiratory syndrome virus (PRRSV) nucleocapsid protein based on live cell imaging. Virology 378:34–47

Yuan X, Shan Y, Zhao Z, Chen J, Cong Y (2005) G0/G1 arrest and apoptosis induced by SARS-CoV 3b protein in transfected cells. Virol J 2:66

Zhu HY, Yamada H, Jiang YM, Yamada M, Nishiyama Y (1999) Intracellular localization of the UL31 protein of herpes simplex virus type 2. Arch Virol 144:1923–1935

Chapter 15
Assembly of Signal Recognition Particles in the Nucleolus

Marty R. Jacobson

15.1 Introduction

Ribonucleoprotein (RNP) particles are RNA–protein machines that participate in many essential cellular processes, including each step of the eukaryotic gene expression pathway (Dreyfus et al. 1988; Collins et al. 2009). Because eukaryotes synthesize RNA and protein in different cellular compartments (nucleus and cytoplasm, respectively), the assembly of RNP complexes requires, at a minimum, that at least one of these components is trafficked between cellular organelles for assembly. For many RNPs, assembly is a complex process, involving intricate pathways coordinating the temporal and spatial steps necessary for RNA maturation and the ordered assembly of protein subunits onto the RNA component (Pederson and Politz 2000; Filipowicz and Pogacic 2002; Gerbi et al. 2003; Fromont-Racine et al. 2003). These processes often involve the translocation and sequestration of subunits and/or precursors into subcellular compartments for ordered assembly.

The nucleolus has long been recognized as a site of RNA processing and RNP assembly. Most notably, the nucleolus is the site of ribosomal RNA (rRNA) synthesis (Perry 1962), pre-rRNA processing, and ribosome assembly (Warner and Soeiro 1967; Liau and Perry 1969). These ribosome biogenesis processes take place in spatially distinct subnucleolar domains, the fibrillar centers, the dense fibrillar component, and the granular component (reviewed by Gossens 1984; Shaw and Jordan 1995; Scheer and Hock 1999). Recently, roles for the nucleolus in processes other than ribosome biogenesis have emerged (Pederson 1998; Olson et al. 2000, 2002; Pederson and Politz 2000; Raska et al. 2006; Boisvert et al. 2007). In this chapter, we review the role the nucleolus plays in the biogenesis of the signal recognition particle (SRP).

M.R. Jacobson (✉)
Saccomanno Research Institute, St. Mary's Hospital & Medical Center,
Grand Junction, CO, USA
e-mail: marty.jacobson@stmarygj.org

M.O.J. Olson (ed.), *The Nucleolus*, Protein Reviews 15,
DOI 10.1007/978-1-4614-0514-6_15, © Springer Science+Business Media, LLC 2011

15.1.1 The Mammalian SRP

The SRP is a ubiquitous and abundant cytoplasmic RNA–protein (RNP) complex that, through a series of orchestrated steps, targets select translating ribosomes to the endoplasmic reticulum for the subsequent cotranslational translocation of secretory and membrane proteins (Lutcke 1995; Bui and Strub 1999; Shan and Walter 2005). The SRP accomplishes this by (1) binding the signal peptide of a nascent secretory or membrane protein as it emerges from the translating ribosome, (2) temporarily arresting or slowing nascent polypeptide elongation, and (3) delivering the SRP–ribosome–nascent polypeptide complex to the cytoplasmic side of the endoplasmic reticulum membrane for subsequent cotranslational translocation of the polypeptide.

The RNA component of the SRP (7SL or SRP RNA) is approximately 300 nucleotides in length (Walter and Blobel 1982) and contains two elements related to the human and rodent *Alu* families of interspersed repetitive DNA sequences connected by a unique sequence termed the S-domain (Li et al. 1982; Ullu et al. 1982; Walter and Blobel 1982). SRP RNA (Fig. 15.1) has an overall secondary structure that has been highly conserved during evolution (Larsen and Zwieb 1991; Strub et al. 1991; Althoff et al. 1994; Zwieb and Larsen 1997). In the mammalian SRP complex, SRP RNA associates with six proteins, SRP9, SRP14, SRP16, SRP54, SRP68, and SRP72 named according to their apparent molecular mass in kilodalton (Walter and Blobel 1980). Although the human genome contains several hundred 7SL-like sequences dispersed in human DNA, only three-to-four 7SL RNA encoding genes are present in the genome (Ullu and Weiner 1984). SRP (or 7SL) RNA is transcribed by RNA Polymerase III (Zieve et al. 1977) as a precursor containing three terminal uridylic acid residues, which are posttranscriptionally removed, followed by the addition of a single 3′-adenylic residue (Sinha et al. 1998).

Mammalian SRP was originally purified from cytoplasmic membrane fractions as an 11S RNA–protein complex (Walter and Blobel 1980). The regions of SRP RNA bound by protein were identified by mild micrococcal nuclease digestion of the complex, which cleaved the SRP into the two subparticles, the Alu domain and the S domain (Gundelfinger et al. 1983). The Alu domain subparticle consists of approximately 100 nucleotides of the 5′-end of SRP RNA base paired with approximately 50 nucleotides of the 3′-end of SRP RNA (Fig. 15.1), bound by the SRP9 and SRP14 proteins. This Alu domain comprises the translational arrest activity of the SRP (Siegel and Walter 1985). The S domain subparticle consists of approximately 150 nucleotides of the core SRP RNA sequence bound by the remaining four proteins SRP19, SRP54, SRP68, and SRP72 (Siegel and Walter 1986; Walter and Johnson 1994). Signal sequence recognition, mediated by SRP54, and protein translocation, mediated by SRP68/SRP72, activities reside within the S domain subparticle (Krieg et al. 1986; Kurzchalia et al. 1986; Siegel and Walter 1986, 1988; Wiedmann et al. 1987).

Purified SRP can be readily disassembled in vitro into its constituents (SRP RNA, SRP19, SRP54, and the two heterodimeric protein complexes SRP9/14 and

15 Assembly of Signal Recognition Particles in the Nucleolus

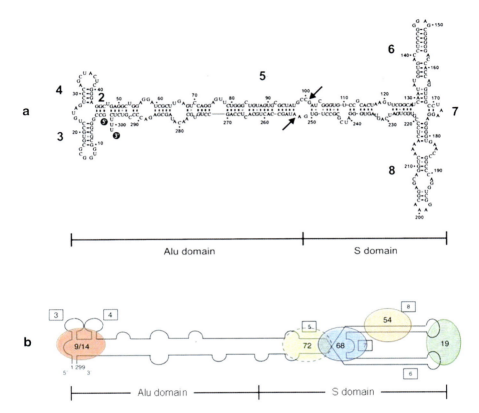

Fig. 15.1 (**a**) Human signal recognition particle (SRP) RNA. Regions denoted by the numerals 2–8 follow the nomenclature of Larsen and Zwieb (1991); helix 1 is present only in archebacterial SRP RNA. The *two arrows* denote sites at which the SRP is cleaved into two particles by mild micrococcal nuclease digestion (Gundelfinger et al. 1983) (reprinted from Jacobson and Pederson (1998) with permission. Copyright (1998) National Academy of Sciences, USA). (**b**) Schematic representation of mammalian SRP. RNA is represented as a *black line*. The *boxed* numbers represent helices following the nomenclature in (**a**). SRP proteins are shown in *ovals*. The Alu and S domains are indicated

SRP68/72), and then reassembled into a functional RNP (Walter and Blobel 1983). Furthermore, SRP protein subunits expressed in vitro from synthetic mRNAs derived from their cDNA clones are also capable of assembling onto SRP RNA and forming a functional SRP (Romisch et al. 1990; Strub and Walter 1990; Zopf et al. 1990; Lutcke et al. 1993; Lingelbach et al. 1998). As recently reviewed by Menichelli and Nagai (2009), the ability to reassemble SRP in vitro has allowed investigators the opportunity to extensively study the dynamic biochemical and biophysical protein–protein and RNA–protein interactions involved in the ordered assembly of a functional SRP complex. In vitro assembly of the Alu domain requires prior heterodimerization of SRP9 and SRP14 before assembly onto SRP RNA (Strub and Walter 1990). Interestingly, accurate 3′-end processing of SRP RNA in vitro also requires prior binding of the SRP9/14 heterodimer to pre-SRP RNA (Chen et al. 1998; Sinha et al. 1999), suggesting a role for Alu domain assembly in SRP RNA processing. In vitro assembly of the S-domain involves the binding of the SRP68/72 heterodimer to the three-way junction formed by helices 5, 6, 7, and 8 of SRP RNA (Fig. 14.1) and the SRP19 subunit to the tips of helices 6 and 8 (Siegel and Walter 1988; Yin et al. 2007). Although SRP68/72 heterodimer disassembled from the SRP remains a stable heterodimer complex (Walter and Blobel 1983; Scoulica et al. 1987), SRP68 and SRP72 expressed in vitro associate inefficiently with each other in the absence of SRP RNA (Lutcke et al. 1993) suggesting prior heterodimerization of the SRP68/72 complex in the absence of SRP RNA does not occur. The assembly of SRP68 and SRP72 onto SRP RNA is a multistep process wherein binding site specificity is provided by SRP68 binding to the three-way junction of helices 5, 6, and 8 of SRP RNA with subsequent SRP72 binding increasing the overall stability of the complex (Maity et al. 2008). SRP54 is the last protein to assemble onto the SRP complex and its correct association requires, at a minimum, prior SRP19 assembly (Yin et al. 2001; Maity and Weeks 2007). In vitro data suggests that the combined addition of SRP68/72 and SRP19 causes a major reorganization of SRP RNA into an SRP54 binding competent state (Menichelli et al. 2007).

Assembly of RNP complexes is generally thought to occur in an ordered and cooperative manner. In vitro studies evaluating the assembly of the SRP S-domain suggest that in the absence of spatial or temporal mechanisms for controlling subunit availability, altered nonnative complexes can form. Binding of SRP19 to SRP RNA is known to be a sufficiently slow, multistep process (Rose and Weeks 2001; Maity et al. 2006) such that if SRP19 and SRP54 are allowed simultaneous access to SRP RNA, the formation of the normal (native) SRP19-SRP RNA complex is prevented and an altered, nonnative SRP19-SRP RNA structure is formed (Maity et al. 2006; Maity and Weeks 2007). Additionally, although both SRP19 and SRP68/72 contribute to the formation of a competent SRP54 binding platform on SRP RNA, prior binding of either subunit to SRP RNA in vitro diminishes the rate of binding of the other subunit suggesting that SRP19 and SRP68/72 assembly onto SRP RNA may be moderately anti-cooperative (Maity et al. 2008). These observations have possible implications on in vivo SRP assembly and emphasize the need for mechanisms in eukaryotic cells that provide the temporal and/or spatial separation required for accurate and ordered assembly of the SRP.

15.2 Nucleolar Involvement in SRP Biogenesis

The nucleolus is perhaps the most prominent structure in the cell nucleus. Over four decades ago, the nucleolus was established as the site of ribosome biogenesis (Perry 1962; Warner and Soeiro 1967; Liau and Perry 1969). Recently, it has become clear that additional RNP related functions besides ribosome biogenesis also reside in the nucleolus (Pederson 1998; Olson et al. 2000, 2002; Pederson and Politz 2000; Lueng and Lamond 2003). Fluorescent RNA cytochemistry (Wang et al. 1991;

15 Assembly of Signal Recognition Particles in the Nucleolus

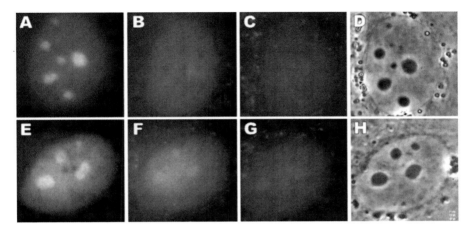

Fig. 15.2 Rapid nucleolar localization of fluorescent SRP RNA after microinjection into the nucleus of NRK cells. Two representative experiments are shown. Fluorescence at 3 min (**a**), 26 min (**b**), and 66 min (**c**) after nuclear microinjection. (**d**) Phase-contrast micrograph of cell in (**a–c**) taken at 66 min. Fluorescence at 6 min (**e**), 33 min (**f**), and 56 min (**g**) after nuclear microinjection. (**h**) Phase-contrast image of cell in (**e, f**) taken at 56 min. Note the early, transient localization in nucleoli (**a, e**) and subsequent cytoplasmic appearance (**b, c** and **f, g**) (photomicrographs reprinted from Jacobson and Pederson (1998) with permission. Copyright (1998) National Academy of Sciences, USA)

Jacobson and Pederson 1997), the microinjection of fluorescently-labeled RNA into living cells is a method that allows the direct visualization of dynamic RNA movement in living cells. Microinjection of fluorescently labeled SRP RNAs into the nucleus of mammalian cells gave the first indication that the nucleolus might also be involved in the biogenesis of the SRP (Jacobson and Pederson 1998).

15.2.1 SRP Subunits Localize in Nucleoli

When fluorescently tagged SRP RNA was microinjected into the nucleus of mammalian cells, it very rapidly localized in nucleoli (Jacobson and Pederson 1998). After this initial nucleolar localization, the fluorescent SRP RNA signal progressively decreased in nucleoli and increased at discrete sites in the cytoplasm (Fig. 15.2). The biological relevance of this observation was indicated by the finding that this rapid nucleolar localization required discrete domains within SRP RNA consisting of the Alu domain and helix 8 (Fig. 15.3). Interestingly, Alu domain and S-domain RNAs are both capable of rapidly localizing to nucleoli (Fig. 15.3a, b, respectively) independent of the other domain. These observations were consistent with findings that a small portion of SRP RNA biochemically fractionates with purified nucleoli (Reddy et al. 1981; Mitchell et al. 1999). Subsequent in situ hybridization

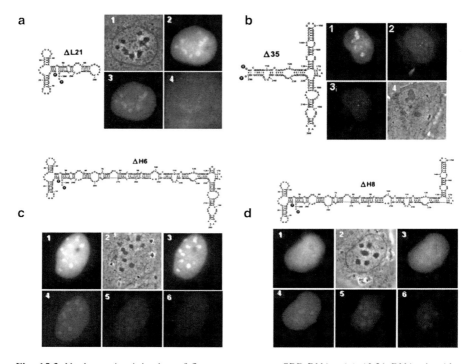

Fig. 15.3 Nuclear microinjection of fluorescent mutant SRP RNAs. (a) ΔL21 RNA, the Alu domain SRP9/14 binding domain. (a1) Phase-contrast image taken at 1 min after nuclear microinjection. Fluorescence at 1 min (a2), 3 min (a3) and 16 min (a4). (b) Δ35 RNA, the SRP S domain, SRP68/72, SRP19, and SRP54 binding domain. Fluorescence at 30 s (b1), 3 min (b2), and 10 min (b3) after nuclear microinjection. (b4) Phase-contrast at 19 min. (c) ΔH6 RNA, SRP RNA lacking helix 6. Fluorescence at 30 s (c1), 3 min (c3), 12 min (c4), 26 min (c5), and 43 min (c6) after nuclear microinjection. (c2) Phase-contrast at 1 min. (d) ΔH8 RNA, SRP RNA lacking helix 8. Fluorescence at 30 s (d1), 3 min (d3), 13 min (d4), 25 min (d5), and 43 min (d6) after nuclear microinjection. (d2) Phase-contrast at 1 min (RNA diagrams and photomicrographs reprinted from Jacobson and Pederson (1998) with permission. Copyright (1998) National Academy of Sciences, USA)

studies (Politz et al. 2000, 2002) confirmed the presence of a portion of endogenous SRP RNA in the nucleoli of mammalian cells (Fig. 15.4) and *Xenopus* oocytes (Sommerville et al. 2005).

The nucleolus is composed of three domains: the fibrillar centers, the dense fibrillar component, and the granular component (see Gossens 1984; Shaw and Jordan 1995; Scheer and Hock 1999). Analysis of endogenous SRP RNA distribution within the mammalian nucleolus revealed very little SRP RNA present in fibrillar centers and the dense fibrillar component, sites of rRNA synthesis, processing, and nascent ribosome assembly (Politz et al. 2002). As shown in the Fig. 15.5c, only minimal colocalization (as indicated by yellow signal) exists between SRP

15 Assembly of Signal Recognition Particles in the Nucleolus 353

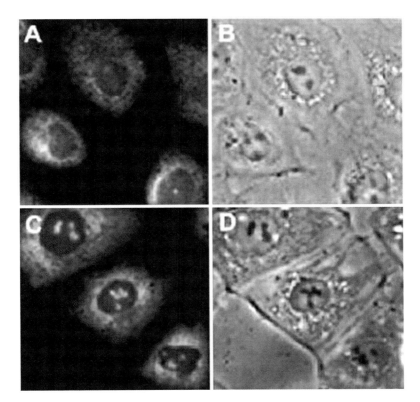

Fig. 15.4 In situ hybridization of endogenous SRP RNA in NRK cells. Hybridization using standard (PO backbone) oligodeoxynucleotides (**a, b**) or peptide nucleic acid (PNA) probes (**c, d**) complementary to human SRP RNA. Fluorescence images (**a, c**). Phase-contrast images (**b, d**). Note that nucleolar signal is present using PO probes but at a very low level of detection, and that the PNA probe enhances SRP RNA detection (photomicrographs reprinted from Politz et al. (2002) with permission from: "©The Rockefeller University Press")

RNA and the granular component marker B23 protein. Using constrained iterative deconvolution to increase the resolution of subnucleolar regions, Politz et al. (2002) further demonstrated that the most concentrated subnucleolar regions of SRP RNA often do not overlap with the most concentrated subnucleolar regions of B23 protein (Fig. 15.5d–h). Similar results were observed for SRP RNA and 28S rRNA colocalization (Fig. 15.5i–m).

Although some SRP RNA is present in the granular component (Fig. 15.5) where late stage ribosomal subunit maturation occurs, a substantial portion of the nucleolar localized SRP RNA is present in what appears to be a previously unidentified region, which extends throughout the nucleolus lacking 28S rRNA and markers for the three classical nucleolar domains (Politz et al. 2002).

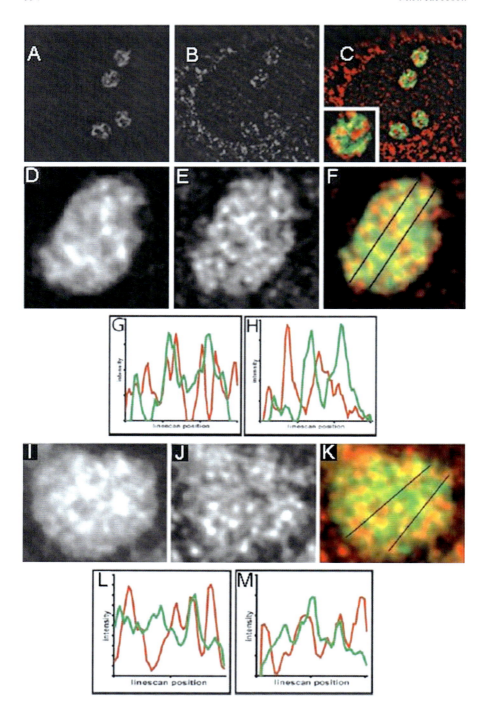

15.2.2 SRP Protein Localization

In yeast, all of the SRP protein subunits with the exception of the SRP54 homologue can be found in nucleoli (Grosshans et al. 2001). In mammalian cells expressing green fluorescent protein (GFP) fusions of SRP19, SRP68, or SRP72 (Politz et al. 2000), the intracellular distribution of each of these SRP-fusion proteins consisted of cytoplasmic, nuclear, and substantial nucleolar localization (Fig. 15.6a–c, e–g). In contrast, GFP-tagged SRP54 (Fig. 15.6d, h), as well as endogenous SRP54, was only detected in the cytoplasm (Politz et al. 2000). These observations are consistent with S-domain in vitro assembly data (Maity et al. 2008) that suggests temporal or spatial compartmentalization is important for ordered SRP19 and SRP54 assembly.

The SRP14 protein subunit has also been identified in nucleoli isolated from cultured human cells (Andersen et al. 2002). The presence of SRP14 in nucleoli coupled with in vitro findings that heterodimerization precedes SRP9/14 assembly onto SRP RNA, and that pre-SRP RNA processing requires SRP9/14 binding suggests that SRP9 is also present in nucleoli (although this has yet to be demonstrated in mammalian cells).

The presence of SRP RNA and the majority of the SRP proteins colocalized in nucleoli strongly suggest that the nucleolus provides part of the compartmentalization necessary for the ordered assembly of the SRP. The rapid nucleolar localization observed after fluorescently-labeled SRP RNA is microinjected into the nucleus of mammalian cells might also suggest that SRP RNA processing occurs within the nucleolus. Alternatively, assembly of the SRP9/14 heterodimer onto SRP RNA with subsequent 3′-end pre-SRP RNA maturation could occur in the nuclear space outside of the nucleolus followed by rapid nucleolar uptake for further RNP assembly. Upon completion of the nucleolar phase of SRP biogenesis, the partially assembled SRP is exported to the cytoplasm for final SRP54 assembly.

Fig. 15.5 Subnucleolar localization of SRP RNA. (**a–c**) SRP RNA signal colocalizes with a portion of the nucleolar granular component. Nearest-neighbor deconvolution of (**a**) GFP-B23 and (**b**) SRP RNA. (**c**) Merged image of (**a**, **b**). *Inset* in (**c**) is a single nucleolus at higher magnification. (**d–h**) Constrained iterative deconvolution of (**d**) GFP-B23 protein and (**e**) SRP RNA signal in a single nucleolus. (**f**) Overlay of (**d**) (*green*) and (**e**) (*red*) images showing regions of similar intensity in *yellow*. (**g**) Linescan of the *left line* in (**f**). (**h**) Linescan of the *right line* in (**f**). Each linescan shows, from *left* to *right* the intensities (arbitrary units) of B23 (*green*) and SRP RNA (*red*) along the line indicated in (**f**), proceeding downward from the top to bottom. Linescans are displayed with the minimal linescan intensity at the origin of the y-axis. (**i–m**) SRP RNA and 28S rRNA distribution in the nucleolus. Deconvolved images of endogenous (**i**) 28S rRNA and (**j**) SRP RNA in a single nucleolus of an NRK cell. (**k**) Color combined images of (**i**) (*green*) and (**j**) (*red*). (**l**) Linescan of the *left line* in (**k**). (**m**) Linescan of the *right line* in (**k**). Each linescan (**l**, **m**) shows, from *left* to *right* the intensities (arbitrary units) of 28S rRNA (*green*) and SRP RNA (*red*) along the line indicated in (**k**), proceeding downward from the top to bottom. Linescans are displayed with the minimal linescan intensity at the origin of the y-axis (photomicrographs reprinted from Politz et al. (2002) with permission from: "©The Rockefeller University Press")

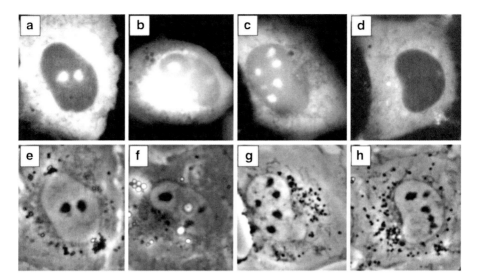

Fig. 15.6 Localization of SRP-GFP fusion proteins expressed in NRK cells. (**a, e**) EGFP-SRP72 protein, (**b, f**) EGFP-SRP68 protein (**c, g**), EGFP-SRP19 protein, and (**d, h**) EGFP-SRP54 protein. Fluorescence images (**a–d**). Phase-contrast images (**e–h**) (photomicrographs reprinted from Politz et al. (2000) with permission. Copyright (2000) National Academy of Sciences, USA)

15.2.3 Nucleolar Export

Nuclear export of the mammalian SRP is dependent on the CRM1 export pathway as well as the guanine nucleotide exchange factor of Ran, RanGEF (Alavian et al. 2004). Inhibition of either CRM1 or RanGEF activity in mammalian cells results in an overall increase in the level of SRP RNA observed in nucleoli. In yeast, nuclear export of SRP to the cytoplasm requires (1) a partially assembled SRP consisting of scR1 RNA associated with the core proteins Srp14p, Srp68p, Srp72p, and Srp21p and (2) NES pathway components, Xpo1p/Crm1p and Yrb2p (Ciufo and Brown 2000). Interestingly, nuclear export of the mammalian 40S and 60S ribosomal subunits (Thomas and Kutay 2003; Trotta et al. 2003) and the yeast 60S ribosomal subunit (Ho et al. 2000; Gadal et al. 2001) is also mediated through the CRM1 export pathway.

15.3 Summary

From the existing in vitro and in vivo data, a model for SRP biogenesis involving nucleolar compartmentalization is emerging. The SRP protein subunits are translated in the cytoplasm and the SRP9, SRP14, SRP19, SRP68, and SRP72 protein subunits are subsequently transported into the nucleus while the SRP54 protein

15 Assembly of Signal Recognition Particles in the Nucleolus

subunit remains partitioned in the cytoplasm. Although SRP9 and SRP14 form a stable heterodimer in the absence of SRP RNA, it is currently unknown whether dimerization occurs in the cytoplasm before nuclear uptake or whether the SRP9 and SRP14 subunits are first imported into the nucleus and then form the SRP9/14 heterodimer. Data from yeast and expression of SRP-GFP fusion proteins in mammalian cells suggest that most if not all of the SRP protein subunits present in the nucleus are further sequestered into nucleoli for staging, RNP assembly, or both.

The SRP RNA is transcribed by RNA polymerase III as a precursor in the nucleus from one-to-four SRP RNA encoding genes. Three terminal uridylic acid residues present on human pre-SRP RNA are posttranscriptionally removed (3′-end processed) and a single adenylic acid residue is subsequently added. Prior binding of the SRP9/14 heterodimer to nascent pre-SRP RNA is required for posttranscriptional adenylation (and perhaps 3′-end processing) of SRP RNA. Following these very early RNP assembly and RNA processing steps in SRP biogenesis, the SRP RNA-SRP9/14 heterodimer complex becomes localized in the nucleolus for further RNP assembly, presumably within a possibly novel subnucleolar domain closely associated with the granular component but lacking both B23 protein and 28S rRNA. The remaining nucleolar sequestered SRP proteins (SRP19, SRP68, and SRP72) are next sequentially assembled onto the S-domain of the SRP RNA-SRP9/14 complex, likely with subnucleolar compartmentalization controlling the ordered assembly of these protein subunits. The partially assembled SRP (containing SRP RNA, SRP19, and the SRP9/14 and SRP68/72 heterodimers) is next exported to the cytoplasm via the CRM1 export pathway where SRP54 is assembled onto the RNP complex creating a functional SRP.

References

Alavian CN, Ritland Politz JC, Lewandowski LB, Powers CM, Pederson T (2004) Nuclear export of signal recognition particle RNA in mammalian cells. Biochem Biophys Res Commun 313: 351–355

Althoff S, Selinger D, Wise JA (1994) Molecular evolution of SRP cycle components: functional implications. Nucleic Acids Res 22:1933–1947

Andersen JS, Lyon CE, Fox AH, Leung AKL, Lam YW, Steen H, Mann M, Lamond AI (2002) Directed proteomic analysis of human nucleolus. Curr Biol 12:1–11

Boisvert FM, van Koingsbruggen S, Navascues J, Lamond AI (2007) The multifunctional nucleolus. Nat Rev Mol Cell Biol 8:574–585

Bui N, Strub K (1999) New insights into signal recognition and elongation arrest activities of the signal recognition particle. Biol Chem 380:135–145

Chen Y, Sinha K, Perumal K, Gu J, Reddy R (1998) Accurate 3′ end processing and adenylation of human signal recognition particle RNA and Alu RNA in vitro. J Biol Chem 273:35023–35031

Ciufo LF, Brown JD (2000) Nuclear export of yeast signal recognition particle lacking Srp54p by the Xpo1p/Crm1p NES-dependent pathway. Curr Biol 10:1256–1264

Collins LJ, Kurland CG, Biggs P, Penny D (2009) The modern RNP world of eukaryotes. J Hered 100:597–604

Dreyfus G, Philipson L, Mattaj IW (1988) Ribonucleoprotein particles in cellular processes. J Cell Biol 106:1419–1425

Filipowicz W, Pogacic V (2002) Biogenesis of small nucleolar ribonucleoproteins. Curr Opin Cell Biol 14:319–327

Fromont-Racine M, Senger B, Savcanu CFF (2003) Ribosome assembly in eukaryotes. Gene 313:17–42

Gadal O, Strauss D, Kessl J, Truppower B, Tollervey D, Hurt E (2001) Nuclear export of 60S ribosomal subunits depends on Xpo1p and requires a nuclear export sequence-containing factor, Nmd3p, that associates with the large subunit protein Rpl10p. Mol Cell Biol 21:3405–3415

Gerbi SA, Borovjagin AV, Lang TS (2003) The nucleolus: a site of ribonucleoprotein maturation. Curr Opin Cell Biol 15:318–325

Gossen G (1984) Nucleolar structure. Int Rev Cytol 87:107–158

Grosshans H, Deinert K, Hurt E, Simos G (2001) Biogenesis of the signal recognition particle (SRP) involves import of SRP proteins into the nucleolus, assembly with the SRP-RNA, and Xpo1p-mediated export. J Cell Biol 153:745–762

Gundelfinger ED, Krause E, Melli M, Dobberstein B (1983) The organization of the 7SL RNA component of the signal recognition particle. Nucleic Acids Res 11:7362–7374

Ho JH, Kallstrom G, Johnson AW (2000) Nmd3p is a Crm1p-dependent adapter protein for nuclear export of the large ribosomal subunit. J Cell Biol 151:1057–1066

Jacobson MR, Pederson T (1997) RNA traffic and localization reported by fluorescent molecular cytochemistry in living cells. In: Richter JD (ed) mRNA formation and function. Academic, New York, pp 341–359

Jacobson MR, Pederson T (1998) Localization of signal recognition particle RNA in the nucleolus of mammalian cells. Proc Natl Acad Sci USA 95:7981–7986

Krieg UC, Walter P, Johnson AE (1986) Photocrosslinking of the signal sequence of nascent preprolactin to the 54-kilodalton polypeptide of the signal recognition particle. Proc Natl Acad Sci USA 83:8604–8608

Kurzchalia TV, Wiedmann M, Grischovich AS, Bochkareva ES, Bielka H, Rapoport TA (1985) The signal sequence of nascent preprolactin interacts with the 54K polypeptide of the signal recognition particle. Nature 320:634–636

Larsen N, Zwieb C (1991) SRP-RNA sequence alignment and secondary structure. Nucleic Acids Res 19:209–215

Li WY, Reddy R, Henning D, Epstein P, Busch H (1982) Nucleotide sequence of 7S RNA: homology to ALU DNA and LA 4.5S RNA. J Biol Chem 257:5136–5142

Liau MC, Perry RP (1969) Ribosome precursor particles in nucleoli. J Cell Biol 42:272–283

Lingelbach K, Zwieb C, Webb JR, Marshallsay C, Hoben PJ, Walter P, Doggerstein B (1998) Isolation and characterization of a cDNA clone encoding the 19 kDa protein of signal recognition particle (SRP): expression and binding to 7SL RNA. Nucleic Acids Res 16:9431–9442

Lueng AKI, Lamond AI (2003) The dynamics of the nucleolus. Crit Rev Eukaryot Gene Expr 13:39–54

Lutcke H (1995) Signal recognition particle (SRP), a ubiquitous initiator of protein translocation. Eur J Biochem 228:531–550

Lutcke H, Prehn S, Ashford AJ, Remus M, Frank R, Dobberstein B (1993) Assembly of the 68- and 72-kD proteins of signal recognition particle with 7S RNA. J Cell Biol 121:977–985

Maity TS, Weeks KM (2007) A three-fold RNA-protein interface in the signal recognition particle gates native complex assembly. J Mol Biol 369:512–524

Maity TS, Leonard CW, Rose MA, Fried HM, Weeks KM (2006) Compartmentalization directs assembly of the signal recognition particle. Biochemistry 45:14955–14966

Maity TS, Fried HM, Weeks KM (2008) Anti-cooperative assembly of the SRP19 and SRP68/72 components of the signal recognition particle. Biochem J 415:429–437

Menichelli E, Nagai K (2009) Assembly of the human signal recognition particle. In: Walter NG, Woodson SA, Batey RT (eds) Non-protein coding RNAs. Springer, Berlin, pp 273–284

Menichelli E, Isel C, Ougridge C, Nagai K (2007) Protein-induced conformational changes of RNA during the assembly of human signal recognition particle. J Mol Biol 367:187–203

Mitchell JR, Cheng J, Collins K (1999) A box H/ACA small nucleolar RNA-like domain at the human telomerase RNA 3′end. Mol Cell Biol 19:567–576

Olson MO, Dundr M, Szebeni A (2000) The nucleolus: an old factory with unexpected capabilities. Trends Cell Biol 10:189–196

Olson MO, Hingorani K, Szebeni A (2002) Conventional and nonconventional roles of the nucleolus. Int Rev Cytol 219:199–266

Pederson T (1998) The plurifunctional nucleolus. Nucleic Acids Res 26:3871–3876

Pederson T, Politz JC (2000) The nucleolus and the four ribonucleoproteins of translation. J Cell Biol 148:1091–1095

Perry RP (1962) The cellular sites of synthesis of ribosomal and 4S RNA. Proc Natl Acad Sci USA 48:2179–2186

Politz JC, Yarovoi KSM, Gowda K, Zwieb C, Pederson T (2000) Signal recognition particle components in the nucleolus. Proc Natl Acad Sci USA 97:55–60

Politz JC, Lewandowski LB, Pederson T (2002) Signal recognition particle RNA localization within the nucleolus differs from the classical sites of ribosome synthesis. J Cell Biol 159:411–418

Raska I, Shaw PJ, Cmarko D (2006) New insights into nucleolar architecture and activity. Int Rev Cytol 255:177–235

Reddy R, Li WY, Henning D, Choi YC, Nohga K, Busch H (1981) Characterization and subcellular localization of 7-8S RNAs of Novikoff hepatoma. J Biol Chem 256:8452–8457

Romisch K, Webb J, Lingelbach K, Gausepohl H, Dobberstein B (1990) The 54-kD protein of signal recognition particle contains a methionie-rich RNA binding domain. J Cell Biol 111:1793–1802

Rose MA, Weeks KM (2001) Visualizing induced fit in early assembly of the human signal recognition particle. Nat Struct Biol 8:515–520

Scheer U, Hock R (1999) Structure and function of the nucleolus. Curr Opin Cell Biol 11:385–390

Scoulica E, Krause E, Meese K, Dobberstein B (1987) Disassembly and domain structure of the proteins in the signal-recognition particle. Eur J Biochem 163:519–528

Shan SO, Walter P (2005) Co-translational protein targeting by the signal recognition particle. FEBS Lett 579:921–926

Shaw PJ, Jordan EG (1995) The nucleolus. Ann Rev Cell Dev Biol 11:93–121

Siegel V, Walter P (1985) Elongation arrest is not a prerequisite for secretory protein translocation across the microsomal membrane. J Cell Biol 100:1913–1921

Siegel V, Walter P (1986) Removal of the Alu structural domain from signal recognition particle leaves its protein translocation activity intact. Nature 320:81–84

Siegel V, Walter P (1988) Binding of the 19-kDa and 68/72-kDa signal recognition particle (SRP) proteins on SRP RNA as determined by protein-RNA "footprinting". Proc Natl Acad Sci USA 85:1801–1805

Sinha KM, Gu J, Chen Y, Reddy R (1998) Adenylation of small RNAs in human cells. J Biol Chem 273:6853–6859

Sinha K, Perumal K, Chen Y, Reddy R (1999) Post-translational adenylation of signal recognition particle RNA is carried out by an enzyme different from mRNA poly(A) polymerase. J Biol Chem 274:30826–30831

Sommerville J, Brumwell CL, Ritland Politz JC, Pederson T (2005) Signal recognition particle assembly in relation to the function of amplified nucleoli of *Xenopus* oocytes. J Cell Sci 118:1299–1307

Strub K, Walter P (1990) Assembly of the Alu domain of the signal recognition particle (SRP): dimerization of the two protein components is required for efficient binding to SRP RNA. Mol Biol Cell 10:777–784

Strub K, Moss J, Walter P (1991) Binding sites of the 9- and 14-kilodalton heterodimeric protein subunit of the signal recognition particle (SRP) are contained exclusively in the Alu domain of SRP RNA and contain a sequence motif that is conserved in evolution. Mol Cell Biol 11:3949–3959

Thomas F, Kutay U (2003) Biogenesis and nuclear export of ribosomal subunits in higher eukaryotes depends on the CRM1 export pathway. J Cell Sci 116:2409–2419

Trotta C, Lund E, Kahan L, Johnson AW, Dahlberg JE (2003) Coordinated nuclear export of 60S ribosomal subunits and NMD3 in vertebrates. EMBO J 22:2841–2851

Ullu E, Weiner AM (1984) Human genes and pseudogenes for the 7SL RNA component of the signal recognition particle. EMBO J 3:3303–3310

Ullu E, Murphy S, Melli M (1982) Human 7SL RNA consists of a 140 nucleotide middle-repetitive sequence inserted in an Alu sequence. Cell 29:195–202

Walter P, Blobel G (1980) Purification of a membrane-associated protein complex required for protein translocation across the endoplasmic reticulum. Proc Natl Acad Sci USA 77:7112–7116

Walter P, Blobel G (1982) Signal recognition particle contains a 7S RNA essential for protein translocation across the endoplasmic reticulum. Nature 299:691–698

Walter P, Blobel G (1983) Disassembly and reconstitution of signal recognition particle. Cell 34:525–533

Walter P, Johnson AE (1994) Signal sequence recognition and protein targeting to the endoplasmic reticulum membrane. Annu Rev Cell Biol 10:87–119

Wang J, Cao LG, Wang YL, Pederson T (1991) Localization of pre-messenger RNA at discrete nuclear sites. Proc Natl Acad Sci USA 88:7391–7395

Warner JR, Soeiro R (1967) Nascent ribosomes from HeLa cells. Proc Natl Acad Sci USA 58:1984–1990

Wiedmann M, Kuzchalia TV, Bielka H, Rapoport TA (1987) Direct probing of the interaction between the signal sequence of nascent preprolactin in the signal recognition particle by specific cross-linking. J Cell Biol 104:201–208

Yin J, Yang CH, Zwieb C (2001) Assembly of the human signal recognition particle (SRP): overlap of regions required for binding of protein SRP54 and assembly control. RNA 7:1389–1396

Yin J, Iakhiaeva E, Menichelli E, Zwieb C (2007) Identification of the RNA binding regions of SRP68/72 and SRP72 by systematic mutagenesis of human SRP RNA. RNA Biol 4:154–159

Zieve G, Benecke BJ, Penman S (1977) Synthesis of two classes of small RNA species in vivo and in vitro. Biochemistry 16:4520–4525

Zopf D, Bernstein HD, Johnson AE, Walter P (1990) The methionine-rich domain of the 54 kd protein subunit of the signal recognition particle contains an RNA binding site and can be crosslinked to a signal sequence. EMBO J 9:4511–4517

Zwieb C, Larsen N (1997) The signal recognition particle database (SRPDB). Nucleic Acids Res 25:107–108

Chapter 16
Relationship of the Cajal Body to the Nucleolus

Andrew Gilder and Michael Hebert

16.1 Introduction

The cytoplasm of eukaryotic cells contains various domains and compartments that provide platforms for specific cellular functions. This functional organization is also present in the nucleus, which contains a menagerie of distinct subcompartments. In contrast to their cytoplasmic brethren, nuclear bodies lack a lipid membrane but are nonetheless structurally stable, and yet, paradoxically, highly dynamic (Matera et al. 2009; Dundr and Misteli 2010). Among the nuclear bodies, the nucleolus is the most easily observable and has been studied for the longest time (since at least 1835), revealing its incontrovertible role in the formation of ribosomal subunits (Pederson 2010). In 1903, Santiago Ramón y Cajal developed a silver staining method that enabled him to detect what he termed the "accessory body" of the nucleolus (Ramón y Cajal 1903). To honor Cajal, this structure is now known as the Cajal body (CB) (Gall et al. 1999). It is likely that Cajal initially called the CB the "accessory body" of the nucleolus because the same staining technique can detect the CB and the nucleolus, rather than implying that these structures were physically associated (Lafarga et al. 2009). Nevertheless, CBs and nucleoli can be physically associated, implying a functional dialogue exists between the two (Gall 2003). In support of this idea, some proteins localize to both nucleoli and CBs (Table 16.1), arguing that their function may take place in both compartments or that the CB may participate in their biogenesis.

Studies of CBs over the last 15 years have demonstrated a role for this structure in ribonucleoprotein (RNP) biogenesis, histone pre-mRNA processing, and telomerase assembly (Gall 2003; Cioce and Lamond 2005; Morris 2008; Matera et al.

M. Hebert (✉)
Department of Biochemistry, The University of Mississippi Medical Center,
Jackson, MS, USA
e-mail: mhebert@umc.edu

M.O.J. Olson (ed.), *The Nucleolus*, Protein Reviews 15,
DOI 10.1007/978-1-4614-0514-6_16, © Springer Science+Business Media, LLC 2011

Table 16.1 Proteins that localize to both CBs and nucleoli

Name	Function	References
NAP57	rRNA processing, implicated in dyskeratosis congenita	Meier and Blobel (1994), Wang and Meier (2004), and Heiss et al. (1998)
EST1A	Telomere replication; part of the dyskerin complex required for CB localization and synthesis of telomeres; implicated in dyskeratosis congenita	Snow et al. (2003) and Venteicher et al. (2009)
Fibrillarin	pre-rRNA processing; interacts with SMN	Tollervey et al. (1993) and Pellizzoni et al. (2001b)
FRG1	Involved in pre-mRNA splicing and facioscapulohumeral muscular dystrophy	Mourelatos et al. (2002), van Koningsbruggen et al. (2004), and Gabellini et al. (2006)
GAR1	H/ACA small nucleolar ribonucleoprotein (H/ACA snoRNP) complex protein involved in ribosome biogenesis and telomere maintenance	Yang et al. (2000), Wang and Meier (2004), Pellizzoni et al. (2001a), and Venteicher et al. (2009)
Gemin4	Part of the SMN complex that functions to assemble Sm proteins on snRNA for pre-mRNA splicing, associated with spinal muscular atrophy	Charroux et al. (2000), Feng et al. (2005), and Shpargel and Matera (2005a)
NHP2	H/ACA snoRNP complex protein involved in ribosome biogenesis and telomerase component linked to dyskeratosis congenita	Henras et al. (1998), Watkins et al. (1998), Yang et al. (2000), Vulliamy et al. (2008), and Venteicher et al. (2009)
NOP10	H/ACA snoRNP complex protein involved in ribosome biogenesis and telomerase component linked to dyskeratosis congenita	Henras et al. (1998), Wang and Meier (2004), and Venteicher et al. (2009)
NOP58	Component of the box C/D snoRNPs, required for pre-18S rRNA processing in yeast	Lyman et al. (1999), Matunis et al. (1998), Gautier et al. (1997), and Bleichert and Baserga (2010)
RAD52	dsDNA break repair	New et al. (1998) and Van Dyck et al. (1999)
NEDD-5 (Septin2)	Cytoskeletal GTPase that maintains actin cytoskeleton; required for mitosis	Spiliotis et al. (2005) and Andersen et al. (2005)
HEST2	Telomerase reverse transcriptase	Nugent and Lundblad (1998)
EXP1	U3 snoRNA transport from CBs to nucleoli	Boulon et al. (2004)
Nopp140	Chaperones protein transport between CBs and nucleoli; possible role in Pol I transcription	Isaac et al. (1998) and Chen et al. (1999)

16 Relationship of the Cajal Body to the Nucleolus

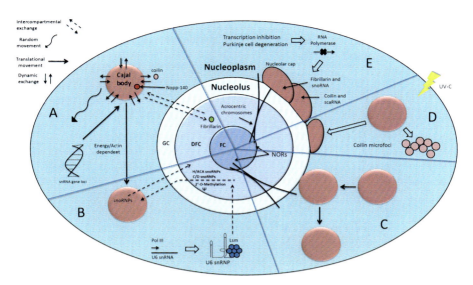

Fig. 16.1 The relationship between CBs and the nucleolus. (**a**) CBs dynamically exchange their contents such as coilin with the nucleoplasm (dynamic exchange), exhibit random movement throughout the nucleoplasm (random movement), exchange proteins such as Nopp140 and fibrillarin with the nucleolus (intercompartmental exchange), and exhibit translational movement and recruit snRNA gene loci (translational movement). (**b**) snoRNPs undergo biogenesis steps in the CB and then shuttle from the CB to the nucleolus. U6 snRNAs are transcribed by RNA polymerase III (Pol III) and assembled onto Lsm (like Sm) proteins to form a U6 snRNP. U6 snRNPs subsequently shuttle to the nucleolus where they are modified by snoRNPs before returning to a CB for assembly into the U4/U6.U5 tri-snRNP. (**c**) CBs translocate to and from the nucleolus. (**d**) UV-C irradiation disrupts CBs and redistributes coilin to numerous microfoci and nucleolar caps in a small subset of cells. (**e**) The components of CBs are relocalized to nucleolar caps in response to transcription inhibition. Caps containing coilin and scaRNAs overlap but are distinct from caps containing fibrillarin and snoRNA on transcription inhibition. In Purkinje cell degeneration-induced DNA damage, CBs are disrupted and coilin localizes to nucleolar caps or in a perinucleolar ring containing fibrillarin. (*Center*) The nucleolus is anchored around nucleolar organizing regions (NORs) on the acrocentric chromosomes. *GC* granular component; *DFC* dense fibrillar component; *FC* fibrillar center. See text for more details

2009; Nizami et al. 2010). These studies were facilitated by the identification of a marker protein for CBs, coilin (Raska et al. 1991; Andrade et al. 1991). Coilin antibodies enabled researchers to determine that CBs were enriched in noncoding RNA-protein complexes called spliceosomal small nuclear RNPs and small Cajal body-specific RNPs (scaRNPs). scaRNAs help guide modifications that take place on the snRNA component of the small nuclear ribonucleoprotein (snRNP) within a CB. scaRNAs also provide another clue about the relationship between the CB and the nucleolus. Within the nucleolus, a closely related guide RNA, snoRNA, facilitates the modification of rRNA nucleotides. These small nucleolar ribonucleoprotein (snoRNP) mediated modifications, like scaRNPs, include 2'-*O*-methylation by box C/D snoRNPs and pseudouridylation by box H/ACA snoRNPs (Fig. 16.1b), but in the nucleolus, these modifications necessarily precede pre-rRNA cleavage. Moreover, in support of this relationship, CBs have been shown to participate in snoRNP maturation (Fig. 16.1b).

Another notable component of CBs is SMN, the survival of motor neuron protein. SMN is mutated in the genetic disease spinal muscular atrophy and plays a crucial role in the cytoplasmic phase of snRNP biogenesis. In the CB, SMN may take part in snRNP regeneration or recycling. CBs are not found in all cell types (Spector et al. 1992; Young et al. 2001), but their presence is correlated with high transcriptional demands. Thus, cells with the greatest need for snRNP and snoRNP resources contain CBs. There are several excellent reviews that explore known CB functions (Gall 2003; Cioce and Lamond 2005; Morris 2008; Matera et al. 2009; Nizami et al. 2010), but a more comprehensive examination of the relationship between CBs and nucleoli is needed. The overarching goal of this chapter is to examine this connection between CBs and nucleoli. Such an analysis is timely, given recent findings showing that the CB and nucleolus play important roles in various stress responses in addition to their more well-known and established activities.

16.2 The Cuerpo Accesorio: Discovered in the Shadow of the Nucleolus

By employing a reduced silver nitrate method with modifications, Cajal was able to illustrate a refined view of the neuronal nucleus that included not only a more detailed view of the nucleolus but also newly described structures such as the "hyaline grumes," which are currently known as splicing speckles and the "accessory body" (cuerpo accesorio, in Spanish) of the nucleolus (Ramón y Cajal 1903, 1910). Drawings by Cajal show what is now known as the CB free in the nucleoplasm, clearly distinct from the nucleolus (some of Cajal's drawings are reproduced in Lafarga et al. 2009; Garcia-Lopez et al. 2010). The shared argyrophilic qualities of the CB and nucleolus observed by Cajal portend a complicated communion between these structures that is still not fully resolved. One reason for this uncertainty is that after their identification by Cajal in 1903, CBs were largely forgotten, with a few notable exceptions, until their rediscovery by electron microscopists 66 years later in 1969. Conversely, nucleolar investigations received a tremendous boost in the 1930s from work showing that the nucleolus forms at a chromosomal locus termed the nucleolus organizer (rRNA genes) (Pederson 2010). This finding provided a key foundational concept that spearheaded the establishment of the nucleolus as the center of ribosome biosynthesis approximately 30 years later (reviewed in Pederson 2010). As the data piles grew from hints, suggestions, and observations to unambiguous facts and dogma with regard to the function of the nucleolus, studies of CBs limped along in the valley between these mountains of information.

Although progress into CB function was slow relative to that obtained for the nucleolus, a similarity was nevertheless drawn between these two structures early on from studies using the germinal vesicle in the oocytes of insects and amphibians. Specifically, spherical structures distinct from nucleoli and what is now thought to be CBs were observed in crickets and newts (reviewed Gall 2003). The work by Joe Gall was particularly notable as the spherical structures were observed in the nucleoplasm and, very interestingly, attached to specific gene loci. Thus, a striking parallel can be made between the nucleolus and the nucleolus organizer to the CB and specific genes. Rather than being formed in response to transcription of the associated gene, as found for nucleoli, however, CBs and specific gene loci are brought together by movement of the gene loci to CBs in an actin-dependent manner (Dundr et al. 2007) (Fig. 16.1a). Later studies conducted by Gall and colleagues demonstrated that CBs were attached to histone gene loci and contained the U7 snRNP that is necessary for histone pre-mRNA processing (Gall et al. 1981; Callan et al. 1991; Gall and Callan 1989; Wu et al. 1991, 1996; Wu and Gall 1993). Other studies have shown that CBs also associate with snRNA gene loci and CBs containing telomerase associate with telomeres during S phase (Frey and Matera 1995; Smith et al. 1995; Gao et al. 1997; Jacobs et al. 1999; Shopland et al. 2001; Tomlinson et al. 2006; Jady et al. 2006).

Strangely, U7 snRNP localization is schizophrenic in that it is found in CBs in mammalian cancer cells, and in another nuclear domain, the histone locus body (HLB), in primary cells, and *Drosophila* (reviewed in Nizami et al. 2010; Matera et al. 2009). In addition to the U7 snRNP, other factors involved in replication-dependent histone gene expression and processing are also found in HLBs, including NPAT and FLASH. Curiously, during S phase, the proteins within CBs and HLBs are colocalized within one structure (Bongiorno-Borbone et al. 2008), clearly indicating that these two subnuclear domains are intimately related in an enigmatic manner. Future studies should endeavor to conduct a systematic analysis across different model systems throughout development and the cell cycle using CB (coilin, scaRNAs) and HLB (NPAT) epitopes to distinguish these structures in order to facilitate this characterization.

Additional insight into the CB and its connection to the nucleolus comes from studies conducted in 1957 by the Barr group (Lafarga et al. 2009). Because they were investigating neuronal cells, these investigators were well aware of the "accessory body" of Cajal, and demonstrated that this structure was DNA negative. These studies also showed that CB size and number decreased on axotomy (severing of axon), which, importantly, is the first indication that CBs are linked to the overall transcriptional vigor of the cell. Finally, an argyrophilic paranucleolar structure observed by this group was hypothesized to give rise to CBs in the nucleoplasm. Following up on this work and concept, Hardin et al. (1969) presented EM data supporting the idea that the "dense component" of the nucleolus leads to the formation of the paranucleolar structure, which separates from the nucleolus to generate CBs that are free in the nucleoplasm (Hardin et al. 1969). The reverse scenario, in which a free CB in the nucleoplasm docks to the nucleolus, was discussed but dismissed as unlikely considering the "known intense biosynthetic activity of the nucleolus and the consequent implication of active export of macromolecules such as ribosomal ribonucleic acid (rRNA) from this structure." In other words, at the time of these studies, it seemed difficult to envisage what a CB could possibly deliver to the nucleolus.

Other interesting tidbits of information from the Hardin study are their observation that paranucleolar structures (i e., CBs docked to the nucleolus) were much more frequent on nucleoli that were free in the nucleoplasm as opposed to nucleoli associated with the nuclear envelope. This selective accumulation of CBs on a subset of nucleoli is somewhat reflective of the nucleolar caps formed by coilin in response to transcription arrest that, for an unknown reason, more heavily accumulate at some nucleoli while others have less accumulation or even lack coilin caps (Raska et al. 1990; Carmo-Fonseca et al. 1992). Why should some nucleoli recruit CBs and coilin while others do not? Clearly this is a topic for future studies. Finally, in their discussion of paranucleolar structures, which are presumably CBs associated with nucleoli, Hardin et al. described them as "being organized into a coiled and possibly branched thread-like structure." This is ironic considering that, 6 days after the Hardin manuscript was received at *Anatomical Record*, the *Journal of Ultrastructure Research* received a manuscript by Monneron and Bernhard describing a novel nuclear subdomain they termed the coiled body because it looked to be composed of "irregularly twisted" "coiled threads" (Monneron and Bernhard 1969). These authors noted that CBs "do not seem to have any topological relationship with the nucleolus," which is somewhat contradictory to the findings from other groups described above, but may be a result of the EDTA staining method Monneron and Bernhard used for their EM work. Interestingly, they did observe that CBs could sometimes be found in close proximity to chromatin, and concluded that these structures contained RNPs, which are two very significant findings that were subsequently corroborated. After these EM studies in 1969, the CB had parallel lives as the "coiled body" and the "accessory body" for over a decade until Lafarga and colleagues definitively showed that they were the same structure (Lafarga and Hervas 1983). Although Lafarga and colleagues suggested that the structure be called the "accessory body," the "coiled body" name stuck until Joe Gall in 1999 proposed that it be renamed the "Cajal body" to honor its discoverer (Gall et al. 1999).

Although the Hardin (1969) study did not find any evidence at the EM level for the recruitment of a CB to the nucleolus, live-cell studies have shown that CBs can move to, associate with, and bud off nucleoli (Boudonck et al. 1999; Platani et al. 2000, 2002) (Fig. 16.1c). The first of these studies utilized a fusion of GFP to U2B″ (a component of the U2 snRNP which is enriched in CBs) that was stably expressed in tobacco BY-2 cells and *Arabidopsis* plants (Boudonck et al. 1999). Movement of CBs in these cells was noted to take place for CBs in the nucleoplasm and in the nucleolus. Dramatic movements of CBs from the nucleoplasm to the nucleolus were also shown to occur, in addition to fusion of CBs at the nucleolar periphery. However, movement of CBs from the nucleolus to the nucleoplasm was not observed in these studies. In contrast, live-cell studies using HeLa cells stably expressing GFP-coilin (the CB marker protein) showed movement of CBs both to and from the nucleolus (Platani et al. 2000) (Fig. 16.1c). As with the plant cell studies, this study also reported that CBs could fuse together, but these fusions could also take place in the nucleoplasm. In one example, a nucleoplasmic CB is shown to split into two CBs, one of which translocates to the nucleolus where it remains for around 30 min, and then returns to fuse to the CB from where it came (Platani et al. 2000). These studies indicate that the conversation between CBs and nucleoli is not, at least in HeLa cells, uni-directional. Another interesting finding from this study is that CBs are heterogeneous given that a subset of CBs can contain fibrillarin (a CB protein that also localizes to the nucleolus), but others do not. These findings indicate that different CBs may have different functions based on the proteins present within them. Follow up studies by this same group demonstrated that CB movement increased when ATP is depleted or transcription is inhibited, and can be described by anomalous diffusion (Platani et al. 2002). FRAP and iFRAP studies on CB components that also localize to the nucleolus, such as fibrillarin and Nopp140 (discussed below), demonstrate that these proteins have significantly slower dissociation kinetics in nucleoli compared to that found in CBs (Snaar et al. 2000; Dundr et al. 2004). This may be due to the fact that CBs lack pre-rRNA transcripts or other interacting partners that are present in nucleoli, thereby accounting for the slow dissociation kinetics of these nucleolar CB proteins in nucleoli compared to CBs.

16.3 Cajal Body Characterization Leads to More Nucleolar Connections

The lack of a protein marker for CBs (to which antibodies could be generated) hindered the classification of CB components until such a marker became available in 1991 in the form of coilin (Raska et al. 1991; Andrade et al. 1991). Prior to this, however, it was found that essential components of snRNPs, Sm proteins, are enriched in CBs (Fakan et al. 1984), confirming the earlier finding by Monneron and Bernhard that suggested CBs contained RNPs (Monneron and Bernhard 1969). The snRNA components of snRNPs were later identified in CBs, demonstrating that CBs contained the essential factors necessary for pre-mRNA splicing (Wu et al. 1991; Carmo-Fonseca et al. 1991, 1992; Huang and Spector 1992; Matera and Ward 1993). However, as CBs are not sites for splicing, the prevailing theory at this time was that CBs played a role in RNP biogenesis in addition to some aspect of histone processing. This theory has been borne out (Morris 2008; Matera et al. 2009).

The development of coilin antibodies facilitated the classification of CB contents in a variety of different species and developmental or cell cycle stage. Intriguingly, many of the proteins identified were also nucleolar, some of which are shown in Table 16.1. It is noteworthy that the function of many of these proteins centers on ribosomal RNA (rRNA) processing (e.g., NAP57 and fibrillarin); however, it is also noteworthy that several of these proteins are implicated in telomerase assembly and telomere maintenance (EST1A, GAR1, NHP2, NOP10, HEST2). The rRNA processing functions are emphasized in this chapter, although future work will undoubtedly shed light on the relationship between CBs and telomeres.

As rRNA is not processed in the CB, it stands to reason that the nucleolar function of the proteins listed in Table 16.1 might be different from their role in the CB.

Alternatively, it could be that these proteins are brought into the appropriate complex and/or matured in the CB, and then delivered (or otherwise make their way to) to the nucleolus. Work investigating box C/D snoRNA trafficking supports the maturation hypothesis. Briefly, studies in yeast have demonstrated that a nucleolar body containing snoRNA can be induced to form under certain conditions (Verheggen et al. 2001). The yeast nucleolar body is functionally equivalent to the CB in that it contains the machinery necessary for U3 snoRNPs maturation, such as the methyltransferase responsible for cap tri-methylation (hTgs1) (Verheggen et al. 2002). In higher eukaryotes, therefore, CBs facilitate snoRNP biogenesis (Fig. 16.1b), but in yeast this function is localized to the nucleolar body.

Further support for the role of the CB in snoRNP maturation comes from studies showing that coilin interacts with Nopp140 (Isaac et al. 1998) and SMN interacts with fibrillarin (Jones et al. 2001; Pellizzoni et al. 2001b). Thus, the relationship between the CB and the nucleolus can be characterized as being supportive given that CBs provide nucleoli a service to mature components necessary for ribosome biogenesis. Supporting this belief is the correlation between the number of CBs and nucleolar organizing regions (NORs). Raska et al. note that the maximum number of CBs in human, mouse, and PtK2 cells lines was 8, 3, and 1, respectively, as compared to 10, 6, and 2 NORs (Raska et al. 1991). So while the presence of CBs is clearly tied to the level of pol II transcriptional activity and consequent snRNP splicing resources needed by the cell, the generation of rRNA by pol I, and needed snoRNP activities, also contributes to CB formation. Given that rRNA accounts for ~80% of the RNA in the cell, it is likely that the presence of CBs, and their number, is more influenced by the level of rRNA compared to pre-mRNA, which accounts for ~5% of the total RNA.

16.4 Characterization of Nucleolar CBs

In 1990 and 1991, three papers from the Tan group described the identification of autoimmune antibodies that detected CBs (Raska et al. 1990, 1991; Andrade et al. 1991). These studies went on to show that the antigen detected by these antibodies was a protein they termed p80-coilin (given its accumulation in "coiled bodies" and apparent molecular weight on SDS-PAGE). A partial cDNA sequence of coilin was isolated and later used to generate coilin polyclonal antibodies (R288, Andrade et al. 1993 and R508, Chan et al. 1994). With the autoantibodies, however, Raska, Andrade, and colleagues demonstrated that CBs, as shown by the Barr group in 1957 and Hardin in 1969, sometimes associate with nucleoli. Specifically, they showed that CBs could be found associated with nucleoli more frequently in rat and brain cells compared to cycling cells, such as HeLa, in which they reported that less than 15% of cells had CBs associated with nucleoli (Raska et al. 1991). In a follow-up study, which actually got published in 1990, Raska et al. further explored the relationship between the CB and nucleolus using the transcription inhibitors actinomycin D and DRB (Raska et al. 1990). They observed that the association between

CBs and nucleoli in cycling cells is greatly increased on transcription inhibition. Furthermore, they demonstrated that low levels of transcription inhibitors in cycling cells can recapitulate the association of CBs with nucleoli as observed in primary neuron cultures without treatment. Additionally, they showed that the accumulation of coilin to nucleoli could be found in both a paranucleolar structure and a thin ring, perinucleolar ring, around the nucleolus. Therefore, it appears that coilin in these transcription-inhibition-induced nucleolar caps forms a distinct subcompartment that partially colocalizes with segregated fibrillarin but not B23 (Fig. 16.1e). The authors conclude from these studies that "it is our belief that the coiled (Cajal) body may be multifunctional, serving both the nucleolus and the nucleoplasm."

Transcription inhibitors and their impact on CB localization were also studied by the Lamond group, who showed that spliceosomal snRNPs no longer accumulate in CBs after actinomycin D or alpha amanitin treatment (Carmo-Fonseca et al. 1992), but instead are retained in speckles. Interestingly, the localization of the U1 snRNP on transcription inhibition differed from the other spliceosomal snRNPs tested given that it formed nucleolar caps that partially overlap with coilin. Therefore, transcription inhibition by DRB, actinomycin D, or alpha amanitin relocalizes CBs to nucleolar caps, although a more precise description would be that a subset of antigens within the CB, such as coilin, IS relocated to caps. Additionally, an exhaustive study of nucleolar caps formed during transcription arrest demonstrated that there are, in fact, several different flavors of caps such as light nucleolar caps (LNC) and dark nucleolar caps (DNC) (Shav-Tal et al. 2005). LNCs were shown to contain fibrillarin, Nopp140, Gar1, U3 snoRNA, and U6 snRNA. This detailed analysis showed that coilin forms distinct caps that contain only scaRNA (Fig. 16.1e). The coilin caps only partially overlap with fibrillarin-containing LNCs, suggesting that, like the nucleolus, components of the CB segregate differently in response to transcription arrest. This functional partitioning appears to fall into coilin/scaRNA complexes and fibrillarin/snoRNA complexes (Fig. 16.1e). This is reminiscent of findings showing that coilin knockout cell lines contain two kinds of residual CBs, one with snoRNP components (Tucker et al. 2001) and the other with scaRNP components (Jady et al. 2003). Coilin, therefore, appears to be the glue that brings these different activities together into one unified structure. In addition to coilin, poly(ADP-ribose) polymerase 1 (PARP1) may also play a role in the formation of CBs as mutants of PARP1 in *Drosophila* fail to recruit fibrillarin to CBs (Kotova et al. 2009). Similarly, SMN is required for Nopp140 accumulation in the CB, and the lack of Nopp140 in the CB correlates with the severity of spinal muscular atrophy (Renvoise et al. 2009).

Importantly, coilin accumulation to nucleolar caps on transcription arrest is not strictly limited to cells that contain CBs. Therefore, it is possible that the segregation of different CB proteins to separate but overlapping caps on transcription arrest facilitates the reformation of a canonical CB after this stress is removed. However, one must consider the fact that primary cell lines that lack CBs, such as WI-38, still have coilin accumulations in caps after actinomycin D treatment (Polak et al. 2003), suggesting that the recruitment of nucleoplasmic coilin is also functional in responding to this stress response. What, exactly, coilin is doing in nucleolar caps is unknown.

Another chemical treatment, besides those causing transcription arrest, also alters coilin and CB localization. Specifically, inhibition of protein dephosphorylation by okadaic acid has been shown to cause CB accumulation in nucleoli (Lyon et al. 1997). Unlike nucleolar caps formed on transcription arrest, nucleolar CBs found in okadaic acid treated cells contain snRNPs. Fibrillarin localization was also altered in treated cells given that it formed rounded structures that surrounded the nucleolar CBs. These findings indicate that one factor which may influence the relationship between the CB and nucleolus is phosphorylation.

16.5 CBs and Nucleologenesis

CBs have been implicated in nucleologenesis during mouse oocyte development (Zatsepina et al. 2003; Pochukalina and Parfenov 2008), where it was revealed that coilin colocalized with RNA polymerase I at the onset of this process. This interaction occurred at the nucleolar surface and, being at an immature developing nucleolus, was associated with an absence of RNA polymerase I activity (Pochukalina and Parfenov 2006). Such a scenario suggests an early and intimate association between the CB and the nucleolus that is gradually diminished as the structures become more specialized in function. Functional clues to the relevance of this early association might be revealed by examining the material that exchanges between the mature forms of these structures as the CB's association with a developing nucleolus would presumably be synergistic. This material includes noncoding RNAs and their associated proteins that collectively operate as rRNA processing factors in addition to other proteins with important nucleolar functions.

16.6 CB Association with the Nucleolus: The Coilin Connection

Characterization and analysis of the CB marker protein coilin has also provided insight into the relationship of CBs with nucleoli. Although coilin has been described as being localized to the nucleoplasm and the CB, a subset of cells in a given population can have nucleolar coilin accumulations, especially in neuronal cells (Raska et al. 1990, 1991) (Fig. 16.2). Nucleolar coilin localization has also been observed in hibernating dormice (Malatesta et al. 1994) and breast cancer cells (Ochs et al. 1994). These findings demonstrate that coilin and CBs normally associate with nucleoli.

Analyses of the human coilin amino acid (aa) sequence have revealed the presence of apparent nucleolar localization signals (Hebert and Matera 2000; Scott et al. 2010). Moreover, mutational analysis of coilin, which is a protein of 576 aa, reinforces the connection between CBs and nucleoli. For example, a coilin fragment comprising aa 1–291 generates large pseudo CBs that accumulate Nopp140 but not other nucleolar proteins (Bohmann et al. 1995). As Nopp140 is a coilin-interacting protein (Isaac et al. 1998) that can also suppress pol I transcription (Chen et al.

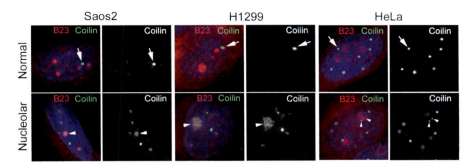

Fig. 16.2 Coilin, the CB marker protein, localizes to the nucleolus in a subset of cells. Saos2, H1299, and HeLa cells were cultured under normal growth conditions, fixed in methanol, permeabilized in 0.5% triton for 10 min, and immunostained for B23 (*red*) and coilin (*green*). DAPI (*blue*) stains the nucleus. The majority of cells in a given population do not have significant amounts of nucleolar coilin (*top*); however, a subset of cells do have coilin accumulation in the nucleolus (*bottom*). *Arrows* show CBs; *arrowheads* show nucleolar coilin

1999), these pseudo CBs may be an attempt by the cell to sequester a potentially deleterious unregulated complex. Like Nopp140, fragments of coilin (aa 94–291) have been shown to inhibit pol I activity, further demonstrating a functional link between CBs and nucleoli (Bohmann et al. 1995). Other truncations of coilin (aa 1–248 and aa 1–315) localize like Nopp140 and fibrillarin to both CBs and nucleoli (Hebert and Matera 2000). Posttranslational modifications of coilin also influence nucleolar coilin localization. Coilin is phosphorylated both on serines and threonines and symmetrically dimethylated on arginines (Hebert 2010). Mutation to alanine of coilin residues that have been shown to be phosphorylated by mass spectroscopic analysis (mimicking constitutive dephosphorylation) results in nucleolar localization (Hearst et al. 2009). Furthermore, hypomethylated coilin has been demonstrated to localize within nucleoli (Tapia et al. 2010). These findings demonstrate that signaling pathways that modify coilin, and possibly other proteins in the CB, impact coilin and CB nucleolar localization.

In addition to coilin modifications, ectopic overexpression of coilin also results in nucleolar coilin accumulations (Hebert and Matera 2000). Coilin's strong propensity to self interact (Hebert and Matera 2000; Shpargel et al. 2003) likely drives this by disrupting CBs and causing coilin to accumulate in nucleoli. This suggests that CB integrity is maintained not just by coilin's self association, but also by its association with other CB proteins, consistent with the model of nuclear body self assembly, where coilin can effectively nucleate a CB (Kaiser et al. 2008). This is reflected in the fact that mislocalizations of other crucial CB components, such as Sm proteins by depletion of PRMT5 or PRMT7 (protein arginine *N*-methyltransferase), results in the distribution of coilin in nucleolar caps (Clelland et al. 2009; Gonsalvez et al. 2007). Collectively, all the above data indicate that there is a nucleolar pool of coilin. This pool is relatively small compared to amount of coilin found in the nucleoplasm and CB, but can be increased in specific situations. The function of this nucleolar pool of coilin is unknown.

16.7 Other Factors That Influence CB Association with the Nucleolus

Besides transcription arrest, phosphatase and methylase inhibition, and coilin mutation or overexpression, alteration of other CB components also results in nucleolar CB/coilin localization. These findings demonstrate that CB integrity and localization are not mediated by just one factor, such as coilin, but instead are responsive to several proteins. For example, recent work has shown that WRAP53 depletion abolishes CBs and results in the accumulation of coilin and SMN inside the nucleolus where it colocalizes with fibrillarin (Mahmoudi et al. 2010). The *WRAP53* gene plays a role in regulating p53 via the production of a p53 antisense transcript (Mahmoudi et al. 2009). This gene also produces the WRAP53 protein (also known as TCAB1 or WDR79), which has been shown to localize to the CB and direct the targeting of scaRNAs and telomerase RNA to the CB (Tycowski et al. 2009; Venteicher et al. 2009). Thus, WRAP53 is an essential factor necessary for proper CB formation, composition, and localization. It should be pointed out that the nucleolar localization of coilin and SMN on WRAP53 knockdown is not similar to what is observed on transcription arrest, which results in coilin localization to nucleolar caps that do not contain SMN. This would suggest that the cell response to WRAP53 depletion and transcription arrest is different, although they both result in distinct nucleolar coilin accumulations.

A similar finding is observed for the reduction of factors required for snRNP biogenesis (hTGS1, SMN, and snurportin) or histone transcription and processing (FLASH), all of which result in altered coilin localization to the nucleolus, either colocalized with fibrillarin or in nucleolar caps (Lemm et al. 2006; Girard et al. 2006; Shpargel and Matera 2005; Barcaroli et al. 2006; Kiriyama et al. 2009). Moreover, inhibition of U snRNA export using leptomycin B also results in the nucleolar localization of coilin that colocalizes with fibrillarin (Carvalho et al. 1999). Collectively, these findings demonstrate that while coilin or CB localization to the nucleolus is a common response to many insults, including transcription arrest and specific protein knockdown, the differing localization of nucleolar coilin indicates that the response to these conditions varies on the basis of the insult.

16.8 The Effect of Viruses and DNA Damage on the Relationship Between CBs and Nucleoli

There is extensive literature detailing the impact of viral infection on the nucleolus (Hiscox 2007; Greco 2009). These findings demonstrate that many viruses commandeer nucleolar protein activity in order complete their viral cycle, or may utilize nucleolar proteins in novel ways. Like the nucleolus, the CB is also targeted by viruses, probably to usurp the RNA processing machinery present within this structure.

For example, adenovirus infection induces coilin microfoci and rosettes that may increase the capacity of viral transcription and splicing (Rebelo et al. 1996; James et al. 2010). In support of this idea, viral yield was reduced after coilin knockdown (James et al. 2010). Herpes simplex virus type 1 (HSV-1) also generates coilin microfoci, but these accumulations correspond to centromeres and may reflect a novel function of coilin in the interphase centromere damage response (iCDR) (Morency et al. 2007). Therefore, as is found for the nucleolus, CB activity may be hijacked by viruses or proteins in the CB may gain additional, novel activity on viral infection.

Another parallel between the CB and nucleolus is found in response to DNA damage induced by UV-C and neurodegeneration (Cioce et al. 2006; Baltanas et al. 2010; Boulon et al. 2010). With regard to UV-C exposure, this treatment induces the formation of coilin microfoci and, in a subset of cells, coilin nucleolar caps (Fig. 16.1d). This coilin redistribution has been linked to the proteasome activator subunit PA28γ, and suggests that the degradation of specific CB components triggers microfoci formation (Cioce et al. 2006). This component could be FLASH, considering that it is degraded by the proteasome after UV-C exposure (Bongiorno-Borbone et al. 2010). Although other types of DNA damage have not been shown to disrupt CBs, we have preliminary evidence indicating that cisplatin and gamma irradiation cause coilin to localize to the nucleolus (our unpublished observations). We have recently shown that coilin interacts with Ku proteins that are involved in the nonhomologous end joining (NHEJ) pathway of DNA repair, and inhibits in vitro NHEJ (Velma et al. 2010). It is possible, therefore, that the relocalization of coilin on DNA damage or HSV-1 infection is a novel response of coilin to monitor the extent of DNA damage.

Strikingly, the tumor suppressor protein p53 localizes to a subset of CBs after activation of its expression or inhibition of its degradation (Young et al. 2002). As it has been shown that SMN interacts directly with p53, it is possible that stress resulting in the accumulation of p53 to CBs may influence p53-depedent apoptosis (Young et al. 2002). In a mouse model of ataxia based on the Purkinje cell degeneration (*pcd*) phenotype, p53 message level as well as detection by immunofluorescence in nuclear spots is increased compared to control mice (Baltanas et al. 2010). Most of the p53 nuclear foci in the *pcd* mice were PML bodies (Baltanas et al. 2010), which are classified as another subnuclear domain responsive to stress that activate p53 (Alsheich-Bartok et al. 2008). Given that p53 is a negative regulator of pol I activity (Zhai and Comai 2000), and PML bodies and CBs are often found in close proximity (Grande et al. 1996; Sun et al. 2005), activated p53 may also disrupt CB activity, which contributes to the degenerative process. Supporting this idea is the finding that both CBs and nucleoli are altered in the *pcd* mice. Specifically, CBs are disassembled and coilin is relocalized to the segregated nucleolus where it forms a ring containing fibrillarin (Baltanas et al. 2010) (Fig. 16.1e). Therefore, as with other types of cellular stresses, the DNA damage-induced neuronal degeneration in the *pcd* mouse model again reinforces the relationship between CBs and nucleoli.

16.9 Conclusions

The organization of the eukaryotic cell nucleus is a manifestation of the vital activities that take place within this organelle, such as transcription, splicing, and ribosome biogenesis. Notable in this regard are the nucleolus, which builds around sites of rDNA transcription, and the CB, which facilitates the organized exchange of material necessary for ribosome biogenesis. Therefore, the coordination of rDNA transcription with rRNA processing underlies the relationship between the CB and the nucleolus. As such, recent findings demonstrating the central role of the nucleolus in response to stress also implicate the CB as a key target of signaling pathways activated by a variety of cellular insults (reviewed in Boulon et al. 2010). Going forward, particular emphasis should be placed on understanding the exact functional consequence of CB alteration in relationship to the nucleolus in response to stress. For example, most of the stress conditions discussed here result in the accumulation of coilin, but not other CB components, to specific nucleolar caps. What coilin is doing in these caps is still not clear and warrants further investigation. In order to fully understand the relationship between CBs and nucleoli, it will also be necessary for future studies to characterize the signaling pathways and components in the CB that are modified by these pathways. Such knowledge will greatly broaden our understanding into the biochemical conversations that take place between these nuclear neighbors.

Acknowledgments Research in the Hebert lab is supported by NIH grant R01GM081448.

References

Alsheich-Bartok O, Haupt S, Alkalay-Snir I, Saito S, Appella E, Haupt Y (2008) PML enhances the regulation of p53 by CK1 in response to DNA damage. Oncogene 27(26):3653–3661. doi:1211036 [pii] 10.1038/sj.onc.1211036

Andersen JS, Lam YW, Leung AK, Ong SE, Lyon CE, Lamond AI, Mann M (2005) Nucleolar proteome dynamics. Nature 433(7021):77–83. doi:nature03207 [pii] 10.1038/nature03207

Andrade LEC, Chan EKL, Raska I, Peebles CL, Roos G, Tan EM (1991) Human autoantibody to a novel protein of the nuclear coiled body: immunological characterization and cDNA cloning of p80 coilin. J Exp Med 173:1407–1419

Andrade LEC, Tan EM, Chan EKL (1993) Immunocytochemical analysis of the coiled body in the cell cycle and during cell proliferation. Proc Natl Acad Sci USA 90:1947–1951

Baltanas FC, Casafont I, Weruaga E, Alonso JR, Berciano MT, Lafarga M (2010) Nucleolar disruption and Cajal body disassembly are nuclear hallmarks of DNA damage-induced neurodegeneration in Purkinje cells. Brain Pathol (in press). doi:10.1111/j.1750-3639.2010.00461.x

Barcaroli D, Dinsdale D, Neale MH, Bongiorno-Borbone L, Ranalli M, Munarriz E, Sayan AE, McWilliam JM, Smith TM, Fava E, Knight RA, Melino G, De Laurenzi V (2006) FLASH is an essential component of Cajal bodies. Proc Natl Acad Sci USA 103(40):14802–14807. doi:0604225103 [pii] 10.1073/pnas.0604225103

Bleichert F, Baserga SJ (2010) Dissecting the role of conserved box C/D sRNA sequences in disRNP assembly and function. Nucleic Acids Res 38(22):8295–8305. doi:gkq690 [pii] 10.1093/nar/gkq690

Bohmann K, Ferreira J, Lamond A (1995) Mutational analysis of p80 coilin indicates a functional interaction between coiled bodies and the nucleolus. J Cell Biol 131:817–831

Bongiorno-Borbone L, De Cola A, Vernole P, Finos L, Barcaroli D, Knight RA, Melino G, De Laurenzi V (2008) FLASH and NPAT positive but not Coilin positive Cajal Bodies correlate with cell ploidy. Cell Cycle 7(15):2357–2367

Bongiorno-Borbone L, De Cola A, Barcaroli D, Knight RA, Di Ilio C, Melino G, De Laurenzi V (2010) FLASH degradation in response to UV-C results in histone locus bodies disruption and cell-cycle arrest. Oncogene 29(6):802–810. doi:onc2009388 [pii] 10.1038/onc.2009.388

Boudonck K, Dolan L, Shaw PJ (1999) The movement of coiled bodies visualized in living plant cells by the green fluorescent protein. Mol Biol Cell 10(7):2297–2307

Boulon S, Verheggen C, Jady BE, Girard C, Pescia C, Paul C, Ospina JK, Kiss T, Matera AG, Bordonne R, Bertrand E (2004) PHAX and CRM1 are required sequentially to transport U3 snoRNA to nucleoli. Mol Cell 16(5):777–787

Boulon S, Westman BJ, Hutten S, Boisvert FM, Lamond AI (2010) The nucleolus under stress. Mol Cell 40(2):216–227. doi:S1097-2765(10)00752-5 [pii] 10.1016/j.molcel.2010.09.024

Callan HG, Gall JG, Murphy C (1991) Histone genes are located at the sphere loci of Xenopus lampbrush chromosomes. Chromosoma 101:245–251

Carmo-Fonseca M, Tollervey D, Pepperkok R, Barabino SML, Merdes A, Brunner C, Zamore PD, Green MR, Hurt E, Lamond AI (1991) Mammalian nuclei contain foci which are highly enriched in components of the pre-mRNA splicing machinery. EMBO J 10:195–206

Carmo-Fonseca M, Pepperkok R, Carvalho MT, Lamond AI (1992) Transcription-dependent colocalization of the U1, U2, U4/U6 and U5 snRNPs in coiled bodies. J Cell Biol 117:1–14

Carvalho T, Almeida F, Calapez A, Lafarga M, Berciano MT, Carmo-Fonseca M (1999) The spinal muscular atrophy disease gene product, SMN: a link between snRNP biogenesis and the Cajal (coiled) body. J Cell Biol 147(4):715–728

Cajal S.R.y (1910) El núcleo de las células piramidales del cerebro humano y de algunos mamíferos. Trab Lab Invest Biol 8:27–62

Chan EK, Takano S, Andrade LE, Hamel JC, Matera AG (1994) Structure, expression and chromosomal localization of human p80-coilin gene. Nucleic Acids Res 22(21):4462–4469

Charroux B, Pellizzoni L, Perkinson RA, Yong J, Shevchenko A, Mann M, Dreyfuss G (2000) Gemin4. A novel component of the SMN complex that is found in both gems and nucleoli. J Cell Biol 148(6):1177–1186

Chen HK, Pai CY, Huang JY, Yeh NH (1999) Human Nopp 140, which interacts with RNA polymerase I: implications for rRNA gene transcription and nucleolar structural organization. Mol Cell Biol 19(12):8536–8546

Cioce M, Lamond AI (2005) Cajal bodies: a long history of discovery. Annu Rev Cell Dev Biol 21:105–131. doi:10.1146/annurev.cellbio.20.010403.103738

Cioce M, Boulon S, Matera AG, Lamond AI (2006) UV-induced fragmentation of Cajal bodies. J Cell Biol 175(3):401–413. doi:jcb.200604099 [pii] 10.1083/jcb.200604099

Clelland AK, Kinnear NP, Oram L, Burza J, Sleeman JE (2009) The SMN protein is a key regulator of nuclear architecture in differentiating neuroblastoma cells. Traffic 10(11):1585–1598. doi:TRA972 [pii] 10.1111/j.1600-0854.2009.00972.x

Dundr M, Misteli T (2010) Biogenesis of nuclear bodies. Cold Spring Harb Perspect Biol 2(12):a000711. doi:cshperspect.a000711 [pii] 10.1101/cshperspect.a000711

Dundr M, Hebert MD, Karpova TS, Stanek D, Xu H, Shpargel KB, Meier UT, Neugebauer KM, Matera AG, Misteli T (2004) In vivo kinetics of Cajal body components. J Cell Biol 164(6):831–842

Dundr M, Ospina JK, Sung MH, John S, Upender M, Ried T, Hager GL, Matera AG (2007) Actin-dependent intranuclear repositioning of an active gene locus in vivo. J Cell Biol 179(6): 1095–1103. doi:jcb.200710058 [pii] 10.1083/jcb.200710058

Fakan S, Leser G, Martin TE (1984) Ultrastructural distribution of nuclear ribonucleoproteins as visualized by immunocytochemistry on thin sections. J Cell Biol 98:358–363

Feng W, Gubitz AK, Wan L, Battle DJ, Dostie J, Golembe TJ, Dreyfuss G (2005) Gemins modulate the expression and activity of the SMN complex. Hum Mol Genet 14(12):1605–1611

Frey MR, Matera AG (1995) Coiled bodies contain U7 small nuclear RNA and associate with specific DNA sequences in interphase cells. Proc Natl Acad Sci USA 92:5915–5919

Gabellini D, D'Antona G, Moggio M, Prelle A, Zecca C, Adami R, Angeletti B, Ciscato P, Pellegrino MA, Bottinelli R, Green MR, Tupler R (2006) Facioscapulohumeral muscular dystrophy in mice overexpressing FRG1. Nature 439(7079):973–977. doi:nature04422 [pii] 10.1038/nature04422

Gall JG (2003) The centennial of the Cajal body. Nat Rev Mol Cell Biol 4(12):975–980

Gall JG, Callan HG (1989) The sphere organelle contains small nuclear ribonucleoproteins. Proc Natl Acad Sci USA 86:6635–6639

Gall JG, Stephenson EC, Erba HP, Diaz MO, Barsacchi-Pilone G (1981) Histone genes are located at the sphere loci of newt lampbrush chromosomes. Chromosoma 84:159–171

Gall JG, Bellini M, Wu Z, Murphy C (1999) Assembly of the nuclear transcription and processing machinery: Cajal bodies (coiled bodies) and transcriptosomes. Mol Biol Cell 10(12): 4385–4402

Gao L, Frey MR, Matera AG (1997) Human genes encoding U3 snRNA associate with coiled bodies in interphase cells and are clustered on chromosome 17p11.2 in a complex inverted repeat structure. Nucleic Acids Res 25(23):4740–4747

Garcia-Lopez P, Garcia-Marin V, Freire M (2010) The histological slides and drawings of Cajal. Front Neuroanat 4:9. doi:10.3389/neuro.05.009.2010

Gautier T, Berges T, Tollervey D, Hurt E (1997) Nucleolar KKE/D repeat proteins Nop56p and Nop58p interact with Nop1p and are required for ribosome biogenesis. Mol Cell Biol 17(12):7088–7098

Girard C, Neel H, Bertrand E, Bordonne R (2006) Depletion of SMN by RNA interference in HeLa cells induces defects in Cajal body formation. Nucleic Acids Res 34(10):2925–2932

Gonsalvez GB, Tian L, Ospina JK, Boisvert FM, Lamond AI, Matera AG (2007) Two distinct arginine methyltransferases are required for biogenesis of Sm-class ribonucleoproteins. J Cell Biol 178(5):733–740. doi:jcb.200702147 [pii] 10.1083/jcb.200702147

Grande MA, van der Kraan I, van Steensel B, Schul W, de The H, van der Voort HT, de Jong L, van Driel R (1996) PML-containing nuclear bodies: their spatial distribution in relation to other nuclear components. J Cell Biochem 63(3):280–291

Greco A (2009) Involvement of the nucleolus in replication of human viruses. Rev Med Virol 19(4):201–214. doi:10.1002/rmv.614

Hardin JH, Spicer SS, Greene WB (1969) The paranucleolar structure, accessory body of Cajal, sex chromatin and related structures in nuclei of rat trigeminal neurons: a cytochemical and ultrastructural study. Anat Rec 164(4):403–431

Hearst SM, Gilder AS, Negi SS, Davis MD, George EM, Whittom AA, Toyota CG, Husedzinovic A, Gruss OJ, Hebert MD (2009) Cajal-body formation correlates with differential coilin phosphorylation in primary and transformed cell lines. J Cell Sci 122(Pt 11):1872–1881. doi:jcs.044040 [pii] 10.1242/jcs.044040

Hebert MD (2010) Phosphorylation and the Cajal body: modification in search of function. Arch Biochem Biophys 496(2):69–76. doi:S0003-9861(10)00075-5 [pii] 10.1016/j.abb.2010.02.012

Hebert MD, Matera AG (2000) Self-association of coilin reveals a common theme in nuclear body localization. Mol Biol Cell 11(12):4159–4171

Heiss NS, Knight SW, Vulliamy TJ, Klauck SM, Wiemann S, Mason PJ, Poustka A, Dokal I (1998) X-linked dyskeratosis congenita is caused by mutations in a highly conserved gene with putative nucleolar functions. Nat Genet 19:32–38

Henras A, Henry Y, Bousquet-Antonelli C, Noaillac-Depeyre J, Gelugne JP, Caizergues-Ferrer M (1998) Nhp2p and Nop10p are essential for the function of H/ACA snoRNPs. EMBO J 17(23):7078–7090. doi:10.1093/emboj/17.23.7078

Hiscox JA (2007) RNA viruses: hijacking the dynamic nucleolus. Nat Rev Microbiol 5(2): 119–127. doi:nrmicro1597 [pii] 10.1038/nrmicro1597

Huang S, Spector DL (1992) U1 and U2 small nuclear RNAs are present in nuclear speckles. Proc Natl Acad Sci USA 89:305–308

16 Relationship of the Cajal Body to the Nucleolus

Isaac C, Yang Y, Meier UT (1998) Nopp 140 functions as a molecular link between the nucleolus and the coiled bodies. J Cell Biol 142:407–417

Jacobs EY, Frey MR, Wu W, Ingledue TC, Gebuhr TC, Gao L, Marzluff WF, Matera AG (1999) Coiled bodies preferentially associate with U4, U11, and U12 small nuclear RNA genes in interphase HeLa cells but not with U6 and U7 genes. Mol Biol Cell 10(5):1653–1663

Jady BE, Darzacq X, Tucker KE, Matera AG, Bertrand E, Kiss T (2003) Modification of Sm small nuclear RNAs occurs in the nucleoplasmic Cajal body following import from the cytoplasm. EMBO J 22(8):1878–1888

Jady BE, Richard P, Bertrand E, Kiss T (2006) Cell cycle-dependent recruitment of telomerase RNA and Cajal bodies to human telomeres. Mol Biol Cell 17(2):944–954

James NJ, Howell GJ, Walker JH, Blair GE (2010) The role of Cajal bodies in the expression of late phase adenovirus proteins. Virology 399(2):299–311. doi:S0042-6822(10)00030-9 [pii] 10.1016/j.virol.2010.01.013

Jones KW, Gorzynski K, Hales CM, Fischer U, Badbanchi F, Terns RM, Terns MP (2001) Direct interaction of the spinal muscular atrophy disease protein SMN with the small nucleolar RNA-associated protein fibrillarin. J Biol Chem 276(42):38645–38651. doi:10.1074/jbc.M106161200 M106161200 [pii]

Kaiser TE, Intine RV, Dundr M (2008) De novo formation of a subnuclear body. Science 322(5908):1713–1717. doi:1165216 [pii] 10.1126/science.1165216

Kiriyama M, Kobayashi Y, Saito M, Ishikawa F, Yonehara S (2009) Interaction of FLASH with arsenite resistance protein 2 is involved in cell cycle progression at S phase. Mol Cell Biol 29(17):4729–4741. doi:MCB.00289-09 [pii] 10.1128/MCB.00289-09

Kotova E, Jarnik M, Tulin AV (2009) Poly (ADP-ribose) polymerase 1 is required for protein localization to Cajal body. PLoS Genet 5(2):e1000387. doi:10.1371/journal.pgen.1000387

Lafarga M, Hervas JP (1983) Light and electron microscopic characterization of the "Accessory Body" of Cajal in the neuronal nucleus. In: Grisolía S, Guerri C, Samson F, Norton S, Reinsoso-Suárez E (eds) Ramón y Cajal's contribution to the neurosciences. Elsevier, Amsterdam, pp 91–100

Lafarga M, Casafont I, Bengoechea R, Tapia O, Berciano MT (2009) Cajal's contribution to the knowledge of the neuronal cell nucleus. Chromosoma 118(4):437–443. doi:10.1007/s00412-009-0212-x

Lemm I, Girard C, Kuhn AN, Watkins NJ, Schneider M, Bordonne R, Luhrmann R (2006) Ongoing U snRNP biogenesis is required for the integrity of Cajal bodies. Mol Biol Cell 17(7): 3221–3231. doi:E06-03-0247 [pii] 10.1091/mbc.E06-03-0247

Lyman SK, Gerace L, Baserga SJ (1999) Human Nop5/Nop58 is a component common to the box C/D small nucleolar ribonucleoproteins. RNA 5(12):1597–1604

Lyon CE, Bohmann K, Sleeman J, Lamond AI (1997) Inhibition of protein dephosphorylation results in the accumulation of splicing snRNPs and coiled bodies within the nucleolus. Exp Cell Res 230:84–93

Mahmoudi S, Henriksson S, Corcoran M, Mendez-Vidal C, Wiman KG, Farnebo M (2009) Wrap53, a natural p53 antisense transcript required for p53 induction upon DNA damage. Mol Cell 33(4):462–471. doi:S1097-2765(09)00073-2 [pii] 10.1016/j.molcel.2009.01.028

Mahmoudi S, Henriksson S, Weibrecht I, Smith S, Soderberg O, Stromblad S, Wiman KG, Farnebo M (2010) WRAP53 is essential for Cajal body formation and for targeting the survival of motor neuron complex to Cajal bodies. PLoS Biol 8(11):e1000521. doi:10.1371/journal.pbio.1000521

Malatesta M, Zancanaro C, Martin T, Chan E, Amalric FL, Vogel P, Fakan S (1994) Is the coiled body involved in nucleolar functions? Exp Cell Res 211:415–419

Matera AG, Ward DC (1993) Nucleoplasmic organization of small nuclear ribonucleoproteins in cultured human cells. J Cell Biol 121(4):715–727

Matera AG, Izaguire-Sierra M, Praveen K, Rajendra TK (2009) Nuclear bodies: random aggregates of sticky proteins or crucibles of macromolecular assembly? Dev Cell 17(5):639–647. doi:S1534-5807(09)00439-0 [pii] 10.1016/j.devcel.2009.10.017

Matunis MJ, Wu J, Blobel G (1998) SUMO-1 modification and its role in targeting the Ran GTPase-activating protein, RanGAP1, to the nuclear pore complex. J Cell Biol 140(3): 499–509

Meier UT, Blobel G (1994) NAP57, a mammalian nucleolar protein with a putative homolog in yeast and bacteria. J Cell Biol 127:1505–1514

Monneron A, Bernhard W (1969) Fine structural organization of the interphase nucleus in some mammalian cells. J Ultrastruct Res 27:266–288

Morency E, Sabra M, Catez F, Texier P, Lomonte P (2007) A novel cell response triggered by interphase centromere structural instability. J Cell Biol 177(5):757–768

Morris GE (2008) The Cajal body. Biochim Biophys Acta 1783(11):2108–2115. doi:S0167-4889(08)00267-X [pii] 10.1016/j.bbamcr.2008.07.016

Mourelatos Z, Dostie J, Paushkin S, Sharma A, Charroux B, Abel L, Rappsilber J, Mann M, Dreyfuss G (2002) miRNPs: a novel class of ribonucleoproteins containing numerous microR-NAs. Genes Dev 16(6):720–728. doi:10.1101/gad.974702

New JH, Sugiyama T, Zaitseva E, Kowalczykowski SC (1998) Rad52 protein stimulates DNA strand exchange by Rad51 and replication protein A. Nature 391(6665):407–410. doi:10.1038/34950

Nizami Z, Derysheva S, Gall JG (2010) The Cajal body and histone locus body. Cold Spring Harb Perspect Biol 2(7):a000653. doi:cshperspect.a000653 [pii] 10.1101/cshperspect.a000653

Nugent CI, Lundblad V (1998) The telomerase reverse transcriptase: components and regulation. Genes Dev 12(8):1073–1085

Ochs R, Stein TJ, Tan E (1994) Coiled bodies in the nucleolus of breast cancer cells. J Cell Sci 107:385–399

Pederson T (2010) The nucleolus. Cold Spring Harb Perspect Biol. doi:cshperspect.a000638[pii]10.1101/cshperspect.a000638

Pellizzoni L, Baccon J, Charroux B, Dreyfuss G (2001a) The survival of motor neurons (SMN) protein interacts with the snoRNP proteins fibrillarin and GAR1. Curr Biol 11(14):1079–1088. doi:S0960-9822(01)00316-5 [pii]

Pellizzoni L, Charroux B, Rappsilber J, Mann M, Dreyfuss G (2001b) A functional interaction between the survival motor neuron complex and RNA polymerase II. J Cell Biol 152(1):75–85

Platani M, Goldberg I, Swedlow JR, Lamond AI (2000) In vivo analysis of Cajal body movement, separation, and joining in live human cells. J Cell Biol 151(7):1561–1574

Platani M, Goldberg I, Lamond AI, Swedlow JR (2002) Cajal body dynamics and association with chromatin are ATP-dependent. Nat Cell Biol 4(7):502–508

Pochukalina GN, Parfenov VN (2006) The nucleolus in oocytes of multilayer mouse follicles: topography of fibrillarin, RNA polymerase I and coilin. Tsitologiia 48(8):641–652

Pochukalina GN, Parfenov VN (2008) Nucleolus transformation in oocytes of mouse antral follicles. Revealing of coilin and RNA polymerase I complex components. Tsitologiia 50(8):671–680

Polak PE, Simone F, Kaberlein JJ, Luo RT, Thirman MJ (2003) ELL and EAF1 are Cajal body components that are disrupted in MLL-ELL leukemia. Mol Biol Cell 14(4):1517–1528

Ramón y Cajal SR (1903) Un sencillo metodo de coloracion selectiva del reticulo protoplasmico y sus efectos en los diversos organos nerviosos de vertebrados y invertebrados. Trab Lab Invest Biol (Madrid) 2:129–221

Raska I, Ochs RL, Andrade LEC, Chan EKL, Burlingame R, Peebles C, Gruol D, Tan EM (1990) Association between the nucleolus and the coiled body. J Struct Biol 104:120–127

Raska I, Andrade LEC, Ochs RL, Chan EKL, Chang C-M, Roos G, Tan EM (1991) Immunological and ultrastructural studies of the nuclear coiled body with autoimmune antibodies. Exp Cell Res 195:27–37

Rebelo L, Almeida F, Ramos C, Bohmann K, Lamond AI, Carmo-Fonseca M (1996) The dynamics of coiled bodies in the nucleus of adenovirus-infected cells. Mol Biol Cell 7(7):1137–1151

Renvoise B, Colasse S, Burlet P, Viollet L, Meier UT, Lefebvre S (2009) The loss of the snoRNP chaperone Nopp 140 from Cajal bodies of patient fibroblasts correlates with the severity of spinal muscular atrophy. Hum Mol Genet 18(7):1181–1189. doi:ddp009 [pii] 10.1093/hmg/ddp009

Scott MS, Boisvert FM, McDowall MD, Lamond AI, Barton GJ (2010) Characterization and prediction of protein nucleolar localization sequences. Nucleic Acids Res 38(21):7388–7399. doi:gkq653 [pii] 10.1093/nar/gkq653

Shav-Tal Y, Blechman J, Darzacq X, Montagna C, Dye BT, Patton JG, Singer RH, Zipori D (2005) Dynamic sorting of nuclear components into distinct nucleolar caps during transcriptional inhibition. Mol Biol Cell 16(5):2395–2413. doi:E04-11-0992 [pii] 10.1091/mbc.E04-11-0992

Shopland LS, Byron M, Stein JL, Lian JB, Stein GS, Lawrence JB (2001) Replication-dependent histone gene expression is related to Cajal body (CB) association but does not require sustained CB contact. Mol Biol Cell 12(3):565–576

Shpargel KB, Matera AG (2005) Gemin proteins are required for efficient assembly of Sm-class ribonucleoproteins. Proc Natl Acad Sci USA 102(48):17372–17377. doi:0508947102 [pii] 10.1073/pnas.0508947102

Shpargel KB, Ospina JK, Tucker KE, Matera AG, Hebert MD (2003) Control of Cajal body number is mediated by the coilin C-terminus. J Cell Sci 116(Pt 2):303–312

Smith K, Carter K, Johnson C, Lawrence J (1995) U2 and U1 snRNA gene loci associate with coiled bodies. J Cell Biochem 59:473–485

Snaar S, Wiesmeijer K, Jochemsen AG, Tanke HJ, Dirks RW (2000) Mutational analysis of fibrillarin and its mobility in living human cells. J Cell Biol 151(3):653–662

Snow BE, Erdmann N, Cruickshank J, Goldman H, Gill RM, Robinson MO, Harrington L (2003) Functional conservation of the telomerase protein Est1p in humans. Curr Biol 13(8):698–704. doi:S0960982203002100 [pii]

Spector DL, Lark G. Huang S (1992) Differences in snRNP localization between transformed and nontransformed cells. Mol Biol Cell 3:555–569

Spiliotis ET, Kinoshita M, Nelson WJ (2005) A mitotic septin scaffold required for Mammalian chromosome congression and segregation. Science 307(5716):1781–1785. doi:307/5716/1781 [pii]10.1126/science.1106823

Sun J, Xu H, Subramony SH, Hebert MD (2005) Interactions between Coilin and PIASy partially link Cajal bodies to PML bodies. J Cell Sci 118(Pt 21):4995–5003

Tapia O, Bengoechea R, Berciano MT, Lafarga M (2010) Nucleolar targeting of coilin is regulated by its hypomethylation state. Chromosoma. doi:10.1007/s00412-010-0276-7

Tollervey D, Lehtonen H, Jansen R, Kern H, Hurt EC (1993) Temperature-sensitive mutations demonstrate roles for yeast fibrillarin in pre-rRNA processing, pre-rRNA methylation, and ribosome assembly. Cell 72(3):443–457. doi:0092-8674(93)90120-F [pii]

Tomlinson RL, Ziegler TD, Supakorndej T, Terns RM, Terns MP (2006) Cell cycle-regulated trafficking of human telomerase to telomeres. Mol Biol Cell 17(2):955–965

Tucker KE, Berciano MT, Jacobs EY, LePage DF, Shpargel KB, Rossire JJ, Chan EK, Lafarga M, Conlon RA, Matera AG (2001) Residual Cajal bodies in coilin knockout mice fail to recruit Sm snRNPs and SMN, the spinal muscular atrophy gene product. J Cell Biol 154(2):293–307

Tycowski KT, Shu MD, Kukoyi A, Steitz JA (2009) A conserved WD40 protein binds the Cajal body localization signal of scaRNP particles. Mol Cell 34(1):47–57. doi:S1097-2765(09)00136-1 [pii] 10.1016/j.molcel.2009.02.020

Van Dyck E, Stasiak AZ, Stasiak A, West SC (1999) Binding of double-strand breaks in DNA by human Rad52 protein. Nature 398(6729):728–731. doi:10.1038/19560

van Koningsbruggen S, Dirks RW, Mommaas AM, Onderwater JJ, Deidda G, Padberg GW, Frants RR, van der Maarel SM (2004) FRG1P is localised in the nucleolus, Cajal bodies, and speckles. J Med Genet 41(4):e46

Velma V, Carrero ZI, Cosman AM, Hebert MD (2010) Coilin interacts with Ku proteins and inhibits in vitro non-homologous DNA end joining. FEBS Lett 584(23):4735–4739. doi:S0014-5793(10)00898-7 [pii] 10.1016/j.febslet.2010.11.004

Venteicher AS, Abreu EB, Meng Z, McCann KE, Terns RM, Veenstra TD, Terns MP, Artandi SE (2009) A human telomerase holoenzyme protein required for Cajal body localization and telomere synthesis. Science 323(5914):644–648. doi:323/5914/644 [pii]10.1126/science.1165357

Verheggen C, Mouaikel J, Thiry M, Blanchard JM, Tollervey D, Bordonne R, Lafontaine DL, Bertrand E (2001) Box C/D small nucleolar RNA trafficking involves small nucleolar RNP proteins, nucleolar factors and a novel nuclear domain. EMBO J 20(19):5480–5490. doi:10.1093/emboj/20.19.5480

Verheggen C, Lafontaine DL, Samarsky D, Mouaikel J, Blanchard JM, Bordonne R, Bertrand E (2002) Mammalian and yeast U3 snoRNPs are matured in specific and related nuclear compartments. EMBO J 21(11):2736–2745

Vulliamy T, Beswick R, Kirwan M, Marrone A, Digweed M, Walne A, Dokal I (2008) Mutations in the telomerase component NHP2 cause the premature ageing syndrome dyskeratosis congenita. Proc Natl Acad Sci USA 105(23):8073–8078. doi:0800042105 [pii] 10.1073/pnas.0800042105

Wang C, Meier UT (2004) Architecture and assembly of mammalian H/ACA small nucleolar and telomerase ribonucleoproteins. EMBO J 23(8):1857–1867. doi:10.1038/sj.emboj.7600181 7600181 [pii]

Watkins NJ, Gottschalk A, Neubauer G, Kastner B, Fabrizio P, Mann M, Luhrmann R (1998) Cbf5p, a potential pseudouridine synthase, and Nhp2p, a putative RNA-binding protein, are present together with Gar1p in all H BOX/ACA-motif snoRNPs and constitute a common bipartite structure. RNA 4(12):1549–1568

Wu C-HH, Gall JG (1993) U7 small nuclear RNA in C snurposomes of the Xenopus germinal vesicle. Proc Natl Acad Sci USA 90:6257–6259

Wu Z, Murphy C, Callan HG, Gall JG (1991) Small nuclear ribonucleoproteins and heterogeneous nuclear ribonucleoproteins in the amphibian germinal vesicle: loops, spheres and snurposomes. J Cell Biol 113:465–483

Wu C, Murphy C, Gall J (1996) The Sm binding site targets U7 snRNA to coiled bodies (spheres) of amphibian oocytes. RNA 2:811–823

Yang Y, Isaac C, Wang C, Dragon F, Pogacic V, Meier UT (2000) Conserved composition of mammalian box H/ACA and box C/D small nucleolar ribonucleoprotein particles and their interaction with the common factor Nopp 140. Mol Biol Cell 11(2):567–577

Young PJ, Le TT, Dunckley M, Nguyen TM, Burghes AH, Morris GE (2001) Nuclear gems and Cajal (coiled) bodies in fetal tissues: nucleolar distribution of the spinal muscular atrophy protein, SMN. Exp Cell Res 265(2):252–261

Young PJ, Day PM, Zhou J, Androphy EJ, Morris GE, Lorson CL (2002) A direct interaction between the survival motor neuron protein and p53 and its relationship to spinal muscular atrophy. J Biol Chem 277(4):2852–2859

Zatsepina O, Baly C, Chebrout M, Debey P (2003) The step-wise assembly of a functional nucleolus in preimplantation mouse embryos involves the cajal (coiled) body. Dev Biol 253(1):66–83. doi:S0012160602908651 [pii]

Zhai W, Comai L (2000) Repression of RNA polymerase I transcription by the tumor suppressor p53. Mol Cell Biol 20(16):5930–5938

Chapter 17
Role of the Nucleolus in HIV Infection and Therapy

Jerlisa Arizala and John J. Rossi

17.1 Introduction

17.1.1 Characterization of HIV

Human immunodeficiency virus (HIV) is classified within lentivirinae – a subfamily of retroviridae viruses in which its genetic composition of RNA utilizes host DNA genomes to proliferate via reverse transcription and chromosomal integration (Fenner 1975; Barre-Sinoussi et al. 1983). The first indication of retroviral particles dates back to 1911, after establishment that transmission of sarcoma in chickens emerged in response to injections of filtered, cell-free tumorgenic homogenates (Rous 1911). Discovery of the first mammalian retrovirus in 1951 – murine leukemia virus (MuLV) – provided an experimental model that further recognized retroviral agents as the cause of leukemia, lymphoma, sarcoma, and other types of cancers (Gross 1951). In 1964, Temin's proviral hypothesis speculated that conversion of normal cells to malignancy after retroviral infection involved genomic alterations caused by the viral RNA (Temin 1964); discovery of reverse transcriptase in 1970 not only supported Temin's theory (Temin and Mizutani 1970; Baltimore 1970) but also characterized the retrovirus family and led to the identification of the first two human retroviruses – human T-cell leukemia virus I (HTLV-I) and II (HTLV-II) (Poiesz et al. 1980; Kalyanaraman et al. 1982).

Knowledge of retroviruses remained ambiguous to the public until 1981 with the emergence of Kaposi's sarcoma as well as immunodeficiency-related illnesses caused by *Pneumocystis carinii*, *Candida albicans*, *Mycobacterium avium* complex, *Cryptococcus neoformans*, *Histoplasma capsulatum*, *Cytomegalovirus*, and papilloma viruses in previously healthy patients (Gottlieb et al. 1981; Masur et al.

J.J. Rossi (✉)
Division of Molecular Biology, Beckman Research Institute of The City of Hope,
Duarte, CA, USA
e-mail: jrossi@coh.org

M.O.J. Olson (ed.), *The Nucleolus*, Protein Reviews 15,
DOI 10.1007/978-1-4614-0514-6_17, © Springer Science+Business Media, LLC 2011

1981; Siegal et al. 1981). These conditions were associated with what was referred to as acquired immuno deficiency syndrome (AIDS), as affected individuals exhibited weakened immune systems due to a significant decrease in T-cell function against opportunistic infectious agents. Although the causative agent of AIDS was unidentified, transmission through contact with infected blood, bodily fluids, and mucosal tissues was understood. The AIDS epidemic climaxed throughout the 1980s worldwide, affecting intravenous drug abusers, blood transfusion recipients, sexually active partners, and infants born to infected mothers. This urgent medical situation prompted identification of the AIDS causative agent.

Retroviruses were immediately assumed as a possible culprit, being that immune deficiencies instigated by cytopathic agents affecting birds (avian leukosis virus – ALV), cats (feline leukemia virus – FeLV), and other mammals are of retroviral infectious origin (Hardy 1985; Gardner and Luciw 1989). To confirm this, the following scientific methodologies were taken: reverse transcriptase detection, viral RNA/ protein complex identification, defective virus rescue, and genome-wide nucleic acid sequencing. The findings of Barre-Sinoussi in 1983 confirmed the retrovirus through identification of reverse transcriptase and proviral particles in infected lymphocytes of AIDS patients; the AIDS retrovirus was initially called lymphadenopathy-associated virus (LAV) (Barre-Sinoussi et al. 1983). LAV accumulated several other names during its characterization, first by Gallo et al. to HTLV-III (Popovic et al. 1984; Gallo et al. 1984; Sarngadharan et al. 1984; Schupbach et al. 1984) and then to AIDS-related virus (ARV) by Levy et al. (1984). Finally in 1986, all AIDS-associated retroviral strains were labeled as HIV (Coffin et al. 1986a, b).

To identify the origination of HIV, an assessment of evolutionary divergence and similarity between *Pol* genes of HIV and its relative simian immunodeficiency virus (SIV) was performed using sequence analysis. Divergence is reflective of a time frame for the existence of viral agents within a population based on gene variation. SIV existed in primates (20% variation of *Pol* gene) longer than HIV existed in humans (7% variation); in addition, *Pol* is 55–60% identical in HIV and SIV strains (Desrosiers et al. 1989). It is currently established that the ancestry of HIV is of an SIV variant (Sharp and Hahn 2008). HIV-1 is genetically similar to SIVcpz, which infects sub-species of chimpanzee of Western and Central Africa (Gao et al. 1999; Santiago et al. 2002), and HIV-2 is closely related to SIVsm, an infectious agent specific to sooty mangabey monkeys (Gao et al. 1992; Bailes et al. 2003; Rambaut et al. 2004). Primate subjects were naturally SIV-infected yet did not develop immunodeficiency-related maladies. This indicated adaptation of primate immune systems with SIV (Sodora et al. 2009). An evolution towards mutual symbiosis would have occured from inability of the immune system to eradicate SIV. This is due to inefficient, error-prone reverse transcription, which lacks a proofreading process to fix mutations induced within the viral DNA. Similarly to HIV, SIV replication is quite active, yet its reverse transcription lacks fidelity, generating a mutation every 1,700–4,000 nucleotides; mutant viruses are hence able to escape detection during antibody surveillance (Preston et al. 1988; Roberts et al. 1988). Investigation of the biological machinery as well as molecular aspects that give rise to a mutual coexistence of primates and SIV could provide novel approaches for treatment of the incurable HIV infection.

17.1.2 A Current Overview of HIV Infection

The AIDS epidemic affects approximately 50 million individuals worldwide, has claimed over 20 million lives, and infects an estimated 7,000 individuals daily (Mystakidou et al. 2009). The most susceptible groups to infection are adolescents between the ages of 15–24. The most common mode of transmission is sexual intercourse, by which mucosal linings of the vagina, cervix, or anorectum are exposed to virus. Two types of HIV isolates exist – HIV type 1 (HIV-1) is dispersed worldwide whereas HIV type 2 (HIV-2) is specific to Western and Central Africa, areas within Europe, and India (Fanales-Belasio et al. 2010). Unlike HIV-1, HIV-2 is less cytopathic, manifests AIDS-related illnesses at a much later stage, and often affects the central nervous system (Lucas et al. 1993; Whittle et al. 1994).

HIV-1 and 2 are genetically similar except that the HIV-2 *Vpx* gene replaces *Vpr* of HIV-1, and the HIV-2 *Vpr* gene replaces *Vpu* of HIV-1 (Fig. 17.1a). Within both viral genomes starting at the 5′ end are *Gag* (group-specific antigen), *Pol* (polymerase), and *Env* (envelope glycoprotein) genes. *Pol* expresses a polyprotein precursor, Pr160^{GagPol}, which is modified into *Pol*-encoded enzymes (p11 or PR), reverse transcriptase (p66/p51 or RT), and integrase (p31 or IN). *Gag* encodes a polyprotein precursor Pr55^{Gag}, which is later cleaved by PR into structural proteins for composition of the viral core (p24 or CA; p7 or NC; and p6), matrix (p17 or MA), and spacer peptides (p1 and p2). The *Env* precursor gp160 is modified by host cellular proteases into the surface glycoprotein (gp120 or SU) and the transmembrane glycoprotein (gp41 or TM) during translocation towards the cell cytoplasmic membrane. Regulatory proteins Tat (*trans*-activator) and Rev (regulator of expression of virion proteins) as well as accessory/auxiliary proteins Vif; Vpr (viral protein R), Vpx of HIV-2; Vpu of HIV-1; and Nef (negative factor) are processed from overlapping genes adjacent to *Gag*, *Pol*, and *Env* (Fanales-Belasio et al. 2010).

The HIV virion is approximately 100 nm in diameter, composed of two copies of 35S single-stranded RNA (Luciw 1996) embedded together with RT, IN, and PR within the protein clusters of NC, CA, and MA. All viral components are enclosed within a lipid/lipoprotein membrane flanked by TM and SU (Fig. 17.1b); such envelope proteins initiate infection through interaction with the CD4 receptor and chemokine coreceptors expressed on host cell membrane surfaces. The CD4 receptor facilitates foreign antigen detection as a component of the major histocompatibility complex class II expressed on T-cells (Miceli and Parnes 1993). HIV targets CD4$^+$ helper T-cells, T-cell precursors within bone marrow, thymus, lymphocytes, monocytes, macrophages, and eosinophils; infection of other cell types with low CD4 expression includes follicular dendritic and microglial cells of the nervous system, and Langerhans cells. Transmission of HIV relies on its concentration within infected biological solutions upon contact and on the vulnerability of nondividing host cells. A successful infection will progress into the following steps: binding and entry, uncoating, reverse transcription, provirus integration, virus protein synthesis and assembly, and budding (Fig. 17.2).

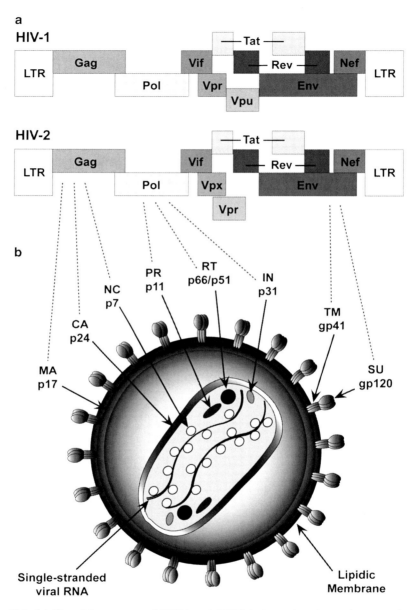

Fig. 17.1 (a) Complete genomes of HIV-1 and HIV-2 isolates give rise to the expression of structural, regulatory, accessory, and packaging proteins that facilitate the viral replication cycle. (b) Structural proteins are organized accordingly around 35S single-stranded HIV RNA during viral packaging to generate an infectious virion

17 Role of the Nucleolus in HIV Infection and Therapy

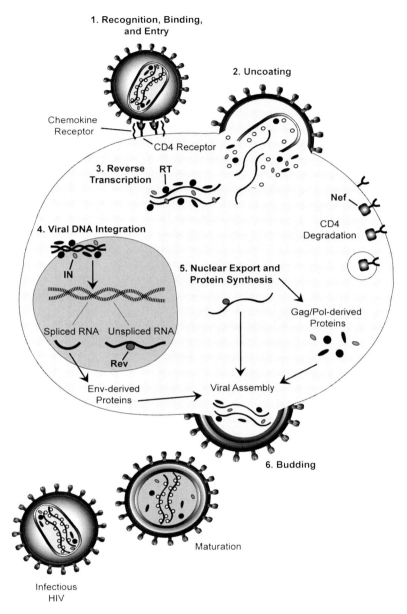

Fig. 17.2 The HIV replication cycle is composed of six steps. Binding and entry of an infectious virion is CD4/chemokine-dependent, leading to reverse transcription and generation of viral DNA. A provirus is formed after integration into the host genome, enabling viral protein synthesis and assembly. Budding and release of an infectious virion concludes the replication cycle

Replication begins on contact of HIV SU with the CD4 coreceptor. A conformational change occurs, exposing chemokine receptors that facilitate HIV tropism. Two chemokines identified in HIV infection are CXCR4, preferred by T-lymphocyte-tropic (T-tropic) X4 viruses and CCR5, preferred by macrophage-tropic (M-tropic) R5 viruses (Broder and Berger 1995). Dual tropic X4R5 viruses are able to bind both CXCR4 and CCR5. Stable interaction of HIV SU with CD4 and the chemokine receptor allows penetration of TM through the membrane lipid bilayer, causing fusion of the HIV virion with the host cell. Entry of the viral capsid into the cytosol takes place, and the reverse transcription complex (RTC) is immediately initiated allowing RT to generate complementary DNA from viral RNA templates. PR degrades the template RNA, leaving behind proviral DNA for transport into the nucleus and integration within the host genome by IN within the preintegration complex (PIC). The host cell is transformed into a reservoir for HIV production (Fanales-Belasio et al. 2010).

HIV infection continues on transcription of the provirus to express Tat and Rev regulatory proteins, which are products of completely spliced proviral transcripts (2 kb). Tat binds the transactivation response (TAR) element site within the 5' long terminal repeat (LTR) of other viral transcripts, initiating transcription for production of full length HIV mRNA. Rev facilitates the nucleocytoplasmic transport of unspliced (9 kb) and partially spliced (4 kb) HIV mRNAs through binding to the Rev binding element (RBE) within the Rev response element (RRE), leading to the expression of accessory, structural, and packaging proteins. Encapsulation of the 35S single-stranded HIV RNA takes place along with membrane localization of *Gag* and *Env*-derived structural proteins for complete viral packaging. Vif is involved in the assembly of mature, infectious virions. Vpr (Vpx in HIV-2) functions in cell cycle arrest during late stages of viral replication. Nef redirects CD4 from the cell membrane to lysosomes for degradation, and Vpu (Vpr in HIV-2) aids the release of viral particles.

17.2   Identification of a Nucleolar Step During HIV Infection

After integration of proviral DNA, the host cell transcriptional machinery produces HIV transcripts for expression of regulatory proteins – Tat and Rev. Tat facilitates transcription of viral genes through association with TAR, found within the 5' LTR of HIV transcripts; Rev translocates between the nucleus and cytoplasm, shuttling HIV mRNA via its multimerization with an intronic *cis*-acting target – RRE – found within the *Env* region. A similarity between both proteins is the nucleolar-specificity in the absence of HIV transcripts. Unlike Tat, which accumulates in both the nucleus and nucleolus, Rev is mainly nucleolar. The purpose of such localization patterns are not fully understood. Tat and Rev are RNA-binding proteins; Tat-induced transcription and Rev-induced nucleocytoplasmic shuttling require the aid of host cell proteins. Perhaps some, if not all, of these proteins associate with Tat and Rev in the nucleolus.

One protein in particular that was discovered to drive nuclear import of Tat and Rev is B23 (nucleophosmin, numatrin, or NO38) (Li 1997; Szebeni et al. 1997; Truant and Cullen 1999) – a nucleolar phosphoprotein involved in ribosomal assembly (see chapter 10 of this volume). B23 is found in nucleoli during cell cycle interphase, along the chromosomal periphery during mitosis, and in prenucleolar bodies (PNB) at the conclusion of mitosis. B23 is speculated to be involved in the shuttling of various nuclear/nucleolar proteins because of its nucleocytoplasmic translocation pattern. In the presence of a B23-binding-domain-β-galactosidase fusion protein, Tat was mis-localized within the cytoplasm and lost its transactivation activity, demonstrating that there is a strong affinity of Tat for B23 (Li 1997). Another study established a Rev–B23 stable interaction in the absence of RRE-containing mRNAs. In the presence of RRE-containing RNAs, Rev dissociated from B23 and bound preferably to the HIV RRE, leading to displacement of B23 (Fankhauser et al. 1991). Although it is possible that HIV regulatory proteins Tat and Rev enter nucleolar compartments via binding B23, the specific location of It is Tat transactivation and the Rev exchange process for HIV mRNA with B23 in the nucleus is unclear. To better understand the nucleolar-specific behaviors of Tat and Rev during HIV infection, both regulatory proteins will be discussed further below.

17.2.1 Tat in the Nucleolus and TAR

Tat is composed of 86 (laboratory-derived variant) or 101 (HIV-1-expressed) amino acids. The amino acid region 22–37, a cysteine-rich motif, participates in interaction with host cell proteins, dimerization with metal ions, and multimerization (Frankel and Young 1998; Bogerd et al. 1993). A structurally superimposable, hydrophobic core of amino acids 37–48 is involved in TAR RNA binding in addition to Tat mul-timerization (Churcher et al. 1993). Mutagenesis of this conserved region revealed Gly44 and Lys41 as important amino acids for interaction with TAR (Churcher et al. 1993). Tat binds TAR with high affinity through a basic domain sequence (amino acids 48–61), which is also the nuclear localization signal (NLS) – RKKKRRQRRRAHQN (Siomi et al. 1990; Berkhout et al. 1989; Hauber et al. 1989; Efthymiadis et al. 1998). Prior to binding, Tat recognizes a tri-nucleotide bulge within the TAR stem-loop through its core sequence (Fig. 17.3). Nucleotides flanking the TAR bulge (G21A22 and G26A27, U38, and U40) additionally collaborate in Tat–TAR interaction (Churcher et al. 1993; Weeks and Crothers 1991; Wang and Rana 1996).

The HIV 5′ LTR region acts as a recruitment center for accumulation of host transcriptional factors that aid in Tat function. The following sequence motifs are documented within the 5′ LTR: two core enhancer elements, three Sp1 transcriptional binding sites, a TATA box, and TAR (Ruben et al. 1989). In preparation for transcription, RNA polymerase is phosphorylated at the carboxy terminal domain (CTD) by TFIIH and interacts with the folded, stem-loop structure of TAR (Wu-Baer et al. 1995). Tat recognition of the TAR RNA bulge allows the interaction of Tat-associated kinase (TAK) (Wei et al. 1998), resulting in phosphorylation of the

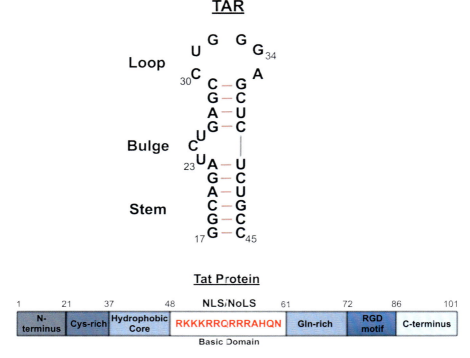

Fig. 17.3 Tat recognizes TAR through the presence of a tri-nucleotide bulge, which initiates transcription of HIV mRNA. A basic domain sequence within Tat achieves three functions – interaction with the TAR stem loop structure within the 5′ LTR of HIV transcripts, nuclear localization, and nucleolar accumulation

CTD of RNA polymerase II (West and Karn 1999). CREB binding protein (p300/CBP) causes dissociation of Tat from TAR via acetylation of Lys50 (Kiernan et al. 1999), leading to the interaction of Tat-TAK with the histone acetyltransferase PCAF (p300/CBP-associated factor). The Tat-TAK-PCAF complex maintains RNA polymerase II in a phosphorylated state during transcription of the entire HIV genome (Karn 1999; Benkirane et al. 1998).

Although transcription of the provirus occurs in the nucleus, Tat intracellular localization is not restricted there. Tat cellular dispersion was observed, implying a variety of functions including the following: transactivation of non-HIV viruses – polyomavirus BK, human papillomavirus (HPV), KS-associated herpesvirus (KSHV), and JC virus (JCV) (Gorrill et al. 2006; Aoki and Tosato 2004; Turner and Palefsky 1998; Stettner et al. 2009); immune suppression of SU-specific T-cell responses (Gupta et al. 2008); down regulation of ERCC1 and IER3 by a TAR miRNA for apoptosis inhibition (Klase et al. 2009) and HIV reverse transcription (Bres et al. 2002; Liang and Wainberg 2002). Tat nuclear localization is conserved and is achieved through the NLS, which also directs nucleolar accumulation (Siomi et al. 1990; Hauber et al. 1989; Kuppuswamy et al. 1989). The last four amino acids of the NLS – AHQN – were demonstrated to function as the nucleolar localization signal (NoLS). Mutational analyses of deletion mutations as well as amino acid substitutions in the Tat NLS interrupted nuclear localization (Siomi et al. 1990; Hauber et al. 1989; Kuppuswamy et al. 1989) and reduced HIV transcriptional activity (Kuppuswamy et al. 1989). The point mutation (R55Q) did not affect TAR binding but delayed HIV replication by 12 days (Neuveut and Jeang 1996). Although the disruption of the Tat NLS on HIV replication is documented, viral replication in response to a damaged Tat NoLS has not been tested. Ponti et al. demonstrated the ability of Tat to interact with fibrillarin and the U3 snoRNA (both are nucleolar factors necessary for pre-rRNA processing), leading to reduced expression of the 80S ribosome (Ponti et al. 2008). Tampering with Tat NoLS may hinder the ability of Tat to manipulate host cell responses during HIV infection.

17.2.2 Rev in the Nucleolus and RRE

The 18-kDa Rev phosphoprotein is a nucleocytoplasmic protein that shuttles, unspliced and partially spliced HIV mRNAs to the cytoplasm. Preliminary studies investigating Rev function involved transient transfections of cells with Rev cDNA expression vectors and Rev-responsive reporter plasmids (proviral and subgenomic) (Chang and Sharp 1989; Ivey-Hoyle and Rosenberg 1990). Such studies demonstrated that Rev is required for cytoplasmic accumulation of intron-containing, 9 and 4 kb HIV mRNAs. In the absence of Rev expression, only fully spliced versions (2 kb) of HIV mRNAs were observed, suggesting the importance of Rev in the prevention of splicing as well as HIV nuclear export.

Rev is documented to accumulate predominantly within the nucleolus; it is suggested that this nucleolar behavior facilitates HIV infection (Felber et al. 1989; Perkins et al. 1989; Malim et al. 1989a; Meyer and Malim 1994). Rev localizes to dense fibrillar components (DFC) and granular compartments (GC) of the nucleolus (Dundr et al. 1995) using an arginine-rich RNA binding motif (ARM) of the NLS/NoLS – RQARRNRRRRWRERQRQ (Cochrane et al. 1990; Kubota et al. 1989). The Rev protein structure is 116 amino acids in length, and residues 34–51 mediate both nuclear and nucleolar localization; this motif has also been shown to bind the RRE (Kjems et al. 1992; Tan et al. 1993). Nuclear magnetic resonance (NMR)-based analyses have shown that four arginine residues (at positions 35, 39, 40, and 44) participate in base-specific contacts with the high-affinity binding site of the IIB and IID RRE stem regions (Battiste et al. 1996), referred to as the RBE. Point mutations incorporating lysine in replacement of arginine decreased the Rev/RRE binding affinity by 3–10-fold (Hammerschmid et al. 1994). As Rev exhibits strong binding affinity to the RRE and has strong nucleolar localization properties, there exists the distinct probability that Rev/RRE complexes accumulate within or traffic through the nucleoli as part of the Rev/RRE-mediated cytoplasmic export machinery.

Rev shuttles HIV transcripts via binding to the *cis*-acting target – the RRE (Hope et al. 1990; Malim et al. 1989b). The RRE lies within the *Env* coding region, which in the absence of Rev binding behaves as an intron (between the 5′-splice site 4 and 3′-splice site 6) and is present in all incompletely spliced viral mRNAs (Pollard and Malim 1998). Studies involving RRE mutagenesis (Heapy et al. 1990), in vitro

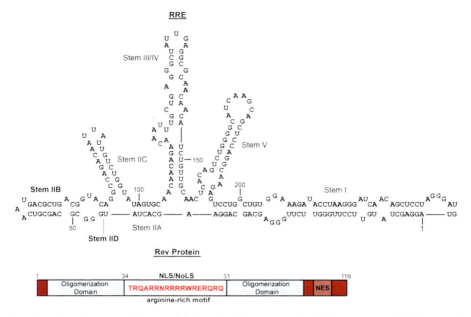

Fig. 17.4 The Rev ARM exhibits triple functionality as an NLS, NoLS, and an RNA-binding motif specific to the RRE intron within unspliced/partially spliced HIV mRNAs. Rev multimerization takes place at stem loops IIB and IID (both stem loops are the RBE of Rev) within the RRE structure

binding (Cook et al. 1991), chemical modification interference (Kjems et al. 1992), and iterative in vitro genetic selection assays (Bartel et al. 1991) mapped the 351 nt RRE/Rev binding sequence to the IIB and IID stem regions (RBE) within unspliced and partially-spliced HIV-1 transcripts (Fig. 17.4). Further characterization of Rev/RRE interactions was accomplished utilizing NMR, which showed that the RRE/Rev complex contained a 34-nt RNA hairpin bound to a 23 amino acid Rev peptide (Battiste et al. 1996). Although the RRE was demonstrated to harbor a single high-affinity Rev binding site, in vitro binding and foot printing studies using full-length Rev and RRE fragments demonstrated multiple binding sites for up to eight Rev proteins on a single RRE (Heapy et al. 1990; Cook et al. 1991; Malim et al. 1990). These observations contributed to the understanding of contact points between Rev and the RRE, and further characterized the Rev-dependent nucleocytoplasmic transport of HIV-1 transcripts.

17.2.3 Significance of Rev Nucleolar Localization

The nucleolar accumulation pattern of Tat is weaker than that of Rev. Using cell fusion assays, Tat nucleocytoplasmic shuttling activity was observed to be 3–4 times slower than Rev; this activity is best explained with the HIV transcriptional process

17 Role of the Nucleolus in HIV Infection and Therapy

requiring minimal amounts of Tat (Stauber and Pavlakis 1998), and the host cellular machinery requiring abundant Tat to cooperate in the HIV infectious cycle. A previous report demonstrated that amino acids 40–45 (NRRRRW) within Rev are sufficient for the maintenance of nucleolar localization (Perkins et al. 1989). Studies by Cochrane et al. further identified residues 45–51 as vital for nucleolar accumulation (Cochrane et al. 1990). Site-directed mutagenesis and indirect immunofluorescence demonstrated that Rev mutants lacking amino acids 48–51 (RQRQ) maintained nucleolar accumulation. However, when deletions were expanded to residues WRE, resulting in complete elimination of WRERQRQ (amino acids 45–51) from the NoLS, nuclear accumulation replaced the nucleolar localization pattern (Cochrane et al. 1990).

The effect of Rev-NoLS mutants on HIV replication is currently under investigation. Rev mutants that lose the ability to enter the nucleus and nucleoli were unable to rescue a Rev-deficient HIV-1 provirus (Arizala and Rossi, unpublished observations). Rev mutants containing single point mutations that replace arginine residues within the NoLS and maintain nucleolar localization were able to rescue HIV production as efficiently as wild-type Rev; however, virions produced by these mutants were defective in infection (Arizala and Rossi, unpublished observations).

Rev is critical in the nuclear export of intron-containing HIV transcripts. It is suggested that the nucleolus acts as an interaction site for Rev with transport factors required in the nuclear-to-cytoplasmic export of HIV transcripts. Fischer et al. demonstrated that the Rev activation domain consists of a nuclear export signal (NES) that redirects RRE-containing viral mRNAs to the cytosol utilizing a non-mRNA export pathway (Fischer et al. 1995). CRM1, an mRNA-independent export factor, was demonstrated to interact with the Rev NES and mediate nuclear export of Rev-bound RNAs (Fukuda et al. 1997; Fornerod et al. 1997). In addition, expression of Rev in human cells resulted in the colocalization of CRM1 and nucleoporins (Nup98 and Nup214) within the nucleolus (Zolotukhin and Felber 1999). A nucleolar step during HIV replication would involve the CRM1-dependent shuttling of Rev-bound HIV transcripts towards the cytoplasm.

Nucleolar localization would be necessary for prevention of spliceosomal assembly along unprocessed HIV transcripts. The oligomeric-bound state of Rev with RRE-containing mRNAs was examined; the Rev NLS blocked spliceosomal assembly by preventing accumulation of the U4·U6, and U5 small nuclear ribonucleoprotein (snRNP) particles along transcripts (Kjems and Sharp 1993). Addition of one, three, or six Rev binding domains in tandem within the RRE-containing intron increased the inhibition of spliceosomal complex formation, prevented splicing, and facilitated rapid mRNA export. It is possible that the Rev NLS/NoLS protects HIV transcripts from being spliced through allowing nucleolar accumulation.

Unsliced/partially spliced HIV transcripts contain premature termination codons (PTCs), and are prone to degradation by nonsense-mediated decay (NMD). Early studies of the Rous sarcoma virus (RSV) indicated that *gag* genes containing nonsense codons resulted in instability of unspliced viral transcripts (Arrigo and Beemon 1988; Barker and Beemon 1991, 1994). A dominant-negative mutant of Upf1 was generated; this mutation disabled the ability of Upf1 to identify transcripts for NMD. In the presence of mutant Upf1, PTC-containing RSV transcripts were stabilized and were not subjected to degradation through the NMD pathway. This observation revealed the susceptibility of unspliced viral transcripts to NMD degradation. In the case of HIV, Rev-directed nucleocytoplasmic shuttling would prevent unpsliced and partially spliced HIV transcripts from undergoing NMD.

17.3 HIV-Specific Targets in the Nucleolus

17.3.1 Nucleolar Ribozyme Against HIV

The nucleolar trafficking of HIV transcripts would depend on its interaction with Tat and/or Rev, both of which accumulate within this subnuclear compartment. Nucleolar trafficking would be necessary for rapid nuclear export of HIV transcripts, facilitated through the CRM1 shuttling complex. To test the hypothesis that HIV RNAs enter the nucleolus during HIV infection, a nucleolar-localizing U16 hammerhead ribozyme fusion was designed to cleave a highly conserved sequence within the U5 region of the HIV-1 LTR (Ojwang et al. 1992; Ratner et al. 1985). The inhibitory function of this ribozyme was investigated in HIV-infected HeLa CD4[+] and CEM T-lymphocytes. The previously studied anti-HIV hammerhead ribozyme (Uhlenbeck 1987; Haseloff and Gerlach 1988; Ruffner et al. 1990) was embedded within the backbone of the U16 snoRNA (Fig. 17.5a, b), which is a nucleolar-localizing, C/D box snoRNA (Fragapane et al. 1993; Weinstein and Steitz 1999). The U16 snoRNA achieves nucleolar localization in the presence of two NoLS sequences – the first NoLS within box C and the second NoLS within box D (Lange et al. 1998; Samarsky et al. 1998). U16 snoRNAs guide and introduce the $2'\text{-}O$-methylation of riboses in pre-rRNA transcripts. Nucleolar localization of

---

Fig. 17.5 U16 snoRNA was used as a backbone for the nucleolar expression of anti-HIV hammerhead ribozyme. (**a**) Ribozyme catalytic activity is specific to the highly conserved position within the 5′ LTR transcription initiation site. Shown within the U16 snoRNA are the C/D boxes, which function as a NoLS. As a control for negative catalytic activity, a single point mutation C to G (depicted with *arrow*) was used. (**b**) Transcription of U16Rz is driven from the human U6 small nuclear RNA promoter. (**c**) In situ hybridization using fluorescently labeled probes complementary to the U16Rz demonstrated a nucleolar localization pattern in HEK293 cells. A probe complementary to endogenous U3 snoRNA was used as a positive control for nucleolar localization. (**d**) As a preliminary test for in vivo efficacy, wild-type and mutant U16Rz were cloned into a retroviral vector, transfected into HeLa CD4[+] cells, and selected by puromycin resistance. RNA

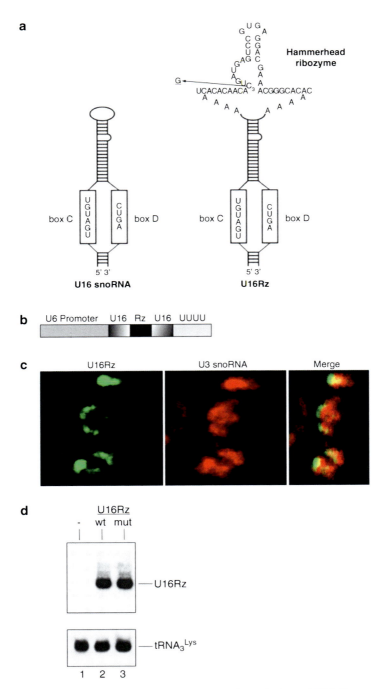

Fig. 17.5 (continued) was collected from the selected cell population and U16 ribozyme expression analyzed. Lane 1 contains extracted RNA from untransfected HeLa CD4[+] cells, lane 2 contains wild-type U16Rz, and lane 3 contains mutant U16Rz. Loading controls were monitored using probes specific for tRNA$_3^{Lys}$

the U16Rz was confirmed using in situ hybridization with fluorescent probes specific to U16Rz; U16Rz was demonstrated to colocalize with the U3 snoRNA control (Fig. 17.5c). The RNA expression from U16Rz was established prior to HIV infectious studies (Fig. 17.5d).

To test the anti-viral activity of U16Rz against HIV-1, CEM T-lymphocytes stably expressing either wild-type or mutant U16Rz were subjected to HIV-1$_{NL4-3}$ infection. Cell media containing infectious HIV-1 were collected 7–25 days after infection and viral production assessed by p24 ELISA antigen capture assay. Unlike the mutant U16Rz, which allowed HIV-1 proliferation up to 25 days of infection, the wild-type U16Rz dramatically suppressed HIV-1 production throughout (Michienzi et al. 2000). In support of the p24 data, indirect immunofluorescence assays utilizing heat-inactivated HIV-1 seropositive human serum (Sandstrom et al. 1985) were used to measure HIV-1 inhibition in the presence of the U16Rz. CEM cells stably expressing the mutant or wild-type U16Rz were infected with HIV-1$_{NL4-3}$ and tested for HIV antigenicity. HIV-1 antigen staining was undetectable in CEM cells expressing the wild-type U16Rz, supporting the p24 antigen results (Fig. 17.6). In contrast, the mutant U16Rz-expressing CEM cells revealed intense HIV antigen staining, indicating HIV-1 proliferation. As a negative control for both the wild type and mutant ribozymes, cells expressing these ribozymes were infected with HIV-2$_{ROD}$, which lacks the ribozyme target sequence. The serum tests were positive for these cell samples.

To test the catalytic strength of U16Rz, stable CEM clones expressing the wild-type or mutant U16Rz were infected with HIV-1$_{NL4-3}$ at an MOI (multiplicity of infection) tenfold greater than the 0.002 used in previous experiments. Total RNA was collected 11 days postinfection and analyzed via Northern blot analyses using a cDNA probe complementary to the Rev mRNA sequence. Splice variants of HIV-1 (9 kb unspliced, 4 kb partially spliced, and 2 kb completely spliced) were detectable in stable CEM clones expressing the mutant U16Rz, whereas unspliced and partially spliced HIV RNA were undetectable in RNA samples from CEM clones expressing the wild-type U16Rz (Fig. 17.7). As expected, fully spliced variants were detected in this sample since they do not contain the RRE and hence would not traffic through the nucleolus. The U16Rz selectively cleaves unspliced and partially spliced transcripts that have traversed the nucleolus where the ribozyme was sequestered. The potent catalytic activity of nucleolar-specific U16Rz against HIV RRE-containing transcripts supports the nucleolar trafficking of HIV-1 transcripts during HIV replication.

17.3.2 HIV in Response to Nucleolar Decoys

The nucleolar localization concept was used as a means of trapping HIV regulatory proteins within the nucleolar compartment by incorporating RBE or TAR elements into a snoRNA backbone. A previous study incorporated the RBE motif into the

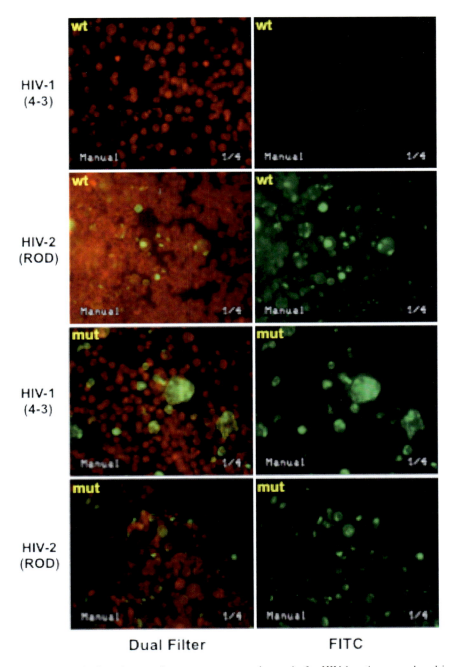

Fig. 17.6 HIV indirect immunofluorescence was used to stain for HIV-1 antigens produced in CEM clones stably expressing either the wild-type or mutant U16Rz after infection with HIV-1$_{NL4-3}$ or HIV-2$_{ROD}$. The *top four panels* depict CEM clones producing wild-type U16Rz. The *lower four* panels are CEM clones producing mutant U16Rz. HIV antigenicity was measured 17 days after HIV infection. The *left panel* of cells was subjected to dual filters (FITC/rhodamine), and visualization of the *right panel* was obtained with a FITC filter. HIV-infected cells fluoresce *green* whereas uninfected cells fluoresce *red*

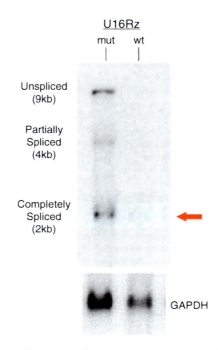

Fig. 17.7 Northern blot analysis was utilized in the examination of U16Rz catalytic activity against HIV at an MOI 10-fold higher than 0.002. After HIV infection of stable CEM clones, total RNA was collected. RNA samples (5 μg) were subjected to 1% agarose/formaldehyde gel electrophoresis and hybridized with a cDNA probe complementary to HIV Rev mRNA for detection of all classes of HIV RNA. The *arrow* depicts fully spliced HIV transcripts. Within the *upper panel*, lane 1 contains total RNA exposed from the mutant U16Rz-expressing cells and lane 2 contains total RNA isolated from the wild-type U16Rz-expressing cells. Endogenous GAPDH mRNA was used as an RNA loading control (*lower panel*)

U16 snoRNA backbone (U16RBE – U16 Rev binding element) for the purpose of capturing Rev within the nucleolus. The U16RBE was an ideal nucleolar decoy to investigate, as the intracellular localization pattern was compartmentalized within nucleoli of HEK293 cell lines (Buonuomo et al. 1999). An eGFP-expression vector containing the U6 promoter-driven, U16RBE Rev decoy was transduced into CD34[+] progenitor cells; eGFP fluorescence was utilized in the sorting of transduced progenitor clones. Isolated CD34[+] clones stably expressing the U16RBE were subjected to challenges with HIV-1 JFRL at an MOI of 0.001. Viral production within the culture supernatant was assessed using p24 ELISA, which revealed the U16RBE Rev decoy as an efficient inhibitor of HIV replication for up to 21 days postinfection (Michienzi et al. 2006).

In addition to the U16RBE decoy, Michienzi et al. utilized the U16 snoRNA backbone as a nucleolar decoy for Tat. A functional TAR motif was placed into the apical loop of the U16 snoRNA (U16TAR - U16 transactivation response element), and its nucleolar localization pattern was demonstrated through colocalization with fibrillarin (Michienzi et al. 2002). Nucleolar-specific fibrillarin is a component of the C/D box snoRNA complexes involved in the processing of ribosomal RNA (Weinstein and Steitz 1999; Venema and Tollervey 1999). The U6-driven vector-expression construct was transduced into CEM cells for the generation of puromycin-resistant CEM clones stably expressing U16TAR. Stable CEM clones were infected with HIV-1$_{NL4-3}$, and HIV-1 reverse transcriptase was monitored 7–21 days postinfection. U16TAR was shown to inhibit HIV-1, indicating an arrest in viral production through the titration of functional Tat within the nucleoli (Michienzi et al. 2002). These results further support a critical role of the nucleolus during HIV-1 replication.

17.4 Conclusion

The potential for novel anti-HIV therapies to develop depends greatly on the full understanding of intracellular methodologies allowing HIV infection. Such research endeavors have led to the characterization of two HIV regulatory proteins – Tat and Rev. The nucleolar behavior of both proteins implicates a nucleolar step during HIV replication, and is expected to function in the presence of other host cell factors. These functions include, yet are not limited to, transactivation and nucleocytoplasmic shuttling of HIV transcripts. Nucleolar-specific ribozymes and decoys demonstrated the inhibition HIV production. This not only confirmed the nucleolar localization pattern of Tat and Rev, but revealed for the first time the ability of HIV RNAs to enter the nucleolus. Therapeutic approaches that manipulate the nucleolus during viral infection are significant in that such concepts can be applied to other viral infectious models that require a nucleolar step for replication.

Acknowledgments This research was supported by grants from the NIH NIAD to JJR.

References

Aoki Y, Tosato G (2004) HIV-1 Tat enhances Kaposi sarcoma-associated herpesvirus (KSHV) infectivity. Blood 104:810–814
Arrigo S, Beemon K (1988) Regulation of Rous sarcoma virus RNA splicing and stability. Mol Cell Biol 8:4858–4867
Bailes E, Gao F, Bibollet-Ruche F, Courgnaud V, Peeters M, et al (2003) Hybrid origin of SIV in chimpanzees. Science 300:1713
Baltimore D (1970) RNA-dependent DNA polymerase in virions of RNA tumour viruses. Nature 226:209–211
Barker GF, Beemon K (1991) Nonsense codons within the Rous sarcoma virus gag gene decrease the stability of unspliced viral RNA. Mol Cell Biol 11:2760–2768
Barker GF, Beemon K (1994) Rous sarcoma virus RNA stability requires an open reading frame in the gag gene and sequences downstream of the gag-pol junction. Mol Cell Biol 14:1986–1996

Barre-Sinoussi F, Chermann JC, Rey F, Nugeyre MT, Chamaret S et al (1983) Isolation of a T-lymphotropic retrovirus from a patient at risk for acquired immune deficiency syndrome (AIDS). Science 220:868–871

Bartel DP, Zapp ML, Green MR, Szostak JW (1991) HIV-1 Rev regulation involves recognition of non-Watson-Crick base pairs in viral RNA. Cell 67:529–536

Battiste JL, Mao H, Rao NS, Tan R, Muhandiram DR et al (1996) Alpha helix-RNA major groove recognition in an HIV-1 rev peptide-RRE RNA complex. Science 273:1547–1551

Benkirane M, Chun RF, Xiao H, Ogryzko VV, Howard BH et al (1998) Activation of integrated provirus requires histone acetyltransferase. p300 and P/CAF are coactivators for HIV-1 Tat. J Biol Chem 273:24898–24905

Berkhout B, Silverman RH, Jeang KT (1989) Tat trans-activates the human immunodeficiency virus through a nascent RNA target. Cell 59:273–282

Bogerd HP, Fridell RA, Blair WS, Cullen BR (1993) Genetic evidence that the Tat proteins of human immunodeficiency virus types 1 and 2 can multimerize in the eukaryotic cell nucleus. J Virol 67:5030–5034

Bres V, Kiernan R, Emiliani S, Benkirane M (2002) Tat acetyl-acceptor lysines are important for human immunodeficiency virus type-1 replication. J Biol Chem 277:22215–22221

Broder CC, Berger EA (1995) Fusogenic selectivity of the envelope glycoprotein is a major determinant of human immunodeficiency virus type 1 tropism for CD4+ T-cell lines vs. primary macrophages. Proc Natl Acad Sci U S A 92:9004–9008

Buonuomo SB, Michienzi Z, Caffarelli E, Bozzoni I (1999) The Rev protein is able to transport to the cytoplasm small nucleolar RNAs containing a Rev binding element. RNA 5:993–1002

Chang DD, Sharp PA (1989) Regulation by HIV Rev depends upon recognition of splice sites. Cell 59:789–795

Churcher MJ, Lamont C, Hamy F, Dingwall C, Green SM et al (1993) High affinity binding of TAR RNA by the human immunodeficiency virus type-1 tat protein requires base-pairs in the RNA stem and amino acid residues flanking the basic region. J Mol Biol 230:90–110

Cochrane AW, Perkins A, Rosen CA (1990) Identification of sequences important in the nucleolar localization of human immunodeficiency virus Rev: relevance of nucleolar localization to function. J Virol 64:881–885

Coffin J, Haase A, Levy JA, Montagnier L, Oroszlan S et al (1986a) Human immunodeficiency viruses. Science 232:697

Coffin J, Haase A, Levy JA, Montagnier L, Oroszlan S et al (1986b) What to call the AIDS virus? Nature 321:10

Cook KS, Fisk GJ, Hauber J, Usman N, Daly TJ et al (1991) Characterization of HIV-1 REV protein: binding stoichiometry and minimal RNA substrate. Nucleic Acids Res 19:1577–1583

Desrosiers RC, Daniel MD, Li Y (1989) HIV-related lentiviruses of nonhuman primates. AIDS Res Hum Retroviruses 5:465–473

Dundr M, Leno GH, Hammarskjold ML, Rekosh D, Helga-Maria C et al (1995) The roles of nucleolar structure and function in the subcellular location of the HIV-1 Rev protein. J Cell Sci 108(Pt 8):2811–2823

Efthymiadis A, Briggs LJ, Jans DA (1998) The HIV-1 Tat nuclear localization sequence confers novel nuclear import properties. J Biol Chem 273:1623–1628

Fanales-Belasio E, Raimondo M, Suligoi B, Butto S (2010) HIV virology and pathogenetic mechanisms of infection: a brief overview. Ann Ist Super Sanita 46:5–14

Fankhauser C, Izaurralde E, Adachi Y, Wingfield P, Laemmli UK (1991) Specific complex of human immunodeficiency virus type 1 rev and nucleolar B23 proteins: dissociation by the Rev response element. Mol Cell Biol 11:2567–2575

Felber BK, Hadzopoulou-Cladaras M, Cladaras C, Copeland T, Pavlakis GN (1989) rev protein of human immunodeficiency virus type 1 affects the stability and transport of the viral mRNA. Proc Natl Acad Sci U S A 86:1495–1499

Fenner F (1975) The classification and nomenclature of viruses. Summary of results of meetings of the International Committee on Taxonomy of Viruses in Madrid, September 1975. Intervirology 6:1–12

17 Role of the Nucleolus in HIV Infection and Therapy

Fischer U, Huber J, Boelens WC, Mattaj IW, Luhrmann R (1995) The HIV-1 Rev activation domain is a nuclear export signal that accesses an export pathway used by specific cellular RNAs. Cell 82:475–483

Fornerod M, Ohno M, Yoshida M, Mattaj IW (1997) CRM1 is an export receptor for leucine-rich nuclear export signals. Cell 90:1051–1060

Fragapane P, Prislei S, Michienzi A, Caffarelli E, Bozzoni I (1993) A novel small nucleolar RNA (U16) is encoded inside a ribosomal protein intron and originates by processing of the pre-mRNA. EMBO J 12:2921–2928

Frankel AD, Young JA (1998) HIV-1: fifteen proteins and an RNA. Annu Rev Biochem 67:1–25

Fukuda M, Asano S, Nakamura T, Adachi M, Yoshida M et al (1997) CRM1 is responsible for intracellular transport mediated by the nuclear export signal. Nature 390:308–311

Gallo RC, Salahuddin SZ, Popovic M, Shearer GM, Kaplan M et al (1984) Frequent detection and isolation of cytopathic retroviruses (HTLV-III) from patients with AIDS and at risk for AIDS. Science 224:500–503

Gao F, Yue L, White AT, Pappas PG, Barchue J et al (1992) Human infection by genetically diverse SIVSM-related HIV-2 in west Africa. Nature 358:495–499

Gao F, Bailes E, Robertson DL, Chen Y, Rodenburg CM et al (1999) Origin of HIV-1 in the chimpanzee Pan troglodytes troglodytes. Nature 397:436–441

Gardner MB, Luciw PA (1989) Animal models of AIDS. FASEB J 3:2593–2606

Gorrill T, Feliciano M, Mukerjee R, Sawaya BE, Khalili K et al (2006) Activation of early gene transcription in polyomavirus BK by human immunodeficiency virus type 1 Tat. J Gen Virol 87:1557–1566

Gottlieb MS, Schroff R, Schanker HM, Weisman JD, Fan PT et al (1981) *Pneumocystis carinii* pneumonia and mucosal candidiasis in previously healthy homosexual men: evidence of a new acquired cellular immunodeficiency. N Engl J Med 305:1425–1431

Gross L (1951) Pathogenic properties, and "vertical" transmission of the mouse leukemia agent. Proc Soc Exp Biol Med 78:342–348

Gupta S, Boppana R, Mishra GC, Saha B, Mitra D (2008) HIV-1 Tat suppresses gp120-specific T cell response in IL-10-dependent manner. J Immunol 180:79–88

Hammerschmid M, Palmeri D, Ruhl M, Jaksche H, Weichselbraun I et al (1994) Scanning mutagenesis of the arginine-rich region of the human immunodeficiency virus type 1 Rev trans activator. J Virol 68:7329–7335

Hardy WD (1985) Feline retroviruses. Adv Viral Oncol 5:1–34

Haseloff J, Gerlach WL (1988) Simple RNA enzymes with new and highly specific endoribonuclease activities. Nature 334:585–591

Hauber J, Malim MH, Cullen BR (1989) Mutational analysis of the conserved basic domain of human immunodeficiency virus tat protein. J Virol 63:1181–1187

Heapy S, Dingwall C, Ernberg I, Gait MJ, Green SM, Kern J, Lowe AD, Singh M, Skinner MA (1990) HIV-1 regulator of virion expression (Rev) protein binds to an RNA stem-loop structure located within the Rev response element region. Cell 60(4):685–693

Hope TJ, Huang XJ, McDonald D, Parslow TG (1990) Steroid-receptor fusion of the human immunodeficiency virus type 1 Rev transactivator: mapping cryptic functions of the arginine-rich motif. Proc Natl Acad Sci U S A 87:7787–7791

Ivey-Hoyle M, Rosenberg M (1990) Rev-dependent expression of human immunodeficiency virus type 1 gp160 in *Drosophila melanogaster* cells. Mol Cell Biol 10:6152–6159

Kalyanaraman VS, Sarngadharan MG, Robert-Guroff M, Miyoshi I, Golde D et al (1982) A new subtype of human T-cell leukemia virus (HTLV-II) associated with a T-cell variant of hairy cell leukemia. Science 218:571–573

Karn J (1999) Tackling Tat. J Mol Biol 293:235–254

Kiernan RE, Vanhulle C, Schiltz L, Adam E, Xiao H et al (1999) HIV-1 tat transcriptional activity is regulated by acetylation. EMBO J 18:6106–6118

Kjems J, Sharp PA (1993) The basic domain of Rev from human immunodeficiency virus type 1 specifically blocks the entry of U4/U6.U5 small nuclear ribonucleoprotein in spliceosome assembly. J Virol 67:4769–4776

Kjems J, Calnan BJ, Frankel AD, Sharp PA (1992) Specific binding of a basic peptide from HIV-1 Rev. EMBO J 11:1119–1129

Klase Z, Winograd R, Davis J, Carpio L, Hildreth R, et al (2009) HIV-1 TAR miRNA protects against apoptosis by altering cellular gene expression. Retrovirology 6:18

Kubota S, Siomi H, Satoh T, Endo S, Maki M et al (1989) Functional similarity of HIV-I rev and HTLV-I rex proteins: identification of a new nucleolar-targeting signal in rev protein. Biochem Biophys Res Commun 162:963–970

Kuppuswamy M, Subramanian T, Srinivasan A, Chinnadurai G (1989) Multiple functional domains of Tat, the trans-activator of HIV-1, defined by mutational analysis. Nucleic Acids Res 17:3551–3561

Lange TS, Borovjagin A, Maxwell ES, Gerbi SA (1998) Conserved boxes C and D are essential nucleolar localization elements of U14 and U8 snoRNAs. EMBO J 17:3176–3187

Levy JA, Hoffman AD, Kramer SM, Landis JA, Shimabukuro JM et al (1984) Isolation of lymphocytopathic retroviruses from San Francisco patients with AIDS. Science 225:840–842

Li YP (1997) Protein B23 is an important human factor for the nucleolar localization of the human immunodeficiency virus protein Tat. J Virol 71:4098–4102

Liang C, Wainberg MA (2002) The role of Tat in HIV-1 replication: an activator and/or a suppressor? AIDS Rev 4:41–49

Lucas SB, Hounnou A, Peacock C, Beaumel A, Djomand G et al (1993) The mortality and pathology of HIV infection in a west African city. AIDS 7:1569–1579

Luciw PA (1996) Human immunodeficiency virus and their replication. In: Fields BN, Knippe DM, Howley PM (eds) Field virology. Lippincott-Raven, Philadelphia, pp 1881–1952

Malim MH, Bohnlein S, Hauber J, Cullen BR (1989a) Functional dissection of the HIV-1 Rev trans-activator–derivation of a trans-dominant repressor of Rev function. Cell 58:205–214

Malim MH, Hauber J, Le SY, Maizel JV, Cullen BR (1989b) The HIV-1 rev trans-activator acts through a structured target sequence to activate nuclear export of unspliced viral mRNA. Nature 338:254–257

Malim MH, Tiley LS, McCarn DF, Rusche JR, Hauber J et al (1990) HIV-1 structural gene expression requires binding of the Rev trans-activator to its RNA target sequence. Cell 60:675–683

Masur H, Michelis MA, Greene JB, Onorato I, Stouwe RA et al (1981) An outbreak of community-acquired *Pneumocystis carinii* pneumonia: initial manifestation of cellular immune dysfunction. N Engl J Med 305:1431–1438

Meyer BE, Malim MH (1994) The HIV-1 Rev trans-activator shuttles between the nucleus and the cytoplasm. Genes Dev 8:1538–1547

Miceli MC, Parnes JR (1993) Role of CD4 and CD8 in T cell activation and differentiation. Adv Immunol 53:59–122

Michienzi A, Cagnon L, Bahner I, Rossi JJ (2000) Ribozyme-mediated inhibition of HIV 1 suggests nucleolar trafficking of HIV-1 RNA. Proc Natl Acad Sci U S A 97:8955–8960

Michienzi A, Li S, Zaia JA, Rossi JJ (2002) A nucleolar TAR decoy inhibitor of HIV-1 replication. Proc Natl Acad Sci U S A 99:14047–14052

Michienzi A, De Angelis FG, Bozzoni I, Rossi JJ (2006) A nucleolar localizing Rev binding element inhibits HIV replication. AIDS Res Ther 3:13

Mystakidou K, Panagiotou I, Katsaragakis S, Tsilika E, Parpa E (2009) Ethical and practical challenges in implementing informed consent in HIV/AIDS clinical trials in developing or resource-limited countries. SAHARA J 6:46–57

Neuveut C, Jeang KT (1996) Recombinant human immunodeficiency virus type 1 genomes with tat unconstrained by overlapping reading frames reveal residues in Tat important for replication in tissue culture. J Virol 70:5572–5581

Ojwang JO, Hampel A, Looney DJ, Wong-Staal F, Rappaport J (1992) Inhibition of human immunodeficiency virus type 1 expression by a hairpin ribozyme. Proc Natl Acad Sci U S A 89:10802–10806

Perkins A, Cochrane AW, Ruben SM, Rosen CA (1989) Structural and functional characterization of the human immunodeficiency virus rev protein. J Acquir Immune Defic Syndr 2:256–263

Poiesz BJ, Ruscetti FW, Gazdar AF, Bunn PA, Minna JD et al (1980) Detection and isolation of type C retrovirus particles from fresh and cultured lymphocytes of a patient with cutaneous T-cell lymphoma. Proc Natl Acad Sci U S A 77:7415–7419

Pollard VW, Malim MH (1998) The HIV-1 Rev protein. Annu Rev Microbiol 52:491–532

Ponti D, Troiano M, Bellenchi GC, Battaglia PA, Gigliani F (2008) The HIV Tat protein affects processing of ribosomal RNA precursor. BMC Cell Biol 9:32

Popovic M, Sarngadharan MG, Read E, Gallo RC (1984) Detection, isolation, and continuous production of cytopathic retroviruses (HTLV-III) from patients with AIDS and pre-AIDS. Science 224:497–500

Preston BD, Poiesz BJ, Loeb LA (1988) Fidelity of HIV-1 reverse transcriptase. Science 242:1168–1171

Rambaut A, Posada D, Crandall KA, Holmes EC (2004) The causes and consequences of HIV evolution. Nat Rev Genet 5:52–61

Ratner L, Gallo RC, Wong-Staal F (1985) HTLV-III, LAV, ARV are variants of same AIDS virus. Nature 313:636–637

Roberts JD, Bebenek K, Kunkel TA (1988) The accuracy of reverse transcriptase from HIV-1. Science 242:1171–1173

Rous P (1911) A sarcoma of the fowl transmissible by an agent separable from the tumor cells. J Exp Med 13:397–411

Ruben S, Perkins A, Purcell R, Joung K, Sia R et al (1989) Structural and functional characterization of human immunodeficiency virus tat protein. J Virol 63:1–8

Ruffner DE, Stormo GD, Uhlenbeck OC (1990) Sequence requirements of the hammerhead RNA self-cleavage reaction. Biochemistry 29:10695–10702

Samarsky DA, Fournier MJ, Singer RH, Bertrand E (1998) The snoRNA box C/D motif directs nucleolar targeting and also couples snoRNA synthesis and localization. EMBO J 17: 3747–3757

Sandstrom EG, Schooley RT, Ho DD, Byington R, Sarngadharan MG et al (1985) Detection of human anti-HTLV-III antibodies by indirect immunofluorescence using fixed cells. Transfusion 25:308–312

Santiago ML, Rodenburg CM, Kamenya S, Bibollet-Ruche F, Gao F et al (2002) SIVcpz in wild chimpanzees. Science 295:465

Sarngadharan MG, Popovic M, Bruch L, Schupbach J, Gallo RC (1984) Antibodies reactive with human T-lymphotropic retroviruses (HTLV-III) in the serum of patients with AIDS. Science 224:506–508

Schupbach J, Popovic M, Gilden RV, Gonda MA, Sarngadharan MG et al (1984) Serological analysis of a subgroup of human T-lymphotropic retroviruses (HTLV-III) associated with AIDS. Science 224:503–505

Sharp PM, Hahn BH (2008) AIDS: prehistory of HIV-1. Nature 455:605–606

Siegal FP, Lopez C, Hammer GS, Brown AE, Kornfeld SJ et al (1981) Severe acquired immunodeficiency in male homosexuals, manifested by chronic perianal ulcerative herpes simplex lesions. N Engl J Med 305:1439–1444

Siomi H, Shida H, Maki M, Hatanaka M (1990) Effects of a highly basic region of human immunodeficiency virus Tat protein on nucleolar localization. J Virol 64:1803–1807

Sodora DL, Allan JS, Apetrei C, Brenchley JM, Douek DC et al (2009) Toward an AIDS vaccine: lessons from natural simian immunodeficiency virus infections of African nonhuman primate hosts. Nat Med 15:861–865

Stauber RH, Pavlakis GN (1998) Intracellular trafficking and interactions of the HIV-1 Tat protein. Virology 252:126–136

Stettner MR, Nance JA, Wright CA, Kinoshita Y, Kim WK et al (2009) SMAD proteins of oligodendroglial cells regulate transcription of JC virus early and late genes coordinately with the Tat protein of human immunodeficiency virus type 1. J Gen Virol 90:2005–2014

Szebeni A, Mehrotra B, Baumann A, Adam SA, Wingfield PT et al (1997) Nucleolar protein B23 stimulates nuclear import of the HIV-1 Rev protein and NLS-conjugated albumin. Biochemistry 36:3941–3949

Tan R, Chen L, Buettner JA, Hudson D, Frankel AD (1993) RNA recognition by an isolated alpha helix. Cell 73:1031–1040

Temin HM (1964) Nature of the provirus of Rous sarcoma. Natl Cancer Inst Monogr 17:557–570

Temin HM, Mizutani S (1970) RNA-dependent DNA polymerase in virions of Rous sarcoma virus. Nature 226:211–213

Truant R, Cullen BR (1999) The arginine-rich domains present in human immunodeficiency virus type 1 Tat and Rev function as direct importin beta-dependent nuclear localization signals. Mol Cell Biol 19:1210–1217

Turner MA, Palefsky JM (1998) HIV-1 Tat protein increases invasion of human papillomavirus type 16 positive keratinocytes. J Acquir Immune Defic Syndr 17:A13

Uhlenbeck OC (1987) A small catalytic oligoribonucleotide. Nature 328:596–600

Venema J, Tollervey D (1999) Ribosome synthesis in *Saccharomyces cerevisiae*. Annu Rev Genet 33:261–311

Wang Z, Rana TM (1996) RNA conformation in the Tat-TAR complex determined by site-specific photo-cross-linking. Biochemistry 35:6491–6499

Weeks KM, Crothers DM (1991) RNA recognition by Tat-derived peptides: interaction in the major groove? Cell 66:577–588

Wei P, Garber ME, Fang SM, Fischer WH, Jones KA (1998) A novel CDK9-associated C-type cyclin interacts directly with HIV-1 Tat and mediates its high-affinity, loop-specific binding to TAR RNA. Cell 92:451–462

Weinstein LB, Steitz JA (1999) Guided tours: from precursor snoRNA to functional snoRNP. Curr Opin Cell Biol 11:378–384

West MJ, Karn J (1999) Stimulation of Tat-associated kinase-independent transcriptional elongation from the human immunodeficiency virus type-1 long terminal repeat by a cellular enhancer. EMBO J 18:1378–1386

Whittle H, Morris J, Todd J, Corrah T, Sabally S et al (1994) HIV-2-infected patients survive longer than HIV-1-infected patients. AIDS 8:1617–1620

Wu-Baer F, Sigman D, Gaynor RB (1995) Specific binding of RNA polymerase II to the human immunodeficiency virus trans-activating region RNA is regulated by cellular cofactors and Tat. Proc Natl Acad Sci U S A 92:7153–7157

Zolotukhin AS, Felber BK (1999) Nucleoporins nup98 and nup214 participate in nuclear export of human immunodeficiency virus type 1 Rev. J Virol 73:120–127

Index

A

Acute myeloid leukemia (AML), mutated NPM1
  4-base insertion, TG^G, 230
  leucine-rich NES motif, generation, 231
  N-terminal oligomerizaton domain, 231
  residues
    Trp288 and Trp290, loss, 221, 230
    tryptophan, NES motif, 231
Alternate reading frame (ARF)
  B23 dependent and-independent roles (*see* B23)
  Mdm2 protein, 290
  tumor suppressor
    p53 pathway, 282–283
    ribosome synthesis and cell cycle progression, 164
Apoptosis
  E9 Tcof1$^{+/-}$embryos, 267
  Nopp140, 266
  ribosomes loss, 268
ARF. *See* Alternate reading frame
Assembly
  nucleolar
    cell cycle, mammalian cells, 20
    eukaryotes, 14–15
    morphological features, 15
    photoactivation (PA), 16
    proteins and snoRNAs, 16
    rDNA transcription, NORs, 15–16
    45S rRNAs, 21
  pre–40S and pre–60S, yeast, 17
  ribosomes, 17

B

B23. *See also* Nucleophosmin 1 (NPM1)
  ARF-dependent and-independent role, 284–286
  nucleolar protein, 282
Binding protein, TTF1, 195

C

Cajal bodies (CBs)
  "accessory body", 357
  antibodies, coilin, 359
  characterization
    autoimmune antibodies identification, 364
    coilin accumulation, 365
    dephosphorylation inhibition, 366
    LNCs, 365
    spliceosomal snRNPs, 365
    transcription inhibitors, 364–365
  coilin connection
    description, 366
    ectopic overexpression, 367
    marker protein, localization, 367
    mutational analysis, 366
    Nopp140, 366–367
    posttranslational modification, 367
  cuerpo accesorio
    argyrophilic paranucleolar structure, 361–362
    argyrophilic quality, 360
    FRAP and iFRAP, 363
    germinal vesicle, oocytes, 360–361
    GFP to U2B" fusion, 362
    histone pre-mRNA processing, 361

404 Index

Cajal bodies (CBs) (*cont.*)
    silver nitrate method, 360
    U7 snRNP localization, 361
  cytoplasm, eukaryotic cells, 357
  influencing factors, 368
  localization, proteins, 358
  nucleolar connections
    components, 363
    correlation, 364
    and NORs, 364
    telomerase assembly and telomere
      maintenance, 363
    U3 snoRNPs maturation, 364
  and nucleologenesis, 366
  relationship, 359
  SMNs, 360
  snoRNPs, 359–360
  viruses and DNA damage, effects
    HSV–1, 369
    p53 accumulation, 369
    RNA processing machinery, 368
    UV-C and neurodegeneration, 369
Cancer, 75–76
CBs. *See* Cajal bodies
Cell cycle
  control, 309
  progression (*See* Ribosome synthesis)
  transcriptional regulation
    FRAP-based survey, 122
    G_1-and S-phase, 121
    HeLa cells, 121
    mitosis, 120–121
    Pol I transcription, 121
    synthesis, rRNA, 120
    UNF phosphorylation, 121–122
Cell proliferation
  GZF1, 190
  nucleolin, 201–203
  REST and phosphorylated nucleolin, 192
Chaperone
  box C/D snoRNPs, 269–270
  snoRNPs, 261–262
Chromatin immunoprecipitation (ChIP), 60
Chromosomal organization, rDNA
  distribution, regulatory elements, 85
  gene promoter, mammalian, 85
  intergenic spacers (IGS), 84
  lymphocytes, human, 86
  repeats, 86
  sequence encoding, 84
  transcriptional terminators, 85–86
  Xenopus, 84–85
Coilin
  antibodies, 359
  CBs and nucleolus, 359, 366–367

cDNA sequence, 364
cisplastin and gamma irradiation, 369
herpes simplex virus type 1 (HSV–1), 369
localization, 368
nucleolar caps, 362
transcription arrest, 366

D

DBA. *See* Diamond–Blackfan anemia
Diamond–Blackfan anemia (DBA)
  diagnosis, 168–169
  p53
    activation, 175
    pathway, 171
  ribosomal proteins, 170
  RPS19 production, 169–170
DNA methylation, 59–60
DNA virus interactions
  herpesvirus-encoded proteins, 328–329
  HIPK2 interaction, 330
  KSHV ORF57 protein, 329
  proteins role, 329
  replication, 329–330
  upstream binding factor (UBF), 329–330
Dynamic proteome, nucleolus
  data, NOPdb
    API, 39
    human nucleolar proteins, 37–38
    interpro motif numbers and gene
      onotology, 38
  nucleolar, 31
  nucleolar proteins analysis
    interpretation, 34–35
    metabolic conditions, 35
    microscopy analysis, 37
    p53, 36
    purification procedures, 34
    ribosome subunit biogenesis, 35
  nucleoli isolation
    characterization, 32
    electron microscopy images, 33
    incorporated BrUTP, 33–34
    purity and intactness, 32–33
Dynamics, ribosome biogenesis
  nucleolar assembly/disassembly, 14–16
  nucleoli and, 5–14
  nucleus, nucleolus, 17–20

E

Electron microscopy (EM)
  DFC, 12
  3D reconstructions, 19
  FCs, 11–12

Index 405

GC, 12–13
HeLa cells nucleolar assembly, 11
hybridization, rRNAs, 12
NIH3T3 nucleus, perinucleolar
heterochromatin, 10
nucleolar organization, 6–7
PtK1 nucleolus, 8
ribosome biogenesis, 4
EM. *See* Electron microscopy
Epigenetics nucleolus
cancer, 75–76
and chromatin features
biochemical analyses, 61–62
ChIP, 60
DNA methylation, 59–60
drosophila, 63
fraction, 60–61
gene expression, 58–59
histone H3K9me2, 61
in human, 59–60
nucleosome positioning, 62–63
roles, 63
rRNA gene copies, 58
silent rRNA genes, 59
functions
assembling DNA repeats, 74
bona fide rRNA genes, 73
establishment heterochromatin,
74–75
heterochromatin structure, 72–73
maintaining, inactive X chromosome, 74
TIP5 depletion, NIH3T3, 73–74
inheritance, rDNA chromatin
analysis, 63–64
chromosomal DNA replication, 64–65
expression, monoallelic, 64
memory, 64
mode of action, 66
NoRC-mediated rDNA, 65
NoRC-pRNA, 66–67
nucleosome remodelling activity, 65
pre-rRNA synthesis, 68
replication, 64
rRNA transcription, 65
spacer promoter, findings, 67–68
regulation, rRNA synthesis
binary unit, 69
CpG methylation, 71–72
eNoSC complex, 70
histone modifications, 70
Pol I, transcription, 68–69
role, UBF, 70–71
schema representation, 71
transcription rates, 69
UBF depletion, 71

rRNA production, 58
transcription, rRNA genes, 57
Evolutionary conserved motif (ECM),
188

F
Fibrillarin
coilin
and scaRNAs caps, 359
truncations, 367
FRAP and iFRAP, 363
localization, 358, 366
transcription-inhibition-induced nucleolar
caps, 365

G
Genotoxic stress, Pol transcription
apoptotic program, 127
mice mutant, 127
nucleolar function, 126
phosphorylation, 125–126
ribosome producing factory, 125
ribotoxic stress, TIF-IA, 126
rRNA synthesis, 125
GFP. *See* Green fluorescent protein
GNL3L. *See* Guanine nucleotide binding
protein-like 3
Green fluorescent protein (GFP)
description, 351
expression, 353
GFP-Nopp140, *Drosophila* embryogenesis,
259
SRP localization, 351, 352
SRP19, SRP68 and SRP72 fusion, 351
Growth factor, rDNA transcription
epidermal growth factor receptor, 118
IGF-IR and IRS–1, 119–120
multiple signalling, TIF-IA, 119
positive effects, 118–119
GTP-binding proteins
MDM2 interaction, 312
MMR1_HSR1, 306
NpLS, 308
Guanine nucleotide binding protein-like 3
(GNL3L)
estrogen receptor-related proteins (ERRγ),
313
expression profile and subcellular
distribution, 311–312
MDM2 protein stabilization,
312–313
nucleostemin depletion, 314
TRF1 and G2/M Transition, 313

H

Herpes simplex virus type 1 (HSV–1), 369
HIPK2. *See* Homeodomain-interacting protein kinase 2
Histone chaperone
  nucleolin, chromatin accessibility, 198–199
  nucleoplasmin and NAP–1, 195
Histone H3K9me2, 61
HIV infection and therapy, nucleolus role
  CD4 receptor, 379
  characterization
    description, AIDS, 378
    error-prone reverse transcription, 378
    identification, sequence analysis, 378
    immunodeficiency-related illnesses, 377–378
    Pol gene, 378
    retrovirus and temin's proviral hypothesis, 377
    scientific methodologies, 378
  epidemic affects and types, 379
  genomes, HIV–1 and 2, 379, 380
  nucleolar step
    B23, 383
    Rev-B23 stable interaction, 383
    Rev nucleolar localization, 386–388
    Rev protein and RRE, 385–386
    significance, Tat and Rev, 382
    Tat, nucleolus and TAR, 383–835
  regulatory proteins, Tat and Rev, 382
  replication cycle, 379, 381
  steps, replication, 382
  transcripts and viral-specific target
    nucleolar decoys, 390–393
    nucleolar ribozyme, 388–390
  virion, 379
Homeodomain-interacting protein kinase 2 (HIPK2), 330
HSV–1. *See* Herpes simplex virus type 1

I

Insulin-like growth factor receptor (IGF-IR), 119
Insulin receptor substrate–1 (IRS–1), 95, 119–120

J

Japanese encephalitis virus (JEV), 237–238, 333
JEV. *See* Japanese encephalitis virus

K

Kaposi's sarcoma associated herpesvirus (KSHV), 239, 329, 384
KSHV. *See* Kaposi's sarcoma associated herpesvirus

L

Light nuclear caps (LNCs), 365
LNCs. *See* Light nuclear caps

M

Mass spectrometry
  nucleoli purification, 30, 31
  PSP1, 37
Mdm2. *See* Murine double minute 2
Mdm2/p53 pathway
  amino acid substitutions, 174
  ARF inhibition, 164
  DBA encode, 175
  proteasome-mediated degradation, 162
  ribosome synthesis, 162
  RPL11 overexpression, 162
Messenger RNA (mRNA), 345
Murine double minute 2 (Mdm2)
  ARF tumor suppressor, 283
  ATM-Chk2 and ATR-Chk1 pathways, 289
  binding sites identification, 290
  cell cycle arrest mediation, 291
  cell growth/protein synthesis, 286
  conserved regions, 289
  C4 zinc finger, 289
  de novo precursor rRNA synthesis, 286
  E3 ubiquitin ligase activity and sequestration, 282
  feedback loop, 288
  mechanisms, 288
  missense mutation, 290
  negative regulator, p53, 281
  and p53
    inactivation, 288–289
    nucleolar stress, 288
  RPL5, RPL11 and RPL23, 287–288
  RP-Mdm2-p53 pathway, 287
  RPS7 and RPL26, 289
  serum starvation and nucleotide depletion, 287
  structure, 290
  UV irradiation, 285

N

Ngp–1
description, 315
structure and subcellular distributions, 302
NLS. *See* Nuclear localization signal
Nopp140
and CBs
nucleoli and, 262–263
SMN/hTGS1 RNAi knockdown, 263
spinal muscular atrophy (SMA), 263–264
coilin-interacting protein, 366–367
coilin truncation, 367
FRAP and iFRAP, 363
LNCs, 365
localization, 358
molecular interaction
C-terminus, 259–260
Drosophila, 259
R-rings, 260
molecular structures
carboxy terminus, 258
CKII, 257–258
LisH motif, 257
mammals, 257
nucleolar locations and associations
GFP-Nopp140, 259
M-phase NORs, 258–259
organismal depletion, 266
orthologs and reports
cDNA, *Drosophila*, 256
135 kDa, 255
Nopp140-RGG isoforms, 257
p130, 255
Srp40, 255–256
vs. treacle and nucleolin proteins, 254
Trypanosome brucei, 256
small nucleolar ribonucleoproteins
(snoRNPs)
C/D box, 261
NAP57, 261
yeast Srp40, 261–262
SMN accumulation, 365
transcription factor
AGP-CAT, 264–265
AGP/EBP, 264
carboxy tail (NOPPC), 265
Pol I transcription, 265
treacle
function, 269–271
and TCOF1, 266–269
Normal rat kidney (NRK) cells
localization, SRP-GFP fusion proteins, 351, 352

nucleolar localization, 347
in situ hybridization, 348, 349
NORs and nucleolar formation, UBF
active *vs.* inactive NORs, 86–87
animal phyla
cinoa, human
96
DNA sequence information, 96
hmol P, yeast, 97
HNG box proteins, 95–96
phylogenetic tree, 96
secondary constrictions, 97
Trichoplax adhaerans, 96
UBF, 96–97
chromosomal organization, rDNA, 84–86
definition, active NORs, 89–90
domain structure and DNA binding, 88
history, 83
nucleolar reformation, 93–94
rDNA repeat (*see* rDNA repeat)
ribosome biogenesis
coordinating, 92–93
regulation, 94–95
Northern blot analysis, 390, 392
NPI46 in *S. cerevisiae*, 271
NPM. *See* Nucleophosmin
NPM2. *See* Nucleophosmin 2
NPM3. *See* Nucleophosmin 3
NRK. *See* Normal rat kidney
Nuclear actin and myosin
cytoplasmic actin, 114
mutants, 114
NMI
and actin function, 116
association, 114–115
observation, 114
Pol I transcription, 115–116
WSTF, 116
Nuclear chaperone
description, 214
NPM1, 213
Nuclear localization signal (NLS)
human treacle, 267
mutation, 333
and nucleolar targeting signals act, 332
Tat, 384–385
Nucleic acids
annealing, 197, 200
binding properties, nucleolin, 188–189
nucleolin interaction
dimethylarginine, 192–193
phosphorylation, 192

Nucleolar cap
  accumulation
    CBs, 362
    coilin, 365
    PRMT5/PRMT7 depletion, 367
    spliceosomal snRNPs, 365
    transcription arrest, 366
    UV-C irradiation, 359
Nucleolar decoy
  fibrillarin, 393
  trapping HIV regulatory proteins, 390, 392
  U16RBE, 392
  U16 snoRNA, 392–393
Nucleolar localization, 308
Nucleolar reformation, 93–94
Nucleolar ribozyme. *See* Ribozyme
Nucleolar stress
  de novo precursor rRNA synthesis, 286
  nucleolar proteins dysfunction, 287
  vs. oncogenic
    DNA damage, 292
    genetic integrity, 292
    multiple stress signals, 292
    p53, 291–292
    ribosome synthesis and DNA
      replication, 292–293
    RPL5 and RPL11, 291
  perturbation, 282
  p53 response, 288, 290
  ribosome biogenesis, 286
  RP-Mdm2-p53 pathway, 287
Nucleolar trafficking, 388
Nucleolin
  cell cycle regulation, cell division
      and proliferation
    marker, 201–202
    phosphorylation and ribosome
      biogenesis, 202
    proteolysis, 202
    siRNA, 202–203
  central region, 187
  C-terminal domain, 187
  DNA metabolism, 199–201
    repair and recombination, 201
    replication, 199–201
  N-terminal domain, 187
  nucleic acid binding properties
    NRE and ECM, 188
    RBD 1 and 2, 188
    3'UTR, 188–189
    5'UTR, GROs and VEGF, 189
  nucleolar functions
    polymerase I transcription, 194–196
    rRNA maturation and pre-ribosome
      assembly, 196–197

pol II transcription
    histone chaperone activity, 198–199
    HPV18-induced cervical
      carcinogenesis, 198
    KLF2, 198
  post-translational modifications
    ADP-ribosylation and glycosylation,
      193
    methylation, 192–193
    phosphorylation, 191–192
  protein-protein interactions
    cell-cycle-dependent, 190
    DNA metabolism, 190
    receptor and subcellular localization, 190
    RGG and N-terminal domains,
      189–190
  regulation, post-transcriptional, 199
  structure and posttranslational
      modifications, 186
Nucleolin recognition element (NRE), 188,
    197
Nucleolus. *see also* Dynamic proteome,
    nucleolus
  assembly/disassembly
    CDK1-cyclinB phosphorylation, 15
    cell cycle, 16
    disruption, 15
    mitosis, 14–15
    PNBs, 16
    rDNA transcription, 15–16
  FCs, 21
  Nopp140 (*see* Nopp140)
  nucleolin, polymerase I transcription
    chromatin structure and function
      regulation, 194
    depletion, nucleoli disorganization,
      195–196
    nucleolin phosphorylation, 194–195
    pre-rRNA synthesis, 194
    rDNA transcription and histone
      deposition, 195
    rRNA gene expression, 196
  nucleus
    and chromatin, 18–19
    envelope relationship, 19–20
    r-proteins, 17
    structure dynamics, 17–18
  rDNAs, 4–5
  ribosome biogenesis, 150
  and ribosome biogenesis
    compartmentation, building blocks,
      9–13
    human HeLa cell and *X. laevis*, 5–6
    organization, 6–9
    production, 5, 13–14

Index

409

RNase MRP functions, 149
rRNAs, 20–21
telomeres and nucleolin, aging, 200–201
and viruses (*see* Viruses and nucleolus)
Nucleolus, stress response
ARF-dependent and-independent roles,
284–286
Mdm2, description, 281
nucleolar *vs.* oncogenic, 291–293
perturbation, 282
ribosomal protein-Mdm2-p53 pathway,
286–291
stressors, 281
tumor suppressor ARF, 282–283
Nucleophosmin 1 (NPM1)
lymphomas and leukemias, gene
translocation
ALCL, 235
chromosomal translocation, 236
NPM1-ALK fusion, 235–236
NPM1 and PML-RARA, 236
NPMc+ AML
cell of origin, 233–234
features, 234–235
mutations, 230–231
putative mechanisms, 232–233
Nucleophosmin 2 (NPM2)
H2A-H2B dimers, 218
human NPM1 and *Xenopus* NPM2, 213
members, nucleophosmin/nucleoplasmin
family, 214
physiological functions, 227–228
posttranslational modifications, 222
tissue distribution, 215
Nucleophosmin 3 (NPM3)
human NPM1 and *Xenopus* NPM2, 213
members, nucleophosmin/nucleoplasmin
family, 214
physiological functions, 227–228
posttranslational modifications, 222
Nucleophosmin (NPM)
A1 and A2 acidic stretches
chaperone activity, 219–220
in vitro replication, 220
N-terminal hydrophobic core region, 219
apoptosis inhibition
BuONa/vanadate, 225
GADD45α, 225
HL–60 cells, 225
hypoxia (HIF1α), 224
PIP3, 224
basic domain, 220
cell cycle regulation, 226–227
C-terminal aromatic domain
aromatic residues, 221

nucleus-cytosol shuttling, 220–221
structure, 221
surface lysin residues, 221
domain organisation, 215
function, alterations and modifications, 222
gene transcription regulation
complex, YY1, 223–224
retinoic acid-induced differentiation,
224
genomic stability, 226
host cell cycle
adenovirus mobilization, 240
posttansfusion hepatitis, 239
ribosome biogenesis, 239
SARS-CoV, 240
suppression effect, YY1, 239–240
intracellular parasites, 236
molecular chaperone, 222
NPM2 (*see* Nucleophosmin 2)
NPM3 (*see* Nucleophosmin 3)
NPM1 alteration in human cancers
(*see* Nucleophosmin 1)
N-terminal core region
CDKN2A gene, 218
"core" domain, 216
NPM1-ARF interaction, 219
nucleo-cytoplasmic shuttling, 219
pentamer-pentamer interface, 218
structure, 217
Xenopus NO38 and nucleoplasmin core
domain, 218
X-ray crystal structure, human, 216
nuclear chaperones, 214
nucleophosmin/nucleoplasmin family
members, 214
physiological functions, NPM2 and NPM3,
227–228
posttranslational modification, 228–229
proliferative and growth-suppressive, 221
ribosome biogenesis, 223
sequence and structural homology, 215
structure and expression
isoforms, 215
tissue distribution, 215
transcript variants, 214–215
tumor suppressors modulation, 225–226
virus-NPM1 interactions
host cell cycle, 239–240
replication cycle, 237–239
Nucleostemin (NS)
dynamic nucleolus, 316
expression profile
cell cycle reentry, 307
TIC/CSC, 306–307
tissue regeneration, 306

Index

Nucleophosmin 3 (NPM3) (*cont.*)
functions
G1/S or G2/M transition, 309
nucleostemin-null embryos, 308
gene and protein structures, 305–306
GNL3, 314–315
GNL3L (*See* Guanine nucleotide binding protein-like 3)
Lsg1 and Mtg1, 316
Ngp–1, 315
nucleolar distribution, 307–308
nucleostemin-interacting proteins
non-nucleolar, association, 311
nucleoplasmic mobilization, MDM2, 310–311
p53, 309–310
TRF1 and RSL1D1, 311
phylogenetic view
GNL3, 305
homology comparison, proteins, 302, 304
Lsg1, Gnl1, and *Mtg1,* 305
MMR1_HSR1, GTP-binding motifs, 302, 303
protein structures and subcellular distributions, 302
tissue regeneration, 317
Nucleus
"cytoplasmic" actin, 114
DNA metabolism, 199–201
HIV–1, 237, 328
JEV core protein, 237
Nopp140, 262
NPM1, 219, 223
nucleolin, 197
nucleolus in
and chromatin, 18–19
nuclear envelope, 19–20
structure, 17–18
Sdo1p, 172
SRP RNAs, mammalian cells, 347
TIF-IA translocation, 125
Tif6p, 173

O
Oncogenes and tumor control, 127–128

P
p53
ARF (*See* Alternate reading frame)
dependent cell cycle arrest, G1 phase, 160–161

ribosomal protein-Mdm2-p53 pathway, 286–291
ribosomal stress, mammalian cells, 162–163
PNB. *See* Prenucleolar body
Pol I transcription regulation
cellular energy supply, 123–124
genotoxic stress, 125–127
growth factor, 118–120
oncogenes and tumor control, 127–128
by reversible acetylation, 122–123
TIF-IA links, 117–118
TOR signalling, 124–125
transcription, cell cycle, 120–122
Prenucleolar body (PNB)
and NDFs, 16
NORs and, 21
NPM/B23 and Nop52, 16
Pre-rRNA processing
ITS2, 150
RNase MRP functions, 149
Proteomics
dynamic, nucleolus (*see* Dynamic proteome, nucleolus)
quantitative
adenovirus *vs.* ActD dataset, 336
analysis, purified nucleoli, 335–336
SILAC experiment, 335
stress sensor, 334–335
Pseudouridylation
bipartite guide sequence, 141
intact hairpins, 144
optimal ribosome function, 137
snR10, 139

R
rDNA chromatin structure
cell type and development stages, 44
genomic organization
nucleolar organizers regions (NORs), 44–45
sequence analysis, 45
SIR1 and FOB1, 45
tandem repeats, 44
intergenic spacers, 43
non-transcribed spacer (NTS), 43
protein and modifications
Arabidopsis, 49
block aberrant transcription, 48
cohesin mutants, 51
condensin distribution, 51
degree of methylation, 50
enzymes covalent modification, 49–50
functions, 48

Index 411

histones, 48
human cells, 50
non-histone proteins, 48–49
Pol I and II transcription, 50–51
promoters, 50
regulation, 51
synthesis, 50
ZmHM-rDNA binding, 49
rDNA repeats, 43–44
roles, nuclear organization, 51
Saccharomyces, 51
S. cerevisiae, 44
sRNA, 43
structural organization
DNA staining levels, 45, 46
epigenetic mechanisms, 48
heterochromatic rDNA, 46
hybridization, fluorescence, 45–46
Miller spreads, 47
in pea, 46, 47
poised transcription, 48
transcriptional active rDNA, 46–47
rDNA repeat
analysis, histone composition, 91
hetrochromatin spreading, 91–92
histone H3, CENP-A, 90
HMGB1, 91
interactors and function, 92
micrococcal nuclease digestion, 91
Pol I and UBF, interactions, 91
promoters, 91
pseudo-NORs, 90
UBF *in vivo* and *xenopus*, 90
XEn elements, 90
Rev
nucleolar localization
cell fusion assays use, 386–387
compartment, 387
mutants, 387
NMD degradation, 388
prevention, spliceosomal assembly, 387
site-directed mutagenesis, 387
protein and RRE
localization, 385
mutagenesis, 385–386
nuclear magnetic resonance (NMR), 386
point mutations, 385
transient transfections, cells, 385
Rev response element (RRE)
B23 displacement, 383
description, 382
nucleolar localization (*see* Rev)
and nucleolar Rev protein (*see* Rev)

Ribonucleic acid (RNA)
fluorescent RNA cytochemistry, 346–347
ribosome biogenesis processes, 343
RNP, 343
SRP
colocalization, 348–349
expression, *in vitro*, 344–346
level, 352
microinjection, 347, 348
nucleolar localization, 351
protein complex, 344
ribosomal subunit maturation, 349
RNP complexes, 346
in situ hybridization, 347–349
structure, 344, 345
Ribonucleoprotein (RNP)
complexes, 346
description, 343
ribosome biogenesis, 346
small (*see* Small ribonucleoproteins)
Ribose methylation, 2'-O-ribose methylations, 135–137, 139, 145
Ribosomal proteins (RPs)
Mdm2-p53 pathway, 286–291
ribosome biogenesis, 282
Ribosomal RNA (rRNA)
chromatin, 18, 19
colocalization, 349
EM hybridization, 12
fibrillar centers and dense component, 348
nucleolar component, 12
nucleolar structure modification, 13
ribosome biogenesis, 4–5
5.8S and 28S rRNAs, 13
32S pre-rRNA into 28S rRNA, 14
45S rRNAs, 15, 16
synthesis, 343
Ribosomal stress
mammalian cell model, 174
p53 activation, 162–163
Ribosome biogenesis
cell proliferation, 202
nucleolin, rDNA transcription regulation, 197
Ribosome synthesis
assembly factors, 165–167
biogenesis factors, mitosis, 167–168
G1 phase and G1/s transition
communication, yeast cells, 158–159
p53 activation, mammalian cells, 162–163
p53-dependent cell cycle arrest, 160–161

412 Index

Ribosomal RNA (rRNA) (*cont.*)
  and human diseases, defection
    DBA and human 5q-syndrome, 168–171
    perspectives, 173–175
    SDS, 172–173
    TCS, 171–172
  tumor suppressor ARF inhibition, 164
Ribosomopathies
  cell cycle arrest and apoptosis, 174
  definition, 168
  gene mutation, 158
Ribozyme
  anti-viral activity and catalytic strength,
    U16Rz, 390
  indirect immunofluorescence assays,
    390, 391
  inhibitory function, 388
  Northern blot analysis, 390, 392
  trafficking, 388
  U16 snoRNA, 388–390
RNA polymerase I-dependent transcription
    cycle
  dynamics, 111
  elongation, 112
  termination, 112–113
  upstream terminator T, 113–114
RNA polymerase I transcription machinery
  components
    basal factors, 109–111
    structure and function, 108–109
  I-dependent cycle
    dynamics, 111
    elongation, 112
    termination, 112–113
    upstream terminator T, 113–114
  mammalian cells, 107
  nuclear actin and myosin, 114–116
  perspectives, 128–129
  regulation
    cellular energy supply, 123–124
    genotoxic stress, 125–127
    growth factor, 118–120
    oncogenes and tumor control, 127–128
    overlapping mechanisms, 116–117
    by reversible acetylation, 122–123
    synthesis, 116
    TIF-IA links, 117–118
    TOR signalling, 124–125
    transcription, cell cycle, 120–122
  structural organization, rDNA, 108
  synthesis, rRNA, 107
RNA virus interactions
  capsid proteins, 330
  molecular mechanisms, 331

  nucleolus, sub-cellular localisation, 331
  poliovirus infection, 331
RNP. *See* Ribonucleoprotein
RPs. *See* Ribosomal proteins
rRNA. *See* Ribosomal RNA
rRNA gene copies, 58

S
SDS. *See* Shwachman–Diamond syndrome
Shwachman–Diamond syndrome (SDS),
    172–173
Signal recognition particle (SRP) assembly
  description, RNP, 343
  mammalian
    Alu and S domain, 344
    description, 344
    expression, SRP68 and SRP72, 346
    proteins, 344
    purification, cytoplasmic membrane
      fractions, 344
    RNP complexes, 346
    structure, RNA component, 344, 345
    in vitro, Alu and S domain, 344–346
  nucleolar involvement, biogenesis
    export, 352
    fluorescent RNA cytochemistry,
      346–347
    localization, 347–350
    protein localization, 351–352
    ribosome biogenesis processes, 343
SILAC. *See* Stable isotope labelling with
    amino acids in cell culture
Small nuclear ribonucleoprotein (snRNP)
  CBs and nucleolus, relationship, 359
  cytoplasmic phase, 360
  histone pre-mRNA processing, 361
  nucleolar caps, 366
  SMN, 360
  snRNA components, 363
  spliceosomal, 365
  U7, localization, 361
Small ribonucleoproteins (snoRNPs)
  box C/D snoRNPs
    archaeal, 145, 146
    architectural models, 147
    Fibrillarin/Nop1, 144
    Nop5-fibrillarin, 145, 146
    Snu13, 148
    Xenopus oocytes, 145
  box H/ACA snoRNPs
    archaeal box, 142–143
    Cbf5/Nap57/dyskerin, 142
    eukaryotic, 143

Index

L7Ae, 144
protein-protein interactions, 142
single and dual-hairpin, 144
conserved secondary structures, nucleolar
box C/D snoRNAs, 141–142
eukaryotes, 140
pseudouridylation, 141
nucleotide modification, rRNAs
base methylations, 135
eubacteria, 136
2'-O-ribose methylation, 137
pseudouridine, 137
pseudouridylations, 135
ribosome function, 136
S. cerevisiae, 136–137
pre-rRNA cleavage events
base pairing interactions, 138
box C/D and H/ACA snoRNA, 139
"processing", 139
U3 snoRNP, 137, 138
RNA helicases
conformational changes, 148
Crick base-pairing interactions, 148
snoRNA-preRNA duplex unwinding,
148–149
RNase MRP snoRNP
archaeal and eukaryotic counterparts, 150
human RNase MRP, 151
protein composition, 150
site-specific endonuclease, 149
snoRNPs. *See* Small ribonucleoproteins
snRNP. *See* Small nuclear ribonucleoprotein
SRP biogenesis, nucleolar involvement
export, 352
fluorescent RNA cytochemistry, 346–347
localization
colocalization, 347–350
compositions, domains, 348
maturation, 349
microinjection, 347, 348
rRNA synthesis sites, 348
in situ hybridization, 347–349
protein localization
colocalization, 351
GFP fusions, 351
heterodimer, 351
subunit, SRP14, 351
RNA polymerase III, 353
Stable isotope labelling with amino acids in
cell culture (SILAC), 91, 335, 336
Stem cells
germline, 315
mesenchymal, 307, 309
neural, 307, 312

Stress sensor, viruses and nucleolus, 334–335
Structure
FC and DFC markers, 13
nucleolar organization
characterization, 8
modifications, 13
nucleic acids, 6
nucleoli, human HeLa nucleus, 9

T
Tat, nucleolus and TAR
cellular dispersion, 384
composition, amino acid, 383
5' LTR region, 383
mutagenesis, 383
nuclear localization, 384–385
TAK interaction, 383–384
TCS. *See* Treacher Collins syndrome
TOR signalling, 124–125
Trafficking
Trafficking, nucleolar
and accumulation, 332
hepatitis delta antigen, 332
NLSs and targeting signals, 332
signals, viral proteins, 331–332
Transactivation response (TAR)
nucleolus (*see* Tat, nucleolus and TAR)
viral genes transcription, Tat, 382
Transcription
factor, Nopp140, 264–265
nucleolin
Pol II, 198–199
polymerase I, 194–196
posttranscriptional regulation, 199
rDNA, 270
Treacher Collins–Franceschetti syndrome
1, 266–269
Treacher Collins syndrome (TCS), 171–172
Treacle
function
box C/D snoRNPs, 269–270
Nop56 proteomic analysis, 269
pre-rRNA processing, rDNA
transcription, 270–271
RPA40 and RPA16, 271
vs. Nopp140 protein, 254
and TCOF1
acidic motifs, 266–267
human carboxy tail, 267
nucleolar stress response, 268–269
p53 gene, 268
Tcof1⁺ᐟ⁻, 267–268
Treacher Collins syndrome, 266

414   Index

U
U16RBE. *See* U16 rev binding element
U16 rev binding element (U16RBE), 392
U16Rz
    catalytic strength, 390, 392
    indirect immunofluorescence, 390, 391
    nucleolar localization and anti-viral
        activity, 390
U16TAR. *See* U16 transactivation response
U16 transactivation response (U16TAR),
    392–393

V
Viruses and nucleolus
    classes and proteins, 322–327
    coupling quantitative proteomic analysis,
        336
    description, 321
    DNA virus interactions, 328–330
    functional relevance, nucleolar interactions
        assembly-activating protein (AAP), 334
        flaviviruses, 333
        maize fine streak virus (MFSV), 334
        nucleolin and plant virus, 333–334
        pathogenesis, nuclear disruption,
            332–333
        point mutations, JEV core protein, 333
        recruitment, cellular proteins, 334
        tomato bushy stunt virus (TBSV), 333

genomes, 321
HIV–1 interaction, 328
pan-virus phenomenon, 321
quantitative proteomics
    adenovirus *vs.* ActD dataset, 336
    analysis, purified nucleoli, 335–336
    SILAC experiment, 335
    stress sensor, 334–335
RNA and positive-strand RNA, 328
RNA virus interactions, 330–331
trafficking, nucleolar
    and accumulation, 332
    hepatitis delta antigen, 332
    NLSs and targeting signals, 332
    signals, viral proteins, 331–332
Virus replication cycle, virus-NPM1
    interactions
    colocalization, HBV core antigen,
        238–239
    HDV, 238
    JEV, 237–238
    KSHV and CDK6 kinase, 239
    NPM1 isoforms, 238
    protein encode, 238
    Rev protein, HIV–1, 237

W
Williams syndrome transcription factor
    (WSTF), 116